National Fire Protection Association

Structural Firefighting
Strategy and Tactics
Second Edition

Bernard J. "Ben" Klaene
Chief of Training/Safety, Cincinnati Fire Department, Retired
Cincinnati, Ohio

Russell E. "Russ" Sanders
National Fire Protection Association
Chief of Department, Louisville Fire Department, Retired
Executive Secretary, NFPA/IAFC Metropolitan "Metro" Fire Chiefs Association
Louisville, Kentucky

JONES AND BARTLETT PUBLISHERS
Sudbury, Massachusetts
BOSTON   TORONTO   LONDON   SINGAPORE

World Headquarters
Jones and Bartlett Publishers
40 Tall Pine Drive
Sudbury, MA 01776
978-443-5000
www.jbpub.com

Jones and Bartlett Publishers Canada
6339 Ormindale Way
Mississauga, Ontario L5V 1J2
Canada

Jones and Bartlett Publishers International
Barb House, Barb Mews
London W6 7PA
United Kingdom

National Fire Protection Association
1 Batterymarch Park
Quincy, MA 02169
www.NFPA.org

Jones and Bartlett's books and products are available through most bookstores and online booksellers. To contact Jones and Bartlett Publishers directly, call 800-832-0034, fax 978-443-8000, or visit our website, www.jbpub.com.

Substantial discounts on bulk quantities of Jones and Bartlett's publications are available to corporations, professional associations, and other qualified organizations. For details and specific discount information, contact the special sales department at Jones and Bartlett via the above contact information or send an email to specialsales@jbpub.com.

Production Credits

Chief Executive Officer: Clayton E. Jones
Chief Operating Officer: Donald W. Jones, Jr.
President, Higher Education and Professional Publishing:
  Robert W. Holland, Jr.
V.P., Sales and Marketing: William J. Kane
V.P., Production and Design: Anne Spencer
V.P., Manufacturing and Inventory Control: Therese Connell
Publisher, Public Safety Group: Kimberly Brophy
Senior Acquisitions Editor, Fire: William Larkin
Managing Editor: Carol B. Guerrero
Editorial Assistant: Justin Keogh

Production Editor: Karen Ferreira
Director of Marketing: Alisha Weisman
Text Design: Anne Spencer
Composition: Graphic World, Inc.
Cover Design: Kristin Ohlin
Associate Photo Researcher and Photographer: Christine McKeen
Cover Image: © RubberBall Selects/Alamy Images
Text Printing and Binding: Malloy
Cover Printing: Courier Stoughton

Copyright © 2008 Jones and Bartlett Publishers, LLC

All rights reserved. No part of the material protected by this copyright notice may be reproduced or utilized in any form, electronic or mechanical, including photocopying, recording, or by any information storage and retrieval system, without written permission from the copyright owner and publisher.

Publication of this work is for the purpose of circulating information and opinion among those concerned for fire and electrical safety and related subjects. While every effort has been made to achieve a work of high quality, neither the publisher, the NFPA, the authors, or the contributors to this work guarantee the accuracy or completeness of or assume any liability in connection with the information and opinions contained in this work. The publisher, NFPA, authors, and contributors shall in no event be liable for any personal injury, property, or other damages of any nature whatsoever, whether special, indirect, or consequential, or compensatory, directly or indirectly resulting from the publication, use of or reliance upon this work. This work is published with the understanding that the publisher, NFPA, authors, and contributors to this work are supplying information and opinion but are not attempting to render engineering or other professional services. If such services are required, the assistance of an appropriate professional should be sought.

Additional illustration and photographic credits appear on page 379, which constitutes a continuation of the copyright page.

Library of Congress Cataloging-in-Publication Data
Klaene, Bernard J.
  Structural firefighting : strategy and tactics / Bernard J. Klaene, Russell E. Sanders.-- 2nd ed.
    p. cm.
  Includes index.
  ISBN-13: 978-0-7637-5168-5
  1. Fire extinction. 2. Fire prevention. 3. Fire departments--Equipment and supplies. I. Sanders, Russell E. II. Title.
  TH9310.5.K54 2007
  628.9'25--dc22
                                    2007027649
6048

Printed in the United States of America
12 11 10   10 9 8 7 6 5 4 3

Brief Contents

CHAPTER 1	Organizing, Coordinating, and Commanding Emergency Incidents	1
CHAPTER 2	Procedures, Pre-Incident Planning, and Size-Up	38
CHAPTER 3	Developing an Incident Action Plan	80
CHAPTER 4	Company Operations	94
CHAPTER 5	Fire Fighter Safety	104
CHAPTER 6	Life Safety	148
CHAPTER 7	Fire Protection Systems	170
CHAPTER 8	Offensive Operations	192
CHAPTER 9	Defensive Operations	232
CHAPTER 10	Property Conservation	248
CHAPTER 11	The Role of Occupancy	258
CHAPTER 12	High-Rise Buildings	296
APPENDIX A	Sample ASETBX Data	347
APPENDIX B	Imperial and Metric Conversions	353
APPENDIX C	FESHE Correlation Guide	357
	Glossary	358
	Index	364
	Photo Credits	379

Contents

CHAPTER 1 Organizing, Coordinating, and Commanding Emergency Incidents....1
- Learning Objectives........................ 1
- Introduction 2
- Evolution of the Incident Management System .. 2
- Command................................ 2
  - National Incident Management System 2
  - Unified Command 3
  - Initial Command 4
  - Command by a Chief Officer....................5
  - Transfer of Command........................5
  - Delegation.................................6
  - Command Post 6
  - Span of Control 7
  - Calling for Additional Resources 8
  - Staging 8
- NIMS Organization and Positions 9
  - Modular Organization........................9
  - Command Staff 9
  - General Staff..............................11
- Intuitive Naming System 21
- Communications 22
- Summary................................ 27
- Key Terms 28
- Suggested Activities 30
- Chapter Highlights....................... 34
- References 37

CHAPTER 2 Procedures, Pre-Incident Planning, and Size-Up 38
- Learning Objectives...................... 39
- Introduction 40
- Developing Standard Operating Procedures .. 40
  - Purpose of Standard Operating Procedures 41
  - Relationship of Standard Operating Procedures to Training and Equipment........41
  - A Word About the Standard Operating Guidelines Controversy...................42
- Evaluating a Specific Property.............. 42
  - Pre-Incident Plans 43
  - What Structures Are Preplanned?...............50
  - Modifying Standard Operating Procedures........51
  - Estimating Life Safety Needs 51
  - Estimating Extinguishment Needs 51
  - Estimating Property Conservation Needs 51
  - Relationship of Pre-Planning to Size-Up..........51
- Analyzing the Situation Through Size-Up...... 53
  - Life Safety/Fire Fighter Safety..................53
  - Structure.................................59
  - Extinguishment 62
  - Property Conservation 65
  - General Factors 66
- Size-Up Chronology 68
  - Standard Operating Procedures and the Pre-Incident Plan 68
  - Shift/Day/Time............................68
  - Alarm Information 68
  - While Responding 68
  - Visual Observations at the Scene 68
  - Information Gained During Continuing Operations ..70
  - Overhaul 71
- Summary................................ 71
- Key Terms 72
- Suggested Activities 73
- Chapter Highlights....................... 77
- References 79

CHAPTER 3 Developing an Incident Action Plan....80
- Learning Objectives...................... 81
- Introduction 82
- Determining Life Safety Needs............. 82
  - Evaluating Structural Conditions 82
  - Estimating Resource Capability and Evaluating Resource Requirements..................83
  - Devising an Offensive or a Defensive Incident Action Plan 83
  - Formulating an Incident Action Plan.............83
  - Deployment...............................84
- Scenario 1: Single-Family Detached Dwelling .. 84
  - Risk-Versus-Benefit Analysis 85
  - Incident Action Plan 86
  - Deployment...............................86
- Scenario 2: High-Rise Apartment Building 87
  - Risk-Versus-Benefit Analysis 87
  - Incident Action Plan 88
  - Deployment...............................88
- Scenario 3: Church Fire 88
  - Risk-Versus-Benefit Analysis 88
  - Incident Action Plan 89
  - Deployment...............................89
- Summary................................ 91
- Key Terms 92
- Suggested Activities 92
- Chapter Highlights....................... 93

CHAPTER 4 Company Operations 94
- Learning Objectives...................... 95
- Introduction 96
- Engine Company Tasks 96
- Truck Company Tasks 97
- Coordinating Company Operations.......... 97
  - A Word About Quint and Quad Companies........99
  - Ventilation................................99
  - Apparatus Positioning 100
- Summary............................... 100
- Key Terms 102
- Suggested Activities 102
- Chapter Highlights...................... 103
- References 103

CHAPTER 5 Fire Fighter Safety 104
- Learning Objectives..................... 105
- Introduction 106
- Fire Fighter Injuries and Fatalities......... 106
- Risk Management Applied to the Fire Ground........................ 111
  - Fire Intensity 111
  - Building Design Loads 112
  - Fuel Load 112
  - Structural Stability 113
- Roof Operations....................... 116
- Pre-fire Conditions and Fire Conditions..... 116
  - Pre-fire Conditions.......................116
  - Fire Conditions Leading to Structural Collapse....117
- Fire Extension 118
- Fire Zones and Perimeters 121
- Relationship of Time, Fire Intensity, and Structural Stability....................... 124
- Time: Ignition to Effective Actions.......... 124
  - Detection/Transmission 125
  - Dispatch Time..........................125
  - Turnout Time...........................125
  - Response/Travel Time 126
  - Set-up Time 126
  - Adequate Number of Personnel 126
- Tactical Reserve 127
- Correlation Between Elapsed Time and Progression to Flashover 128
- Fire-ground Operations.................. 129
  - Communications 129
  - Command and Control 130
  - Accountability 130
  - Safety Officer 132
  - Alternative Egress 132
  - Rapid Intervention Crews 134
  - Opposing Fire Lines.....................136
  - Personal Protective Clothing..............137
- Overhaul........................... 138
- Rehabilitation 139
- Critical Incident Stress Management....... 141
- Summary........................... 141
- Key Terms.......................... 142
- Suggested Activities 143
- Chapter Highlights.................... 144
- References 146

CHAPTER 6 Life Safety.................148
- Learning Objectives..................... 149
- Introduction 150
- Evaluating the Probability of Extinguishment...................... 150
- Assessing the Ventilation Profile 151
- Analyzing the Available Rescue Options 152
  - Interior Stairs..........................152
  - Fire Escapes 152
  - Aerial Ladders and Elevated Platforms.......153
  - Ground Ladders........................153
  - Elevator Rescues 154
  - Rope Rescues 154
  - Helicopter Rescues 154
- Classifying Evacuation Status 154
- Estimating the Number of People Needing Assistance....................... 158
- Surveying Floor Layout and Size 159
- Prioritizing Rescues by Location/Proximity.. 160
- Evaluating the Medical Status of Victims 162
  - Evaluating Victims in Mass-Casualty Incidents....163
  - Triage, Prioritizing, and Transport..........163
- Evaluating the Need for Shelter 164
- Estimating Life Safety Staffing Requirements 164
- Summary........................... 165
- Key Terms.......................... 166
- Suggested Activities 166
- Chapter Highlights.................... 169

CHAPTER 7 Fire Protection Systems 170
- Learning Objectives..................... 171
- Introduction 172
- Sprinkler Systems 172
- Working at a Sprinkler-protected Building with No Signs of Fire 175
  - Gaining Entry..........................175
  - Checking the Main Control Valve...........175
  - Checking the Fire Pumps 176
  - Checking the Building for Fire and/or Sprinkler Operation 176
  - Supplying the Fire Department Connection 177
- Working at a Sprinkler-protected Building with Evidence of Fire Showing from the Exterior......................... 177
  - Gaining Entry..........................177
  - Checking the Main Control Valve and Fire Pump..............................177
  - Supplying the Fire Department Connection 177
  - Letting the System Do Its Job 177
  - Backing Up the System 177
  - Ventilating............................178
  - Performing Property Conservation Tasks 178
  - Placing the System Back in Service.........178
- Working at a Property Protected by a Deluge System.................... 178
  - Checking the Control Valve and Fire Pump.....178
  - Operating the Deluge Valve...............178
  - Checking Interlocks 178
  - Letting the System Do Its Job 179
  - Backing Up the System 179
- Working in a Building Equipped with a

v

Standpipe System **179**
   Checking Fire Pumps and Main Control Valves....180
   Supplying Fire Department Connections 180
   Providing Standpipe Equipment 181
   Connecting to the Standpipe Discharge 182
Non-water-based Extinguishing Systems.... **183**
   Foam Systems................................183
   Carbon Dioxide Systems.......................184
   Halon and Other Clean Agents 184
   Dry and Wet Chemical Systems 185
Operating in Areas Protected by Total
   Flooding Carbon Dioxide or Clean
   Agent Systems..................... **185**
   Letting the System Do Its Job 185
   Final Extinguishment and Rescue 185
   Manual Activation 185
   Checking Interlocks..........................185
   Checking Agent Supply 186
   System Restoration 186
Operating in Areas Protected by Local Application
   Carbon Dioxide, Clean Agents, Dry Chemical,
   or Other Special Extinguishing Agents ... **186**
   Letting the System Do Its Job 186
   Checking the Interlocks 186
   Manual Activation 186
   Backing Up the System 186
   System Restoration 187
A Word About Responses to Building Fire Alarm
   Systems........................... **187**
Summary............................ **187**
Key Terms............................ **188**
Suggested Activities **188**
Chapter Highlights..................... **189**
References **190**

CHAPTER 8   Offensive Operations............ **192**
Learning Objectives.................... **193**
Introduction **194**
Calculating Rate of Flow................. **194**
   Indirect Application Theory 195
   The Royer/Nelson Formula 195
   The U.S. National Fire Academy Formula 196
   Sprinkler Rate-of-Flow Calculations..............197
   Estimating the Size of the Largest Area 197
   Estimating the Percent of Area on Fire 200
Comparing Rate-of-Flow Calculations....... **201**
Which Rate-of-Flow Calculation Is Best? **203**
   First Floor 204
   Second Floor................................204
   Third Floor..................................204
Selecting the Attack Hose Size............ **204**
Selecting the Nozzle Type................ **210**
Selecting the Stream Position **211**
Estimating the Number of Attack Lines...... **214**
Evaluating Exposures **215**

Estimating Backup Needs................ **216**
Estimating the Number of Hose Lines Needed
   Above the Fire **216**
Evaluating Other Hose Lines Needed **216**
Estimating Water Supply Needs........... **216**
Estimating Ventilation Needs **218**
Calculating Staffing Needs............... **220**
   Engine 1 (First-Arriving Engine Company) 220
   Engine 2 (Second-Arriving Engine Company) 221
   Truck 1 (First-Arriving Truck Company) 221
   Chief Officer 220
   Engine 3 (Third-Arriving Engine Company).......222
   Engine 4 222
   Heavy Rescue, Truck 2, and Engine 5 223
Determining Apparatus Needs............ **224**
A Word About Class A Foam.............. **225**
Summary............................ **225**
Key Terms............................ **226**
Suggested Activities **226**
Chapter Highlights..................... **230**
References **231**

CHAPTER 9   Defensive Operations **232**
Learning Objectives.................... **233**
Introduction **234**
Classifying as a Defensive Attack **234**
   Establishing a Collapse Zone 234
   Evaluating Exposures 236
   Evaluating a Direct Attack.....................236
   Estimating the Number and Type of Master
      Streams Needed 237
   Estimating Water Supply Needs 238
   Calculating Staffing Needs 238
   Determining Apparatus Needs 239
   Conflagrations and Group Fires................239
Classifying as a Non-attack **242**
Summary............................ **243**
Key Terms............................ **244**
Suggested Activities **244**
Chapter Highlights..................... **247**
References **247**

CHAPTER 10   Property Conservation.......... **248**
Learning Objectives.................... **249**
Introduction **250**
Classifying as an Offensive Attack **250**
   Estimating Indirect Damage 250
   Calculating Staffing Needs 254
   Evaluating Property Conservation Needs 254
   Evaluating Overhaul Needs 255
A Word About Fire Investigation........... **255**
Summary............................ **255**
Key Terms............................ **256**

Suggested Activities 256	
Chapter Highlights 257	
References 257	

CHAPTER 11 The Role of Occupancy 258
- Learning Objectives 259
- Introduction 260
- Classifying the Occupancy Type 260
- Assembly Occupancies 261
  - Churches 262
  - Eating and Drinking Establishments. 263
  - Sports Arenas 265
  - Convention Centers 266
  - Theaters 266
- Educational Occupancies 267
  - Elementary Schools 268
  - Middle, Junior High, and High Schools. 270
  - Colleges and Universities 270
- Health Care Occupancies 270
  - Hospitals. 271
  - Nursing Homes 271
  - Limited Care Facilities 272
  - Ambulatory Care Facilities 272
- Residential Board and Care Occupancies. . . . 273
- Detention and Correctional Occupancies 274
- Residential Occupancies 275
  - One- and Two-Family Dwellings 277
  - Apartment Buildings. 277
  - Dormitories 278
  - Hotels and Motels 279
- Mercantile Occupancies 280
  - Shopping Centers 280
  - Enclosed Shopping Malls 281
  - Lifestyle Centers .281
  - "Big Box" Stores .282
  - Multi-Level Department Stores 282
- Business Occupancies 282
- Storage Occupancies. 283
- Industrial Occupancies 285
- Multiple-and Mixed-Occupancy Buildings 286
- Buildings Under Construction, Renovation, or Demolition 287
- Renovated Buildings 287
- General Occupancy Considerations 288
- Estimating the Number of Potential Victims . . 289
- Summary. 289
- Key Terms 290
- Suggested Activities 290
- Chapter Highlights 293
- References 295

CHAPTER 12 High-Rise Buildings 296
- Learning Objectives 297
- Introduction 299
- Developing and Revising High-Rise Standard Operating Procedures 299
  - Fire Fighter Safety 300
- Fire Fighters' Use of Elevators 301
- Stairway Support 303
- Life Safety 304
- Rescuing and Evacuating Occupants 304
  - Helicopter Rescues .304
  - Partial or Sequential Evacuation 305
  - Emergency Voice/Alarm Communications System (EVACS). .306
- Extinguishment 306
- Command Post Location. 308
- Developing Building-Specific High-Rise Pre-Incident Plans 308
- Analyzing the Situation Through Size-Up. 310
  - Smoke Movement 312
- Developing and Implementing an Incident Action Plan. 314
  - Company Operations 315
- Applying NIMS to a High-Rise Fire. 317
  - Communications 317
  - Tactical Worksheets .318
  - Base (Exterior Staging for High-Rise Fires) 320
  - Staging (Interior) 320
  - Lobby Control 321
- High-Rise Case Histories 321
  - The One Meridian Plaza Fire 321
  - The First Interstate Bank Building Fire 322
  - Comparing the One Meridian Plaza and First Interstate Bank Building Fires 322
  - The Peachtree Plaza Fire 323
  - Cook County Administration Building Fire. 324
  - Terrorist Attacks at the World Trade Center. 325
  - The MGM Grand Fire 326
- Summary. 327
- Key Terms 328
- Suggested Activities 328
  - Las Vegas Hilton Fire 328
  - RGB High-Rise Scenario 332
  - Sample Answer to the RGB High-Rise Problem . . .338
  - Incident Action Plan .341
- Chapter Highlights 344
- References 345

APPENDIX A Sample ASETBX Data 347

APPENDIX B Imperial and Metric Conversions . . .353

APPENDIX C FESHE Correlation Guide 357

Glossary 358

Index 364

Photo Credits 379

Chapter Resources

Structural Firefighting: Strategy and Tactics, Second Edition

The National Fire Protection Association and Jones and Bartlett Publishers are pleased to bring you the *Second Edition* of *Structural Firefighting: Strategy and Tactics*. This new edition builds on the extremely successful first edition by providing both fire officers and professionals in training with the tools they need to become skilled incident commanders. It guides readers through all phases of strategic and tactical planning so they can manage any incident regardless of its complexity. The *Second Edition* brings you features from the last edition, such as fallacy/fact boxes, as well as new features, such as detailed learning objectives. The next pages walk you through the features of this textbook.

Learning Objectives
New to this edition, detailed learning objectives specifically outline what you will learn in each chapter.

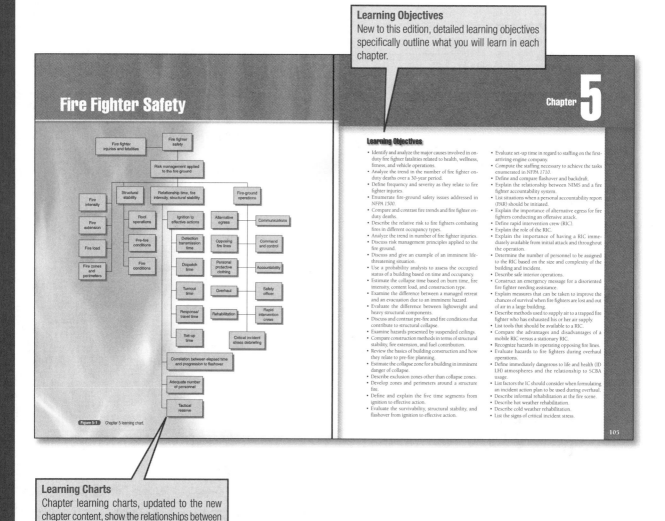

Learning Charts
Chapter learning charts, updated to the new chapter content, show the relationships between topics covered in the chapter.

Margin Notes
These emphasize important points.

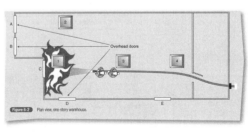

Fallacy/Fact Boxes
These boxes defuse myths and explain the actual facts.

Case Summaries
These case summaries discuss and analyze recent structural fires of note, such as the Worcester, Massachusetts, warehouse fire and the events of September 11, 2001, as well as cases covered in the last edition.

---

112 STRUCTURAL FIREFIGHTING: STRATEGY AND TACTICS

Case Summary

In Worcester, Massachusetts, six career fire fighters died in a vacant, six-floor, maze-like, cold storage and warehouse building while searching for two homeless people who accidentally started the fire and escaped prior to the arrival of the fire department. Approximately 30 minutes after the first alarm was struck, two fire fighters searching for victims and checking for extension sounded an emergency message. A personal accountability report (PAR) confirmed two missing fire fighters. The search-and-rescue operation was expanded to find the two missing fire fighters and the homeless people thought to be in the building. Four additional fire fighters became disoriented during this part of the operation. The fire was believed to have 30 to 90 minutes of pre-burn time. After more than an hour on the scene, a company on the interior notified command of structural problems and an arson investigator on the exterior reported the fire was venting from the roof. A defensive attack was ordered approximately 1 hour and 45 minutes after the initial alarm. The six missing fire fighters perished.

Source: National Institute for Occupational Safety and Health (NIOSH), Fire Fighter Fatality Investigation 99 F-47, Six Career Fire Fighters Killed in Cold Storage Warehouse Building Fire—Massachusetts.

Figure CS5-1 The combination of a massive fire load and the maze of rooms in this Worcester, Massachusetts, warehouse contributed to the death of six fire fighters.

failure. No one can determine exactly when flashover will occur or when the building will fail. However, there are useful approximations. In post-flashover fires, the chance of occupant survival is minimal within the flashover compartment.

In the past, the time from ignition to flashover was given as 10 minutes. The actual time can vary significantly, depending on a number of variables, including the following:
- Compartment size
- Ventilation
- Ignition source
- Fuel supply
- Fuel geometry
- Distance between fuel cells
- Location of the fuel
- Heat capacity of the fuel
- Geometry of the enclosure

There are several computer programs that can provide rough estimates of when flashover will be reached in a compartment. However, given the wide range of variables that must be entered into these computer programs, it is very difficult to provide a reliable time frame that they can use on the fire ground. Generally, one assumption can be made is that the larger the volume of the enclosure where the fire is located, the longer the time required to reach flashover.

Building Design Loads

Loads imposed on a building are categorized as <u>live loads</u>, <u>dead loads</u>, and seismic, wind, snow, and ice loads. Building loads affect structural stability. Unusually high building loads can result in premature collapse. Loads placed on lightweight roof structures are particularly hazardous to fire fighters.

Fuel Load

The fire or <u>fuel load</u> consists of fuels provided by the contents and combustible building materials. Most of the construction materials used in wood frame buildings will burn, thus creating a large fuel load. Combustible building materials are very limited in fire-resistive and

---

Vocabulary Terms
Vocabulary terms appear in red and underlined, and are defined at the end of the chapter and in the glossary at the end of the book.

CHAPTER 10 Property Conservation 253

remain for a considerable time and these absorbent materials may expand, placing pressure against outside walls, which can also affect the structural integrity of the building. Water weighs 8.33 lb/gal (1 kg/L). When multiple master stream appliances are flowing water into a structure, the weight of the water can rapidly increase the collapse potential. Depending on conditions, the water may flow out of the building or be contained. **Table 10-1** assumes that materials stored in the building or fire debris is containing the water being discharged from large master stream appliances.

It is difficult to estimate the amount of water remaining inside a building. Water streaming out of a building may represent only a fraction of the total amount of water being applied.

5. *Covering valuable property.* It is important for the IC, as well as all other members on the scene, to be aware of the need for property conservation. By placing salvage covers over exposed property, an alert fire fighter can often prevent water damage to valuable property. When using salvage covers, crews should group together furniture and other items to protect them from flowing water. This will allow several items to be protected by a single cover.

6. *Moving or removing valuable property.* Water flows downward through a building, following the path of least resistance. Stairs, elevator shafts, and drains provide paths of least resistance. At times the water is blocked by debris, or the volume simply overwhelms the natural flow out of the building. Even when water is flowing through the ceiling, the path of least resistance will be followed. This is generally around openings in the ceiling, such as light fixtures. If a ceiling is holding water run-off for a period of time, the weight of the water is likely to cause the ceiling to collapse. This ceiling collapse can injure fire fighters assigned to the area and increase damage to the contents below. Using a pike pole to drain the ceiling can alleviate this problem.

As soon as possible, property should be moved away from natural flow paths, grouped together, and covered with salvage covers.

On occasion, property is completely removed from the building by fire crews to save it from water damage **Figure 10-4**. This is a very labor-intensive

Figure 10-4 Museum employees look at art that was removed by fire fighters.

Figures
Illustrations and photos embellish concepts to enhance the reader's understanding.

Tables
Tables present information in an easy-to-read format.

TABLE 10-1 Weight of Water from Master Streams

Number of Master Streams at 1000 GPM (63 L/sec) Each	Weight Added Each Minute at 8.33 lb/gal (1 kg/L)	Total Weight at 30 Minutes, lb (kg)	Total Weight at 60 Minutes, lb (kg)
1	8330 lb (3785 kg)	249,900 lb (113,550 kg)	499,800 lb (227,100 kg)
3	24,990 lb (11,355 kg)	749,700 lb (340,650 kg)	1,499,400 lb (681,300 kg)
10	83,300 lb (37,850 kg)	2,499,000 lb (1,135,500 kg)	4,998,000 lb (2,271,000 kg)

End of Chapter Wrap-Up

Key Terms: The chapter's vocabulary terms are defined here.

Suggested Activities: In-depth, critical thinking activities ask students to analyze scenarios, pre-plan structures in their local areas, design incident action plans, and much more.

Chapter Highlights: This bulleted list summarizes important points from the chapter.

References: Chapter references are listed at the end of the chapter.

Instructor Resources

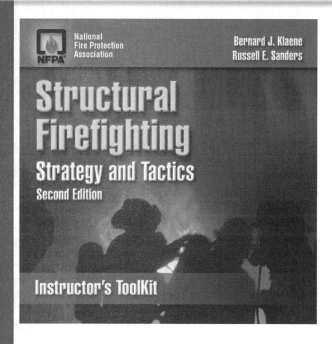

Instructor's ToolKit CD-ROM
ISBN 13: 978-0-7637-5604-8

Structural Firefighting: Strategy and Tactics, Second Edition is accompanied by a complete Instructor's ToolKit CD-ROM to facilitate teaching strategy and tactics. Contents include:

- **Adaptable PowerPoint Presentations**
  These slides follow the textbook's chapter content and include images to enhance the student's classroom experience. Slides can be modified and edited to meet your needs.
- **Lecture Outlines**
  Designed to fit hand-in-hand with the PowerPoint presentations and the textbook chapters, these lesson plans provide additional notes for instructor presentations. These Word documents can be modified and customized to fit your course. The lecture outlines also include discussion points for the Suggested Activities from the end of each textbook chapter.
- **Electronic Test Bank**
  The multitude of multiple-choice and scenario-based questions offered in this test bank can be edited and organized specifically for your course. Select only the questions you want. Tests and quizzes can be printed along with an answer key, which includes page references to the text.
- **Image and Table Bank**
  This resource provides you with the most important images and tables from the text. Use them to import more graphics into your PowerPoint presentation, make handouts, or enlarge specific images for further discussion.
- **Investigation Reports**
  All National Fire Protection Association (NFPA), National Institute for Occupational Safety and Health (NIOSH), and U. S. Fire Administration (USFA) fire fighter fatality investigation reports discussed in the text are included for additional discussion and analysis in the classroom.
- **Course Administration**
  The resources of the Instructor's ToolKit CD-ROM have been formatted so that you can seamlessly integrate them onto the most popular course administration tools.

Acknowledgments

Advisory Committee

Kelly Joe Aylor
Battalion Commander
Florence, Kentucky

Gregory B. Cade
Fire Chief, EFO, CFO, MIFirE
Virginia Beach Fire Department
Virginia Beach, Virginia

Kelvin J. Cochran
Chief, Shreveport Fire Department
Past President, Metropolitan Fire
 Chiefs Association
1st Vice President, International
 Association of Fire Chiefs
Shreveport, Louisiana

Rebecca F. Denlinger, CFO
Chief, Cobb County Fire and
 Emergency Services
Past President, Metropolitan Fire
 Chiefs Association
Marietta, Georgia

Luther L. Fincher, Jr.
Chief, Charlotte Fire Department
IAFC/NFPA Board Representative
 to the Metropolitan Fire Chiefs
 Association
Charlotte, North Carolina

Thomas Lakamp
District Chief, Cincinnati Fire
 Department
Cincinnati, Ohio

Otis J. Latin, Sr., MBA, EFO
Director of Homeland Security and
 Emergency Management
City of Austin
Chief (retired), Fort Lauderdale Fire
 Rescue
Austin, Texas

William J. McCammon
Chief (retired), Alameda County Fire
 Department
Immediate Past President,
 Metropolitan Fire Chiefs Association
San Leandro, California

Robert Ojeda
Chief, San Antonio Fire Department
Secretary, Metropolitan Fire Chiefs
 Association
San Antonio, Texas

Christopher Pope
Chief, Concord Fire Department
Concord, New Hampshire

Keith Richter
Chief, Contra Costa County Fire
President, Metropolitan Fire Chiefs
 Association
Pleasant Hill, California

Tim Ronayne
Chief, Canton Fire Department
Canton, Massachusetts

Mario Rueda
Deputy Chief, Los Angeles City Fire
 Department
Los Angeles, California

Wes Shoemaker
Associate Deputy Minister, Emergency
 Management British Columbia
Chief (retired), Winnipeg Fire and
 Paramedic Services
Past President, Metropolitan Fire
 Chiefs Association
Victoria, British Columbia

William Stewart
Chief, Toronto Fire Services
Vice President, Metropolitan Fire
 Chiefs Association
Toronto, Ontario

Timothy P. Travers, EFO, CFO
Chief of Department
Whitman, Massachusetts Fire-Rescue
Whitman, Massachusetts

Reviewers

Don Abbott
Arizona State University
West Valley Community College
Glendale, Arizona

William P. Alexander
Chief
Pierre Fire Department
Pierre, South Dakota

Raul A. Angulo
Captain
Seattle Fire Department
Seattle, Washington

Greg Burroughs
Program Chair, Fire Protection
 Technology
Southeast Community College
Lincoln, Nebraska

Jack Collie
Chief of Training
Topeka Fire Department
Topeka, Kansas

Bryn Crandell
USAF Fire Protection
USAFA, Colorado

Ronald W. Endle
Division Chief (Retired)
Buffalo Fire
New York State Academy of Fire
 Science
Cheektowaga, New York

Mike Gagliano
Captain
Seattle Fire Department
Seattle, Washington

Acknowledgments, cont.

Reviewers, continued

Mike Garcia
Assistant Chief
Director of Training
Long Beach Fire Department
Long Beach, California

Bill Hepburn
Assistant Chief
Seattle Fire Department
Seattle, Washington

Ronnie Holton
Coastal Carolina Community College
Department of Fire Technology
Jacksonville, North Carolina

Daniel G. Klein, MA
Firefighter/EMT
Cologne, Minnesota

Shawn C. Koser, AAS
Captain
Columbus Ohio Division of Fire
Columbus, Ohio

Richard Kosmoski
President, New Jersey Volunteer Fire Chiefs Association
Ex-Chief
Middlesex County Fire Academy
Parlin, New Jersey

Kurt Leben
Battalion Chief
Bismarck Fire Department
Bismarck, North Dakota

Tim Linke
Instructor
Lincoln Fire and Rescue Training Division
Southwest Fire and Rescue
Lincoln, Nebraska

Jerry Marrison, Sr.
MFETI
Deltona, Florida

Paul R. Martin
Deputy District Chief
Chicago Fire Department
Chicago, Illinois

Michael S. Mayers
Director, Urban Search and Rescue Program
South Carolina Emergency Response Task Force
Captain, Hilton Head Island Fire and Rescue Department
Hilton Head Island, South Carolina

Ike McConnell
West Georgia Technical College
Hampton, Georgia

Gregg Moore
Driver/Operator, Instructor
Richmond Fire Department
Richmond, Indiana

Jim Moreland
City of Westminster Fire Department
Westminster, Colorado

Ken Nelms
Emergency Service Training Institute
Goodlettsville, Tennessee

Michael Richardson
Louisville Metro
Jefferson County Fire
Louisville, Kentucky

Douglas Rohn
Madison Area Technical College
Fire Service Education Center
Madison, Wisconsin

Holly A. Scribner, AAS, AS, IC, FF, EMT-P
Waldo County Technical Center
Cushing, Maine

William Stipp, EFO
Goodyear Fire Department
Estrella Mountain Community College
Goodyear, Arizona

Gary "Skip" Tinagero
Captain
Albuquerque Fire Academy
Albuquerque, New Mexico

H. Jeffrey Turner, MSM
Mohave Community College
Bullhead City, Arizona

David Walsh
Dutchess Community College
Poughkeepsie, New York

Brent Willis
Captain
Martinez Columbia Fire Rescue
Georgia Fire Academy Adjunct
Grovetown, Georgia

The authors have many people to thank. Dr. Rita Fahy, Manager of NFPA Fire Data Bases and Systems, researched and provided much of the statistical information contained in this text. Carl Peterson, Director of the NFPA Public Fire Protection Division, carefully reviewed the manuscript, drawings, and photographs. And, Robert Duval, NFPA Senior Fire Investigator, provided background information related to fire investigations. A very special "thank you" goes to Carol Guerrero, our Managing Editor, who patiently endured a constant barrage of questions from us and guided copy editors, compositors, and others throughout the process.

Preface

The goal of this textbook is to explain proven tactics and strategies used at structure fires. The guiding objective throughout this book is preparing the fire officer to take command at structure fires, fully utilizing available resources in a safe and effective manner.

We assume that the reader has a basic understanding of firefighting and associated tasks as outlined in *Fundamentals of Fire Fighter Skills* and such NFPA documents as:

- *NFPA 1001, Standard for Fire Fighter Professional Qualifications*
- *NFPA 1002, Standard for Fire Apparatus Drive/Operator Professional Qualifications*
- *NFPA 1021, Standard for Fire Officer Professional Qualifications*

or other similar firefighting manuals and standards. Qualified fire fighters with basic training and some level of experience who are now seeking positions within a fire department as company and/or chief officers are the intended audience for this book.

Many fire departments adopted the first edition of *Structural Firefighting* as part of their reading list for promotional examinations. Some of these departments, as well as participants at the many seminars we presented over the years, believe that the book's primary audiences are company and battalion level officers. To ensure consistency throughout a fire department, we believe that it's best when all department members use the same strategy and tactics textbook. The lessons learned in *Structural Firefighting* apply to all members of the department and it is an excellent text for providing consistent, department-wide strategy and tactics training.

The basis for an incident action plan that leads to a safe and effective fire-ground operation is an operational priority list. This textbook is designed around a three-point list:

1. Life safety
2. Extinguishment
3. Property conservation

Several other priority lists exist, but each list basically prioritizes life safety first, followed by a logical sequence of activities aimed at reducing property loss. It must be stated that crossover exists among the listed priorities, and quite often, a lower priority item provides the means to achieve a higher priority. For example, when the fire is extinguished, occupants and fire fighters are in less danger. Thus, extinguishment becomes an essential part of the life safety priority. The importance of extinguishment cannot be overemphasized and is a recurrent theme throughout this book.

An important goal of any fire strategy and tactics textbook, including this one, is giving the reader the necessary tools to achieve maximum productivity under adverse fire-ground conditions. Fire-ground experience is necessary to effectively use the tools explained in this textbook. This book will enable readers to learn fire-ground procedures at an accelerated pace, thus reducing the cost in lives and property associated with learning by experience only. In addition to updating information in this second edition of *Structural Firefighting*, to further clarify various concepts we incorporated frequently asked questions from students, seminar participants, and fire officers who studied the text. Further, explanations of many of the topics covered in the first edition have been expanded, and the text addresses the recent Department of Homeland Security mandates related to the use of the National Incident Management System. We think you will find this second edition even more useful than the first.

The development of an effective incident commander involves several steps over a period of time. Training, education, and experience all play a role in the developmental process. A successful incident commander's background should include the following:

- Basic fire fighter training (as outlined in *NFPA 1001, Standard for Fire Fighter Professional Qualifications*)
- Experience as a fire fighter
- Advanced fire fighter training
- Company officer training (*NFPA 1021, Standard for Fire Officer Professional Qualifications*)
- Experience as a company officer
- Education in tactics, building construction, command, and organization
- Experience as a command staff officer (especially safety officer) and as a line officer
- Experience as an incident commander

This list is not all inclusive, and the mix of experience, education, and training will vary in the developmental process. Experience is especially difficult to categorize. The needed experience may not be provided by serving several years with a department or company that has few structure fires. In turn, the assignment of a fire fighter or officer to a very busy unit does not guarantee diverse learning. For example, a fire fighter on a busy company may always be in the attack position with a 1 3/4-in (44-mm) hose line. This fire fighter will be very proficient at advancing hose lines, but may not be knowledgeable about other aspects of firefighting such as truck company operations.

While it is not possible (nor desirable!) to totally eliminate experience from the process of becoming an effective incident commander, training and education do reduce the experience necessary to become proficient. Of particular value is live-fire training and realistic simulations. Some level of experience is necessary to "read the fire ground" and to understand the capabilities and limitations of resources and the impact of lead time. Ongoing training and education play a key role in the development of even the most experienced incident commander.

By applying the principles described in this text, an incident commander can utilize his or her experiences more effectively at the scene of structure fires. The incident commander who relies entirely on experiential learning essentially places the lives and property of others at a greater risk while he or she learns the trade. Sadly, many departments do not provide formal structural firefighting training. After the attack on the World Trade Center on September 11, 2001, the NFPA conducted a fire department needs assessment for the United States Fire Administration. The study found, in part, that many fire departments do not provide structural firefighting training to all members. A follow-up report, *Four Years Later—a Second Needs Assessment of the U.S. Fire Service*, found that 42% of departments still lacked formal training programs in structural firefighting. We hope that some of these departments will use this book to provide this essential training and education.

Learning charts are used to show the relationship of topics within each chapter. The learning charts, located on the first page of each chapter, group like topics much as a NIMS organizational chart groups like functions on the fire ground. The learning charts provide an overview of the chapter and relate similar topics, but do not necessarily follow the order of topics in the chapter.

Case summaries are used in this textbook to provide the opportunity to learn from the experience of others. Since some types of fires may only occur once in a lifetime (conflagration, large area fire, major high-rise fire), incident commanders relying strictly on experience will be at a loss when confronted with these rare incidents. Collectively, the fire service has learned many lessons about strategies and tactics that were applied at these major events. Only by reviewing case studies and promoting ongoing education can most fire officers learn about these most challenging incidents.

Case studies are available from many different sources. This book uses numerous case summaries within the text and references many other case histories discussed in fire service periodicals, NFPA Fire Investigations, NIOSH Firefighter Fatality Investigation and Prevention Program, and the Technical Report Series from the U.S. Fire Administration. Sometimes it is difficult to share our failures, but through sharing we can prevent future tragedies.

Pre-incident planning and standard operating procedures (SOPs) are essential elements of successful operations. Pre-incident planning is particularly important in large and complex structures. The authors of this book continually reiterate the importance of pre-incident planning to successful operations. Advance knowledge of some of the factors leading to strategic decisions before an incident occurs will reduce the information processing needs of the incident commander in the crucial early moments of an operation. Pre-designating initial operations using SOPs allows the incident commander to process information while a predictable course of action takes place.

While a great deal of science is associated with fighting fires in structures, the application of this science requires a less-than-scientific approach (some degree of art). As a matter of fact, the rigid application of scientific principles sometimes leads to theories that are technically correct but empirically fallacious.

As you progress through this text, you will find several places where a fallacy is introduced but then disputed. Some commonly held fallacies are described in this text for the purpose of disproving misconceptions. Fallacy/Fact boxes are used to alert you to commonly held beliefs that are fallacious.

This book is designed to be used in a distance-learning format with an instructor providing additional insight and practice materials as well as in a classroom environment. Suggested Activities are included at the end of each chapter to assist the student and instructor in applying the information contained in it. Discussion points for all the Suggested Activities are included on an Instructor's ToolKit CD-ROM along with PowerPoint presentations, lecture outlines, an electronic test bank, and an image and table bank. Given the distance-learning design, the book is ideal for self-study.

Fire is an integral part of our society. We depend on it for warmth and convenience. Fire allowed humans to progress from cavemen to the computer age. Yet uncontrolled fire is feared, and rightfully so, as it continues to destroy life and property. The modern fire department utilizes public safety education and code enforcement to help prevent and reduce the impact of fire. When these proactive efforts fail, fire fighters are forced into a reactive suppression mode. Even where considerable effort is expended to prevent fires, they do occur. In this book fire is viewed as a formidable adversary, and understanding "the enemy" is crucial if we are to be successful. Firefighting is as much an art as a science, but applying the available science can aid us in our war against fire.

Many have compared fire-ground management to commanding a military battle or coaching a football game. These analogies are correct and useful, but with one significant difference: The military commander and football coach face an unpredictable enemy as opposed to the incident commander who, if he or she has done effective pre-incident planning, has a very predictable enemy. Fire is bound by the laws of chemistry and physics. The surprises that may occur are a result of incomplete information or failing to understand basic natural laws governing the fire, building construction, and the impact of the fire on the structure.

This textbook is by no means the only one a fire officer endeavoring to be a proficient fire department incident commander should read. Although the modern fire department may respond to hazardous materials releases, medical emergencies, technical rescues, wildland fires, and a wide variety of other emergencies, the scope of this book is limited to structural firefighting. The incident commander should be a qualified safety officer and be well versed on fire fighter safety laws and standards. The National Fire Protection Association (NFPA), the Occupational Safety and Health Administration (OSHA), and the National Institute of Occupational Safety and Health (NIOSH) have developed standards related to fire fighters' health and safety that should be familiar to the command officer. Furthermore, incident commanders and safety officers must have an in-depth knowledge of current building and fire codes and standards.

Dedication

This book is dedicated to our colleagues

who gave their lives serving others.

Hopefully, by sharing our many years

of fire service training and experience,

we can prevent future tragedies.

Organizing, Coordinating, and Commanding Emergency Incidents

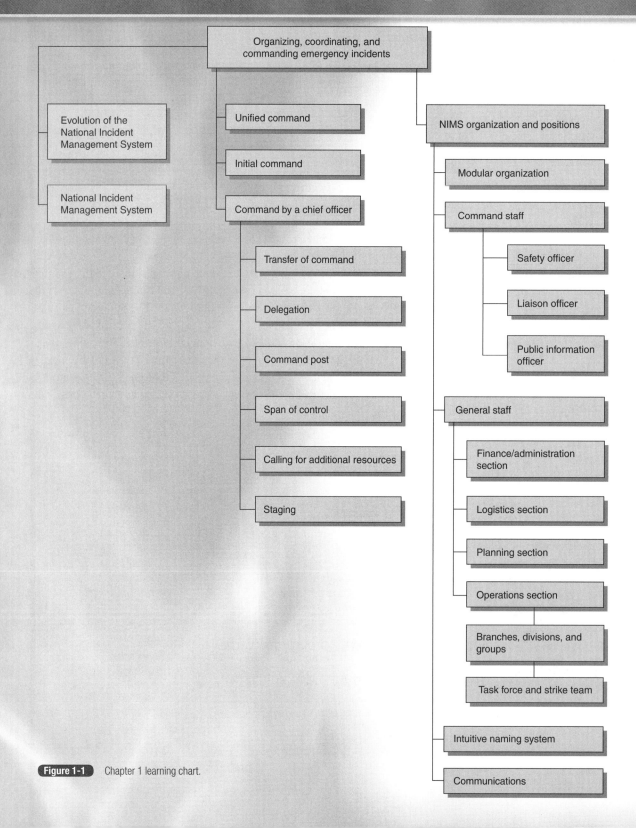

Figure 1-1 Chapter 1 learning chart.

Chapter 1

Learning Objectives

- Identify and define the main functions within the National Incident Management System (NIMS) and how they interrelate during an incident.
- Given different scenarios, organize an operation using NIMS.
- Discuss and contrast fire-ground management compared to administrative management.
- Discuss the history and evolution of incident management systems including the development of the National Incident Management System.
- Define unified and single command listing the advantages and disadvantages of each.
- Compare command modes available to the first arriving officer determining situations where each mode would be appropriate.
- Develop an initial report.
- Explain the importance of and develop a status report.
- Analyze the command transfer process discussing when and how command should be transferred.
- Define and list the problems associated with freelancing.
- List the attributes of a good command post.
- Define and explain the importance of maintaining a reasonable span of control.
- Describe and enumerate the importance of staging.
- Compare a staged company to a parked apparatus.
- Define incident commander (IC).
- Identify, define, and place command staff positions on a NIMS organization chart.
- Identify, define, and place the four sections on a NIMS organization chart.
- Describe the position of and function of a chief's aide.
- Define and describe the functions of branches, divisions, groups, task forces, and strike teams.
- Explain the two-in/two-out rule.
- Organize an operation using geographical and functional sectoring and describe when each should be used.
- Given a fire situation apply an intuitive naming system for various tactical level management units.
- Recognize and articulate the importance of fire-ground communications.
- List general rules for incident scene communications.
- Define and explain unity of command.
- List and compare various means of communications that could be used at the incident scene.
- Develop a communications network that supports a NIMS organization.
- Explain methods that can be used to reduce radio communications to and from the incident commander (IC).

Introduction

This chapter discusses the importance of adopting and implementing the National Incident Management System (NIMS) for use at all structure fires. Use of NIMS is essential to enable the **incident commander (IC)** to safely and effectively manage incident resources. NIMS is instrumental in ensuring that fire fighter safety, as well as the three operational priorities of life safety, extinguishment, and property conservation, are addressed at the fire scene. Because even the best organization is at a distinct disadvantage without good communications, which is probably the most frequently cited deficiency at actual emergencies and exercise scenarios, we also address communications in this chapter.

Decision making at the incident scene is different than day-to-day administrative decision making. The IC must decide on the proper course of action with limited information available in a relatively short period of time. This process is referred to as **recognition primed decision making (RPD)**. Administrative decision making is less time sensitive; therefore, input should be included from diverse sources, and careful analyses of all options should be considered. This is referred to as **rational decision making (RDM)**. Proficient ICs provide rational decisions within the time and information constraints of the fire ground by taking advantage of pre-incident planning and matching the situation at hand to similar experiences at previous fires.

The relationship of topics within chapters in this book is graphically represented by learning charts, which are similar to the NIMS organizational charts that ICs use during an incident **Figure 1-1**. Learning charts are designed to give the reader an overview of chapter topics and show the relationship between topics. These charts appear on the opening page of each chapter.

The adoption of an incident management system meeting the criteria outlined in *NFPA 1561: Standard on Emergency Services Incident Management System*[1] is a must. Such a system will allow for the expansion of forces while maintaining a reasonable span of control. The authors highly recommend the use of the **National Incident Management System (NIMS)**.[2]

Fire-ground operations have often been compared to military operations. A military commander uses a battle plan much as an IC uses a pre-incident plan. Standard military maneuvers (the equivalent of fire department standard operating procedures, or SOPs) are developed in advance. The military has a very rigid organizational structure that limits the span of control to a relatively small number. A general trying to give orders to each of the thousands of men and women under his or her command would soon be overwhelmed and lose control. This general may have success with skirmishes in which few soldiers are actually involved in battle, but he could not hope to win a major campaign.

On the fire ground, it is a mistake for a chief officer to attempt to control the entire incident. Like the general, he or she may be successful in handling a skirmish (e.g., a one-alarm fire involving only three or four companies). However, a chief who tries to apply a micromanaging style to a major incident will quickly become overwhelmed.

NFPA 1500: Standard for Fire Department Occupational Safety and Health Program[3] requires the use of an IMS, as do current Occupational Safety and Health Administration (OSHA) regulations dealing with hazardous materials response.[4] The implementation of NIMS, which the authors strongly recommend and which uses the original Incident Command System (ICS) terminology and structure, satisfies these requirements.

Evolution of the Incident Management System

A significant amount of work has gone into developing systems to manage emergency incidents of all types. Forest fires in California in 1970 prompted the development of the FIRESCOPE Incident Command System (ICS).[5] FIRESCOPE greatly improved operations at large wildland fires, and its use by fire departments of all sizes in southern California provided the testing ground that led to the urbanization of the FIRESCOPE Incident Command System. As use of the system expanded, improvements were made based on the experience gained at actual incidents. The result was an extremely useful tool that is available to the fire service and others involved in emergency response. This is the system used to coordinate resources within NIMS.

The fireground command system[6] was developed in Phoenix, Arizona, for use on structure fires. This system gained widespread acceptance as a very practical urban system and has been modified to manage larger incidents.

Command

National Incident Management System

It is essential that the entire response community, not just the fire service, be familiar with and trained in using NIMS. Agencies such as police, health department, and local disaster agencies, as well as mutual aid departments, should all work under the same system. The use of NIMS provides common terminology and operational assignments for all agencies at the incident scene. This uniformity can greatly facilitate the successful conclusion of an incident. Homeland Security Presidential Directive (HSPD)-5 establishes NIMS as the national system, requiring its use by federal agencies and making its adoption mandatory when the federal government is providing preparedness assistance.

NIMS should be used for *all* incidents, regardless of their size. If personnel are using NIMS for smaller, routine incidents, they will be better prepared to use it for large-scale incidents. It is a mistake to wait until an incident grows to a certain size before implementing NIMS; by then, it is too late to start trying to put together the organizational segments to support the incident. It is much easier to let NIMS grow with the incident and integrate other agencies and jurisdictions as needed.

The command system should be regional in scope and capable of handling large numbers of resources from beyond the local jurisdiction. NIMS is national in scope, and because of its flexibility and adaptability, it is capable of handling the largest imaginable structural fire incident. The focus of this book is on structure fires, but NIMS is an all-hazard system capable of handling any emergency situation that the IC may encounter.

Unified Command

NIMS addresses situations in which more than one jurisdiction or agency has responsibility by establishing a **unified command**. Unified command provides an invaluable method for controlling very large incidents in which multiple agencies or jurisdictions are involved. In a unified command scenario, different jurisdictions and/or agencies share responsibility for developing the incident action plan.

A word of caution about unified command is in order. Whenever possible, a **single command** (one person is designated as the overall IC) is preferred to a shared or unified command. However, some situations dictate the use of a unified command system, such as when the incident priority is not clearly defined as being one of fire or law enforcement.

Structure fires are clearly under the jurisdiction of the local fire department unless the structure spans more than one jurisdiction. Much has been written about the difficulty in communications and coordination between fire and police at the World Trade Center on September 11, 2001. Some feel this was a case for a unified command. The police department played a major role, but the emergency phase of the operation clearly should have been under the jurisdiction of the fire department. It is not unusual for

Case Summary

When the Alfred P. Murrah Federal Building in Oklahoma City was blown up by a terrorist's bomb on April 19, 1995, a myriad of agencies became involved in many different stages of the incident. Overall command of the incident was taken and held by the Oklahoma City fire chief, Gary Marrs. However, as the incident evolved, other jurisdictions became involved in the cause and origin determination, search for evidence, and other law enforcement functions. Despite the size and complexity of the incident, the fire department retained command of the incident, coordinating all operations with the different jurisdictions involved.

police to assist in evacuating or isolating the area during a structure fire; they also control traffic around the incident scene. There could be a large contingent of police officers at the fire scene, but the fire department should retain control, with a police department supervisor coordinating police activities from the fire department incident command post. The police department supervisor should be equipped with a radio capable of communicating with all police personnel, enabling the IC to communicate through the police supervisor. Though the World Trade Center events on September 11, 2001, may have been a case for unified command, seldom would a unified command be warranted for a structure fire.

A unified command can also evolve after the emergency has been mitigated and the incident has moved into the cleanup phase. For example, once the life safety issues have been resolved, several agencies might become engaged in a fire involving hazardous materials.

NIMS recognizes the potential problems in having various agencies with different priorities implementing an incident action plan. In using the unified command structure, plan development is shared, but one operations chief should direct all field units.

Initial Command

Major incidents typically begin with response by one or two fire companies. It is critically important that the person serving as the initial IC get the operation off to a good start by establishing command and following department procedures. It is very difficult to recover once an operation becomes chaotic.

It is certainly desirable for the first-arriving company officer to establish a formal command post outside of the structure and concentrate on directing operations. However, few fire departments have sufficient staff to allow the company officer to establish a stationary command post on arrival. It is important to recognize that the officer of a first-in unit, who is performing task-level functions and commanding the operation, is at a distinct disadvantage over the chief officer, who is at a stationary command post

Fallacy
Whenever more than one agency responds or the fire spreads over more than one jurisdiction, establish a unified command.

Fact
Whenever possible, establish a single command with one incident commander.

with a single function. If the first-in officer delays his or her actions to establish a stationary command post, the fire may make substantial progress during this delay.

It is essential that someone always be in command. Most departments respond to structure fires as fire companies, with a company officer or member in charge of the company. However, there are some situations where an individual fire fighter or officer arrives at the scene prior to the arrival of a fire company. In this case, this person is the initial IC, and most departments allow three command options:

- *Investigation:* Generally, nothing showing on arrival, no need to take immediate action to save lives or protect property
- *Fast attack:* Obvious signs of fire or some other dangerous conditions that require immediate action
- *Command:* A situation of such a magnitude or complexity that the company officer must assume a command position and not become directly involved in the operation so that he or she can coordinate the actions of personnel and other incoming units

Standard operating procedures (SOPs) allow the operation to continue while the company officer is busy managing the system.

Company officers must be reminded of the importance of establishing command at fires and other emergencies. To ensure fire fighter safety, there should always be a strong command presence. Establishing command during training does much to remind the company officer of this important concept. Post-incident critiques and tactics training are also necessary if company officers are expected to establish a stationary command post when circumstances permit or require it.

Some departments adhere to the strict interpretation of establishing command by announcing command. Others are satisfied with carrying out the duties, without an announcement of command, when the officer is in the investigation or fast attack mode. Each has its advantages. Announcing command reinforces the fact that this first-arriving officer is in command. Others believe that the announcement is unnecessary, as command is assigned to the first-in officer per SOPs. Regardless, there must always be someone in command of every incident, no matter how large or small. The expected initial report must be included in the department SOPs. The first arriving officer must report conditions upon arrival at the scene. Some of the common items included in an initial report are:

- Confirm address
- Confirm command

- Command mode (investigation, fast attack, or command)
- Brief description of building or specific name of well-known structure
- Occupancy
- Conditions—e.g., heavy smoke with occupants at windows
- Actions being taken
- Resources needed

The information required in the initial report should match department needs and resources. As an example, it may be necessary to include information regarding water supply if water supply requirements are not specific in the SOPs, or when parts of the jurisdiction do not have water mains and hydrants.

Use good communications techniques when using the radio:
- Take a deep breath.
- *Think* about what you are going to say before transmitting.
- Key the radio, followed by a very short delay.
- Speak slowly and distinctly.

It is imperative that all members who could possibly be in charge of the first-arriving unit practice issuing an initial report. Suggested activities at the end of this chapter provide an opportunity to do just that.

Command by a Chief Officer

The three command modes (investigation, fast attack, and command) apply only to company-level operations. When a chief officer assumes command, it is critical that a stationary command post be established. As higher ranking chief officers arrive, they have the option of accepting the command role and reassigning the previous command officer to a position that may be mobile (e.g., the safety officer), to an interior division, or to another general and/or command staff position. This should occur only after a formal transfer of command.

The chief officer assuming command must provide an updated status report covering the items listed for the initial report along with current progress. Part of the size-up and status report process involves reconnaissance from units at the scene. This is most often done using the radio system to request a progress report from operating units. The IC will also need to use the communications network to relay orders to tactical level management units or individual companies depending on the size and complexity of the incident. It is important to first identify the unit you are calling—e.g., "Command to Engine 1"—and when Engine 1 replies, issue brief, specific, clear orders. The unit being assigned should repeat the order. ICs must remember that completing a task will take time, and on occasion the unit assigned to complete the task will be unsuccessful. However, when an order is issued, the IC will assume that the objective is being successfully completed unless the company given the directive advises otherwise. Once the IC gives an order, the companies assigned the task(s) must complete the objective(s) and notify the IC that the assignment is complete or advise the IC that they are unable to complete the assignment and reasons why.

Transfer of Command

Command transfer must be addressed in department SOPs. In larger departments where there is a multi-level rank structure, it is essential that later arriving company officers know whether they are required to assume command or if they have other options. The first-arriving company officer may not be at a command post. When command is transferred to another officer, it should be formalized at a stationary command post.

Unless there is a compelling reason, command should not be transferred between company officers. The situations where command transfer between company officers is permitted should be clearly outlined in department SOPs. Multiple command transfers in the initial stages of the operation generally results in confusion and could reduce the number of people engaged in hands-on life safety and extinguishment activities. However, an officer who finds an unsafe operation in progress upon arrival must immediately take command. In this case, a formal command post should be established and immediate corrective actions taken.

When command is transferred, there is the possibility of information being lost in the process. It is necessary for the person assuming command to communicate with the previous IC to determine both situation and resource status of the operation in progress. While this communication is taking place, arriving units should be awaiting orders. The longer it takes to transfer command, the greater the chance of **freelancing**. A strong command presence and efficient transfer process will ensure a smooth transition and eliminate independent actions. Keeping everyone focused on achieving the tactical objectives outlined in the incident action plan is imperative.

The IC assuming command must evaluate the safety and effectiveness of the operation in progress. If the operation is unsafe or is not focused on accomplishing the incident priorities of life safety, extinguishment, and property conservation, the IC assuming command must reorganize the operation and reassign operating units as needed. It is very difficult to reorganize and redirect operations in the heat of battle, but it is sometimes necessary. If the current operation is being conducted safely and effectively, the task of assuming command is much easier. The person assuming command should improve the operation by updating the size up and augmenting operations in need of additional resources. Information available to the IC improves as companies relay critical information from the interior or remote positions and more time is available to review pre-incident plans and other information sources.

If department procedures give a more senior or higher ranking officer the option of taking command or allowing the present IC to continue serving in the position, numerous and unnecessary command changes may be eliminated. However, the higher ranking officer, especially chief officers, face a dilemma in allowing a lower ranking officer to retain command: If there is a major problem in resolving the incident, the highest ranking officer present on-site will be held accountable, even if he/she did not assume command. Remember, you can delegate authority, but you cannot delegate responsibility! Thus, the department and its officers must balance the possible efficiency gained in retaining command with the fact that higher-ranking officers are always held responsible and accountable.

The need for command transfer between chief officers other than the chief of the department is less obvious. Command transfers tend to disrupt the continuity of operations, but department culture or rules may require command changes.

On rare occasions, command of a structure fire could be transferred to another agency. Fires and explosions are used as weapons of terror. This could be a case where law enforcement would assume command from the fire department after the fire situation is controlled. The February 26, 1993, explosion and fire at the World Trade Center would be an example of a structure fire where command might eventually be transferred to police authorities.[7]

Fires involving hazardous materials can also result in transferring command to another agency. The accidental fire involving a large quantity of solvents at the BASF plant in Cincinnati is an example. Once the fire was brought under control and all victims removed, the Ohio EPA assumed command and maintained a presence at the scene for five months during cleanup operations.

Delegation

Establishing control over available resources entails the delegation of authority. The IC develops the strategy while branch, division, and group supervisors develop tactics within the overall strategy by assigning and coordinating tasks for units that are working under their supervision. Information exchange through good communications is the key to managing the fire ground.

Command Post

The first-arriving company officer will be in command but generally will be working inside the building during an offensive fire attack. However, fires that are obviously defensive operations from the beginning are normally handled by the first-arriving officer establishing a stationary command post and assigning other first-in companies.

A good command post will:
- Be established in a location that is known and easily found
- Be outside the hot zone, in an area where personal protective clothing is not required (cold zone)
- Provide a view of the two most important sides of the building
- Never hinder apparatus movement

In most cases, companies and agencies will report to the command post for instructions and information. Therefore, the command post should be easily found. When establishing a command post, the IC should communicate the location using the street name for an exterior command post (e.g., First Street Command) or the building name when using a command center within a building (e.g., Sears Tower Command). There are times when some degree of isolation is preferred. Companies can be assigned to report directly to group and division supervisors or branch directors. Outside agencies should

Fallacy: Responsibility can be delegated.

Fact: Responsibility cannot be delegated, but authority can be delegated.

report to a liaison officer, and the media should be managed by the public information officer.

Positioning the command post so that two sides of the building can be viewed is usually good practice. There is a distinct advantage in being able to see the effects of tactical decisions. However, there are times when seeing the scene can be a distraction, causing the IC to focus on the visible while failing to deal with higher priorities that cannot be seen. Generally, the larger and more complex the situation, the farther away and more isolated the command post should be. Isolation from distractions at the scene can be an important factor in developing good incident action plans. Most often the command post will be located at or near the scene of a structure fire. When confronted with a large-scale operation involving very large buildings or multiple buildings it is sometimes a good idea to place the command post in a nearby building with good lighting and communications. For a catastrophic incident, such as the September 11, 2001, attack on the World Trade Center, a command post should be established at a location well away from the incident. Most city and county governments have pre-planned Emergency Operations Centers (EOCs), which could be used for this purpose. The command post is at the location of the IC; even if an EOC is functioning for a structure fire, when the IC is at the incident scene, then so is the command post.

Managing an incident requires the IC's undivided attention. The command post should be in a location that supports command, control, and coordination of all incident activities. Good communication is critically important to the command function at the command post.

Most fires are managed by an IC with a few command staff, sections, or tactical level management units. ICs at these everyday incidents generally use an automobile or sports utility vehicle as a command post. Some ICs prefer to establish command by establishing an outside position away from their vehicle or on the outside at the rear of their vehicle. Many command vehicles have status boards and other command post equipment in the rear vehicle compartment or trunk. Other departments require or strongly suggest that the IC stay inside the vehicle and place command equipment in the front seat area to support this policy. Sports utility vehicles are particularly well suited for placing command post materials and equipment in the front seat or rear compartment. Working from the vehicle generally provides better communications, as vehicle radios are usually more powerful and the vehicle engine maintains a constant charge. Working inside the vehicle also affords a measure of security and safety for the IC, as well as providing climate control. Commanding from an exterior position in high winds, heavy snow, torrential rains, or other types of inclement weather can be a distraction or even dangerous. The larger and more complex the incident, the greater the need for command post facilities and space.

> Wherever the Incident Commander is located—that is the command post.

Span of Control

Safety factors, as well as sound management planning, both influence and dictate span-of-control considerations. The **span of control** is the number of people reporting to a supervisor, and for any individual with emergency management responsibility, the span of control should range from three to no more than seven, with five being the rule-of-thumb average.

The type of incident, the nature of the task, hazard and safety factors, and the distance separating tactical units will influence span-of-control decisions. An important consideration in span of control is to anticipate change rather than react to it. This is especially true during rapid buildup of the command structure, when good management is made difficult because of numerous reporting units.

Large-scale operations where the span of control exceeds manageable limits become chaotic and unsafe. Span-of-control limitations are generally in the area of five subordinates reporting to a single supervisor under emergency conditions. NIMS recommends a span of control ranging from three to seven people reporting to a supervisor. A one-to-one span of control serves no useful purpose, making the NIMS organizational structure more complex than necessary. The larger and more complex the NIMS organization, the more difficult it is to control. An important rule to remember is "keep it simple." Form tactical level management components as necessary to maintain a reasonable span of control and to coordinate specific geographic and functional operations. However, remember that each organizational layer places the IC farther away from having direct contact with operating units. In some large, complex incidents, this is necessary and desirable; however, at most incidents it is not.

A question sometimes arises regarding a two-to-one span of control. If there is a need to sector an operation to maintain a reasonable span of control, and only two units are needed in a geographic area or to perform a specific function, a two-to-one span of control may be acceptable **Figure 1-2**.

Calling for Additional Resources

Some would recommend a piecemeal approach for requesting help to avoid being overwhelmed by units reporting for assignments. A much better approach is to stage arriving resources until assignment needs are identified. Companies or crews can be sent directly from staging to a group, division, or other assignment. Remember, it is best to call for additional help before it is needed. If you call for additional resources after the need is obvious, they will arrive too late. The need for additional units must be anticipated and requested well in advance of when the units actually should be in position.

Staging

Staging areas should be established to locate resources that are not immediately assigned a task. Staging allows the IC to control access to an incident scene, while deploying resources in a safe and effective manner. A staging area can be located anywhere mobile equipment can be temporarily parked while awaiting assignment. It is best to stage units two or more blocks from the actual incident to avoid the temptation to freelance into action and to ensure that staged units are in a safe area. When establishing a staging area, locate it far enough away to avoid obstructing or slowing access to the incident scene, but close enough to allow companies to quickly arrive once summoned. When weather conditions are extreme, such as freezing temperatures, shelter should be provided for companies that could be staged for a long period of time.

The best time to identify good staging locations is during pre-incident planning—not in the middle of the night during the heat of battle. Predetermining the location of the staging area is especially important for large, complex properties. Depending on factors such as property size, access issues, and weather conditions, the staging area could be an open-area parking lot, a nearby firehouse, or a predetermined location that would provide some measure of security and protection from the elements. When a chief officer arrives on the scene, there is an immediate need to re-evaluate the situation and complete the incident action planning process. During these early stages of an operation the IC can be easily distracted by information overload from several sources, including radio communications from both on-scene and incoming units. If the IC has not completed the incident action planning process, including the determination of how responding fire forces are to be deployed to carry out the plan, incoming units will have a tendency to freelance into action. Further, an IC who is already overwhelmed with radio traffic is often reluctant to call for additional assistance, even when the need is obvious. Establishing a staging area early on, and directing all responding companies to that location unless otherwise ordered, allows the IC to better manage

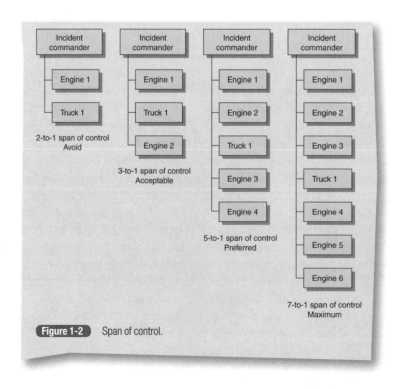

Figure 1-2 Span of control.

> Staging is an important management tool that increases incident scene safety and efficiency by providing a means to immediately deploy forces to achieve identified tactical objectives while preventing dangerous freelancing.

on-scene units, establish a ready tactical reserve, and eliminate potentially dangerous freelancing. Establishing a tactical reserve is an important command consideration. If all units at the scene are committed to the operation, a tactical reserve is needed to cover unanticipated problems and provide relief to operating crews.

The staging area can also be used as a parking area for out-of-service apparatus. During a large-scale offensive operation, later arriving units often are needed only to deliver additional staff. The apparatus pumps, ladders, and tools are not needed; therefore, the unneeded apparatus are parked out of the way in the staging area. It is important to distinguish between an out-of-service apparatus parked in staging and a staged unit that is part of the tactical reserve. A staged unit is a fully staffed unit, such as an engine company, truck company, or medic unit that is ready to respond immediately to the incident scene. Apparatus without adequate staffing are classified as out of service, rather than staged.

SOPs should outline how the staging area is managed, the duties of the staging officer and of the company officers whose units are assigned to staging, as well as the physical characteristics of an effective staging location. During large-scale incidents a staging officer is assigned to manage staging and a more formal staging area is established in an area that provides adequate space and access to the incident scene. The staging officer is responsible for managing all incoming resources and dispatching resources at the request of the IC or operations chief. Equipment and staffing in staging areas must be ready for immediate response. Crews should remain intact and available for immediate deployment. Span of control is not a problem in staging. The staging officer can manage numerous companies because of their inactive status.

Interior staging is typically used for a fire on an upper floor of a high-rise building. High-rise staging procedures are addressed in Chapter 12.

NIMS Organization and Positions

As the need for resources increases, there is a need for a larger and more complex organization. NIMS is a tool used to organize an operation. NIMS is not a tactical objective, but a means to command and control an incident to achieve incident action plan objectives. With this in mind the NIMS organization should be as simple as possible.

> The only NIMS position that must be established at every incident is incident commander.

Modular Organization

The NIMS organizational structure develops in a modular fashion based on the type and size of the incident. First and foremost, there must always be an IC. Other line and staff positions are assigned by the IC according to incident priorities.

The specific organizational structure that is established for any given incident will be based on the management needs of the incident. If one individual (the IC) can simultaneously manage all of the major functional areas, no further organization is required. However, if one or more of the areas require independent management, then an individual should be assigned to manage each of the necessary areas. If the IC does not specifically delegate the incident functions, then he or she retains responsibility for all that are not delegated. The NIMS positions should be thought of as job descriptions for the IC. The IC decides whether to do the task or to have someone else perform that function. NIMS positions are divided into command staff and general staff (sometimes referred to as line officers).

Command Staff

All command staff positions report directly to the IC. Command staff positions are established to assume responsibility for key activities that are not included in the line organization. NIMS identifies three command staff positions:

1. Incident safety officer
2. Liaison officer
3. Public information officer

These positions are generally placed in "staff" position on the NIMS organizational chart as shown in Figure 1-3.

Incident Safety Officer

The **incident safety officer** is one of the key positions on the fire ground that can play a critical role in ensuring fire fighter safety. This position should be staffed by an experienced

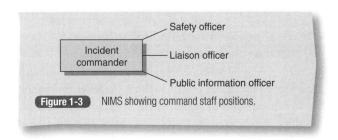

Figure 1-3   NIMS showing command staff positions.

officer who meets the requirements outlined in *NFPA 1521: Standard for Fire Department Safety Officer*.[8] This standard outlines situations when the position should be formally established on the fire ground:

> **6.1.1** The incident safety officer shall be integrated with the incident management system as a command staff member, as specified in *NFPA 1561: Standard on Fire Department Incident Management System*.
>
> **A-6.1.1** Incident scene safety needs to be carried out at all incidents. It is the responsibility of the incident commander (IC) who cannot perform this function due to the size or complexity of the incident to assign or request response of an incident safety officer to this function. There are, however, incidents that require immediate response or appointment of an incident safety officer. This type of incident should be defined in the fire department's response policy or procedures to ensure that the incident safety officer responds. Likewise, some situations require an incident safety officer to respond after members are on the scene, such as a working fire or at the request of the incident commander.
>
> **A-6.1.2** A fire department should develop response procedures that ensure that a predesignated incident safety officer independent of the incident commander responds automatically to predesignated incidents. Examples could be as follows:
>
> **(1)** Commercial fire
> **(2)** Multiple-alarm fire
> **(3)** Serious member injury or member transported for treatment
> **(4)** Hazardous materials incidents
> **(5)** Technical rescue incident
> **(6)** At the request of the incident commander

Safety is the command staff position that should be staffed most often. The IC should be at the command post, but the incident safety officer must monitor all areas where fire fighters are conducting operations. Therefore, it is virtually impossible for one person to effectively handle both functions at a large or long-term incident. Some incidents may require assistant safety officers. When more than one safety officer is assigned, safety officers could be assigned responsibility for specific areas, or multiple assistant safety officers could be assigned to the incident safety officer responsible for coordinating all incident safety activities.

It is important that incident safety officers focus on the overall operation and major risks to fire fighters. There is a tendency for safety officers to spend too much time concentrating on minor safety infractions while failing to recognize potentially deadly hazards. As an example, the safety officer checking protective clothing could fail to evaluate structural stability.

Liaison Officer

Liaison is the point of contact for agencies that are not assigned to operations functions. If more than one person or unit from an agency responds to the incident, one person from that agency should be appointed to communicate with the **liaison officer** (or the IC if liaison is not staffed).

This person would represent the agency and be responsible for coordinating other people and units from the agency that are working at the scene. In most cases, the police department reports to liaison if this command staff position is staffed. However, if law enforcement personnel are playing a major role in meeting incident action plan objectives, they would then report to a general staff position within the operations section or directly to the IC. If the IC does not staff the liaison position, representatives from all responding agencies will communicate directly with the IC. Most structure fires do not require separate staffing of the liaison officer position. However, when many agencies are assisting and all of them are reporting to the IC, it will become extremely difficult for the IC to effectively communicate with emergency resources operating at the scene. When this is the case, the IC should staff the liaison position.

Public Information Officer

The **public information officer (PIO)** disseminates information to the public, usually via the media. This officer provides critical information to the community regarding protective actions that need to be taken as well as information of general interest. An efficient PIO will monitor communications at the command post and keep abreast of the current status of the operation. It is essential that all communications to the public, concerning emergency operations related to the incident, be from the IC or PIO in consultation with the IC. All information relayed to the public must be cleared by the IC. This is particularly critical when there are injuries or fatalities, or when there is suspicion that a crime has been committed.

If there is a high level of public interest in the incident, the media will require a substantial amount of time, thereby making it difficult for the IC to maintain communications with operating units and the media. In this case, a PIO should be appointed or other general NIMS positions handed off to allow the IC time to communicate with the media.

Some question arises as to whether the command staff positions should be counted in the span of control for the IC. A valid argument can be made for not counting these positions against the recommended three to seven reporting positions in the IC's span of control, as they serve roles similar to those of command aides or adjuncts.

Some departments preassign staff officers to command staff positions. For example, the training chief may function as the safety officer at the scene. There is value in preassignment. However, preassignments may delay the staffing of important positions. To ensure that positions such as safety and liaison are filled, it is often necessary to make temporary assignments until the preassigned personnel arrive at the scene. Whoever is assigned to fill one of these positions must be thoroughly trained and qualified to carry out the responsibilities.

At times it is possible to combine assignments. One person may be able to manage both liaison officer and PIO, provided that the workload is not too great. However, when delegated, the safety officer should always be a separate assignment. Command staff officers and section chiefs are frequently located at the command post, whereas personnel who are assigned as safety officers are usually mobile.

General Staff

General staff positions provide a pyramid-structured hierarchy capable of coordinating and controlling the incident. The IC is at the top of the pyramid with five possible organizational layers between the IC and individual responders **Figure 1-4**. Rarely would it be necessary to use all NIMS hierarchical levels and all subordinate units at a structure fire. Positions are staffed when incident conditions require separate management of the functions of that position.

Sections

Section chiefs report directly to the IC **Figure 1-5**. As priorities are established, four separate sections can be assigned (finance, logistics, operations, and planning), and

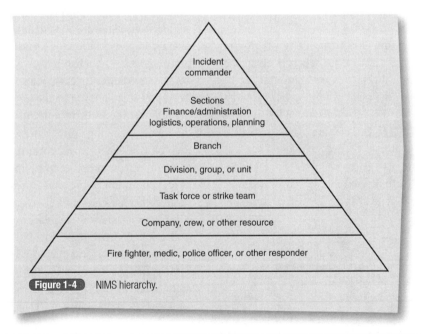

Figure 1-4 NIMS hierarchy.

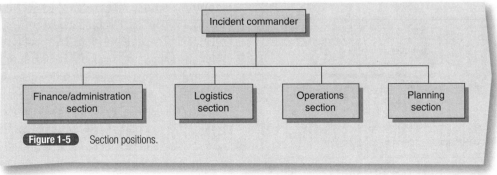

Figure 1-5 Section positions.

> If you do not practice and implement NIMS, your operation will likely be a FLOP.

each section can have several subordinate units. Only in the most extreme circumstances would all section positions be staffed at a structure fire. When established, section chiefs can further delegate management authority for their area of responsibility as required. If necessary, functional units may be established within the section by the section chief. Similarly, each functional unit leader can assign individual tasks within the unit as needed. Units subordinate to each section are staffed depending on incident requirements.

The four main sections of the NIMS organization under the direction of the IC are as follows:
- Finance/Administration section
- Logistics section
- Operations section
- Planning section

Intelligence is recognized as a possible fifth section under NIMS.

It is important to note that these positions are *sections*, not sectors.

Finance/administration section. The **finance/administration section** is least likely to be separately staffed at the scene of a structure fire. Finance/administration, as the name indicates, manages financial matters and/or provides administrative services; the functions provided by this section are shown in **Figure 1-6**. There are some cases where the fire department is able to recover expenses, such as during hazardous materials–related incidents. For instance, the Cincinnati Fire Department was able to recover expenses after the previously referenced BASF chemical company fire. However, it is important to remember that, even at smaller incidents where the finance/administration section isn't staffed, these duties are important and remain the IC's responsibility. In these instances, finance/administration duties might be limited to completing a company incident report that could be later used by the property owner and/or insurance company. However, at larger, more complex incidents, detailed documentation might be necessary to secure reimbursement for damaged equipment, payroll for responding or back-fill personnel, personal injuries, or many other reasons. At these incidents the finance/administrative position should be staffed.

Logistics section. The **logistics section** locates and provides the materials, equipment, supplies, and facilities required to support incident operations. The logistics chief can be thought of as the supply sergeant or quartermaster. Most of the resources needed to support incident operations are already on the scene or provided by additional fire companies; therefore the logistics section is seldom staffed at small structure fires. However, there are circumstances where supplies or facilities are needed. As an example, a fire situation where companies are on the scene for an extended period of time would require rehabilitation with food, water, sheltered areas, and possibly toilet facilities. The logistics chief would be responsible for bringing these resources to the scene.

One of the most important logistics section units is the communications unit. This unit assists in setting up the communications network, as well as providing and maintaining communications equipment. Some departments have a communications person respond to the scene with spare radio batteries and chargers. The need to staff the communications unit increases as the communications network becomes more complex.

The ground support unit would provide items such as fuel at the scene of a structure fire. Logistics section subordinate units are shown in **Figure 1-7**. Some charts further subdivide the logistics section into service and support branches.

Planning section. The **planning section** chief is the information manager. Planning should be the most frequently implemented section position. The planning chief gathers information about the incident, tracks resources, and assists the IC in developing the incident action plan and alternatives. This should be one of the first sections staffed during a major incident. When transferring command, consider assigning the previous IC as the planning section chief, as he or she should know the status of the operation as well as the number and location of incident resources. A major part of the planning

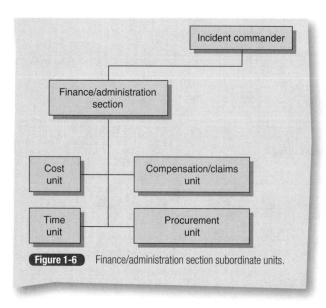

Figure 1-6 Finance/administration section subordinate units.

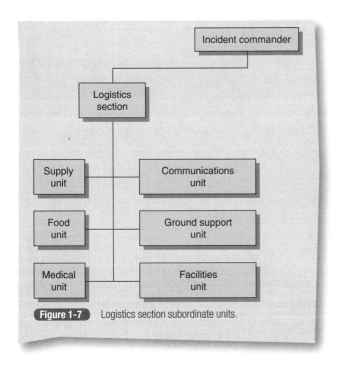

Figure 1-7 Logistics section subordinate units.

The documentation unit would collect incident information, which is important to a formal incident evaluation. The documentation unit becomes more important when there will be a formal investigation or study following the incident and the finance/administration section is not staffed.

Most technical specialists report to the planning section, but they could report to other sections, command staff, or the IC as needed.

In most cases, the intelligence function would be handled by the planning section. However, an intelligence section can be established if there is a need to analyze a large quantity of information. Separate staffing of the intelligence section is most important when there is sensitive or classified information being processed. Intelligence can also be established as a planning or operations section unit, as well as a command staff position.

A word about the chief's aide position. Closely related to planning is the position of chief's aide. At one time most large city fire departments provided chief officers with a chief's aide, but due to budget constraints, many cities eliminated this position. Some departments called the chief's aide a driver or chauffeur, which made it more difficult to defend retaining the position. As administrators examined fire departments in search of ways to cut the budget, the "driver" position was a likely candidate for elimination.

The chief's aide drives the chief's vehicle. Driving is a small but important part of the duties assigned to an aide. Having an aide drive the chief officer to the fire scene

section chief's responsibilities is tracking and documenting incident status and on-scene resources, oftentimes referred to as "SITSTAT" and "RESTAT" or Situation Status and Resource Status, as shown in **Figure 1-8**.

The demobilization unit prepares and implements a plan to return personnel and resources to service as an incident is being brought under control. This is an important planning section function when large areas have been stripped of resources.

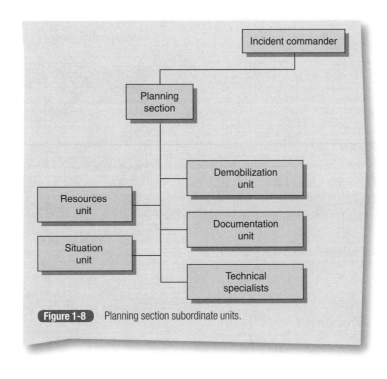

Figure 1-8 Planning section subordinate units.

allows the chief officer to concentrate on communications as well as accessing pre-incident plans and other sources of information to start the size-up process. At the scene of an incident, the chief's aide can manage command tasks so the chief officer can concentrate on developing an incident action plan and the deployment of forces. The chief's aide can also be assigned to do a complete walk-around of the fire building, handle communications, manage the accountability system, maintain the NIMS organizational chart, and research information sources.

The aide's duties could be described as the planning section or as a planning section subordinate unit, such as Resource Status (RESTAT) or Situation Status (SITSTAT). Departments who have chiefs' aides strongly defend keeping the position and usually refer to the chief's aide by a name more suitable to the duties they perform, such as Field Incident Technician or Command Adjunct. In the City of Los Angeles, the aide is called a Staff Assistant and is part of the Battalion Command Team.

In the Louisville (Kentucky) Fire Department, the chiefs' aides assigned to battalion and assistant chiefs are typically assigned to a planning role on the incident scene, where the aide assigned to the Chief of the Department serves as the public information officer. Under this structured system, the aide's role directly supports the chief officer's primary duties. RESTAT and SITSTAT, which are part of the planning section function managed by battalion and assistant chief aides, provide the operations chief and/or IC with information critically important to ensuring fire-ground efficiency, effectiveness, and safety. The aide assigned to the Chief of the Department works directly with the chief to ensure that the media receives timely and accurate updates. Even if the officers change roles, i.e., when the chief assumes command, the aides continue in their assigned planning and public information roles. This ensures a smooth transition during change of command.

A chief's aide can assist the IC in organizing and coordinating a safe and effective operation. Departments who do not have members assigned as chiefs' aides often assign responding personnel to these duties at the incident scene. This temporary assignment is highly recommended as an alternative. However, this temporary assignment is not equal, as command assistance will be delayed and lack the team concept developed between a chief officer and a regularly assigned aide.

Operations section. The <u>operations section</u> manages all tactical operations, such as search and rescue, extinguishment, and providing medical care to victims. When the IC hands off the operations section, the operations chief will make all tactical assignments and control all resources working to resolve the incident. Delegating the operations section allows the IC to focus on the overall strategy and other command functions. However, it is important to note that staffing the operations section means the IC no longer has direct control over operating companies, so there is another layer of organization between the IC and operating units. Typically, the IC delegates authority for operations only at large-scale incidents. Senior, ranking officers sometimes assign the person they are relieving of command to the operations section to provide the lower ranking officer with experience in managing larger operations, while the senior officer maintains overall control as the IC.

The operations section generally has the most resources, and includes a complex organization that may require several hierarchical levels to effectively maintain a reasonable span of control. The hierarchical levels within the operations section are shown in **Figure 1-9**.

The operations section is capable of controlling as many as 625 companies or crews with a five-to-one span of control **Figure 1-10**. At the maximum allowable span of control (seven-to-one) a total of 2401 companies

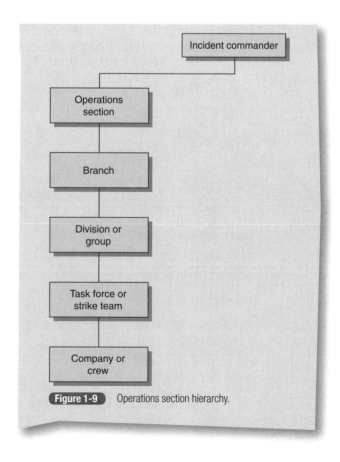

Figure 1-9 Operations section hierarchy.

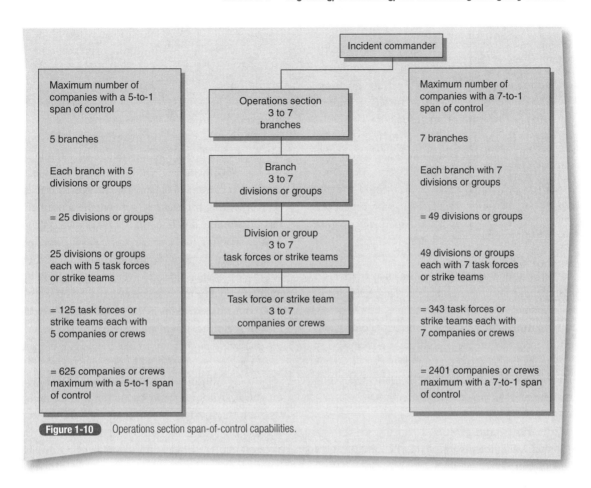

Figure 1-10 Operations section span-of-control capabilities.

or crews can be effectively controlled. Numerous large-scale disasters, including hurricanes, wildland fires, and earthquakes, have demonstrated that when properly implemented, NIMS includes all of the tools necessary to effectively manage the largest and most complex operations. Obviously, NIMS has the capacity to manage the largest imaginable structure fire.

Which sections the IC should hand off depends on the incident. Some officers may be reluctant to hand off the operations section, as they lose all direct contact with tactical operations. Some departments require the highest ranking officer at the scene to assume command. Other departments give a higher ranking officer the option to assume command of the incident, to fall into the command structure under the direction of the present IC, or to serve as a senior advisor.

Allowing a lower ranking officer to retain command shows confidence in the officer and provides experience in commanding a large-scale situation. However, in the end, the senior officer will be responsible for the outcome of the incident, regardless of whether he or she formally assumed command. By handing off the operations to the lower ranking officer, for example, the higher ranking officer can assume command and direct the general strategy while the lower ranking officer gains experience in handling the dynamic, tactical assignments.

Command staff and section leaders make up the **incident management team**, which provides the necessary staff and line functions for an incident. Incident management teams for most structure fires are made up of members from the responding department or from departments that normally provide mutual aid assistance. However, the Federal Emergency Management Agency (FEMA) is encouraging the development of all hazard incident management teams on a regional and state level. Regional, state, or national incident management teams would include members certified to function in the various command staff and section positions, much like the "Red-Card" system used by the forest service when combating large-scale wildland fires. The U.S. Fire Administration provides incident management team training to facilitate certification as an all-hazard member of an incident management team. Some regional teams establish training and certification standards identified within the region. Regional teams are particularly valuable when the responding

> **Two-in/two-out rule:** Fire fighters working inside the hazard area must work in crews of at least two people (two-in). During the initial stages of an operation the two people working in the hazard area must be backed up by at least two people outside the hazard area (two-out) who are properly equipped and immediately available to come to the aid of the inside crew.

department does not have personnel specifically qualified to staff a command staff or section position. As an example, members of regional departments could attend training and educational sections to qualify as safety officers. A qualified safety officer could then be dispatched to any report of a working structure fire within the region.

A more detailed description of each section's responsibilities and subordinate units is provided in the U.S. Homeland Security publication, *National Incident Management System*.[2]

Throughout an incident, from inception to conclusion, the duties of each section are the ultimate responsibility of the IC. The IC chooses to retain the duties or delegates sections that will assist in managing the incident, depending on the incident and the resource requirements.

Branches, divisions, and groups. In using NIMS, the first management assignments by the IC will normally be to hand off geographic and functional areas of responsibility at the **division/group** level. A division is in charge of a geographic area, whereas a group is in charge of a functional area.

As the fire progresses from a first-alarm response to extra alarms, there may be a need to sector the operation to maintain a reasonable span of control or to better manage specific functions of the operation. For example, suppose a fire department responds to an apartment fire with three engine companies and one truck company. The command structure would be as shown in **Figure 1-11**.

Engine 1 represents the first-arriving engine company, which arrives nearly simultaneously with Truck 1 from the same fire station. Given the fact that an interior operation is indicated, the **two-in/two-out rule** must be followed with at least two members assigned as the rapid intervention crew.

Another example is an apartment fire extending into the hallway and the floor above. Rescue operations are needed on all five floors of the building. The IC would normally assign forces working in the same geographic area to a division to reduce the span of control. For example, suppose the apartment building has a fire on the first floor with three engine companies and a truck working that area. An engine company and a truck company are working on the second floor, and an engine company is assigned to each of the remaining three floors. The incident could be managed in several different ways, but with nine companies working at the scene, there is a definite need to start sectoring. There are many correct ways to organize the incident; one of many correct methods is depicted in the organization chart in **Figure 1-12**.

In the authors' opinion, all of the companies working on the first floor should be under the supervision of a single branch, division, or group. Some companies are performing search-and-rescue tasks and therefore could be assigned to the search and rescue group. However, having units in the same general area reporting to different supervisors is not a good practice. The authors of this book prefer geographic rather than functional assignments whenever possible. Most ICs share this preference because communications are improved and the status of the assignment can be personally monitored by the division supervisor, thus eliminating the need for status reports within the division.

The three companies working on the upper floors in the apartment building example could report directly to command, but functional sectoring helps the IC to manage this important component of the incident action plan. The supervisor in charge of the search and rescue group will manage his or her assignment on all three floors, reassigning or requesting assistance as needed to complete the search-and-rescue operations for the upper floors. The IC's incident action plan calls for removing the occupants. Therefore, the search and rescue group supervisor applies the tactics necessary to achieve command's objectives and makes the necessary task-level assignments to companies within the search and rescue group.

For example, suppose the company assigned to the third floor finds very little smoke and can quickly complete

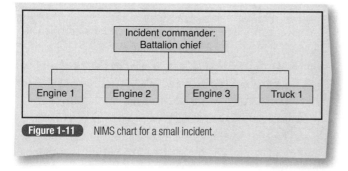

Figure 1-11 NIMS chart for a small incident.

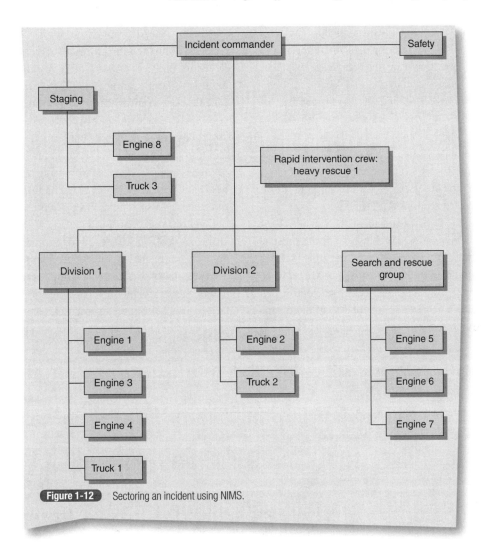

Figure 1-12 Sectoring an incident using NIMS.

the primary search on that floor. There are many occupants on the top floor who need assistance, and the primary search is going very slowly there. The group supervisor could reassign the company from the third floor to assist personnel assigned to the top floor. After the primary search is complete, the group supervisor could assign companies to the secondary search, avoiding repeat searches by the same companies.

All tactical operations within this group are managed by a single supervisor, who organizes and coordinates all activities assigned to his or her group. The IC is regularly informed of progress being made within the search and rescue group but does not need to assign or reassign companies, as these responsibilities are usually assigned to the branch, group, or division supervisors.

<u>Branches</u> can be used in place of divisions or groups, but this is not the recommended way to initially reduce the span of control. Branches are reserved for operations beyond the span of control of a single division or group, or when a contingent of units from another agency is working together, such as a police branch. For example, if nine companies were being used on the first floor of the apartment building, division 1 would become branch 1 with a search-and-rescue group or groups and a suppression group reporting to the branch director.

A large medical or police contingent could also dictate the need to establish branches. If medical services are provided by a separate agency, a branch assignment is even more likely for the medical component of the operation.

Nine companies working on the first floor under a single branch, as well as the entire medical operation under a medical branch, are shown in the organizational chart in Figure 1-13.

A division is used to identify a geographic area of responsibility, while a group designates a functional responsibility.

The BASF Chemical Plant Case Summary (see the case summary box) describes a structure fire where branches, groups, and divisions are used, as well as implementing three of the four NIMS sections Figure 1-14.

In 1990 when the BASF fire occurred, ICS terminology included the term "finance section" instead of the present "finance/administration section." The IC began by sectoring the BASF operation geographically, but there were 15 companies working on the Dana Avenue side of the fire. The IC recognized that the recommended span of control was being exceeded; therefore, the Dana Avenue side was further subdivided into the Dana Division East and Dana Division West. Even with this subdivision, there were eight companies reporting to Dana Division East, which exceeds the recommended seven-to-one maximum span of control. To reduce the span of control, the Dana Avenue side could have been assigned to a branch with three or more subordinate tactical level management units. Also worth noting is that not everyone reporting to a branch needs to be at the division/group level. Individual companies or task forces could also report to a branch. A revised version of the organization chart for the Dana Avenue side of this fire is shown in Figure 1-15. There are many correct ways to organize this operation, but the geographic sectoring allowed division supervisors, many of whom were company officers, to manage the operation in their area of responsibility using face-to-face communications.

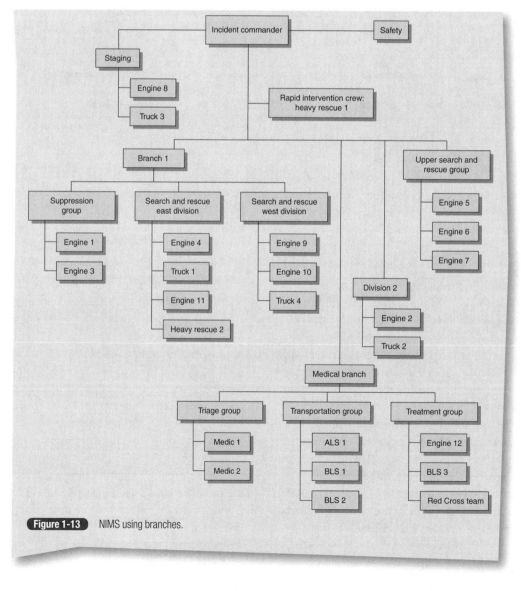

Figure 1-13 NIMS using branches.

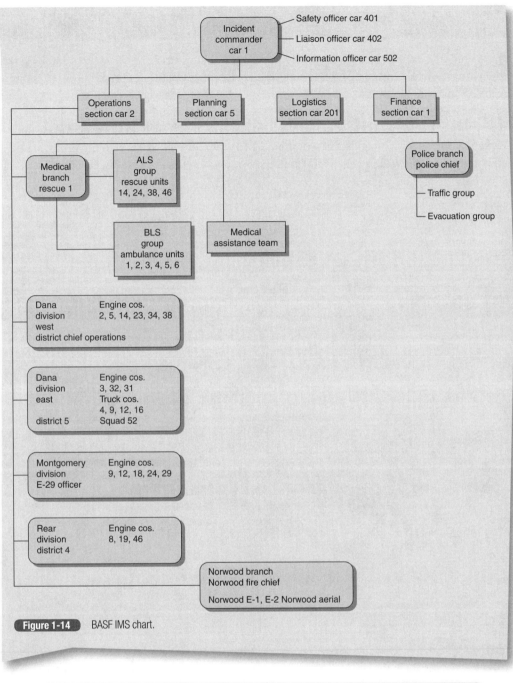

Figure 1-14 BASF IMS chart.

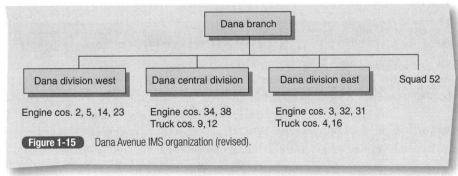

Figure 1-15 Dana Avenue IMS organization (revised).

Case Summary

On the afternoon of July 19, 1990, an explosion and fire at the BASF Chemical Plant in Cincinnati, Ohio required the services of 25 fire companies, EMS units, police, and other agencies. The ICS was used to coordinate and control activities at the scene. The rear of the plant bordered Norwood, Ohio. As companies arrived, they found severely injured employees, a massive fire, and continuing explosions. Two plant employees were killed, and 88 people were injured. The 16 buildings on the BASF complex were heavily damaged or destroyed; 161 non-BASF buildings were damaged. Command was transferred to the Ohio EPA on the evening of July 20, 1990. Cleanup efforts with fire companies in a stand-by mode continued for nearly five months.

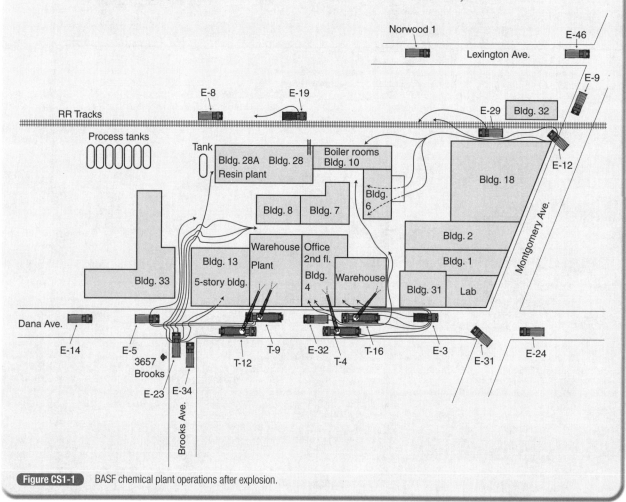

Figure CS1-1 BASF chemical plant operations after explosion.

A Word About Sectors

Prior to the formal implementation of NIMS, the term "sector" was commonly used to identify geographic or functional tactical level management units at the division/group level. The current edition (2005) of NIMS does not recognize the term "sector." *NFPA 1561: Standard on Emergency Services Incident Management System*[1] has also removed the term sector from the standard.

Task Force and Strike Team

The use of **task forces** and **strike teams** is an additional way to reduce the span of control, thus reducing the communications load at an incident Figure 1-16. Task forces can be any combination of resources, whereas strike teams must be resources of the same type. Strike teams are a common way to organize large numbers of mutual aid companies or private sector resources. Some states define strike teams as a specific number of companies to be used as a statewide resource. A strike team could be any number of units of the same type, but the jurisdiction defining the strike team must predetermine the number of units. As an example, most state and local governments assign five engine companies to an engine company strike team. They also address minimum staffing, communications, and other factors. Task forces are much more flexible and are more likely to be formed at the time of the incident. A common application of the task force concept would be two engine companies and a truck company working as a team.

Intuitive Naming System

Notice in Figures 1-12, 1-13, and 1-14 that areas of responsibility are designated by using intuitive naming whenever possible. In Figure 1-13, Branch 1 is in charge of operations on the first floor. The upper search and rescue group is distinguished from the search and rescue group on the first floor.

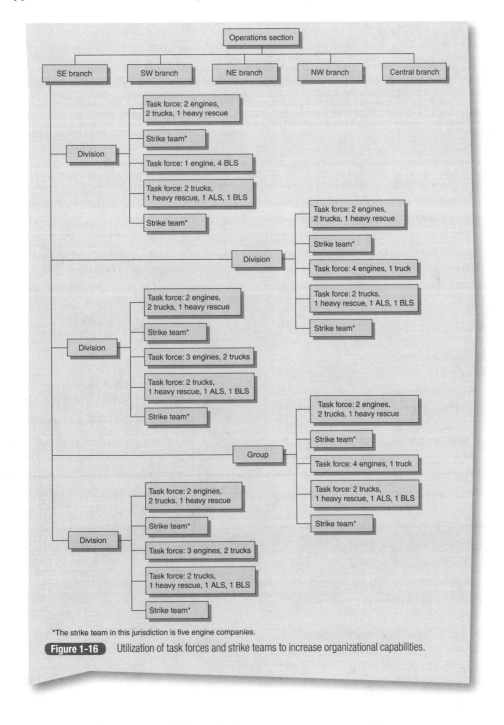

Figure 1-16 Utilization of task forces and strike teams to increase organizational capabilities.

It is best to use intuitive naming systems whenever possible. Many departments use an alphanumeric system to designate geographic assignments. To avoid confusion during emergency operations, the system used should be addressed in department SOPs. When using an alphanumeric system, each floor level is named using the floor level. As an example, Division 21 would be supervising operations on the 21st floor of a high-rise building. Naming tactical level management units by floor number is intuitive.

An example is illustrated in **Figure 1-17**. Here, the IC has established the 13th Street side as Side A to avoid confusion. Sides B, C, and D follow in a clockwise manner. Side A may not be obvious at every fire, so this system is not the most intuitive way to name sides of a building. If a building faces only one street, the alphabetic naming system is simplified. Even then, the front of the building could simply be called the front, with left, rear, and right sides.

Another, more intuitive, approach to identifying the sides of a building is simply to use street names combined with directions. The sides of the hotel shown in Figure 1-17 could be the 13th Street side, the left (or office) side, the rear side, and the Walnut Street side. The ICS organization chart for the BASF fire shown in Figure 1-14 is a good example of an intuitive naming system. Many jurisdictions use compass directions to identify the sides of the buildings. This system works effectively if the building is oriented in a north/south–east/west manner and units operating at the scene know the geographic directions. If the building is not situated at a direct north/south/east/west geographic orientation, then using compass points may not be obvious.

There are numerous ways to identify geographic locations. When you are working with mutual aid departments, everyone should agree on a common system, include it in the SOPs, and use it during joint training exercises. Whatever system you select, use it *consistently* and at *every* incident to make sure everyone is comfortable with the labeling/identifying system.

Communications

Communication is the lifeblood of any command system. It is impossible to coordinate and control a successful operation without effective communications.

Some general rules for communication at the incident scene include the following:

- Use face-to-face communication whenever possible.
- Provide mobile communication to units that are remote from the command post.
- Ensure that all operating units have some form of communication that ultimately relays information to the command post.
- Place representatives of agencies on different frequencies at the command post to handle communications within their agency.
- Follow the command organization structure, facilitating **unity of command**.
- Keep the number of radio channels used by any supervisor to an absolute minimum, preferably no more than two.
- Do not clutter radio channels with unnecessary transmissions.
- Use standard terminology.
- Use clear English; don't use **ten-codes**.

Within the NIMS structure is a communications unit that reports to the logistics section chief. At large-scale incidents, establishing a communications unit is critical. This unit is responsible for establishing a communications plan and installing, procuring, and maintaining the communications equipment at the scene. At incidents where a formal communications unit is not established, preplanned resources should be used.

All radio communications at the incident should be in *plain English*. Codes should be avoided because they can be easily misunderstood. Furthermore, not all personnel operating at large-scale incidents from different jurisdictions or agencies will be familiar with the codes.

All communications should be confined to essential messages. It is important not to clog the airways with

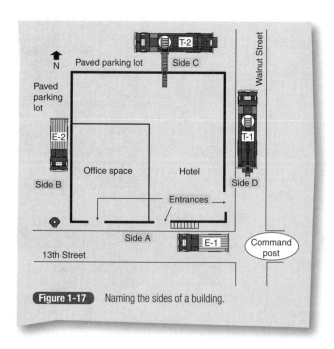

Figure 1-17 Naming the sides of a building.

nonessential messages and to keep the frequency available for vital transmissions. Each transmission should be brief, not a long narrative of what is transpiring within the area.

Radios are the most common mobile communications tool for emergency operations, but other communications devices offer advantages and should be considered during long-term or communications-intensive situations. In addition to radios, the following means of communication can be used at the scene:

- Face-to-face communication
- Messengers
- Telephones (cellular, satellite, and hard wire)
- Public address systems
- Computers/Mobile Data Terminals/Mobile Data Computers
- Fax

The most effective form of communication is face to face; however, running an entire large-scale incident using only face-to-face communications would be all but impossible. Cellular telephones have gained popularity and are often used to contact outside resources. One problem with cellular telephones, however, is that the system can be easily overwhelmed or damaged during a major disaster. In some cases, media representatives and others at the scene may tie up the available cell phone channels by locking in open lines. During an extended incident, the local cellular telephone company might be able to provide additional channel capacity and resources. Satellite telephones are now being used, which tend to reduce the problems of overload and inoperability due to damaged cell towers.

Computers can provide nonverbal communications that free "airtime," as well as provide additional information especially in graphic form. Mobile Data Terminals and Status Message Terminals are two types of systems used to relay digital information from field units to dispatch. These terminal-based systems are dependent on a data network. Mobile Data Computers (MDCs) and other available devices combine dispatch communications with on-board data storage, thus reducing the dependence on a wireless network and the dispatch center.

For purposes of radio communications the IC is simply referred to as "command," a designation that should be reserved solely for the IC. Some departments refer to each division, group, or branch using the term "command," such as "rear command" or "interior command." The authors of this book strongly discourage the use of multiple command designations because of the confusion that it may cause.

The IC is designated as "command" whether this person is the chief of the department or a fire fighter working as an acting officer. Department SOPs must define the term "command" and specify how it is to be used and by whom. By using well-defined terminology, it is possible to eliminate confusion during command transfers. Operating units, divisions, and branches might not remember that Chief Jones assumed command, but they do recognize the importance of orders being issued by

Case Summary

After the Oklahoma City bombing, the local cellular phone providers were able to provide additional capacity in the area of the Alfred P. Murrah Federal Building by bringing in portable cellular transmitters called cells on wheels. They also provided the many agencies operating at the scene with cellular telephones at no cost. These were programmed with priority numbers that could grab a channel. Several locations were established where telephone personnel were available around the clock to repair phones and provide charged batteries. Use of the cellular system was instrumental in coordinating the overwhelming logistics that this incident involved.

Source: Edward R. Comeau and Stephen Foley, "Oklahoma City, April 19, 1995," *NFPA Journal,* July/August, 1995.

Figure CS1-2 The rescue and recovery operations at the Alfred P. Murrah Federal Building went on around the clock. There were numerous government agencies and private contractors on the site that needed to communicate regularly and reliably.

command and the need to communicate through channels to command.

It is crucial that the communications network support unity of command within the NIMS organization. An example of a communications network following the NIMS structure is shown in Figure 1-18. It is important to keep the communications network as simple as possible. It is not necessary to designate scores of radio channels even if you have the equipment to do it.

In Figure 1-18, the IC has retained Operations and is using Channel 2 to communicate with all field units. The IC is using only two radio channels: one to communicate with dispatch, and the other for field communications.

With the many divisions and fire companies using Channel 2, radio discipline would be imperative, and this single channel would be very crowded. If multiple channels were available, the communications network could be expanded, with each division using a separate channel to communicate with units under their supervision and Channel 2 or another channel to communicate with the interior branch. For example, command could communicate with the interior branch on Channel 2, and the interior branch could use Channel 4 to communicate with Divisions 1, 2, 3, and 4, as shown in Figure 1-19. Each division could then be assigned a different channel. More airtime is now available for units communicating with the interior branch; however, the communications network is much more complex, and there are more communications layers between the IC and companies reporting to the divisions within the interior branch. Units subordinate to the interior branch can no longer listen to messages being transmitted by command.

Interoperability is a concern at incidents involving multiple agencies or more than one jurisdiction. Technology

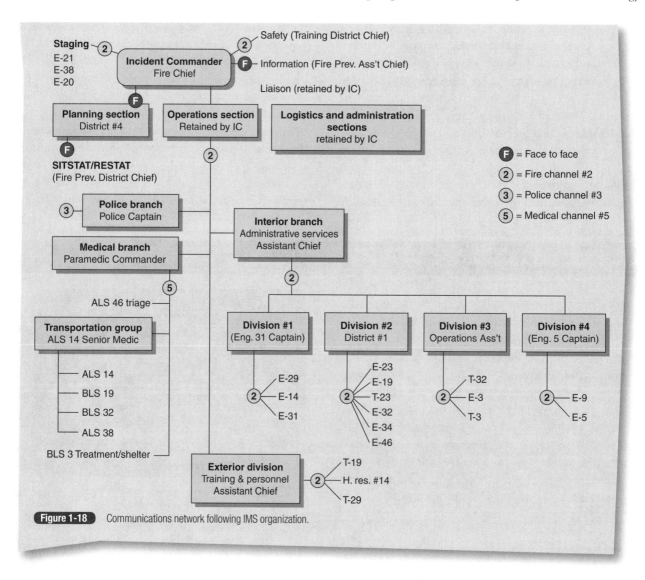

Figure 1-18 Communications network following IMS organization.

provides a means of communicating on many radio channels. With proper planning and equipment, it is possible to communicate with all agencies at the scene, as well as agencies providing support from off-site. Unfortunately, not every department has the communications equipment necessary to communicate with everyone who might respond to an emergency. Even if a department has the capability of communicating with all responding agencies, communicating via multiple radio channels is problematic. It is essential that the IC and all units that report to the operations section limit the number of radio channels used to reduce the possibility of missing critical messages from operating companies or crews. Although it is possible to monitor several radio channels using radios as scanners, it is very difficult to conduct two-way communications on more than two channels. This problem can be solved in several ways as follows:

- Placing a representative of the assisting agency with its communications equipment at the command post. The agency representative manages and communicates with responders from their agency.
- Assign a liaison officer to communicate with other agencies.
- Assign the logistics section to communicate with agencies that are providing equipment and supplies.
- Direct communications to technicians who re-transmit critical messages to the IC. This could be part of the communications unit within the logistics section.

Also consider using alternative communications methods such as cell or satellite telephones when communicating with other agencies.

Common terminology is the cornerstone of effective interoperability. It is essential that common terminology be established for any management system, especially

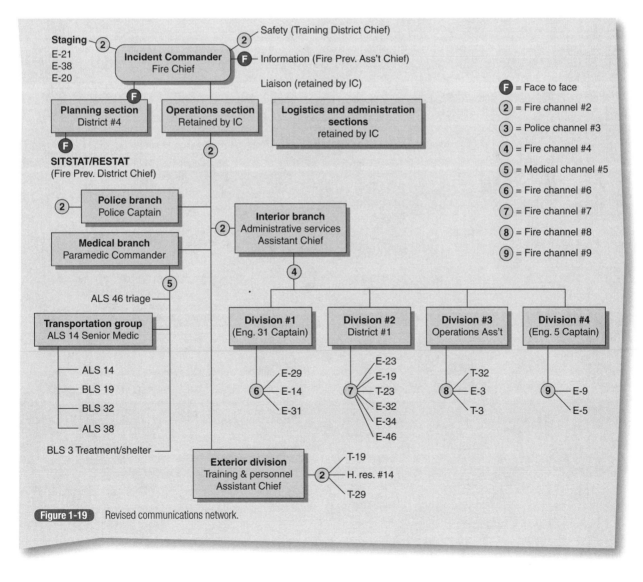

Figure 1-19 Revised communications network.

one that will be used in joint operations by numerous and diverse participants under emergency conditions. This commonality of terms is particularly important in the areas of organizational functions, resources, and facilities.

Positions within the NIMS organizational framework are named, using the same names for the function regardless of the type of emergency. Line and staff positions remain the same, even though incident situations and objectives may differ significantly.

Summary

The fire ground can present complex challenges. A tremendous amount of information must be processed rapidly and accurately to ensure the safety of the fire fighters and victims. The only safe and effective way to manage fire-ground information and make the proper command decisions is to use NIMS from the beginning of the incident to its conclusion. This allows the IC to maintain a proper span of control and ensure accountability while safely and efficiently accomplishing the objectives listed in the incident action plan.

Wrap-Up

Key Terms

branches Immediately subordinate to section in NIMS hierarchy. These units are subordinate to the logistics, finance/administration, and planning sections. Divisions and groups are subordinate to these in the operations section. These are used to reduce the span of control at very large operations or to manage a particular function/agency.

division Tactical level management unit in charge of a geographic area.

finance/administration section The section that tracks and provides financial and administrative services required to compensate people or organizations providing goods and services at the incident scene.

freelancing Performing tasks outside the incident organization structure.

group Tactical Level Management Unit in charge of a function.

incident commander (IC) The person in command of the entire incident and all related activities.

incident management team An incident management team is the incident commander (IC) and appropriate command and general staff personnel assisting the IC in managing an incident.

incident safety officer The command staff position assigned to monitor the scene for safety hazards or unsafe operations, enforce safety practices, and establish a safety plan.

liaison officer The command staff position responsible to communicate and coordinate with agencies not directly involved in meeting objectives identified in the incident action plan.

logistics section The section that obtains needed supplies, equipment, and facilities.

National Incident Management System (NIMS) A Department of Homeland Security system designed to enable federal, state, and local governments and private-sector and nongovernmental organizations to effectively and efficiently prepare for, prevent, respond to, and recover from domestic incidents, regardless of the cause, size, or complexity, including acts of catastrophic terrorism. Adopted by the U.S. Department of Homeland Security, this system uses Incident Command System (ICS) organizational terminology and structure, thus the terms ICS and NIMS can be used interchangeably when referring to organizational structure and terminology.

operations section The section that manages all tactical units deployed at an incident scene.

planning section The section that gathers and evaluates information, assists the incident commander (IC) in developing the incident action plan and tracks progress; also tracks resource status.

public information officer (PIO) The command staff position responsible for relaying information to the public.

rational decision making (RDM) A form of decision making in which input is obtained from diverse sources and careful analyses of all options is considered.

recognition primed decision making (RPD) A form of decision making in which the incident commander (IC) must decide on the proper course of action with limited information available in a relatively short period of time.

single command One person is designated as the incident commander (IC). This person is responsible for the development and implementation of the incident action plan. The IC can delegate staff and command positions as needed to assist in command and control functions.

span of control The number of people reporting to a supervisor. The span of control should not exceed seven people reporting to a single supervisor under emergency conditions. As an example, a fire captain supervising an apparatus operator and three fire fighters would have a four-to-one span of control.

strike teams A set number of resources of the same kind and type that have an established minimum number of personnel, e.g., five, four-person engine companies. Strike teams always have a leader (usually in a separate vehicle) and have common communications among resource elements.

task forces Any combination of resources that can be temporarily assembled for a specific mission. All resource elements within a task force must have common communications and a leader. Task forces should be established to meet specific tactical needs and should be demobilized as single resources. A typical task force would be two engine companies and a truck company under the supervision of a chief officer.

ten-codes A numeric code used to communicate predefined situations or conditions. For example "10-4" generally means "finished communicating." The use of ten-codes is discouraged.

two-in/two-out rule Fire fighters working inside the hazard area must work in crews of at least two people (two-in). The two people working in the hazard area must be backed up by at least two people outside the hazard area (two-out) who are properly equipped and immediately available to come to the aid of the inside crew.

unified command Application of the National Incident Management System when there is more than one agency with incident jurisdiction or when an incident crosses political jurisdictional boundaries. Agencies work together through the designated members of the unified command, often the senior person from an agency participating in the unified command, to establish a common set of objectives making up the incident action plan.

unity of command A pyramidal command system ensuring that no one reports to more than one supervisor.

Suggested Activities

1. Develop an organizational chart for a room and content fire in a two-story, single-family, residential building. Assume that all first-alarm units are on the scene performing the following tasks:
   - Engine 1 – Per SOP, has secured a water supply using a 5" (12.7 cm) forward lay and split into two crews. Crew #1 is the Initial Rapid Intervention Crew (IRIC) stationed outside the structure and consists of the hydrant fire fighter and apparatus operator. The officer and other fire fighter make up Crew #2 and are attacking the fire with a 1¾" (44-mm) line.
   - Engine 2 – Relieves Engine #1's IRIC with all four members forming the Rapid Intervention Crew. Engine #1 hydrant fire fighter joins the officer and other fire fighter on the interior and the apparatus operator continues operating the pump.
   - Engine 3 – Secures a second source of water with the three remaining members placing a hose line on the second floor.
   - Truck 1 – Establishes two crews: Crew #1 is conducting a primary search in the second floor bedrooms; Crew #2 sets up positive pressure ventilation before conducting a primary search of the first floor.
   - Battalion Chief 1 assumes command at a command post on the A/D corner of the building after driving past the A/B side.

2. Develop an organizational chart showing all operating units for a large fire on the second floor of a two-story warehouse. The building is tightly secured and unoccupied at the time of the fire. Pre-incident plans indicate that the second floor is undivided with dimensions of 150 ft (45.7 m) wide by 50 ft (15.2 m) deep and a 25-ft (7.6-m) ceiling.
   - Engine 1 – Per SOP and pre-incident plan instructions, secured a water supply using a 5" (12.7 cm) forward lay and split into two crews. Crew #1 is the Initial Rapid Intervention Crew (IRIC) stationed outside the structure and consists of the hydrant fire fighter and apparatus operator. The officer and other fire fighter make up Crew #2 and are attacking the second floor fire with a 1¾" (44-mm) line.
   - Engine 2 – Relieves the IRIC with all four members forming the Rapid Intervention Crew. Engine #1 hydrant fire fighter joins the officer and other fire fighter on the interior and the apparatus operator continues operating the pump.
   - Engine 3 – Secures a second source of water with the three remaining members operating a 2½" (64-mm) hose line on the second floor.
   - Engine 4 – Connects a 2½" (64-mm) line to Engine 1's pumper and the four-member company is operating a second 2½" (64-mm) hose line on the second floor.
   - Truck 1 – Establishes two crews: Crew #1 is assisting units on the second floor; Crew #2 is venting the roof.
   - Truck 2 – Gains access to the first floor and is checking the first floor for fire extension.
   - Battalion Chief 1 assumes command at a command post on the A/B corner of the building after driving past the A/D side and orders a second alarm assignment.
   - Engine 5 – Connects a 2½" (64-mm) line to Engine 2's pumper and the four-member company is operating a third 2½" (64-mm) hose line on the second floor.
   - Engine 6 – Connects a 2½" (64-mm) line to Engine 1's pumper and the four-member company advances a fourth 2½" (64-mm) hose line into the second floor.
   - Engine 7 – Connects a 2½" (64-mm) line to Engine 1's pumper and the four-member company is standing by with 2½" (64-mm) back-up hose line.
   - Engine 8 – Staged
   - Truck 3 – Staged
   - ALS 1 – Is managing the medical function with two primary responsibilities: providing REHAB and medical treatment for fire fighters; providing medical transportation as needed.
   - ALS 2 – Assigned to REHAB, which is being set up in a nearby building
   - BLS 1 and BLS 2 are assigned to transportation.

3. A large six-story warehouse is fully involved in fire, and the IC has decided on a defensive attack for the warehouse. There are occupied exposure buildings on each side of the warehouse, as shown in **Figure 1-20**. The IC has assigned companies to protect and evacuate the exposures on both sides of the warehouse. Develop an IMS organizational chart for this operation as it is presently being conducted.

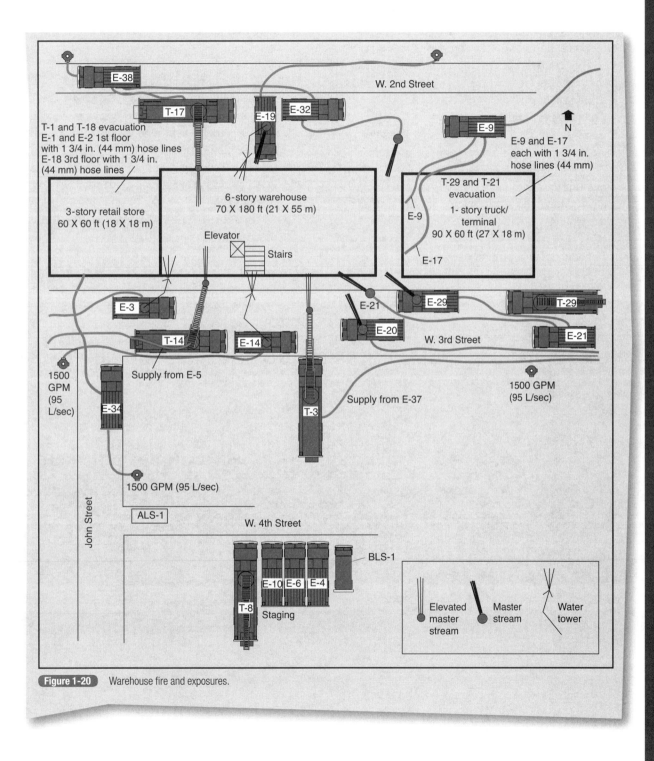

Figure 1-20 Warehouse fire and exposures.

Wrap-Up, continued

4. Assume the role of the first-arriving company officer for a fire in the building pictured in **Figure 1-21**. The fire was reported as 11 West Third Street, but the correct address is for the Showcase Cinema at 111 West Third Street. Develop a report to dispatch including:
   - Address confirmation
   - Command confirmation
   - Command mode (offensive, defensive, investigative, etc.)
   - Brief description of building (or name of well-known building)
   - Occupancy (if not obvious in the building description)
   - Conditions on arrival (nothing showing, occupants at windows, etc.)
   - Resource needs

5. Assume the role of the first-arriving chief officer (Battalion 1) with an engine and truck company on the scene for the fire pictured in **Figure 1-22**. The fire was reported as 1234 Main Street, which is the correct address. Develop a status report to dispatch, including:
   - Address confirmation
   - Command confirmation
   - Brief description of building (or name of well-known building)
   - Occupancy (if not obvious in the building description)

Figure 1-21 Showcase Cinema 111 West Third Street.

- Conditions on arrival (nothing showing, occupants at windows, etc.)
- Progress (or lack of progress)
- Resource needs

6. Incident action plan objectives for the fire shown in Question #5 include assigning a truck company to gain access to the attic by pulling the ceiling on the second floor. Develop a message assigning the next arriving truck to perform this task.

7. Assume the role of the first-arriving engine company for the fire pictured in Question #5. You are on the floor with a 1¾″ (44-mm) hose line encountering heavy smoke conditions, but you have not found the fire. Battalion 1 requests a status report. "Command to Engine 1, what is your status?" Develop a status report including:
- Unit number ("Engine 1 to Command," wait for Command reply)
- Your location
- Progress/conditions
- Resources
- Any safety issues

8. Analyze incident communications by securing and listening to audio media from actual incidents. The Learning Resource Center at the National Fire Academy has audiotapes as part of their major fire reports. Most dispatch centers have the capability of taping and duplicating audio.

Figure 1-22   Fire at 1234 Main Street.

Wrap-Up, continued

Chapter Highlights

- The incident commander (IC) must manage resources, ensure fire fighter safety, and target priorities.
- Good communications are essential to decision making.
- National Incident Management System (NIMS) enables the IC to delegate tasks and make decisions at both major and minor incidents.
- FIRESCOPE Incident Command System was developed to help manage California forest fires in the 1970s, and was expanded for urban and nationwide use.
- NIMS should be used by the entire response community, not just fire departments.
- NIMS provides common terminology and operational assignments for all agencies.
- NIMS should be used for *all* incidents, regardless of size.
- NIMS is the required incident management system for federal agencies and for departments using federal preparedness funding.
- A unified command can be used when more than one jurisdiction or agency has responsibility.
- Although a single command is preferable, some situations require a unified command.
- Unified command is seldom warranted for the emergency phase of a structure fire, but may evolve during the cleanup phase.
- When using a unified command structure, plan development is shared, but one operations chief should direct all field units.
- The first-arriving company officer is in command and should follow department procedures, but in most cases will adopt a fast attack or investigative command mode.
- Three command options are available:
  - *Investigation:* no obvious hazard is evident on arrival; no obvious need for immediate action to save lives or protect property.
  - *Fast attack:* obvious signs of fire or another dangerous condition that require immediate action.
  - *Command:* large or complex incident requiring that the company officer coordinate the actions of personnel and other incoming units.
- Standard operating procedures (SOPs) allow the operation to continue while a chief officer develops and implements an incident action plan.
- Strong command presence is critical to fire fighter safety.
- The initial report should confirm incident location, type, and conditions.
- Good communications and clear initial reporting are essential to command.
- When a chief officer assumes command, a stationary command post must be established.
- Officers assuming command must provide an updated status report based on observation and reconnaissance from units at the scene.
- Communications between command and field units must be established through the communications network.
- Command transfer must be addressed in department SOPs so company and chief officers know whether they are required to assume command.
- When command is transferred to another officer, it should be formalized at a stationary command post.
- To avoid confusion, command should not be transferred between company officers unless there is a compelling reason to do so.
- An officer who finds an unsafe operation in progress upon arrival must immediately take command, establish a formal command post, and take immediate corrective action.
- The person assuming command must communicate with the previous IC to determine both situation and resource status of the operation in progress.
- Avoid freelancing by making command transfers infrequent yet efficient.
- The highest ranking officer at the scene is held responsible and accountable for incident operations.
- Command may be transferred to another agency in specific circumstances (terrorist attack, hazardous waste spill).
- The IC develops the incident strategy.
- Branch, division, and group supervisors develop tactics within the overall strategy.
- Information exchange and good communications are the keys to a safe and effective operation.
- A good command post is located in a known or easily found place, within the cold zone, in a location that

- supports command, control, and coordination of all incident activities.
- In larger, more complex incidents, the command post should be farther from the scene.
- The command post is wherever the IC is located.
- The span of control is the number of people reporting to a supervisor, and should include from three to no more than seven people reporting to a single supervisor.
- The type of incident, the nature of the task, hazard and safety factors, and the distance separating tactical units will influence span-of-control decisions.
- Tactical level management components should maintain a reasonable span of control and coordinate specific geographic and functional operations.
- Anticipate the need for additional units and request them well in advance.
- Staging areas should be established for resources not immediately assigned to a task.
- Staging allows the IC to control access to an incident scene while deploying resources in a safe and effective manner.
- Staging areas should be far enough away to avoid impeding access to the scene or freelancing into action, but close enough to allow companies to arrive quickly once summoned.
- Early establishment of a staging area allows better management of on-scene units, establishes a tactical reserve, and eliminates freelancing.
- Apparatus without staffing that are parked in staging are out-of-service and not staged.
- Management of staging areas should be part of SOPs.
- NIMS is a tool used to organize an operation—a means of providing command and control at an incident to achieve incident action plan objectives.
- NIMS organization should be as simple as possible.
- NIMS organizational structure develops in a modular fashion based on the type and size of the incident
- IC is the one position that must always be staffed within a NIMS organization
- IC determines organizational structure based on the management needs of the incident.
- Three command staff positions in NIMS include:
  - *Incident safety officer:* monitors fire fighter safety
  - *Liaison officer:* coordinates outside agencies
  - *Public information officer:* coordinates with the media/public information systems
- Some NIMS positions may be combined at smaller incidents.
- NIMS has a pyramid-structured hierarchy with the IC at the top.
- Command staff and line positions are staffed when incident conditions require separate management of the functions of that position.
- NIMS includes four possible sections, each with its own section chief as follows:
  - *Finance/Administration* documents equipment expenses, payroll for responding or back-fill personnel, personal injuries, and other financial costs for reimbursement.
  - *Logistics* locates and provides materials, equipment, supplies, communications, and facilities required to support incident operations.
  - *Operations* manages all tactical operations such as search and rescue, extinguishment, and medical care.
  - *Planning* gathers information about the incident, tracks resources, and assists the IC in developing the incident action plan and alternatives.
- A fifth section, intelligence, is rarely needed or implemented.
- A division is in charge of a geographic area, whereas a group is in charge of a functional area.
- Use geographic rather than functional assignments whenever possible to improve communications and eliminate the need for status reports within the division.
- Branches are for operations beyond the span of control of a single division or group, or when a contingent of units from other agencies are working together.
- Task forces and strike teams can be used to reduce the span of control and the communications load at an incident.
- Task forces combine different resources.
- Strike teams combine a set number of resources of the same type.

Wrap-Up, continued

- Strike teams can be used to organize large numbers of mutual aid companies or private sector resources.
- Task forces are more flexible and more likely to be formed at the time of the incident.
- Areas of responsibility are designated by using intuitive naming whenever possible.
- The intuitive naming system used should be addressed in department SOPs.
- Standard rules for maintaining good communications should be observed.
- A communications unit reporting to the logistics section chief is important at large or complex incidents.
- All radio communications should be in plain English, not codes, and confined to essential messages.
- Make use of other available forms of communication in addition to the radio to facilitate communications and planning.
- Communications system must support unity of command.
- When multiple agencies are responding, coordination of communications should be streamlined to avoid information conflicts.

References

1. National Fire Protection Association, *NFPA 1561: Standard on Emergency Services Incident Management System.* Quincy, MA: NFPA, 2005.
2. U.S. Department of Homeland Security, *National Incident Management System.* Accessed July 3, 2007 at www.nimsonline.com/nims_3_04/index.htm.
3. National Fire Protection Association, *NFPA 1500: Standard for Fire Department Occupational Safety and Health Program.* Quincy, MA: NFPA, 2007.
4. Code of Federal Regulations, Title 29, Section 1910.120, *Superfund Amendments and Reauthorization Act of 1986. Hazardous Waste Operations and Emergency Response.* Accessed July 2, 2007 at http://www.osha.gov/pls/oshaweb/owadisp.show_document?p_table=standards&p_id=9765.
5. National Interagency Incident Management System, *Incident Command System: Operational System Description* (ICS 120-1). Stanford, CA: FIRESCOPE Program, 1981.
6. Alan Brunacini, *Fire Command.* Quincy, MA: NFPA, 1985.
7. Michael S. Isner, Thomas J. Klem, *NFPA Fire Investigation Report: World Trade Center Explosion and Fire*, New York, NY, February 26, 1993. Accessed July 2, 2007 at http://www.fireox-international.com/fire/NFPA1993WTCIncidentReport.pdf.
8. National Fire Protection Association, *NFPA 1521: Standard for Fire Department Safety Officer.* Quincy, MA: NFPA, 2008.

Procedures, Pre-Incident Planning, and Size-Up

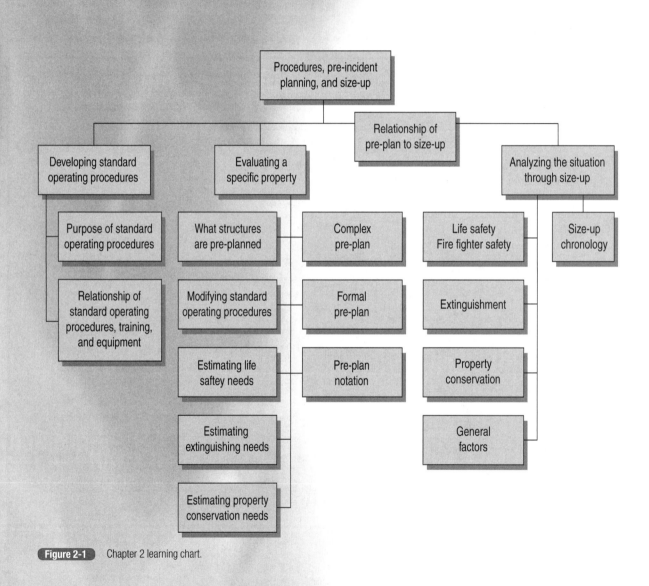

Figure 2-1 Chapter 2 learning chart.

Chapter 2

Learning Objectives

- List the kinds of operations that should be covered by standard operating procedures.
- Explain the importance of standard operating procedures.
- Discuss the relationship between standard operating procedures, pre-incident plans, and size-up.
- Examine the relationship between standard operating procedures, equipment, and training.
- Compare standard operating procedures to standard operating guidelines, explaining the role of a "reasonable person" clause.
- Articulate the main components of prefire planning and identify steps during a prefire plan review.
- Recall major steps taken during size-up and identify the order in which they will take place at an incident.
- Recognize the relationship between pre-incident planning and construction characteristics common to a community.
- Analyze construction methods during everyday responses and while surveying buildings under construction and demolition.
- Explain how pre-incident plan information is gathered using pre-formatted forms, as well as methods for storing and retrieving pre-plan information.
- Examine compatibility issues and usefulness of computer programs in pre-incident planning.
- Construct a priority chart of buildings to be pre-planned by occupancy type.
- List factors to be considered during size-up and briefly define and explain the significance of each factor.
- Demonstrate (verbally and in writing) knowledge of fire behavior and the chemistry of fire.
- Recall the basics of building construction and how they interrelate to prefire planning and size-up.
- Define and explain the difference between occupancy, occupant, and occupied.
- Explain the size-up process in the chronological order in which information is received.
- Evaluate a specific fire department's standard operating procedures.
- Prioritize occupancies to be pre-incident planned in a specific jurisdiction.
- Create a pre-incident plan drawing and narrative.
- Perform an initial size-up based on limited information.
- Apply size-up factors to a fire situation and categorize factors as primary or secondary.

Introduction

This chapter will discuss the importance of <u>pre-incident plans</u> and <u>standard operating procedures (SOPs)</u> and their relationship to size-up. Every strategist recognizes the importance of gaining intelligence. Military officers go to great lengths to understand the factors that are involved in all operations, as well as the particulars of the specific battleground and enemy. In the fire service, factors that apply to fire-ground operations can and should be outlined in SOPs. The battle particulars are a result of pre-incident planning. As with military operations, the better the pre-incident plan and SOPs, the fewer decisions that will need to be made in the heat of battle, which allows the incident commander (IC) to focus on important incident-related factors when developing an incident action plan.

SOPs, pre-incident plans, and incident-specific information are interrelated and are important components of the size-up. The IC who is faced with a large or complex building and no pre-incident plan or applicable SOPs is at a great disadvantage.

The learning chart on the first page of this chapter shows the relationships between pre-incident planning, SOPs, and size-up, all of which are discussed in this chapter **Figure 2-1**. SOPs and a good size-up are necessary prerequisites in the development of an incident action plan. For larger and more complex properties, pre-incident planning is essential.

Important information about specific buildings can be obtained in advance through pre-incident planning. This information is critically important to the IC when making strategic and tactical decisions at the incident scene. Some things will be known only after the incident occurs, as is shown in a later discussion of pre-planning and size-up (see *Analyzing the Situation Through Size-Up*). Pre-fire information and incident-specific information are evaluated in terms of factors related to the operational priorities of life safety, extinguishment, and property conservation. Structural conditions, resources needs, contents, occupancy type, access limitations, water supply, and special challenges are addressed as sub-topics.

The entire, complex process of evaluating an incident and developing an incident action plan must take place in a few minutes. To accomplish this, it is necessary to focus on the major (primary) factors during the initial plan development process. As more information becomes available and the IC has time to reevaluate, the size-up information and the incident action plan should improve.

Developing Standard Operating Procedures

SOPs are general guidelines to be used at all structure fires or fires in similar occupancies. SOPs recognize similarities, covering areas such as those listed in **Table 2-1**.

TABLE 2-1 Areas Covered in SOPs

Command	• Implementing the incident management system • Identifying who is in charge • Establishing a command post • Command transfer • Communications • <u>Rapid intervention crew (RIC)</u> • Accountability
Staging	• Exterior staging • Interior staging
Water supply	• Attack pumper • Forward hose lay • Pumper/hose wagon • Tanker shuttle • Water relay
Special occupancy operations	• Places of assembly • Educational occupancies • Health care occupancies • Board and care occupancies • Hotels • Dormitories • Detention facilities • Mercantile • Business • Storage • Industrial • High-rise • Large area buildings
Company operations	• Truck company • Engine company • Quint/Quad company • Squad unit • Rescue unit
Special operations	• Hazardous materials • Natural disasters • Electrical fires • Civil disturbances • Terrorism
Using private fire protection	• Standpipe • Sprinkler • Alarm systems

SOPs address any operation that can be handled using a standard approach. SOPs will vary among departments. The types of property to be protected, resources available, equipment, and training, among other factors, guide the promulgation of SOPs. This does not mean that fire departments should not share procedures. Instead, it is necessary to learn all that can be learned from the experiences of others. However, the final procedure must reflect department-specific needs. For example, the SOP for an automobile fire would be similar in a small city and in a large urban area. However, a high-rise SOP would be of no value to a small, rural department with no high-rise buildings within its jurisdiction or mutual aid response area.

Also, as the challenge increases, the allowances for local resources become larger. Large, urban departments can quickly summon hundreds of fire fighters to the scene of a large structure fire. A smaller city would, by necessity, handle a fire in the same size and type of structure differently, by calling on surrounding communities through mutual or automatic aid agreements. SOPs for handling the same type of fire may look very different from department to department.

Although SOPs must be written specifically for the department, there is a need for regional planning in writing procedures. It makes little sense to have specific procedures for the incident management or accountability systems limited to just one department when there is a good probability that these systems will be critically important during large-scale incidents when other regional departments are on the scene providing mutual aid.

Purpose of Standard Operating Procedures

Good department SOPs and pre-incident plans take the guesswork out of those first few precious moments on the fire ground. Company officers have pre-designated assignments that allow them to take immediate action within the overall plan or, in some cases, to stage, awaiting orders from command.

In the absence of SOPs and pre-incident plans, the IC is so busy assigning companies that little time is left to collect the information necessary for the development of an action plan. Freelancing may also occur while the IC gathers the intelligence necessary for the formulation of an incident action plan, including the development of strategy and tactics. Either case results in inefficient and possibly dangerous operations.

With this in mind, it is best to provide specific SOPs for the first-arriving engine company. As an example, the first-arriving engine company is assigned to secure a source of water and advance a hose line into the fire area. The SOP for the second-arriving unit could be specific or general in nature, depending on whether the first-arriving unit is directed to secure a water supply. If SOPs do not require the first-arriving engine company to secure a water supply, water supply is generally assigned to the second-arriving engine company. Departments may assign the second-arriving engine to staging, water supply, or as the rapid intervention crew (RIC), or the department should provide general instructions if the second-arriving company is not so assigned. The first-arriving truck company will generally choose from a menu of possible tasks, as enumerated in Chapter 4.

During large-scale operations the IC is faced with numerous, complex decisions. SOPs provide a structure for the decision-making process, including answering the questions of who makes what decisions, at what level of command, and from where. NIMS does an excellent job of establishing a command structure and describing the roles of various players at the incident scene.

Relationship of Standard Operating Procedures to Training and Equipment

In regard to SOPs, *NFPA 1500: Standard on Fire Department Occupational Safety and Health Programs*,[1] states:

> **4.1.2** The fire department shall prepare and maintain written policies and standard operating procedures that document the organization structure, membership, roles and responsibilities, expected functions, and training requirements, including the following:
>
> **(1)** The types of standard evolutions that are expected to be performed and the evolutions that must be performed simultaneously or in sequence for different types of situations
>
> **(2)** The minimum number of members who are required to perform each function or evolution and the manner in which the function is to be performed
>
> **(3)** The number and types of apparatus and the number of personnel that will be dispatched to different types of incidents
>
> **(4)** The procedures that will be employed to initiate and manage operations at the scene of an emergency incident
>
> **5.1.2** The fire department shall provide training, education, and professional development for all department members commensurate with the duties and functions that they are expected to perform.
>
> **5.1.6** The fire department shall provide all members with training and education on the fire department's written procedures.
>
> **5.1.10** Training programs for all members engaged in emergency operations shall include procedures for the safe exit and accountability of members during rapid evacuation, equipment failure, or other dangerous situations and events.

NFPA 1500 makes a definite statement that training must be commensurate with SOPs. If new equipment is placed in service, the availability of the equipment makes a statement that fire fighters are expected to use the equipment properly and safely. Likewise, it becomes necessary that SOPs outline how and when the equipment is to be used.

There is a relationship between SOPs, equipment, and training, as illustrated in Figure 2-2. Any time new equipment is introduced or a new procedure is written, the entire cycle must be completed.

Consideration should be given to SOPs when developing new apparatus and equipment specifications. As an example, if department SOPs specify that the first-arriving pumper should pull past the main entrance to allow room for the first-arriving truck company, pre-connected hose lines should be specified in locations that permit efficient deployment to the rear of the apparatus.

A Word About the Standard Operating Guidelines Controversy

Within the fire service there is some discussion about whether to refer to *standard operating procedures as standard operating guidelines, general operating guidelines*, or another, similar name. The first point that needs to be made is that it is far more important to have written procedures or guidelines than it is to argue about the name. Semantics can be important, but not in this case. Many in the fire service believe that calling procedures "standard operating guidelines" or "general operating guidelines" somehow changes their strength, in that a guideline may be considered less stringent than a procedure. In reality, procedures are guidelines, and guidelines become procedures through practice. If a task is always performed the same way, that method becomes the procedure for doing that task.

Likewise, most procedures should be taken as general guidelines. Situations will arise in which the SOP does not fit the circumstance, and the officer or IC would be expected to modify his or her actions accordingly. For example, an SOP that requires the first-arriving truck company to ventilate would not be appropriate if the fire had self-vented or was obviously a defensive operation. Some SOPs are always in effect, such as an SOP stating that fire fighters *must* wear self-contained breathing apparatus (SCBA) in contaminated atmospheres.

The solution to this problem is a "reasonable person" clause in the procedures manual. Essentially, a statement is made that procedures are to be followed, but that fire fighters should follow a reasonable course of action when confronted with a situation in which modification of the procedure is appropriate. When a decision is made to modify procedures, the noncompliant unit or member is required to communicate the variance and be prepared to justify the modification.

Call them procedures or guidelines, but write them down, train to them, and use them consistently.

Evaluating a Specific Property

During an official visit to a facility, it makes sense to gather information about all kinds of problems that may occur at a property. If the facility falls under the requirements of Title III of the <u>Superfund Amendments and Reauthorization Act</u>

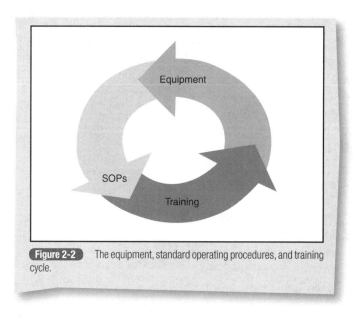

Figure 2-2 The equipment, standard operating procedures, and training cycle.

> There should be a procedure describing the pre-incident planning system.

(SARA), then hazardous materials planning is mandated by law.

Establishing SOPs is the first step in the size-up. Pre-incident planning is step two. After SOPs are developed, individual properties should be examined for specific hazards and characteristics. Pre-incident plans are a natural extension of SOPs. For example, a fire department should have a written procedure for operations conducted in buildings protected by sprinkler systems. These procedures would not be repeated in the pre-incident plan, but any deviations from the normal operation would be outlined. For example, if the water supply at a property is reliable and no off-site water supply exists, the pre-incident plan might modify the typical SOP at this sprinkler-protected building by not automatically requiring the sprinkler system to be supplied at the fire department connections.

Pre-Incident Plans

Formal pre-incident plans include both a narrative and drawings. Narratives are best written in outline form with extremely important information highlighted, color coded, or otherwise identified by some method to draw attention to critical information.

Pre-incident plans can take various forms. A pre-incident plan that includes both a narrative and a drawing would be a formal pre-incident plan. On the other end of the spectrum is a building where a simple notation is made about a particular problem, such as holes in the floor due to a previous fire.

NFPA 1620: Recommended Practice for Pre-Incident Planning outlines the steps involved in developing, maintaining, and using a pre-incident plan.² An *NFPA 1620* chart showing the process used to develop and maintain a pre-plan is shown in **Figure 2-3**.

There are also recommended practices to be applied to specific occupancies. These occupancies include:
- Assembly
- Educational
- Health care
- Detention and correctional
- Residential
- Residential board and care
- Mercantile
- Business
- Industrial
- Warehouse and storage
- Special outdoor locations, such as transformer sub-stations

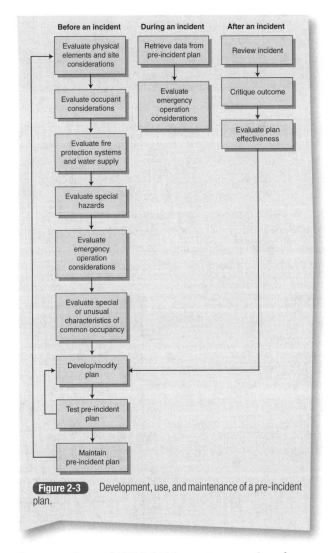

Figure 2-3 Development, use, and maintenance of a pre-incident plan.

The appendices of *NFPA 1620* contain examples of pre-incident plans as well as case histories of several warehouse fires.

Types of Pre-Incident Plans

There are many ways to describe and categorize pre-incident plans. This text considers three levels or types of pre-incident plans:

1. <u>Complex pre-incident plan:</u> A plan of a property with more than three buildings or when it is necessary to show the layout of the premises and relationship between buildings on the site. Complex pre-incident plans are used to identify building and fire protection features as well as hazards for each building. They also provide an overview of the complex to assist in locating buildings or areas within the complex. A pre-incident plan of this type may have several drawings of specific floor areas but also includes a drawing

showing an overview of the complex and the position of various structures by name or number, as in **Figure 2-4**. If Building 20 (Figure 2-4) is given as the fire location, fire fighters use the overview drawing to locate Building 20. Then the IC uses the formal pre-incident plan for Building 20 to assist in sizing up the situation and formulating the incident action plan. Often, buildings are numbered or named according to their function or when they were constructed, rather than by using a pattern that is readily apparent to those who are unfamiliar with the facility.

2. <u>Formal pre-incident plan:</u> A property with a substantial risk to life and/or property would need a formal pre-incident plan. This formal pre-incident plan would include a drawing of the property, specific floor layouts and a narrative describing important features. Several formal pre-incident plans could be included within a pre-incident plan for a single property within a complex pre-incident plan. One large, complex property could easily fill a three-ring binder with drawings of specific areas as well as the relationship between buildings.

The formal pre-incident plan drawing showing the interior layout for two connected buildings is detailed in **Figure 2-5**. These buildings are located within the complex shown in Figure 2-4. Only items of importance to the IC are shown on the formal pre-incident plan. The narrative portion of this pre-incident plan would further describe

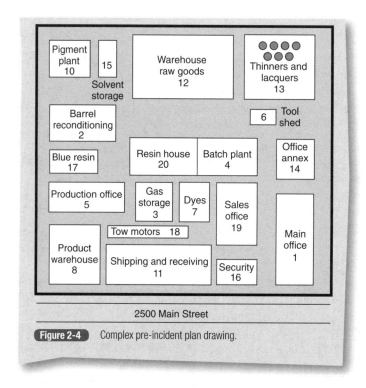

Figure 2-4 Complex pre-incident plan drawing.

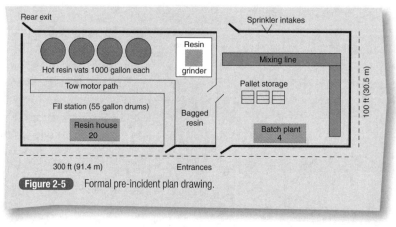

Figure 2-5 Formal pre-incident plan drawing.

the sprinkler system and hazards associated with the different processes taking place in these two buildings.

3. <u>Notation:</u> A simple notation may be made about the premises, such as the building has damage from a previous fire. Computer-aided dispatch systems usually have provisions for making a special notation for specific addresses. In this case, the dispatch center is the obvious location for the information. However, other pre-incident plan information should be available on the scene. Relaying information from a dispatch center could cause delays and misinterpretation of information. Another form of a notation pre-incident plan would include marking the exterior of dangerous buildings, for instance by painting a large "X" at the top level of a building that should not be entered because of structural problems. The State of New Jersey uses such an exterior marking system to indicate truss construction.[3]

A fourth category of pre-incident planning by occupancy could be described as either a training issue or the topic of an SOP. For example, church fires have common traits that should be taken into consideration (see Chapter 11). Most fire companies/districts have a predominance of buildings of similar construction. More can be learned about properties in your jurisdiction during EMS or other responses. This is particularly true for buildings that are not formally pre-planned, such as single-family residential buildings. In many cases, buildings in the same area were built as a subdivision and share common traits, such as having in-ground basements, walk-out basements, cellars, or no basements.

Buildings constructed during a given period will also be similar. Frame buildings built prior to 1940 may be **balloon-frame construction**, whereas in most areas frame buildings built later will probably be **platform-frame construction**. Similarly, frame structures built prior to 1940 will not have **wood truss** roofs.

Fire companies should survey their district on a regular basis, paying particular attention to buildings being demolished and under construction. Buildings being demolished will show the common construction characteristics for an era and provide a look inside the building's skeleton. When a building is under construction, the interior framework and other features can be seen **Figure 2-6**. Many fire fighters have discovered construction methods and materials that they did not know existed when

Figure 2-6 Building under construction.

> Two types of wood frame construction are balloon-frame and platform. In balloon-frame construction, the exterior wall studs extend the height of the building instead of stopping at each floor. This creates a channel through which a fire in the basement could travel inside the walls and emerge in the attic space. In platform construction, each floor is built as a platform on top of the lower floor, thus creating a barrier at each floor level. However, penetrations through the platform for plumbing, wiring, and the like can compromise the ability of this type of construction to stop the spread of fire within the walls.

visiting a building under construction. This is a good time to take photographs and share discoveries with the other crews.

Pre-plan Incident Checklist and Drawings

Pre-incident planning includes planning for special occupancies or specific types of buildings. Some departments prefer to gather and distribute information using a standard pre-incident plan form. The use of forms has advantages and disadvantages. The advantages include being able to find specific information in a predictable location on each pre-incident plan and having defined categories, which ensure the gathering of data for items listed on the form. The disadvantages include having large amounts of "not applicable" space on the form and failing to gather information about items that are not listed on the form. The preformatted pre-incident plan tends to be several pages long because of its generic formats, and much of it may be irrelevant. If a building does not have a sprinkler system, there is really no need to state that it is not sprinkler protected.

The authors of this book recommend using a detailed format based on the checklist presented in our later discussion of pre-planning and size-up (see *Analyzing the Situation Through Size-Up*). The pre-incident plan narrative for Memory Lane Apartments lists items from the size-up/pre-incident plan checklist that can be known in advance **Figure 2-7**. When a factor is assumed or covered in the SOP, it is not included. As an example, we show the factor "Additional Hose Lines Needed" and state "nothing special (SOP calls for back-up and line on floor above the fire)"; in reality, this would not be shown on the pre-plan as it is part of the SOP. Factors that do not apply are not listed. Memory Lane Apartments is not equipped with a standpipe or sprinkler system; they are therefore not addressed in the narrative.

The size-up/pre-incident plan checklist lists 76 items, not counting titles where subtitles fully address the factor. If you closely examine the Memory Lane Apartment pre-incident plan narrative and drawing, you can see that 27 of these factors are directly addressed, five are addressed in a general way, and two do not apply. Property conservation, in this instance, would follow SOPs and be typical; therefore, the 10 factors listed under property conservation are also addressed in a general way. In other words, information about 44 of 76 of the size-up/pre-incident plan factors can be known or partially known in advance. In some buildings, more or fewer factors may be known in advance, and incident-related factors, such as time and weather conditions will not be known until the day of the fire or upon arrival at the fire scene.

In addition to addressing the size-up/pre-plan factors, the building name, address, and owner/manager/agent name with telephone numbers and emergency contact information should be included in the narrative.

The IC and responding units do not have time to read a novel about the property. They should be able to quickly find and easily read the needed information. A sample pre-incident plan narrative is shown in Figure 2-7.

Memory Lane Apartments
25 Memory Lane

Life Safety/Fire Fighter Safety
Occupancy type
- 89-unit apartment building
- Most occupants are elderly, some with severe handicaps
- 7 small retail stores on first floor facing Main St.
- Video Rental Store (50 ft × 45 ft [15.2 m × 13.7 m])
- Joe's Barber Shop (24 ft × 45 ft [7.3 m × 13.7 m])
- Hair by Gloria Beauty Parlor (24 ft × 45 ft [7.3 m × 13.7 m])
- Old Tyme Antiques (12 ft × 45 ft [3.6 m × 13.7 m])
- Nail Boutique Manicure Salon (12 ft × 45 ft [3.6 m × 13.7 m])
- Song Shop CDs (24 ft × 45 ft [7.3 m × 13.7 m])
- Thrift Shop (24 ft × 45 ft [7.3 m × 13.7 m])
- Basement

Figure 2-7 Sample pre-incident plan narrative.

Occupancy type (continued)
- Bingo hall (180 ft × 45 ft [55 m × 13.7 m])
  - Bingo hall posted for 500 maximum occupants
  - Exits for Bingo hall stairs to Memory Lane and Main St. sides

Estimated number of occupants
- Bingo hall could have 500 on Tuesday and Thursday evenings
- Occasionally used as a party hall
- Most apartments occupied by a single resident
- Shops vary, seldom more than 10 per shop

Primary and alternative egress routes
- Three stairways and fire escape (see drawing)
- One-story roof to rear could provide emergency or secondary egress from second floor

Access to building exterior
- Access streets to front (Memory Lane) 50 ft (15 m) setback
- Right side (Main St.)
- Driveway on left side
  - Narrow, but apparatus accessible
  - Too close for aerial use
  - U-shaped driveway to front (may not support apparatus without damage)
  - Rear – one-story roof – good access to second floor

Construction type
- Ordinary construction (brick exterior wooden floor/ceiling assembly)

Roof construction
- Built-up flat tar roof supported by 2″ × 12″ (0.6 m × 3.7 m) beams

Condition
- Well maintained – no noted damage

Live and dead loads
- Average load for residential property

Enclosures and fire separations
- Only open areas in basement and stores
- Metal fire doors from apartments to hallways
- Hallway not separated, open from end to end

Extension probability
- Stairways are not enclosed
- Utility openings penetrate floors

Concealed spaces
- Inaccessible crawl space between top floor and roof

Age
- Built in 1932

Height and area
- Four-story building
- Basement to rear of building above grade level
- 220 ft × 180 ft (67 m × 55 m) U-shaped (see drawing)

Complexity and layout
- U-shaped hallway with center hallway (see drawing)
- Access to basement down stairs
- Stores all face Main St. with direct access to street

Extinguishment
Probability of extinguishment
- Basement only significant problem

External exposures
- College of Design on left side (Side B) 15′ (4.6 m)

Internal exposures
- Johnny's Restaurant attached to rear (Side C)
- Open stairways and floor to floor via walls/ceilings

Fuel load
- Average residential
- Stores and bingo hall average

Calculated rate of flow requirement
- Rate of flow for apartments and stores within the capacity of standard pre-connect and back-up line
- Bingo hall 810 GPM (51 L/sec) required

Number and size hose lines needed for extinguishment
- Standard pre-connect except bingo hall where 2½″ (64-mm) lines may be required

Figure 2-7 Sample pre-incident plan narrative.

Continues

Additional hose lines needed
- Nothing special (SOP calls for back-up and line on floor above the fire)

Water supply
- Hydrant located to front of building and on Main Street
- Additional hydrants on Main St.
- Hydrant flow on Memory Lane 1000 GPM (63 L/sec) per hydrant
- Hydrant flow on Main St. 1500 GPM (95 L/sec) per hydrant
- Water supply on grid system 8" (203 mm) main on Memory Lane; 20" (508 mm) main on Main St.
- System flow in area is approximately 12,000 GPM (757 L/sec)
- System can be cross-tied to two other supply systems for a total 30,000 GPM (1892 L/sec)

Property conservation
- Ordinary household items in apartments
- Stores have property of moderate value

General Factors

Staffing/Apparatus on Alarm Card

Alarm	Engine Companies	Truck Companies	Others
First	23, 19, 32	23, 19	District Chief 1
Second	3, 9, 5	32, 3	District Chief 4
Third	46, 34, 29	29	H. Rescue 9, ALS 31, Deputy Chief
Fourth	14, 31, 12	None	Chief of Department
Fifth	21, 28, 20	None	Training Chief, Prevention Chief

Additional company and command staff available
- 11 engine companies
- 8 truck companies
- 1 heavy rescue company
- 4 ALS, 6 BLS units
- Mutual aid

Staging/Tactical reserve
- Suggested staging area – parking lot off Main St. and Taft Roads

Utilities (water, gas, electric)
- Gas, water, and electric shut-offs in basement utility room just left of Bingo hall

Figure 2-7 Sample pre-incident plan narrative, continued.

It is said that a picture is worth a thousand words, and nowhere is this more true than on the incident scene. Including a drawing in a pre-incident plan is extremely useful. A sample pre-incident plan drawing is shown in **Figure 2-8**.

The authors of this book highly recommend using intuitive drawing symbols on pre-incident plans. Computer programs can easily print out map symbols. The CAMEO software package used for hazardous materials incidents includes intuitive symbols.[4] A symbol system that does not require a legend is highly desirable. It is also possible to photograph the actual items and then convert the image to a computer graphic for use in pre-incident plan drawings.

Drawings should include the following:
- Entry and exit doors
- Fences, gates, and other exterior security features
- Areas of safe refuge
- General floor layout
- Building and area dimensions
- Stairways
- Fire escapes
- Elevators
- Fire separations
- Utility shutoffs
- Alarm panels
- Lockbox
- Fire protection system and sectional control valves
- Fire protection system intakes
- Fire protection system manual actuation
- Fire protection system agent supply
- Hose outlets for standpipe systems
- Fire pumps
- Fire hydrants
- Potential staging areas
- Hazardous materials locations
- Emergency vents and controls
- Other

An important feature of a good pre-incident plan is its usefulness in the field. A pre-incident plan that is much more than two double-spaced typewritten pages in outline format plus a drawing page tends to be less useful during initial operations. An inclination to simulate a fire during the pre-planning process provides good training in preparing for fires in a building, but rarely should this type of information be included on the pre-incident plan. SOPs ensure that a company will supply the sprinkler intakes and that the IMS will be implemented; however, a tendency to repeat SOP items unnecessarily and include too many minor details on drawings weakens the plan. The IC does not always care about the location of bathrooms, plumbing, and other details included in architectural drawings. However, there are occasions when this information is useful.

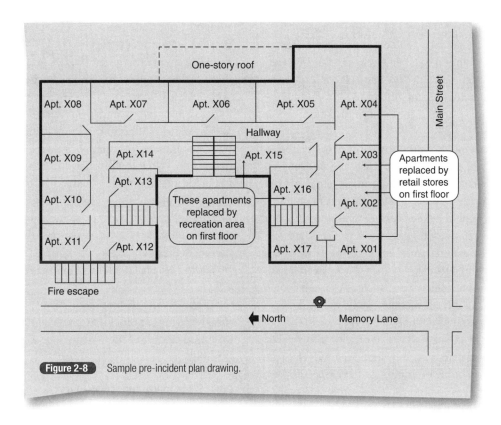

Figure 2-8 Sample pre-incident plan drawing.

For example, to remove water from the upper floors of a building, crews can remove the toilets from the floors, creating a 4″ (102-mm) drain.

There are excellent examples of pre-incident plans in Appendix E of *NFPA 1620: Recommended Practice for Pre-Incident Planning*.

When preparing a pre-incident plan, one of the most common errors is preassigning companies to respond to specific locations. This is an acceptable practice up to a point. For example, in the absence of visual signs of fire, a company could be assigned to check the alarm system with other companies standing by, or a second truck company could be assigned to the rear of the building. However, it is usually poor practice to pre-select a specific entry point or specific tactics. If the actual fire occurs in a location other than the pre-planned area, companies will be in the wrong locations. The pre-incident plan would then have made the situation worse. Including a tactical consideration on the pre-incident plan provides a reminder of possible problems or tactics but does not constitute a preassignment.

Preparation and Time Involved in Pre-Incident Planning

Pre-incident planning is a time-intensive activity that requires extensive data collection, then entering the data into a computer or onto paper. In addition, a building's occupancy, layout, and contents may change frequently, and pre-incident plans must be kept up to date. These changes to the interior of a building can greatly affect safety and tactics. In some cases a building that contained a fairly light **fuel load** may have had a change of occupants, with a subsequent change in the life hazard and/or fuel load. For example, a storage warehouse that was storing noncombustible commodities could have changed occupants, who may now be storing rubber tires in open racks, greatly increasing the fuel load. If this storage warehouse is protected by a sprinkler system, the new fuel load could exceed the design limitations of the sprinkler system.

Keeping pre-incident plans current is at least as important as creating the initial pre-incident plan. Bad information can be worse than no information. This revision process is also time intensive, especially if plans

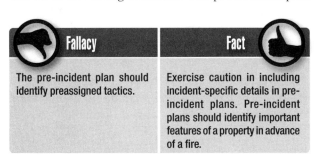

must be completely redrawn and the information must be reentered. This is where the computer can be used to great advantage: minor details can be changed and the new information printed out with little effort. If computers are available at the incident scene, the process of finding the right pre-incident plan and the right information can be greatly simplified.

Imagine having hundreds of pre-incident plans, with thousands of pages and accompanying information (material safety data sheets, hydrant maps, etc.), filed in three-ring binders. Finding the right property and material safety data sheet could involve considerable effort. If your department has relatively few pre-incident plans, a three-ring binder or card system may be sufficient. However, if you have large numbers of properties in the pre-incident plan file, a computerized system, such as the one shown in Figure 2-9, is highly recommended.

Many computer programs are available that allow easy navigation from page to page or provide information when the cursor is placed over the item. There are specific pre-incident plan software programs developed for fire service use, as well as common software programs such as PowerPoint™ that can be used. Easy navigation provides the means to quickly access needed information. Most architectural drawings produced today are drawn with AutoCad™ software. When considering new pre-incident plan software, it is best to purchase a program that can import and modify AutoCad™ drawings. If the property being pre-incident planned has architectural drawings, you may be able to simply load the drawings and add your symbols. AutoCad™ and most other architectural drawing programs are done in layers, allowing you to eliminate unwanted drawing information and add layers with other information. Cities or counties may have Geographic Information Systems (GIS) that include maps with valuable information such as the location of fire hydrants and water mains, aerial photographs or plot plans of buildings, flood zones, and underground pipelines. Fire department pre-incident plan information can be added to most GIS programs. If a GIS is available, the fire department should take full advantage of the data by adding its pre-incident plan information to the local GIS.

> Good pre-incident plans do not repeat standard operating procedures but recommend modifications where necessary.

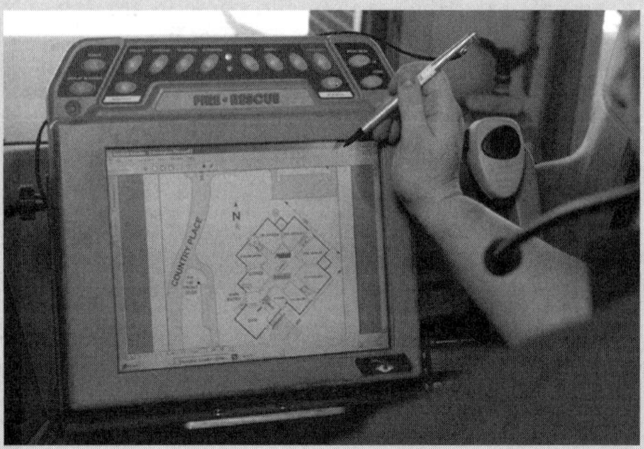

As an alternative to computerization—or as an interim measure while pre-incident plans are being converted to on-board computer files—it may be a good idea to place printouts in the facility lock-box if one is available.

What Structures Are Preplanned?

Just as some SOPs are department-specific, a decision regarding what properties to pre-incident plan depends on the jurisdiction being protected. Few communities have the resources necessary to pre-plan all buildings or even all commercial buildings. The operational priority list (life safety, extinguishment, property conservation) provides direction regarding what buildings may need to have pre-incident plans. If there is a high life hazard (including fire fighter safety issues), a particularly difficult extinguishment problem, or high-value property, then there is a need to prepare a pre-incident plan. High life-hazard properties should be given the highest priority for planning. Pre-incident planning deals more with potential demand (properties with the potential for a large loss of life or property) than it does with realized demand (the actual number of fires in a building or type of building).

Buildings that present an extraordinary challenge in terms of life safety, extinguishment, and salvage should be pre-planned. Other buildings that are generally pre-planned, whether or not they present a high life safety, extinguishment, or property conservation problem, include the following:

- Buildings protected by fire protection systems (with the exception of small restaurants with a hood system)

Figure 2-9 Onboard computerized pre-incident plan.

- High-rise buildings
- Industrial complexes

Modifying Standard Operating Procedures

SOPs describe a standardized method for addressing predictable operational circumstances. Pre-incident planning addresses what is different or unusual. Pre-incident plans are building specific, while SOPs are general. If the specific property fits the general category of operations, the SOP is assumed to be in effect. However, there can be special circumstances in which the SOP is not the most effective way to go to work.

For example, department SOPs require the first-arriving engine company to secure a source of water. The specific building being planned is small and located a considerable distance from the nearest hydrant. Based on information obtained while pre-planning the building, a recommendation is made to make an exception to the department water supply procedure. It is determined that the extra time involved in supplying water would result in complete destruction of the structure, and the rate of flow indicates that the apparatus tank has more than sufficient water capacity to extinguish the fire. Since the pre-planned tactical consideration is different than the department SOP, it is very important that this recommended tactic be approved for this property and then shared with other responding companies.

Estimating Life Safety Needs

As mentioned, any building that poses an unusually high risk to fire fighters or occupants must be included in the pre-planning process. Occupancies with long-span roofs present a hazard to fire fighters but could also be pre-planned due to the large rate of flow required to extinguish a fire. Heavy roof loading or unusual building features that could lead to partial or full collapse should be noted. A notation-type pre-incident plan may be enough for buildings that have previously been damaged by fire or weather.

Nursing homes, hospitals, places of assembly, schools, churches, and other places holding large numbers of people or a significant number of disabled people should be pre-planned.

Estimating Extinguishment Needs

A building with compartments requiring more than two standard pre-connected hose lines, as calculated by using the volume of the fire compartment divided by 100, should be pre-planned. (Rate of flow will be covered in more detail in Chapter 8.) If special or hazardous materials are present in quantity, such as flammable liquids, plastics, rubber tires, and idle wood pallets, smaller areas may require a pre-incident plan.

The estimated rate of flow should be included on the pre-incident plan when applicable. Where very large areas with heavy fuel loads exist, a decision may be made in advance of the fire regarding defensive or non-attack situations for fires beyond a certain percentage of involvement.

Estimating Property Conservation Needs

Any number of items could be considered high-value contents. Furs, jewelry, electronic equipment, and many other items could justify a pre-planning effort. Identifying the location of floor drains is generally beyond the scope of a pre-incident plan. However, if information of this type would be of value to the IC, then include it in the pre-incident plan. When using drawing software, it is possible to include information such as floor drains on a separate layer. The IC can then turn on the floor drain layer if this information is needed. Some property is more susceptible to damage, and the factors increasing its vulnerability should become part of the pre-incident plan. Stock that is piled on floors, rather than in racks or on pallets, and food products are more likely to sustain water damage and may require special attention after a fire.

Relationship of Pre-Planning to Size-Up

Size-up is a natural extension of the SOP and pre-planning process. SOPs get the operation off to a predictable start while pre-incident plans provide specific information about the building in advance of the fire. Whenever possible it is

Fallacy	Fact
All standard operating procedures are always in effect.	Standard operating procedures are not etched in stone. They can be modified during pre-incident planning or whenever situations dictate a different course of action.

> Size-up is a continuous process.

Size-Up/Pre-Plan Checklist

Life Safety/Fire Fighter Safety
- Smoke and fire conditions
  - Fire location
  - Direction of travel
- Ventilation status
- Occupancy type*
- Occupant status
  - Estimated number of occupants*
  - Evacuation status
  - Occupant proximity to fire
  - Awareness of occupants*
  - Mobility of occupants*
  - Occupant familiarity with building*
  - Primary and alternative egress routes*
  - Medical status of occupants*
- Operational status
  - Adherence to SOPs
  - Fire zone/perimeter
  - Accountability*
  - Rapid intervention*
  - Organization and coordination*
  - Rescue options*
  - Staffing needed to conduct primary search
  - Staffing needed to conduct secondary search
  - Staffing needed to assist in interior rescue/evacuation
  - Staffing needed for exterior rescue/evacuation
  - Apparatus and equipment needed for evacuation
  - Access to building exterior*
  - Access to building interior* (forcible entry)
- Structure
  - Signs of collapse
  - Collapse zone*
  - Construction type*
  - Roof construction*
  - Condition*
  - Live and dead loads*
  - Water load
  - Enclosures and fire separations*
  - Extension probability*
  - Concealed spaces*
  - Age*
  - Height and area*
  - Complexity and layout*

Extinguishment
- Probability of extinguishment*
- Offensive/defensive/non-attack
- Ventilation status
- External exposures*
- Internal exposures*
- Manual extinguishment
  - Fuel load*
  - Calculated rate of flow requirement*
  - Number and size hose lines needed for extinguishment*
  - Additional hose lines needed*
  - Staffing needed for hose lines*
  - Water supply*
  - Apparatus pump capacity*
  - Manual fire suppression system*
- Automatic fire suppression equipment*

Property Conservation
- Salvageable property*
- Location of salvageable property*
- Water damage
  - Probability of water damage*
  - Susceptibility of contents to water damage*
  - Water pathways to salvageable property*
  - Water removal methods available*
  - Water protective methods available*
- Smoke damage
  - Ventilation status
  - Probability of smoke damage*
  - Susceptibility of contents to smoke damage*
- Damage from forcible entry and ventilation*

General Factors
- Total staffing available versus staffing needed*
- Total apparatus available versus apparatus needed*
- Staging/tactical reserve*
- Utilities (water, gas, electric)*
- Special resource needs*
- Time
  - Time of day
  - Day of week
  - Time of year
  - Special (e.g., holiday season)
- Weather
  - Temperature
  - Humidity
  - Precipitation
  - Winds

* This factor can be at least partially known in advance of the fire through pre-planning.

Figure 2-10 Size-up/pre-incident plan checklist.

best to gather as much information in advance as possible; therefore the list of size-up factors is also used as a pre-incident plan list. Use the size-up list from Figure 2-10 and description of each factor as a guideline when pre-incident planning.

A brief explanation of what is meant by each of the size-up/pre-incident plan factors is provided in this chapter. Many of these factors are further explained in other chapters of this book.

Analyzing the Situation Through Size-Up

Size-up factors, which may change from incident to incident, are difficult to categorize in terms of relative importance. Incident conditions will determine which size-up factors are most important. Factors related to life safety are most likely to be critical. These most important factors are known as **primary factors**. The initial size-up analysis is limited to evaluating primary factors. Less important factors are categorized as **secondary**. Incident conditions will dictate which conditions are primary. As an example, weather can be categorized as critically important (primary factor) when at the extreme and as a less important secondary factor during moderate conditions. The IC must first sort through information and evaluate available data related to primary factors.

Size-up actually begins before the incident with the development of SOPs and pre-incident planning. When an alarm occurs, the IC considers what is already known about the specific property and, generally, about the type of property and area. Important factors such as weather conditions, time of the alarm, and day of the week will be known as the response begins. Other incident-related information begins with the dispatcher's information about the location of the fire, method of alarm, and other conditions reported to dispatch. In many situations, at least one fire company will arrive before the dispatched chief officer. Communications from the scene should add to or confirm dispatch information.

Once on the scene, the IC will add to what is known through personal observation, communications with fire companies and building personnel, and reconnaissance. Once the on-scene information is processed, the IC must quickly evaluate the action that is taking place and the condition of the building.

The amount and quality of information improve with time, and the IC has time to better evaluate the situation once initial assignments are made. Size-up continues throughout the incident and into the overhaul phase.

Life Safety/Fire Fighter Safety
Smoke and Fire Conditions

Smoke and fire conditions are directly related to occupant survival and fire fighter safety and are primary factors at a structure fire. Heavy, dark, pressurized smoke and visible fire conditions may necessitate a defensive attack, whereas light smoke with no fire evident indicates a high probability of saving occupants still inside the building. It must be remembered that reading smoke and fire is not an exact science, and hidden fire can cause a rapid increase in the volume of smoke and fire when it breaks out of containment. **Flashover** is a critical indicator at a structure fire. Occupants inside post-flashover compartments have a very low probability of survival. Smoke and fire conditions can also provide a warning of an impending **backdraft**.

Experienced ICs learn to evaluate pressure, smoke characteristics, and other factors in determining the intensity of the fire; however, interior reconnaissance is generally the best way to realistically determine fire intensity when an interior attack is possible Figure 2-11.

Figure 2-11 Smoke and fire conditions.

David W. Dodson explains methods of reading smoke in terms of volume, velocity, density, and color.[5] All of these factors provide clues about the possibility of flashover, location of the fire, and survivability inside the burning building. Smoke and fire conditions will not be known until the time of the fire, but the compartment size and tightness, as well as fuel content provide pre-fire indicators of the relative fire potential.

Fire Location. Knowing the fire's location is necessary to successfully combat the fire. Extinguishment is the primary life safety tactic as well as an operational priority. In most cases, company officers will be able to determine the location of the fire on arrival. The information received from dispatch, alarm systems, and information from occupants as well as visual clues assist in finding the fire location. The first-in crews typically will be attacking or setting up to attack the fire when the chief officer arrives. There are times when the fire location is not obvious or when dispatch information is wrong.

When information is incomplete or incorrect, finding the fire can be very difficult especially in a large building. High-rise buildings can hide a fairly large interior fire, and smoke alarms may be sounding far away from the actual fire location. A fairly small fire located at a window may appear to be much larger, whereas a large fire toward the core of the building may not look like much of a problem from the exterior. Locating the fire and its extension paths are critical to the operational plan.

Open flaming is a definite indicator of fire location, but the origin of the fire could be from another location usually below the visible flame. The old adage "where there's smoke, there's fire" is true, but the primary goal here is to find the seat of the fire. David W. Dodson also explains that smoke velocity will be greatest near the main body of fire. Understanding the basic chemistry and physics of fire is helpful in learning to read smoke. All of matter, including smoke and heat, naturally try to reach equilibrium. When openings are available, smoke will try to reach the same temperature as the temperature on the other side of the opening. In most cases, smoke will go to the nearest and largest opening unless wind conditions reverse the venting. Likewise, all gases, including fire gases, are governed by the physical properties of gases, which state that a given volume of gas will expand as the temperature of the gas rises. If the gas is contained under conditions where expansion is not possible, the heated gas will undergo an increase in pressure. This can be calculated using the ideal gas law. But the fire ground is no place for precise calculations; it is enough to understand that smoke rapidly flowing from an opening is a clue that the fire may be nearby.

Direction of Travel. Knowing where the fire is most likely to spread is important to life safety and extinguishment tactics. Search teams with hose lines need to protect and evacuate occupants in areas where the fire is likely to extend. Basic chemistry and physics of fire determine the direction of fire travel. The products of combustion travel upward until reaching a barrier; they will then travel horizontally to fill the top of the compartment, and finally they will travel downward if contained. Fire, heat, and smoke will, therefore, travel upward via the path of least resistance, before traveling horizontally or downward. Fire will most likely travel up stairways and other available openings to the floors above the fire. The upward path could also be through concealed wall areas (balloon-frame construction is particularly noted for this). Fire and smoke entering a wall will travel to the attic unless it meets a solid barrier. Nearly all buildings have vertical paths to the top of the building, although some are much better protected than others. Fire in an open wall, such as in balloon-frame construction, will travel unimpeded to the attic. In platform-frame construction and other types of construction with fire barriers, the path may not be as direct, but there are generally smaller pathways via utility openings for drains, water pipes, electrical wiring, etc.

Although it is not possible to know the exact means of fire travel during pre-incident planning, it is possible to know the probability of fire travel. Rated and non-rated fire compartments tend to limit fire spread. Construction methods and alterations affect fire spread and can be known in advance.

Ventilation Status

Notice that ventilation status is listed under all three operational priorities (life safety, extinguishment, and property conservation). Ventilation is a key factor during all phases of the operation. Venting for life safety involves moving the fire away from occupants and fire fighters. Understanding smoke and fire movement, as discussed previously, is essential to good venting. As mentioned earlier, the fire will follow the path of least resistance upward, then horizontally, and finally downward. The vent opening should pull the fire away from occupants and fire fighters. A common life safety venting tactic is to remove a scuttle or other cover over a built-in opening at the top of a stairway in a multi-storied building. The intended purpose of this tactic is to clear the stairway of smoke. However, this tactic can result in the unintended consequence of channeling fire, heat, and smoke into the stairway being vented. An illustration of this unintended consequence is shown in **Figures 2-12, 2-13** and **2-14**. In this scenario, the first-arriving engine company is advancing a hose line to the fire floor, while the first truck company is split into two crews. One truck crew is gaining access to the roof and the other is in the stairway making entry into the floor above the fire to check for extension and begin the primary search as well as facilitating evacuation (see Figure 2-12).

When the engine company opens the door to attack the fire an unvented fire will be released from confinement and will enter the stairway (see Figure 2-13). This will tend to make smoke conditions in the stairway worse, even if there is no vent opening at the top of the stairway. If an opening is made at the top of the stairway fire and smoke could be pulled into this "chimney-like" path of least resistance making conditions much worse in the stairway (see Figure 2-14). Opening the scuttle at the top of the stairs can be a good tactic if the fire is controlled, but can cause serious injury or death if done under the wrong circumstances.

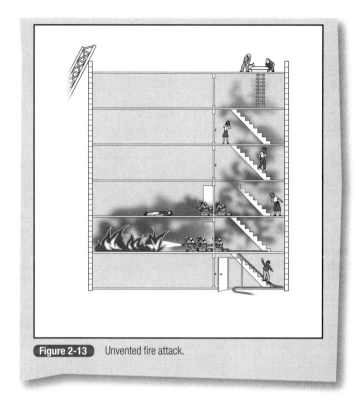

Figure 2-13 Unvented fire attack.

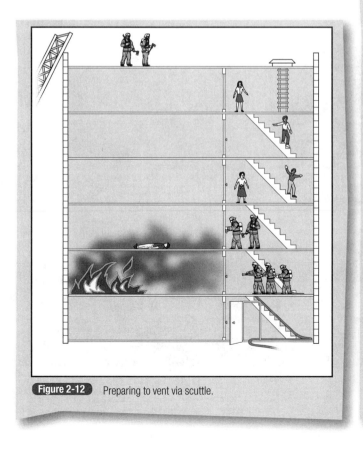

Figure 2-12 Preparing to vent via scuttle.

Figure 2-14 Fire vented at top of stairway. Fire is entering the stairway at the second floor, causing heavy smoke above.

Ventilation possibilities can be known through pre-incident planning. Roof openings such as scuttles and vents may be present. Ventilators and skylights such as those shown in Figure 2-15 could be removed to vent the area below. Areas where aerial apparatus can be positioned to reach the roof can be known in advance. Department SOPs may determine what types of roofs are to be vented, as well as how and when positive pressure venting is to be done.

Occupancy Type

The building's use will determine how likely the building is to be occupied at the time of the fire, the number of occupants, the fuel load, fuel type, value of the contents and many other essential facts. Major occupancies can and should be identified in advance of the fire through pre-incident planning. Smaller occupancies can often be readily identified, e.g., a single-family detached residence. However, some buildings are used for purposes other than their original intended use, e.g., a single-family residence being used as a real estate office. See Chapter 11 for a complete discussion of various occupancies and tactics related to occupancy.

Estimated Number of Occupants. Estimating the number of occupants in a large building is difficult at best. However, the building's intended use (occupancy) provides evidence of relative occupant density. *NFPA 101: Life Safety Code* establishes the maximum occupant load per square foot by occupancy type, e.g., one person per 100 ft^2 (9 m^2) in an office building (business).[6] Places of public assembly may be posted with the maximum number of occupants. In hotels and apartment buildings, the number of available rooms/apartments provides an indicator of maximum occupant load. All of these provide a starting point for estimating the number of people that could be in the building. Time factors, e.g., hours of operation, combined with the maximum occupant load and evacuation status,

Figure 2-15 Fixed roof vents.

help determine the number of people endangered by fire and the possible need for some level of assistance.

Evacuation Status. Estimating the number of people still inside the building is the next logical step after estimating how many people may have occupied the building upon discovery of the fire. Some occupancies routinely conduct fire drills and account for occupants when a fire alarm sounds, e.g., primary and secondary schools. The NFPA recommends that each home have an evacuation plan following the Exit Drill In The Home (EDITH) educational program. Many times, family members account for other family members; unfortunately, this is not always the case, and information from occupants is not always accurate. Most buildings do not have an occupant accountability system.

Consider a fire in a large apartment building. Some of the residents probably would not be home when the fire occurred, others would evacuate and leave the scene, and some occupants may have guests. For all these reasons and more, the only way to be sure that the building has been evacuated is to conduct a primary search, and even then, fire fighters need to verify the search with a secondary search. Information from occupants who have escaped would be less reliable in a larger building without a formal accountability system than in a school or home.

If a pre-planned building has an occupant accountability system, the specifics of the plan should be noted on the pre-incident plan. Of particular interest is determining who will report to the fire department and where this person will be located (including an alternate site). Also note where occupants will be assembled.

Occupant proximity to fire. Obviously being close to a well-involved fire places occupants in imminent danger. Chapter 6 explains the life safety priorities for occupants on various levels of a multi-story building. Knowing the location of occupied areas through pre-incident planning can be very useful during the size-up process.

Awareness of occupants. People who are awake and alert are more likely to hear an alarm or sense the products of combustion, and then take action to evacuate the building. Occupants who are asleep or mentally incapacitated may not be aware of the fire. Awareness is related to occupancy and will be further explained in Chapter 11. Identifying the occupancy type during pre-incident planning provides an indication of occupant awareness, e.g., a nursing home fire at night where occupants are likely to be sleeping and possibly under the influence of medications.

Mobility of occupants. Most large buildings will contain persons unable to fully evacuate on their own. As with other factors, knowing the occupancy of the building can be very helpful in determining whether occupants can successfully escape on their own.

> Occupancy: Building's primary use, e.g., school, hospital, nursing home.
> Occupant: Person who could be in the building.
> Occupied: Occupant(s) are in the building.

Occupant familiarity with building. In most cases, the people inside a place of public assembly are not familiar with the building layout or alternate exit facilities. Many large, loss-of-life fire reports in places of assembly address unfamiliarity as a major problem. People who live or work in a building would be expected to know the building layout and location of exits. However, people will be most familiar with their regular ingress path, e.g., a person working on an upper floor of a high-rise office building likely enters the lobby and takes the same elevator to his/her office every workday. Unless the building management regularly conducts fire drills, many of the employees on upper floors will not use the stairway or even know its location.

Primary and alternative egress routes. This factor is related to the mobility factor insofar as most occupants will escape unassisted if there is sufficient egress. Codes specify egress facilities and pre-plan drawings should show the location of all exits. Some occupancies, such as hospitals, designate a defend-in-place strategy for occupants.

Many public buildings also have **areas of refuge** where immobile occupants wait to be helped. It is essential that pre-incident plans address these special facilities. Fire fighters must know the location, how many occupants could be at the area of refuge, and what the occupant and building managers expect to happen in an emergency. Often fellow occupants or security personnel are designated to assist people in these locations. The fire department must always check areas of refuge during a fire emergency.

Medical status of occupants. This factor requires a twofold analysis. What is the normal medical status of people occupying the building, and what effect is the fire having on building occupants? Whenever people are still in the building upon arrival, emergency medical service units need to be summoned to the scene. A medical branch is needed if there are large numbers of potential victims.

Operational Status

SOPs provide a standard way of going to work at a fire scene. This allows the operation to get started, following a predictable plan. However, there are situations that require a different approach. The IC continually evaluates the safety and effectiveness of the standard operation and determines whether a nonstandard attack would be more effective.

Safety is the IC's most important consideration. If the building is in danger of early collapse or the situation necessitates a transition from an offensive to a defensive operation for other reasons, the IC must order all personnel out of the building.

The effectiveness of the operation must constantly be evaluated. The IC must consider the following questions: Is the search being conducted in a systematic manner? Are the occupants who are in the most danger being rescued? Has the fire been properly vented to control fire spread? Is progress being made in controlling the fire? Are rate-of-flow requirements being met? Has salvage been considered?

It is critical that the IC continually reevaluate the situation in terms of risk management. The question that should regularly be asked is whether the desired benefit is worth placing fire fighters at risk.

Most of the operational status factors apply to an IC assuming command of an operation that is in progress.

Adherence to SOPs. This factor is evaluated when command is transferred. Are on-scene units following SOPs? Is the operation progressing as expected? Safety is a major part of this analysis. Is the operation being conducted safely? Are SCBAs and turnouts being used as prescribed? Is there a need to change from an offensive to a defensive strategy? This is one of few size-up factors about which very little can be known in advance of the fire through pre-incident planning. However, SOPs that are developed in preparation for fire emergencies provide an advanced indicator.

Fire zone/perimeters. Fire zones and perimeters will be further discussed in Chapter 5. However, the fire zone generally refers to an area where a specified level of protective clothing is required or possibly a safe area where no protective clothing is needed. The fire perimeter is set up to keep non-response people out of the area. The IC assuming command should evaluate the present zones/perimeter to determine whether they are sufficient. General rules for fire zones/perimeters should be part of the department SOPs. In addition, pre-incident plans for large secured properties may designate specific perimeters.

> The fire department accountability system accounts for all fire fighters at the incident scene. It does not account for building occupants.

Accountability. The primary accountability system is NIMS. A properly organized incident provides the first level of fire fighter accountability. NIMS is a safety system. It is important that freelancing be avoided and that units working in the hazard area work as a unit (usually a fire company at a structure fire) within the command system. A formal accountability system is required by *NFPA 1500: Standard on Fire Department Occupational Safety and Health Programs*. The IC must ensure that the local system is being used. Chapter 5 includes a detailed description of accountability systems.

Rapid intervention. *NFPA 1500* also requires a rapid intervention crew (RIC). A team of fire fighters must be immediately available to rescue fellow fire fighters who need assistance. Chapter 5 also includes a detailed description of rapid intervention.

Organization and coordination. NIMS, as described in Chapter 1, is the acceptable method of organizing an incident. All units on the scene must be included in the organizational structure. Activities of units must also be coordinated. The first-arriving company begins this process by organizing the initial attack and communicating orders to incoming units. When command is transferred the person assuming command must be sure all units are working toward common tactical objectives within the overall strategy.

Rescue options. The IC first evaluates the ways occupants can be removed from the building, then selects the safest and most efficient option. The pre-incident plan can identify primary means of egress, which are the preferred evacuation pathways, but should also address alternatives such as fire escapes and ladder positions. See Chapter 6 for a more detailed discussion of rescue options.

Staffing needed to conduct primary search. The size of the area to be searched, smoke conditions, rescue methods available, and the condition of the occupants determine how many crews should be assigned to the primary search. In a single-family dwelling, a single company or possibly one company per floor conducts the primary search. In a large area with obstacles and poor visibility, several companies may be required to complete the primary search of a single floor. If victims must be physically removed, there will be a need for additional staffing.

Staffing needed to conduct secondary search. Quite often the same teams that perform the primary search will conduct a secondary search in an area other than the area they originally searched. The secondary search generally will not involve increased staffing.

Staffing needed for interior rescue/evacuation. This could be the same staffing required for the primary search, or additional teams may be assigned to remove victims who are found during the primary search. In many cases, occupants will be attempting to evacuate on their own, and staffing may be needed to assist and direct them to safety.

Staffing needed for exterior rescue/evacuation. If ladders or other exterior rescue methods are warranted, additional staffing will be needed. Most exterior rescues require more staffing per rescue than removal via the interior stairs.

Apparatus and equipment needed for rescue/evacuation. In most situations, an aerial apparatus or ground ladder will be the apparatus/equipment needed for rescue. There is other rescue equipment that is rarely used, e.g., rope rescue equipment.

Access to building exterior. When pre-incident planning a building it is important to note apparatus entry points for secured properties. Quite often, special arrangements are made with building security for these properties. Street or road access around the structure should be included on the pre-incident plan drawing. Also note aerial access points and roadways that are not safe or are inaccessible to fire apparatus.

Access to building interior (forcible entry). In the post–September 11th era, there is more security on the exterior and interior of buildings. Some buildings that were formally open to the public are now blockaded. Be aware of changes to buildings in your jurisdiction. The need for forcible entry can significantly delay search and rescue operations as well as the initial fire attack.

Note the times when a building is secured or when special security features such as internal locks and gates impede entry.

Structure

A thorough risk-versus-benefit analysis determines whether the operation will be offensive or defensive. A major risk-versus-benefit consideration involves evaluating structural conditions. If the structure is in imminent danger of collapse, no expected benefit is worth the risk posed by entry into the structure. If the building is being damaged but adequate time remains before a potential collapse, an offensive attack posture to save lives could be warranted. A decision may be made to contain the fire and rescue occupants, with a switch to a defensive attack once the primary search and rescue are complete. The effect of the initial operations should be reevaluated once occupants are removed from the structure.

The necessary risk-versus-benefit analysis is somewhat subjective. However, experience and training can greatly improve its accuracy.

Signs of Collapse

Collapse (failure of a supporting structure, with complete or nearly complete destruction of the building) can be catastrophic. However, there are many partial collapses that can injure and kill, such as a floor collapse where fire fighters are struck by the floor or fall into a burning lower level. See Chapter 5 for a list of possible collapse cues, but be aware that many collapses occur with no perceptible warning, especially in truss-constructed buildings.

Collapse Zone

In Chapter 5, we recommend a collapse zone the height of the structure plus an allowance for debris scatter. In most cases this translates to a collapse zone 1½ times the height of the building. It is very difficult to estimate the height of the building looking up from the street level. It is much easier to measure or estimate building height during pre-incident planning. Other pragmatic decisions can be made during the pre-planning tour. Determine if the width of the streets is less than the predicted collapse zone. In such a case, a decision could be made not to place apparatus in the street if there is an imminent collapse hazard. Prior to the fire is also the time to determine the reach of master streams: Can water be applied to the building from distances beyond the collapse zone? Determining a collapse zone provides a good example of how the same size-up factor can be primary or secondary. When confronted with a well-involved fire in a Type II structure, determining the collapse zone is a primary factor. On the other hand, it would be a secondary factor for a minor fire in a Type I building.

> The IC must always consider structural conditions as part of the size-up.

Construction Type

We highly recommend that all members of fire departments become familiar with building types and common building problems. Few fire fighters would be capable of an engineering analysis to determine structural stability, but they can recognize potential structural problems during pre-incident planning and by studying past fires in different types of construction. *Brannigan's Building Construction for the Fire Service* is a "must read" for anyone who might take command at a structure fire, including the first-arriving officer.[7]

During pre-incident planning the building should be classified by construction type:

- **Type I construction**: Fire-resistive
- **Type II construction**: Non-combustible
- **Type III construction**: Ordinary
- **Type IV construction**: Heavy timber
- **Type V construction**: Frame

See *NFPA 220: Standard on Types of Building Construction* for specific fire-resistive requirements for each type of construction.[8] Some buildings defy classification, particularly buildings that have been renovated. The original construction might be classified as fire-resistive construction, but removal of protective coatings on steel trusses could reduce the building to non-combustible status. *NFPA 5000: Building Construction and Safety Code* would classify a building so modified as the weakest form of construction (non-combustible in this example).[9] There can also be different types of construction in various areas. An older section of a building may be constructed using heavy wooden beams while a newer section may have wood or steel truss construction.

The fire-resistive (Type I) building, as the name indicates, is superior to all other building types in regard to structural stability under fire conditions. Some might argue whether wood frame or non-combustible is most likely to collapse. Wood frame is generally of limited size, whereas non-combustible construction is used on very large buildings. Most large non-combustible buildings are sprinkler protected. Chapter 5 discusses the relative safety of the five major types of construction.

Roof Construction

The roof covering can be important, particularly if it is combustible, such as wood shingle. However, the structure that supports the roof is generally most important to fire fighters in this era of nearly universal non-combustible roof coverings. Roof collapse is a killing mechanism in its own right, but also a precursor to catastrophic collapse. The roof tends to tie in the walls of the building. When walls damaged by fire lose roof support, they often fail. Of particular concern are truss and other lightweight supporting structures. Most modern buildings with large open areas will have a truss roof. These long spans are particularly dangerous. The roof structure should be identified and noted during pre-incident planning. If a truss roof is present, especially a long span truss roof, it should be highlighted on the pre-plan narrative and a symbol added to the drawing indicating the truss roof. Truss roofs have been responsible for many fire fighter fatalities. The truss roof collapse mechanism and safety issues will be further discussed in Chapter 5.

Condition

A building in poor repair or one that is previously damaged presents an extra hazard for fire fighters. Abandoned buildings often have structural damaged caused by weather, previous fires, or unauthorized occupants. Even if the building does not warrant formal pre-incident planning, a notation should be made warning responding fire fighters of structural damage or hazards such as holes in the floor. The IC must consider serious pre-fire damage as a critical factor when making an offensive/defensive decision.

Live and Dead Loads

The **dead load** is the load imposed on structural members by the building and permanent attachments. The type of construction will be a major factor in determining the overall dead load. Of special consideration are heavy roof loads such as roof mounted equipment, particularly if the roof is supported by unprotected truss construction. Heavy roof loads should be noted on the pre-incident plan.

Live loads are those loads produced by the building contents. The live load will vary from very light to extra hazard. Warehouse buildings tend to carry very heavy live loads with a possible added impact load due to materials handling equipment. If the materials being stored are combustible, they also add to the fuel load. Extraordinary live loads should be noted on pre-plans. Knowing a building's live load is important when determining incident-specific tactics.

Fire Suppression Water Load

The obvious water load is the weight of the water fire fighter's discharge into the building during fire operations. Each gallon of water weighs 8.33 pounds (3.8 kg). Therefore, a 1000 GPM (63 L/sec) master stream operating into the building is adding 8330 pounds (3778 kg) to the building each minute. Given the fact that defensive operations often require several master streams operating over a prolonged period of time, it is easy to see how this would affect a building's structural integrity. Fortunately, most buildings will hold only a small percentage of the water from these master streams, as the water will run down vertical openings to the basement or outside. It is very difficult, if not impossible, to accurately determine how much water is being retained in any area of the building. Here again, pre-plan information can be invaluable. If the storage is stacked directly on the floor and is absorbent, a large percentage of water will be retained. On the other hand, storage on skids tends to allow water to run off. Other factors can also affect run-off. When a building is severely damaged, structural members, wall materials, and contents can fall on the floor, forming dams that retard water run-off. When pumping large quantities of water into a building, observe the run-off and consider the possibility that the weight of the water could facilitate structural failure.

Enclosures and Fire Separations

The type of construction and occupancy, both of which can be known before the fire, will be major clues in determining extension probability. Fire-resistive construction contains fire-rated compartments. However, the size of the compartments will vary by occupancy. A business occupancy such as the World Trade Center may have large, open office spaces, which will allow the fire to spread freely within the compartment. Residential high-rise buildings may have a few large common areas, but will mainly consist of small rooms, which hinders fire extension.

Non-combustible (Type II) construction may also have large open areas, but the fire-resistive qualities between areas will generally be inferior to fire-resistive barriers.

The balloon-frame building is noted for a lack of compartmentation having no rated fire assemblies to prevent fire spread. However, even non-rated doors and walls will retard fire spread for a period of time.

In a multi-story building, the floor/ceiling assembly provides a barrier between floors. This barrier will be much more fire-resistive in fire-resistive construction than in other types of construction. Floor/ceiling assemblies are seldom perfect barriers due to penetrations made for wiring, plumbing, air movement equipment, and other building services.

Extension Probability

Extension probability is directly related to the enclosures/fire separation factor above and the concealed space factor discussed next. Fire spreads through concealed spaces and from area to area when enclosures are missing or weak. The IC must consider how fire spread will endanger fire fighters and occupants during size-up.

Concealed Spaces

Most buildings (except for heavy timber, Type IV construction) contain multiple concealed spaces. Of special note are attics, particularly when several buildings or units share a common attic. Fire separations are often damaged or removed in buildings that were originally constructed with attic fire separations. Sometimes the only separation between the attic and useable space below is a suspended ceiling. Suspended ceilings create a false space where subsequent fire travel can result in fire cutting off fire fighter egress paths. As fire fighters make entry, ceiling tiles should be lifted or removed to determine if fire is in the false ceiling area. If the concealed space is involved in fire and contains a truss roof or floor assembly, expect rapid collapse. Immediately communicate this finding to the IC. The time to discover concealed spaces and truss roofs/floors is during pre-incident planning before the fire. Smoke conditions may not allow an examination during the fire. Many ICs have been surprised by a fire that breaks through the roof or extends to a distant apartment or store.

Age

The age of the building can have positive and negative effects. Most older buildings will have heavier, more fire-resistive construction. However, buildings that have been renovated may contain new, lightweight construction in the renovated areas. Not everyone who remodels secures a building permit, but renovations can be discovered as you tour your response area. It is more likely that a building permit will be secured when larger commercial buildings are remodeled. The fire department should be part of the permit approval process, or at the very least be notified when such construction is taking place. Renovated buildings using lighter construction can be an extra hazard because fire fighters expect heavy construction methods in older buildings. The negative side of the building's age is that the building may be getting weaker as it ages. This depends on the construction materials and building maintenance. The year the building was constructed should be noted on the pre-incident plan.

Height and Area

It is impossible to have a large structure fire in a small building; thus, the building's size will at least partially dictate the total volume of fire. Rate of flow formulas are based on the size of the fire compartment. Rate of flow is thoroughly discussed in Chapter 8. The location and size of large undivided areas of buildings should be noted on pre-incident plans.

The height of the building determines the effectiveness of ground-based fire apparatus. When pre-incident planning a multi-story building, determine the maximum reach of available aerial apparatus in terms of floors, e.g., aerial will reach the sixth floor. There may be areas around the building where aerial apparatus cannot be used due to obstructions or lack of access. Aerial access points should be identified in advance of the fire.

The height of the building will also affect the number of possible occupants, the type of construction, fuel load, and other factors. The height of the building should be included in the pre-incident plan. All high-rise buildings should be pre-planned.

Complexity and Layout

The pre-incident plan should include a general floor layout for the building along with any information that might affect firefighting operations or fire fighter safety. The layout of some buildings is maze-like, particularly large buildings with multiple additions Figure 2-16 . When visiting a large, complex building, take a moment to orient yourself. Try to determine your location in relation to the stairway, street, standpipe, and other critical landmarks. Could you find your way out under heavy smoke conditions? Storage occupancies may have aisles and cross aisles that could be confusing under low visibility conditions. Large open areas in business occupancies may have cubicles separating work areas. The layout of these cubicles can be extremely confusing and very difficult to search under fire conditions.

Extinguishment

Probability of Extinguishment

The probability of extinguishment is also important to life safety. When the fire is extinguished, the probability of death and injury are greatly diminished. Extinguishment is nearly always a primary factor at a working structure fire. Rate of flow is discussed later in this chapter, as well as in Chapter 8. Knowing flow requirements and the extent of fire are key factors in determining whether a rapid, lifesaving extinguishment is possible.

Offensive/Defensive/Non-Attack

Deciding the overall attack strategy is critically important. The entire operation hinges on this early decision. Whenever command is transferred, the person assuming command must consider whether the present strategy is correct. Critical factors change as tactical objectives are achieved.

The most important objectives are those related to life safety. Once the building is cleared of occupants, the reason for continuing an offensive attack is reduced; therefore, a new risk-versus-benefit analysis should be conducted to determine if an offensive attack should continue.

Ventilation Status

Finding and extinguishing the fire is much easier when the fire has self-vented in a building that is still safe to enter. If the fire is unvented, it should be vented as soon as possible. Venting for extinguishment involves making openings to pull smoke and heat away from the hose crew to support a quick, efficient fire attack. Part of the pre-incident planning process involves identifying available built-in vents and evaluating the roof structure for roof operations.

External Exposures

Are nearby structures (not connected), vehicles, and other property threatened by the fire? Protecting exposures can be done in several ways as discussed in Chapters 8 and 9. The pre-incident plan drawing should show the location of nearby exposures. The narrative part of the pre-plan should identify the occupancy and any special considerations regarding the exposures.

Internal Exposures

Other parts of the fire building or connected structures are categorized as internal exposures. Identifying fire pathways is best done prior to the fire. At the time of the fire the IC evaluates whether the fire is likely to follow one of the pre-identified pathways. Tactics for protecting internal exposures are also discussed in Chapter 8.

Manual Extinguishment

If there is no automatic fire suppression system or the system does not control the fire, manual extinguishment will be necessary.

Fuel load. Fuel load varies as to quantity, type of fuel, geometric orientation and other factors. Fuels identified as "extra-hazard" in sprinkler calculations would generally require more water to extinguish than fires in ordinary combustibles. While fuel loads tend to change, in some buildings the maximum and average fuel load can be evaluated during pre-incident planning.

Calculated rate of flow. The big question in calculating rate of flow is whether the standard attack line with a back-up line can extinguish expected fires. It is impractical to calculate rate

Figure 2-16 Large, complex building.

of flow at the time of the fire. If the fire compartment is larger than what two standard pre-connects can extinguish, or if the fuel load will create a fire situation beyond the control of two standard attack lines, the rate of flow should be pre-calculated for these areas and noted in the pre-incident plan. Chapter 8 explains the formulae used to calculate rate of flow.

Number of hose lines needed for extinguishment. The number of hose lines needed for extinguishment is an extension of rate of flow. Once the rate of flow and the flow provided by department hose lines are known, the required number of hose lines needed to extinguish an advanced fire can be estimated.

Additional hose lines/master streams needed. Extinguishment is the primary purpose for using hand-held hose lines and master streams, but there is also a need to cover internal and external exposures, provide a back-up line, and protect critical egress routes.

Staffing needed for hose lines. In most cases, a full fire company should be assigned to each hose line except when the apparatus operator is needed at the pump panel (two members are assigned as the initial RIC) or when there are more than five people assigned to the apparatus. Departments should operate nozzles and hose lines under simulated fire conditions during training. This training should include advancing the hose up stairways, through doorways, and around obstacles to determine the actual staffing needs for each hose/nozzle combination. Keeping company members together also results in better accountability.

Water supply. Once a fire occurs, it is too late to evaluate the water supply capability. Some areas are fortunate to be protected by a high-flow water supply. Most areas have weaknesses in the water supply system, and others have no public water supply.

Growing communities often outstrip their water supply, resulting in reliability issues. These communities are likely to experience lower pressures and lower flows during business hours as demand for domestic water increases on a marginal water system. Even well-established, high-volume water supplies will contain problem areas if hydrants are on dead-end or small mains. If the system pressure is supplied by gravity, and the protected area is hilly, pressures and flows will be different at various locations depending on the difference in height between the hydrant and the water supply tank as well as the distance from the water supply (friction loss).

Some public and nearly all private water systems have a limited supply. These limitations must be understood when deploying large flow appliances. Private hydrants are likely to have a separate supply that may be connected to automatic fire suppression systems. Sometimes connecting to these hydrants will have a negative effect on the automatic fire suppression system.

Large municipal water systems are often supplied by multiple water systems. These multiple systems can usually be cross-tied to supply water from adjoining systems, thus increasing the total flow available at the scene of a major emergency. If the water system has this capability, members should know how to open valves (or how to get assistance in opening valves) connecting water systems.

Some communities or sections of communities do not have a public water system and rely on ponds, rivers, swimming pools, etc. as a water supply. As with the municipal water system, the reliability of the water source must be known in advance. Ask questions, including:
- Will the source freeze during cold months?
- Is there access for apparatus assigned to draft water?
- Are dry hydrants or other water supply equipment available?
- Are water sources dry for part of the year?

Even departments with a strong water supply should plan for situations where the water supply could be reduced or unavailable. After natural disasters such as Hurricane Katrina, water supplies were unavailable, requiring large municipal departments to use alternative water supplies.

Apparatus pump capacity. Apparatus specifications should consider available water supply and fire hazards when pumps are specified. Individual apparatus pump capacity is a known factor. However, the total flow needed for a given situation is incident dependent. Seldom is the apparatus pump capacity challenged during an offensive attack. During defensive operations, the need for water could be greater than the apparatus pump capacity. In many cases, the pump capacity exceeds the available water supply.

Manual fire suppression systems. In most cases, manual fire suppression system refers to a standpipe system. The IC must make a decision whether the system is to be used in a one-story, low-rise, or lower floor of a high-rise building. Usually, it is best to use the standpipe because its use reduces the work required to advance a hose line. However, some standpipe systems are supplied by the same water supply as the sprinkler system. When there is a chance that the sprinkler system will be ineffective because hose lines are using water needed to supply the automatic sprinkler system, it is best to avoid using the standpipe. The water pressure and flow are pre-established for standpipe systems, although pressure and flow can usually be increased by pumping into the fire department connection. If the standpipe has a severely limited flow or very low pressures, it may be best to use hose taken directly from the apparatus. The reliance on the standpipe is directly related to the height of the fire floor; the higher in the building the fire, the greater the reliance on the standpipe.

The location of control valves (main and sectional), pumps, fire department connections, and hose outlets should be shown on pre-incident plans. Some systems have pressure or flow-reducing valves. The pre-incident plan should explain the use of these valves, whether they can be adjusted, and what equipment is needed to adjust the pressure/volume. SOPs and pre-incident plans should specify hose, nozzles, and equipment needed during standpipe operations. Standpipe systems will be further discussed in Chapters 7 and 12.

Automatic Fire Suppression Equipment

There may be more than one sprinkler system in a large building. Systems may or may not be interconnected. Just as with the standpipe system, the location of control valves (main and sectional), pumps, fire department connections, and hose outlets should be shown on the pre-incident plan for sprinkler protected buildings. When a building is equipped with an automatic sprinkler system, the primary tactic involves letting the system do its job while fire fighters support the system and move in for final extinguishment. The sprinkler system becomes the first line of defense.

Other automatic systems may be found in buildings; they must be addressed in the pre-incident plan. See Chapter 7 for more information on automatic fire suppression systems and tactics to be used at protected properties.

Property Conservation

Property conservation is the third operational priority. Seldom will factors related to property conservation take on the urgency associated with life safety and extinguishment. Information about property value and location within the building can usually be known in advance; pre-planning property conservation measures can significantly reduce the overall loss.

Salvageable Property

Nearly every property has some form of salvageable property. During pre-incident planning, it is important to determine what there is to salvage and the value of the property. Remember that property might also have non-monetary value. Most fires occur in residential occupancies that contain personal property, such as heirlooms and photographs, that is considered invaluable by its owners.

Location of Salvageable Property

In the residential occupancy, salvageable property can be found throughout the building. In other occupancies, there may be concentrations of high-value property, such as computer rooms in a school or business occupancy. Pre-incident plans should identify these locations and possibly provide guidance in handling the salvage effort, e.g., do not de-energize computers.

Water Damage

The primary extinguishment tactic involves applying water to the burning material; thus, there will be some quantity of water in the building. Water will migrate through various openings and possibly damage property on floors where water is discharged and below.

Probability of water damage. Some buildings tend to hold water, while others have drains or pathways that allow water to harmlessly drain off. Property in water pathways below the fire will have a high probability of water damage.

Susceptibility of contents to water damage. Susceptibility of contents to water damage has to do with how easily the property is damaged. Some property is highly vulnerable to water damage, e.g., paper files.

Water pathways to salvageable property. Water will flow downward through paths of least resistance. If the floor contains openings or holes, water will flow through the openings or holes to the floor below. Otherwise, it will flow to stairways, elevators, drains, or other routes.

Water removal methods available. Preferred methods involve using built-in features that drain the water out of the building. Otherwise, water removal will involve pushing or directing water to channels leading out of the building, or possibly removing it with water vacuums.

Protective measures available. When property cannot be moved out of harm's way, the most common way to protect property from water damage is by placing covers over the exposed contents, starting with the property that is in the water pathway and most susceptible to water damage. At times it is possible to move contents above floor level, e.g., placing furniture on skids or placing more valuable property on top of less valuable property. Water removal, as mentioned above, is also a means of protecting property from flowing water.

Smoke Damage

Smoke can infiltrate the entire building, especially through ventilation systems. However, the most common pathway for smoke is upward; therefore, most smoke damage occurs on the fire floor and floors above.

Ventilation. The best way to reduce smoke damage is to ventilate the building. Some ventilation methods that are ill-advised prior to full fire containment are acceptable after extinguishment. As an example, as mentioned under life safety factors, venting the stairway could place fire fighters and occupants in danger. Later in the operation, it may be possible to control access to the stairway; once the fire threat no longer exists, this may be a good option for reducing smoke damage.

Property that is susceptible to smoke damage on the fire floor and above is most likely to be damaged. Materials that absorb smoke are more susceptible to damage, as is sensitive electronic equipment.

Forcible entry and ventilation are often necessary to achieve tactical objectives, but they can be overdone. If a building is closed and locked, forcible entry will nearly always result in some damage, as will ventilating. However, damage should be limited to what is needed to achieve these tactical objectives.

General Factors

Total Staffing Available Versus Staffing Needed

The implementation of an incident action plan will require resources, and until the actual plan is developed, the exact number of resources needed is unknown. However, approximations need to be made during the early size-up. Staffing is generally the most important and difficult resource to obtain when initiating an offensive attack.

It has been estimated that it takes a minimum of 11 to 13 fire fighters to safely and efficiently combat a working structure fire in a multilevel building with a life hazard. This estimate is for a single-family dwelling where one or two hose lines will be sufficient and the areas to be searched are limited. As the size and complexity of the property increase, the required staffing will also increase. If the calculated fire flow requires multiple 2½″ (64-mm) lines to attack the fire, more staffing will be needed to meet the rate of flow.

The *NFPA Fire Protection Handbook* lists some suggested guidelines for staffing depending on the type of occupancy:[10]

1. *High-hazard occupancies* (schools, hospitals, nursing homes, explosives plants, refineries, high-rise buildings, and other high-life-hazard or large-fire-potential occupancies): At least four pumpers, two ladder trucks (or combination apparatus with equivalent capabilities), two chief officers, and other specialized apparatus as may be needed to cope with the combustible involved; not fewer than 24 fire fighters and two chief officers.

2. *Medium-hazard occupancies* (apartments, offices, and mercantile and industrial occupancies not normally requiring extensive rescue or firefighting forces): At least three pumpers, one ladder truck (or combination apparatus with equivalent capabilities), one chief officer, and other specialized apparatus as may be needed or available; not fewer than 16 fire fighters and one chief officer.

3. *Low-hazard occupancies* (one-, two-, or three-family dwellings and scattered small businesses and industrial occupancies): At least two pumpers, one ladder truck (or combination apparatus with equivalent capabilities), one chief officer, and other specialized apparatus as may be needed or available; not fewer than 12 fire fighters and one chief officer.

4. *Rural operations* (scattered dwellings, small businesses, and farm buildings): At least one pumper with a large water tank (500 gal [1.9 m^3] or more), one mobile water supply apparatus (1000 gal [3.78 m^3] or larger), and such other specialized apparatus as may be necessary to perform effective initial firefighting operations; at least 12 fire fighters and one chief officer.

This is further reinforced in *NFPA 1710: Standard for the Organization and Deployment of Fire Suppression Operations, Emergency Medical Operations, and Special Operations to the Public by Career Fire Departments,* where a staffing requirement of at least 14 fire personnel are needed on the scene—15 if an aerial device is being used.[11] Staffing requirements enumerated in *NFPA 1710* will be further discussed in Chapter 5.

Searching large areas, physically removing victims, larger rate of flow requirements, or areas beyond a fixed water supply will require additional staffing.

Available water resources become a high-priority issue in areas without hydrants or when the required rate of flow is large. Fire departments working in rural areas can sometimes summon large numbers of fire fighters to the scene, but the immediately available water supply is usually limited.

Offensive operations are typically labor intensive. Conversely, defensive operations require fewer fire fighters operating apparatus and equipment that require large-volume water supplies. In any case the IC needs to estimate the staffing resources required.

Once an estimate of staffing requirements is made, the IC needs to consider how the responding resources match these requirements. Augmenting resources could take the form of requesting additional alarms or mutual aid. If resource capabilities cannot match the needs of an offensive attack, attempting an offensive operation will place fire fighters in extreme danger. If the needs of neither an offensive nor a defensive attack can be met, the IC's only safe option may be to assign available resources to exposure protection. Once the IC matches incident requirements with available resources, the offensive/defensive attack decision can be made.

Total Apparatus Available Versus Apparatus Needed

Offensive operations are staffing intensive; the initial response generally provides an adequate number of apparatus. Defensive operations are apparatus-intensive and require more apparatus for water supply and master streams. Each master stream will require a minimum of one apparatus. If the apparatus does not have pumps, such as an aerial truck without pumps, an additional apparatus will be needed to supply water.

Staging/Tactical Reserve

SOPs should address staging in a general way. Some properties, especially large complexes, may have specific staging opportunities that should be identified when pre-incident planning.

When all staffing or all apparatus are being used and the incident is not resolved, a tactical reserve is needed. For a small-scale incident, this is usually at least one engine company and one truck company. Larger and more complex incidents require a larger tactical reserve.

Utilities (Water, Gas, Electricity, Other)

Some departments routinely shut down the electrical supply in a residential property. This can provide a safer environment and reduce chances of re-ignition. However, this can also be a very dangerous operation, even in a residence. Fire fighters should not attempt to disrupt the power supply in larger properties with high-voltage service. If the department routinely interrupts the power supply at residential properties, SOPs and pre-incident plans should specifically state that fire fighters should not disconnect power supplies at properties with high energy service.

Residential gas or fuel supplies can be shut down if they are causing a problem or if they are leaking. Utility company personnel are better equipped to shut down electrical and piped gas supplies. Pre-incident plans should show the location of gas and electric shut-offs with particular attention to hazard areas.

Special Resource Needs

While not always considered a special resource, police and EMS are often needed at the scene. The IC should assess the need for these resources as part of the size-up. In many areas police and EMS routinely respond to a reported structure fire. Other resources could also be required, such as air trucks, hazardous materials teams, the utility company, Red Cross, and others.

Time

In examining the size-up/pre-incident plan factors above, much can be known in advance of the fire through pre-incident planning. Of course time will not be known until a fire occurs.

Time of Day. Time of day can be a factor in determining the likelihood of people being in the building and their degree of awareness of the fire. An apartment building fire at noontime generally presents a much different scenario than a fire at midnight. A church fire on Sunday morning could involve an extreme life hazard. A fire in the same church late on a Monday night might gain considerable headway before being discovered, but probably would not involve an extreme life hazard. Some properties are continuously occupied, e.g., nursing homes.

Day of Week. As with time of day, occupancies are generally fully occupied on different days of the week. A school would normally be occupied during the week during the school year and minimally occupied at other times, except when a special event is taking place, e.g., an evening basketball game.

Time of Year. The time of year should also be considered. In the northern United States, winter is a time when people tend to stay inside more and use heating systems, which can serve as additional ignition sources.

Special Times. Lastly, special times that should be considered would include holidays. An office building early on Christmas day would contain few occupants, whereas this same building could house thousands of people on a normal workday. Likewise, a Christian church would be expected to contain a maximum number of occupants early on Christmas day.

Weather

The IC must consider the effect of the present weather conditions on the operation. The variety of weather conditions is nearly endless, but the IC should concentrate on extremes. High heat and humidity take a toll on fire fighters. Extreme cold also affects fire fighters and their equipment, as well as victims who are being removed from a structure. High wind affects fire spread and ventilation, especially in high-rise structures or at large fuel fires, such as those at lumberyards.

The general kinds of weather that an area experiences can be known in advance, but the weather at the time of the incident is unknown until the alarm is transmitted. Fire fighters in Miami, Florida, expect to deal with high temperatures and humidity several months a year, just as fire fighters in Fairbanks, Alaska, expect to work in extremely cold temperatures. Of course, a fire could occur in Fairbanks during the summer, with heat being a potential problem for fire fighters that are unaccustomed to working in warmer temperatures.

Extremely cold or hot weather conditions may require occupant shelter and fire fighter rehabilitation (REHAB) stations. Very cold weather can also result in frozen fire hydrants, frozen pumps on apparatus, and damaged apparatus and equipment.

Humidity is particularly important during hot weather extremes, where it increases the fatigue factor. Humidity can also affect smoke movement. On a hot, humid day, smoke may stratify, creating heavy smoke conditions outside the building, requiring apparatus to be positioned farther away, or requiring apparatus operators to don SCBA. High humidity and heat conditions can also cause smoke stratification within high-rise buildings.

Precipitation in the form of rain is less problematic than frozen precipitation, which slows response and makes it more dangerous, as well as creating hazardous on-scene operating conditions.

High winds can push the fire toward exposures or affect venting. High winds are particularly problematic during high-rise fire operations.

A list of Size-Up/Pre-Incident Plan factors is included in Figure 2-10. An attempt has been made to include major factors affecting size-up, but invariably, one or more factors will be omitted from such a list. The three top priorities (life safety, extinguishment, and property conservation) are the basis of the Size-Up/Pre-Incident Plan checklist. All of these factors are interrelated at some level. As an example, weather conditions have an effect on fire fighter and occupant safety. Weather also affects ventilation and thus extinguishment. Structural stability affects the attack method and almost every other tactic at the incident scene.

Size-Up Chronology

Size-up follows a chronological sequence:

1. SOPs
2. Pre-incident plan
3. Shift/day/time
4. Alarm information
5. En route
6. Visual observations at the scene
7. Information gained during continuing operations
8. Overhaul

Standard Operating Procedures and the Pre-Incident Plan

These size-up activities are done well in advance of an alarm but play a significant role in developing an incident action plan. They give the IC a head start on the size-up process and actually become part of the incident size-up.

Shift/Day/Time

As the IC reports for a shift, unusual weather conditions, day of the week, and related factors are considered. On occasion, special measures are taken, such as adding an extra company to certain locations because of snow-covered streets. The time of the alarm may also be very important. The life hazard in a high-rise office building will be much greater at 2:00 PM on a workday than at 2:00 PM on a weekend.

Special information may change from day to day. The water system may be out of service, a parade may be scheduled, or the area might have been experiencing extremely hot, dry weather for a prolonged period. These are all important factors to consider.

Alarm Information

When the alarm is received, the time-of-day factor is immediately processed along with dispatch information. Dispatch information can include the following:

- Building location/address
- Fire location
- Fire intensity (numerous calls usually mean big fires)
- Occupant status (report of an occupant trapped on the second floor, for example)

Additional information may also be known about the area, or the specific building, or the dispatch file may contain a pre-incident plan notation.

While Responding

The dispatcher may be able to provide additional information to the responding units. Companies arriving on the scene should give status reports. There may, for example, be visual indications of a working fire, such as visible smoke or flames. During the response, all previous information is reconsidered, such as what effect the high winds will have on the operation. Initial information received from units on the scene is critical to formulate an incident action plan.

Visual Observations at the Scene

Many questions are answered as the IC gets the first look at the situation. When conditions permit, it is a good idea to get a view from several angles. By driving up to the building and going past the front to set up a command post on the far side of the building, the IC may be able to view three or even four sides of the building before establishing the command post. However, this must be done quickly so that an incident action plan can be developed and implemented.

Case Summary

In Pangnirtung, a small community in a remote area just below the Arctic Circle in the Northwest Territories of Canada, a fire in a school completely destroyed the building. The school was a one-story structure with a combination of wood frame construction and unprotected steel construction. A crawl space 34″ (860 mm) high was located under a major portion of the building. This crawl space contained utilities such as the domestic water tanks, fire protection water tanks, and sewage tanks as well as piping and wiring for various utilities.

The roof and exterior wall assemblies throughout the building were composed of 4′ × 8′ (1200 mm × 2400 mm) sheets of plywood ⅝″ (16 mm) thick, covered with two layers of 6″ (152-mm) thick solid polystyrene insulation, covered by another layer of ⅝″ (16-mm) thick plywood. Cedar siding was placed over the final plywood layer on the exterior of the walls, and a built-up roofing system was placed on the plywood on the roof.

Interior nonbearing walls were constructed of wood studs measuring 2″× 4″ nominal (58 mm × 89 mm), or 2″ × 6″ nominal (58 mm × 140 mm), located 16″ nominal (400 mm) on center. The interior walls were covered by gypsum wallboard ½″ (12.7-mm) thick or ⅝″ (16-mm) thick.

The school was being renovated at the time of the fire and was scheduled to be turned over to the client within two weeks. As part of the renovation, a sprinkler system was installed in the occupied space, in the combustible void spaces, and in the combustible crawl space located underneath the building. However, a combustible void space at the roof level near the gymnasium was not properly protected by an automatic fire sprinkler system.

Fire protection in the community is provided by a 25-person volunteer fire department with only one pumper truck. There is no municipal water supply, and all water (for both domestic use and firefighting) is provided by tankers.

During the winter months, Pangnirtung is accessible only by air. Therefore, the town is completely reliant on its own firefighting resources if a fire occurs.

The fire started at approximately 1:30 PM on Sunday, March 9, 1997. An electrician and a plumber were working in the crawl space under the industrial arts classroom. They heard the fire alarm system activate, and very shortly afterward, the lighting in the crawl space automatically shut off. (The electrical system was designed so that when the fire pump activated, nonessential circuits were shut down, including lighting in the crawl space.) The plumber used a lighter to provide illumination, and both men exited the space through an access hatch.

At the same time, an alarm was automatically transmitted to the Pangnirtung fire station which was located directly across the street from the school building.

Two fire fighters entered the building, advancing a hose line through the building's south door. They ran out of hose before reaching a point where they could attack the fire and began to exit the building through the south corridor. While exiting, they came across a fire fighter in the corridor who had collapsed for an unknown reason. They removed this fire fighter from the building. No further interior firefighting operations were attempted.

The firefighting personnel did not succeed in opening the roof in the vicinity of the fire, and a defensive firefighting operation was attempted. Since Pangnirtung has no municipal water supply, water had to be trucked in with domestic water tankers that were not designed to supply water for firefighting operations in the volumes required. In addition, the discharges on the front-mounted pump on the fire engine froze because of the extremely low temperatures (−40°F [−40°C]), further impeding the fire-ground operations. The fire eventually spread to the roof structure, resulting in destruction of the building and its contents.

The fire marshal for the Northwest Territories determined that a combustible void space near the gymnasium was the area of origin for the fire. Either this void space was not adequately protected by the sprinkler system, or the sprinkler system had failed to control the fire, due to failure of the nonmetallic sprinkler pipe. This failure may have resulted in a loss of water supply throughout the system. The exact cause of the fire is undetermined.

Source: Edward R. Comeau, *School Fire, Pangnirtung, Northwest Territories, Canada, 3/3/97*, NFPA Fire Investigation Report. Quincy, MA: NFPA.

The combination of visual information and reconnaissance from companies working on the interior or other unseen locations provides the basis for the initial incident action plan. This may be the first opportunity to evaluate structural conditions and the effectiveness of action being taken. The big question is whether the current incident action plan will accomplish the desired objectives.

The quality and quantity of information will greatly increase with time. The IC goes from a dearth of information to what may be information overload. At this point, it is important to prioritize information as primary and secondary and spend time reevaluating the incident action plan on the basis of the information received.

Information Gained During Continuing Operations

If the operation is successful, good news such as "the primary search is complete and the fire is extinguished" will reach the IC. If on-scene resources are unable to accomplish tactical objectives, either bad news or no news will be communicated to the IC. Bad news requires an adjustment in plans. On occasion the entire action plan is changed, such as shifting from an offensive attack to a defensive attack.

Overhaul

Overhaul should be planned and deliberate. The emergency phase is over, and extreme caution should be taken to avoid injuring fire fighters. At this point, there is no justification for rushing to make decisions that could put fire fighters at risk.

Summary

By establishing procedures in advance of an incident and by identifying target hazards, as well as developing pre-incident plans, the IC's job is simplified. By having this information available, coupled with a thorough size-up, a safe and effective incident action plan can be developed. Fire fighter safety can be addressed, as well as the three operational priorities—life safety, extinguishment, and property conservation—by applying the principles outlined in this chapter.

Wrap-Up

Key Terms

area of refuge A floor area with at least two rooms separated by smoke-resisting partitions in a building protected by a sprinkler system, or a space located in an egress path that is separated from other building spaces.

backdraft Occurs when oxygen (air) is introduced into a superheated, oxygen-deficient compartment charged with smoke and pyrolytic emissions, resulting in an explosive ignition.

balloon-frame construction An older type of wood frame construction in which the wall studs extend vertically from the basement of a structure to the roof.

complex pre-incident plan A plan of a property with more than three buildings or when it is necessary to show the layout of the premises and relationship between buildings on the site.

dead load The weight of a building; consists of the weight of all materials of construction incorporated into a building, including but not limited to walls, floors, roofs, ceilings, stairways, built-in partitions, finishes, cladding, and other similarly incorporated architectural and structural items, as well as fixed service equipment.

flashover An oxygen-sufficient condition where room temperatures reach the ignition temperature of the suspended pyrolytic emissions, causing all combustible contents to suddenly ignite.

formal pre-incident plan A plan for a property with a substantial risk to life and/or property; includes a drawing of the property, specific floor layouts, and a narrative describing important features.

fuel load Fuels provided by a building's contents and combustible building materials; also called fire load.

live load The weight of the building contents, people, or anything that is not part of or permanently attached to the structure.

notation A piece of information about the premises, such as the building has damage from a previous fire. This information may accompany a pre-incident plan or may be available when the building does not have a pre-incident plan.

platform-frame construction Construction technique using separate components to build the frame of a structure (one floor at a time). Each floor has a top and bottom plate that act as fire stops.

pre-incident plan A written document resulting from the gathering of general and detailed information to be used by public emergency response agencies and private industry for determining the response to reasonably anticipated emergency incidents at a specific facility.

primary factors The most important factors, assessed during size-up, which change from incident to incident and depend on specific incident conditions.

rapid intervention crew (RIC) A minimum of two fully equipped personnel on-site, in a ready state, for immediate rescue of injured or trapped fire fighters.

secondary factors Less important factors in an incident, which change from incident to incident and depend on specific incident conditions.

standard operating procedures (SOPs) Written rules, policies, regulations, and procedures intended to organize operations in a predictable manner.

Superfund Amendments and Reauthorization Acts (SARA) Enacted in 1986, Title III of this federal law is also known as the Emergency Planning and Community Right to Know Act. Title III requires businesses that handle or store hazardous chemicals in quantities above specific limits to report the location, quantity, and hazards of those chemicals to the State Emergency Response Commission (SERC), Local Emergency Planning Committee, and the local fire department. Some of this information has been classified since the attack on the World Trade Center on 9/11/2001.

Type I construction Construction in which the structural members, including walls, columns, beams, girders, trusses, arches, floors, and roofs, are of approved noncombustible or limited-combustible materials and which have the highest level of fire resistance ratings.

Type II construction Construction in which the structural members, including walls, columns, beams, girders, trusses, arches, floors, and roofs, are of approved noncombustible or limited-combustible materials but the fire resistance rating does not meet the requirements for Type I construction.

Type III construction Construction in which exterior walls and structural members that are portions of exterior walls are of approved noncombustible or limited-combustible materials, and interior structural members, including walls, columns, beams, girders, trusses, arches, floors, and roofs, are entirely or partially of wood of smaller dimensions than required for Type IV construction or of approved noncombustible, limited-combustible, or other approved combustible materials.

Type IV construction Construction in which exterior and interior walls and structural members that are portions of such walls are of approved noncombustible or limited-combustible materials. Other interior structural members, including columns, beams, girders, trusses, arches, floors, and roofs, are of solid or laminated wood without concealed spaces. Wood columns supporting floor loads are not less than 8″ (203 mm) in any dimension; wood columns supporting roof loads only are not less than 6″ (152 mm) in the smallest dimension and not less than 8″ (203 mm) in depth. Wood beams and girders supporting floor loads are not less than 6″ (152 mm) in width and not less than 10″ (254 mm) in depth; wood beams and girders and other roof framing supporting roof loads only are not less than 4″ (102 mm) in width and not less than 6″ (152 mm) in depth. Specifics for other structural members are required to be large dimension lumber as well.

Type V construction Construction in which exterior walls, bearing walls, columns, beams, girders, trusses, arches, floors, and roofs are entirely or partially of wood or other approved combustible material smaller than material required for Type IV construction.

wood truss An assembly made up of small dimension lumber joined in a triangular configuration that can be used to support either roofs or floors.

Suggested Activities

1. Review your department's SOPs.
   A. Are they up to date? When was the last revision?
   B. Does the department have SOPs addressing:
      i. Use of NIMS including command transfer, establishing a command post, and specifying the IC?
      ii. Apparatus management—staging, positioning, collapse zones?
      iii. Structural firefighting, including water supply, initial attack, and company duties?
      iv. Operating in buildings with built-in fire equipment?
      v. Special hazard operations, such as hazardous materials?
      vi. High-rise, large-area buildings and other special occupancies?
2. Develop a list and prioritize occupancies that should be pre-incident planned in your jurisdiction. Explain the prioritization in terms of life safety, extinguishment, property conservation and other factors, using the sample list in Table 2-2 as a guide.
3. Develop a pre-incident plan (narrative and drawing) for one of the following occupancies in your community using the format shown in Figures 2-7 and 2-8 or a local pre-incident plan form:
   - Public assembly
   - Educational

TABLE 2-2 Prioritization Factors

Occupancy Type	Priority	Rationale
Educational Primary School	3	Large number of children during school hours, expect orderly evacuation, ground level exits from each room
Assembly Night Club	2	Large number of guests during peak hours of operation, intoxicated occupants possible
Assembly Church	4	Large number of people with a high level of awareness, but occupied a very limited number of hours
Nursing Home No Sprinkler System	1	Large number of people 24/7, many immobile and incapacitated occupants
Mercantile Big Box	5	Large number of people, large undivided area, truss construction, automatic sprinkler protection
High-Rise Office	6	Large number of people, positioning problems for aerial apparatus, automatic sprinkler protection

The above sample list would be incomplete for most communities. The priorities could be different depending on many factors.

Wrap-Up, continued

- Institutional
- Residential (should be a complex or large building)
- Business
- Industrial
- Manufacturing
- Storage
- Mercantile

Follow the process shown in Figure 2-3 when developing and evaluating the pre-incident plan.

4. Assume the role of the first-arriving engine company officer responding to the building pictured in **Figure 2-17**. It is Friday, June 29, 1039 hrs., 90°F with 90% humidity. There is a report of an automatic alarm from the 7th floor of the South Tower in the office building at 25th and Peachtree.

   You do not have a building pre-plan available.

   A. Develop an initial report!

   B. What is your command mode (fast attack, investigative, or command)?

   Assume you decided on a fast attack mode. Further, assume this building is not sprinkler protected, but does have a standpipe system.

   C. How would you access the fire floor (elevator or stairs)?

   D. Without a pre-incident plan you do not know the exact location of the elevator or stairs. Where would you enter the building to find the elevator and/or stairway?

   E. Where do you think the standpipes are located?

   F. Where is the fire department connection for the standpipe?

   G. Which side is the South Tower?

   H. Assume that you decided to begin operations by checking the alarm panel. Where do you think the alarm panel is located?

   These questions should reinforce the need to pre-incident plan large buildings.

Figure 2-17   25th and Peachtree building.

This scenario is based on the USFA Technical Report 033.[12]

Below is a list of critical information that could be known had this building been pre-incident planned:

- This structure is actually two buildings built at two different times (North Tower 1962; South Tower 1969) with a black glass covering to make it appear to be one building.
- The two towers are connected on some levels, but a solid fire wall separates the towers on other levels.
- Elevators are not equipped with fire fighter service or an automatic grounding feature.
- There is a common plenum in the space above the suspended ceiling with walls terminating at the suspended ceiling level.
- Each tower has an area of 19,000 ft² (1765 m²) per floor.

5. Using the 25th and Peachtree building in Question #4, take the role of a chief officer assuming command from the first-arriving company officer. You now have a pre-incident plan available and discover the fire is on the 6th floor, but smoke has extended to the 7th floor via a closet containing an electrical bus duct. This bus duct is routed through closets running the entire height of this 10-story building. The electrical closet is in the common hallway of the South Tower as shown in **Figure 2-18**.

Upon arrival you observe people at the window on the Peachtree side of the West Tower as shown in

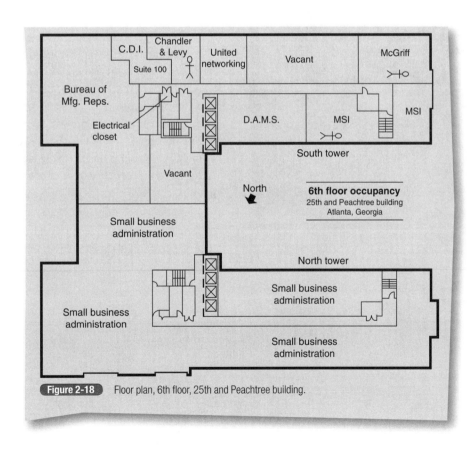

Figure 2-18 Floor plan, 6th floor, 25th and Peachtree building.

Wrap-Up, continued

Figure 2-19 Visible occupants in the 25th and Peachtree building.

Figure 2-19. The first alarm response is three engine companies, two truck companies, an EMS unit, and a battalion chief. Because of a large number of calls, a rescue unit and manpower squad were added to the initial alarm.

The first-arriving company took the stairway to the 5th floor, where they connected to the standpipe, and they are advancing a hose line up the stairway to the 6th floor. The two truck companies are involved in aerial rescues. Truck company members also notice two victims at windows to the rear of the building, who cannot be reached by an aerial ladder. Truck company members advance to the 7th floor to rescue these victims. A victim (jumper) is also found in the alley on the south side of the building. Some occupants are trying to escape via the stairways.

The first-arriving engine has controlled the main body of fire, but is now forced to retreat due to low air supply.

This fire occurred in 1989 and there is no mention of accountability or rapid intervention.

Note: Company operations are not fully explained in the report; therefore some details are added to allow a complete size-up.

Use the size-up/pre-incident plan checklist to size-up this structure fire. Analyze each size-up factor including the importance of the factor. Also categorize the factors as primary or secondary.

6. Obtain a copy of the report USFA Technical Report 033, High-Rise Office Building, Atlanta, Georgia. Compare your size-up with the actual situation. This report is available through the U.S. Fire Administration.[12]

Chapter Highlights

- SOPs, pre-incident plans, and incident-specific information are interrelated and are important components of the size-up.
- When pre-incident plans are available for a property their use in evaluating the situation is crucial.
- A proper size-up is crucial to developing effective firefighting strategies and the incident action plan.
- Due to time constraints the IC must focus on primary size-up factors when developing an initial incident action plan.
- The incident action plan should adapt to changes in circumstances or when additional critical information indicates a need to change tactics.
- Structural firefighting SOPs are guidelines to be used at all structure fires except when circumstances indicate the need for a different approach.
- SOPs establish command structure and address staging, water supply, operations in special occupancies, company operations, special operations, and using private fire protection.
- SOPs must address department-specific training and equipment, but should also reflect regional procedures to facilitate operations with other jurisdictions and agencies.
- SOPs allow first-arriving units to take immediate action or stage while awaiting orders from command.
- SOPs help prevent freelancing while the IC is gathering information and developing an incident action plan.
- SOPs specify action to be taken, but should be flexible enough to allow fire fighters to follow a reasonable course of action when confronted with a situation where modification of the procedure is appropriate.
- Routine evaluation of properties enables fire departments to evaluate specific hazards and characteristics that might affect fire-ground operations.
- For some buildings, hazardous materials planning is mandated by Title III of the Superfund Amendments and Reauthorization Act (SARA).
- Occupancies with a potential for a large loss of life or property (e.g., public buildings, warehouses, healthcare facilities) should be addressed in SOPs, pre-incident planning, and training.
- Three types of pre-incident plans:
  - *Complex*: used for a property with three or more buildings, including layout of buildings and fire protection features or hazards.
  - *Formal*: used for a property with substantial life or property risk, including floor plans and narrative description of features and hazards.
  - *Notation*: systems used to note additional information about a property's special risks to fire fighters (e.g., note at the communications center indicating holes in the floor).
- A building's construction type, age, and current usage are important considerations in the pre-incident plan.
- Pre-incident plan drawings should include information about entrances, security features, floor layout, stairways and other methods of egress, fire protection features, and areas of safe refuge.
- Pre-incident plans must be kept up to date.
- If possible, pre-incident plans should be computerized to allow easier access to pre-plan files.
- Pre-incident plans are needed when there is a high life hazard, a difficult extinguishment problem, or high-value property.
- Pre-incident planning commonly includes buildings protected by fire protection systems, high-rise buildings, and industrial complexes.
- Long-span roofs, heavy roof loading, or unusual building features that could lead to collapse should be noted as potential life hazards on pre-incident plans.
- Occupancies intended for large numbers of people or disabled people should be pre-planned.
- Pre-incident plans should estimate the required rate of flow for large undivided areas, note any hazardous materials, and evaluate fuel loads.
- If high-value contents are present, pre-incident planning can assist the IC in developing strategies to reduce smoke and water damage.
- Size-up builds upon the pre-planning information to develop strategy and tactics appropriate under specific circumstances.
- Incident conditions will determine which size-up factors are more important (primary) or less important (secondary).

Wrap-Up, continued

- Size-up begins with the development of SOPs and pre-incident planning, and is further developed based on the specific conditions of the incident when it occurs.
- SOPs and pre-incident plans identify many size-up factors in advance of a fire.
- Size-up continues throughout the incident and into the overhaul phase.
- Smoke and fire conditions are primary factors at a structure fire because of their effect on life safety; these conditions can be evaluated through reading smoke characteristics, interior reconnaissance, and knowledge of building construction and contents.
- Direction of fire travel, ventilation status, occupancy type, the number and mobility of occupants, and the operational status are important life safety factors.
- Different types of rescue methods (e.g., stairs, ladders, etc.) have different staffing needs that must be considered.
- Building materials, roof construction, age and overall condition, fuel load, live loads, dead loads, and layout are key factors in determining likelihood of fire extension and building collapse.
- Knowing flow requirements and the extent of fire are key factors in determining whether a rapid, life-saving extinguishment is possible.
- Overall attack strategy may be offensive (interior operation) or defensive (exterior operation), depending on the risk-versus-benefit analysis.
- Factors related to property conservation are rarely as urgent as life safety and extinguishment factors.
- The goal of property conservation measures is to mitigate smoke and water damage.
- Staffing and equipment requirements vary depending on the type of occupancy and the nature of the operation.
- It is important to maintain a tactical reserve of apparatus and staff unless the incident is immediately contained.
- Outside resources, such as utility personnel, the Red Cross, police, and EMS, should be called upon as needed to handle special tasks (utility shut-off, medical, sheltering).
- Factors such as weather conditions, time of day, day of the week, season, and type of occupancy can affect the overall action plan.
- Information is gained in advance of a fire through SOPs and pre-incident plans, then the IC considers the specific size-up conditions in chronological sequence as an alarm is received, during response, and upon arrival at the scene.
- Alarm information usually includes the building address and sometimes other information, such as the location and intensity of the fire and the status of occupants.
- Initial information is updated during response by additional messages from dispatch and upon arrival through visual observations and status reports.

References

1. National Fire Protection Association, *NFPA 1500: Standard on Fire Department Occupational Safety and Health Programs*. Quincy, MA: NFPA, 2007.
2. National Fire Protection Association, *NFPA 1620: Recommended Practice for Pre-Incident Planning*. Quincy, MA: NFPA, 2003.
3. Department of Health and Human Services/Centers for Disease Control and Prevention, *NIOSH Alert: Preventing Injuries and Deaths of Fire Fighters due to Truss System Failures*. DHHS (NIOSH) Publication No. 2005-132. Cincinnati, OH: NIOSH-Publication Dissemination, 2005. Accessed July 5, 2007 at www.cdc.gov/niosh/docs/2005-132/#sum.
4. National Safety Council, *CAMEO for Windows*. Chicago, IL: National Safety Council, 1996.
5. Donald W. Dodson, *The Art of Reading Smoke*. Saddlebrook, NJ: Fire Engineering, September 2005.
6. National Fire Protection Association, *NFPA 101: Life Safety Code*. Quincy, MA: NFPA, 2006.
7. Francis Brannigan, National Fire Protection Association. *Brannigan's Building Construction for the Fire Service*, 4th edition. Sudbury, MA: Jones and Bartlett Publishers, 2007.
8. National Fire Protection Association, *NFPA 220: Standard on Types of Building Construction*. Quincy, MA: NFPA, 2006.
9. National Fire Protection Association, *NFPA 5000: Building Construction and Safety Code*. Quincy, MA: NFPA, 2006.
10. John Granito, *NFPA Fire Protection Handbook*, 19th edition. Quincy, MA: NFPA, 2003.
11. National Fire Protection Association, *NFPA 1710: Standard for the Organization and Deployment of Fire Suppression Operations, Emergency Medical Operations, and Special Operations to the Public by Career Fire Departments*. Quincy, MA: NFPA, 2004.
12. Charles Jennings, *USFA Technical Report 033: Five-Fatality High-Rise Office Building Fire, Atlanta, Georgia (June 30, 1989)*. United States Fire Administration National Fire Data Center. Accessed July 5, 2007, at www.interfire.com/res_file/pdf/Tr-033.pdf.

Developing an Incident Action Plan

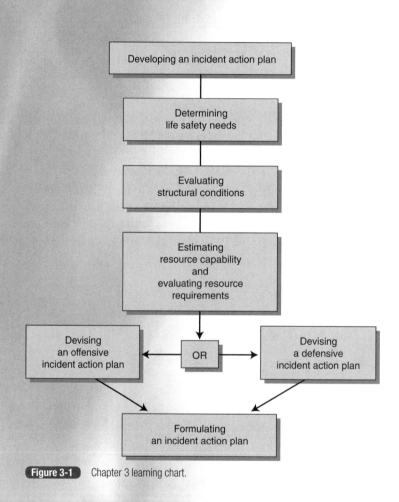

Figure 3-1 Chapter 3 learning chart.

Chapter 3

Learning Objectives

- Describe how extinguishment is both an operational priority and tactical objective with an emphasis on the relationship between life safety and extinguishment.
- Evaluate conditions leading to an offensive or defensive operation.
- Compare probability of occupant survival to fire and building conditions.
- List situations when a written incident action plan is needed.
- Use a case study or actual fire to develop an incident action plan based on a risk-versus-benefit analysis.

Introduction

An <u>incident action plan</u> is derived from an analytical approach to information gained through size-up. The result should be a straightforward, easy-to-understand plan outlining major tactical objectives. Size-up is a continuous process, as conditions constantly change; therefore, the incident action plan must remain flexible. Often, the tactics necessary to accomplish a strategy are modified as conditions change. The incident action plan provides the central focus for operations. All activities should lead to completion of major objectives identified in the incident action plan.

The primary strategic considerations are always life safety, extinguishment, and property conservation. However, the three priorities are not mutually exclusive. When resources are limited—and they usually are in the beginning stages of an operation—the incident commander must take action related to life safety first, extinguishment second, and property conservation third. In most cases, however, extinguishment and life safety are closely related. If the fire is extinguished, rescue often takes care of itself, and the overall operation is much safer.

Extinguishment is normally, but not always, the most important life safety tactic. Extinguishment is both an operational priority and a tactical objective. There are times when there are only a few, easily rescued victims. In these instances, it may be best to rescue the victims first. Even in these situations, extinguishing the fire and venting the smoke could alleviate the problem.

When developing an incident action plan, it is best to categorize tactical objectives in terms of their relative importance as primary and secondary (and possibly tertiary) objectives. As an example, extinguishing the fire would normally be a primary objective, as would the primary search and rescue operation for occupants in dangerous positions. Most property conservation measures would fall under secondary objectives. The idea is to provide an effective plan for deployment, with primary objectives being assigned prior to secondary objectives.

Once there are available resources on the scene, all priorities can be handled simultaneously. If the fire operation is going to continue for more than a few minutes, the incident commander (IC) should request enough staffing and equipment to handle the immediate life safety, extinguishment, and property conservation activities plus an allowance for a tactical reserve.

Remember, the key to a successful fire-ground operation is keeping it simple. Incident action plans should be simply stated and concise.

The learning chart in **Figure 3-1** shows the development and implementation of an incident action plan. This chapter emphasizes the development of an incident action plan leading to offensive or defensive tactics. Three scenarios (a dwelling, a high-rise residence, and a church) are used to provide examples of incident action plans.

Determining Life Safety Needs

Chapter 2 discussed the importance of life safety during pre-incident planning and size-up. The primary objective of structural firefighting is saving lives; therefore, life safety is the first consideration in developing an incident action plan. For this reason, the IC is willing to expose fire fighters to greater risk when lives are threatened.

Evaluating Structural Conditions

Structural conditions bear heavily on the offensive/defensive decision. Even with sufficient resources, an interior attack should not be conducted in an unsafe building. A building that is in danger of imminent collapse should not be entered for any reason. However, rescue efforts may be justified in a building that is currently structurally sound but may ultimately fail as continuing damage from fire undermines the structure. An offensive attack is conducted to assist occupants from the building with a reevaluation of the structure and fire conditions once the building is evacuated. If extinguishment is being accomplished during the initial offensive attack, offensive tactics will normally be continued. However, if the fire has not been contained or extinguished during the initial offensive attack, but the rescue effort has been accomplished, the expected benefit

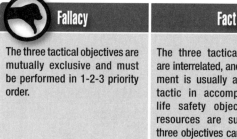

Fallacy
The three tactical objectives are mutually exclusive and must be performed in 1-2-3 priority order.

Fact
The three tactical objectives are interrelated, and extinguishment is usually an important tactic in accomplishing the life safety objective. When resources are sufficient, all three objectives can be accomplished simultaneously.

has changed from life safety to saving property. Therefore, there is less reason to stay in a building of questionable stability, as the **risk-versus-benefit analysis** has changed. Proficient ICs are always weighing risk against expected benefits.

Estimating Resource Capability and Evaluating Resource Requirements

Comparing resource capability to incident requirements is part of the size-up process. The resources that are required to fight a fire offensively are different from those needed for a defensive battle. Generally speaking, it takes more staffing to conduct an offensive operation. Lives and property are best saved by conducting an offensive attack. However, a lack of necessary resources could lead to a defensive decision, even when an offensive attack is clearly the better approach.

The IC must consider what will be needed to conduct an offensive attack early in the plan development process. In making the offensive-versus-defensive decision, the IC must apply sound risk management principles to ensure fire fighter safety. **Figure 3-2** charts time/resources versus occupant probability of survival in terms of structural stability, temperature, and smoke layer. At the beginning of the incident, there is a very high probability of saving occupant lives and extinguishing the fire. As the temperature increases in an unvented structure, the smoke layer begins to fill the enclosure while the structure weakens; the probability of saving lives decreases, and extinguishment is more difficult. In other words, the probable benefit is lower and the risk is higher. In the early stages of the fire, no fire department personnel would be on the scene. As time passes, fire department resources become available, but the probable benefits of life safety and extinguishment decrease.

Each fire will be different; thus, Figure 3-2 is a general depiction rather than an actual fire model.

Devising an Offensive or a Defensive Incident Action Plan

The entire operation is governed by the offensive/defensive decision. A casual observer ought to be able to determine whether the tactics being applied are interior (offensive) or exterior (defensive). However, it may not always be this clear. Large master stream appliances can be used to support rescue efforts in large, complex structures with fire separations. Master streams can also be used to push fire away from critical evacuation routes, such as a fire escape, or to cover exposures. When master streams are being used during an offensive operation for any reason, coordination through command is critical. An operation that begins as an offensive attack is sometimes changed to defensive, but the actions taken during either attack must be coordinated as offensive or defensive; they must never be both.

Whenever it is safe to do so, an offensive attack should be initiated.

Formulating an Incident Action Plan

The IC sets the objectives, decides on the tactics necessary to achieve those objectives, and finally assigns units to complete the tasks associated with each objective and tactic. The incident action plan is the focus of the entire operation. All tactics are directed toward completing the objectives, and each objective is directed toward accomplishing the overall incident action plan. The incident action plan should be simple and understandable.

Every incident needs some form of incident action plan. For small incidents of short duration, the plan need not be written. However, in some situations a written plan will provide a central focus, eliminate confusion, and reduce disputes. The following are examples of situations when a written action plan should be used:

- Resources from multiple agencies are being used.
- Several jurisdictions are involved.
- The incident will require a transfer of command beyond the initial changes from company officer to higher ranking chief officers.

The IC will establish objectives, making strategic determinations for the incident based on the requirements of the

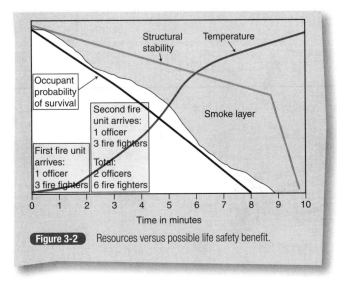

Figure 3-2 Resources versus possible life safety benefit.

jurisdiction. In the case of a unified command, the incident objectives must adequately reflect the policies and needs of all the jurisdictional agencies Figure 3-3 . The incident action plan should cover all tactical and support activities required during the operational period.

An incident action plan becomes more important as the incident grows in size. However, there should be a plan for every structure fire, and units operating on the scene should be coordinated using a common plan.

Deployment

Writing a list of tactical objectives is useless unless you assign sufficient resources to accomplish the objectives. Experienced ICs know that, after making assignments, they must follow up and request status reports. Sometimes an assignment is made but not achieved. Ordering it done and getting it done are not always the same. Experienced company-level officers provide status reports, as appropriate, without being prompted. However, when reports are not received in a timely manner and the assigned company has had ample time to complete the assignment, it is incumbent upon the IC to request a status report. If status reports are requested too frequently, the status report communications will interfere with efficient operations.

Scenario 1: Single-Family Detached Dwelling

Fires in the home make up the vast majority of fires in the United States. Sometimes department SOPs spell out company duties for the entire first alarm assignment. When this is the case, the IC may need to modify operations only when the SOPs are not being followed or when the procedures do not fit the circumstances. It could be said that the IC is on autopilot when SOPs are followed, but going on autopilot can be dangerous. No fire should ever be considered routine. On the other hand, there is no need for unnecessary interference in an operation that is going well. Most ICs feel comfortable handling a small fire, yet the process that is used to develop an incident action plan is the same as that for a very large and complex fire. Even when the fire is going well, the IC should focus on completing the tactics necessary to meet strategic objectives while simultaneously developing Plan B in case the situation changes quickly.

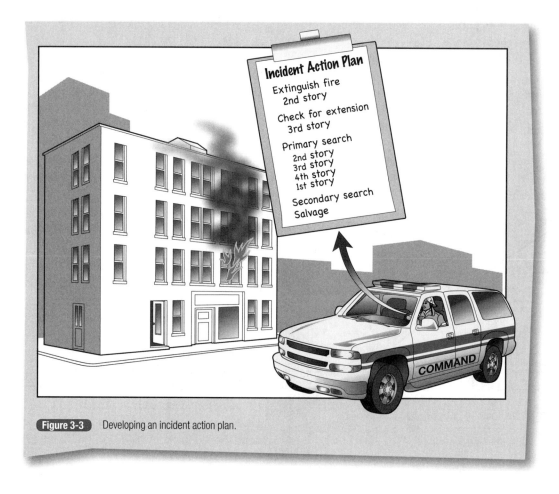

Figure 3-3   Developing an incident action plan.

CHAPTER 3 Developing an Incident Action Plan

Case Summary

At the Oklahoma City bombing site, the incident management team met twice daily to review the progress that had been made in the previous 12 hours and to establish plans for the next 12 hours. These plans were written, copied, and distributed immediately to everyone involved in the operation. By clearly stating the operational period's objectives, the plans let everyone know what the focus of their tasks were in relation to the overall objective.

Source: Edward R. Comeau and Stephen Foley, "Oklahoma City, April 19, 1995," *NFPA Journal*, July/August 1995.

For this scenario, assume a point of origin in a first floor living room of a two-story, balloon-frame constructed home. There are no external exposure problems. It is 6:30 AM on Saturday, and the first-arriving engine company reports a working fire on the first floor **Figure 3-4**.

Risk-Versus-Benefit Analysis

Expected Benefits
Saving the lives of the occupants and saving property. There is a good chance that someone will be home and asleep, possibly in second floor bedrooms.

Expected Risk
Moderate. A life safety benefit is the primary reason for conducting an offensive operation. Risk is associated with any offensive operation, but the risk to fire fighters is usually less in a single-family dwelling than it would be in a larger, more complex occupancy.

Structural Stability
Primary failure not expected until near full involvement. A frame building is a fairly unstable structure and contributes significant fuel to the fire. However, the combustible characteristics of the

Figure 3-4 Scenario 1: Fire in a single-family detached dwelling.

building provide stability clues. Frame structures will have considerable fire involvement before collapse, with the exception of truss roof and floor construction. For this reason, if the fire is in a modern residential building, extreme caution should be used if the IC decides that it is necessary to vent the roof. When possible, it is a good idea to provide a stable platform using an aerial, elevated platform, or a roof ladder **Figure 3-5**.

> Benefit as used here is the expected or potential benefit to occupants or owners. Rescuing occupants would be a life safety benefit.

Offensive/Defensive Decision

Offensive. Given the probable life safety benefit and moderate risk to fire fighters, a decision is made to initiate an offensive attack.

Resource Needs

Initial needs are forcible entry, advancement of an attack line, and simultaneous, coordinated ventilation. In most operations, two fire fighters initially accomplish each of these tasks by laying a preconnected 1¾" (45-mm) hose line, forcing the door, and opening windows. As the operations progress, anticipate the need for back-up hose lines, a hose line on the second floor, search-and-rescue crews, and more complete ventilation.

> Risk as used in the risk-versus-benefit analysis refers to the risk to fire fighters, not the risk to occupants.

Incident Action Plan

- Force entry and extinguish the fire on the first floor.
- Vent heat, smoke, and gases.
- Search and rescue on first and second floors.
- Advance a hose line to the second floor and check for extension on second floor and attic.
- Establish a rapid intervention crew (RIC).

Deployment

- *First engine:* Establish a water supply, force entry, and extinguish fire on first floor; conduct initial horizontal ventilation. Apparatus operator and hydrant fire fighter are the initial RIC. The establishment of a RIC before the first-arriving team makes entry is not mandated here due to the high probability of an imminent life hazard. However, it is good practice to establish a RIC whenever staffing permits.
- *First truck:* Search and rescue and conduct horizontal ventilation.
- *Second engine:* Advance a hose line to second floor; check for extension on second floor. The truck company should assist in checking for extension and complete ventilation. The second engine could be assigned as the RIC.

As you can see, it would be difficult to accomplish all of these tasks with only 11 people (including the IC). Also, owing to a shortage of resources, the IC would be unable to cover other critical assignments, such as assigning a team to a back-up line. If the first-arriving engine company is successful in extinguishing the fire, the entire operation is favorably changed, and the 11-person response will be adequate. This is the reason for placing "extinguish fire on the first floor" as the highest strategic priority, thus accomplishing the life safety objective by extinguishing the fire and ventilating the smoke, heat, and gases.

> Roof ventilation would normally not be required if the fire in scenario #1 were confined to the first floor.

Figure 3-5 Venting a roof from a roof ladder.

Scenario 2: High-Rise Apartment Building

High-rise buildings pose extreme life hazards. Office buildings may contain thousands of people, and large residential buildings can house hundreds of families. The mere fact that the fire could be located several floors above grade level adds to the complexity and difficulty of extinguishment. In addition, many floors above the fire could be occupied, making it necessary to conduct search-and-rescue operations on many different levels. The actual fire conditions could be similar to those in the single-family detached dwelling. However, the challenges that are presented will be much more complex. Fires in residential high-rise buildings will typically be limited to a small area in the building unless the fire occurs in a common area. The smoke, however, can spread far from the fire, endangering a large number of residents. For this scenario, the fire occurs in the living room area, just as in the previous example. Most apartments in residential high-rise buildings have one-floor layouts. Instead of having sleeping occupants above the fire, it is assumed that the bedrooms are on the same floor level as the living room, but with other apartments above.

For this scenario the fire is on the sixth floor in apartment 608 (X8), as illustrated in the floor plan in **Figure 3-6**. The structure is a nine-story, high-rise apartment building of fire-resistive construction. The building, shown in **Figure 3-7**, has ten apartments per floor and primarily houses elderly occupants. The fire originates in the living room, and it is 6:00 AM Saturday morning. A standpipe is in the center hallway leading to the center stairs.

Risk-Versus-Benefit Analysis
Expected Benefits
Saving the lives of the occupants and saving property. There is a good chance that most residents will be home and asleep.

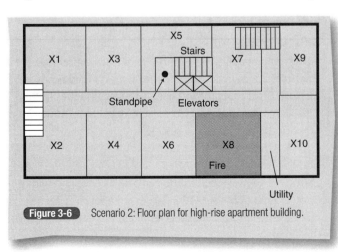

Figure 3-6 Scenario 2: Floor plan for high-rise apartment building.

Figure 3-7 Scenario 2: Fire in a high-rise apartment building.

Occupants of the three floors above the fire and those on the fire floor are in imminent danger.

Expected Risk
Moderate to high. The risk in this building is somewhat greater than that in the single-family detached dwelling, due to the number and location of occupants who are beyond the reach of aerial ladders and towers and the difficulty in ventilating the building. The layout of this building is fairly straightforward in comparison to some high-rise buildings. The fire-resistive construction and multiple stairs reduce the risk to some extent.

Structural Stability
Failure not expected. The structure is fire-resistive construction; therefore, it is unlikely that a moderate fire would seriously threaten the stability of this building.

Offensive/Defensive Decision
Offensive. Given the high likelihood of many occupants being endangered on four floor levels, a decision is made to initiate an offensive attack.

Resource Needs
Resource needs are much greater than those in the single-family detached dwelling in Scenario 1, as it will take longer to reach the fire floor, the floor layout is much larger, and there are three occupied floors above the fire instead of one. The back-up hose line is also more critical in this situation. The early estimation is for three companies to be assigned to the fire floor, two to the floor above, and one each on the eighth and ninth floors. Property conservation, especially

below the fire, is also important, but it is a secondary objective.

Incident Action Plan
- Force entry (if necessary) and extinguish the fire on the sixth floor.
- Search and rescue and ventilate the sixth to ninth floors.
- Advance a hose line to the seventh floor and check for extension and smoke conditions.
- Evacuate occupants and arrange for shelter (as needed).
- Implement property conservation on the fifth floor.

Deployment
- *First engine*: Establish a water supply, supply standpipe, force entry, and extinguish fire on sixth floor; begin venting.
- *Second engine*: RIC.
- *First truck*: Assist with forcible entry, search and rescue, and ventilation on the sixth floor.
- *Third engine*: Advance a hose line to seventh floor and ventilate (if possible).
- *Fourth engine*: Advance a back-up hose line to the sixth floor.
- *Rescue group*: Perform search-and-rescue operations on the seventh to ninth floors and ventilate if possible.
- *Additional fire companies*: Initiate an evacuation and complete ventilation, forcible entry, and property conservation.

If the first-arriving engine company is successful in extinguishing the fire, the entire operation is favorably changed, and the typical five- or six-company response (20 to 24 fire fighters) is adequate. This accomplishes the life safety objective by extinguishing the fire and ventilating the heat, smoke, and gases. If the fire is brought under control in a short period of time, the additional units listed under the rescue group and additional fire companies may not be needed.

High-rise residences come in all sizes, including buildings that are much larger than the one in this example. As the size and complexity of the building increase, the danger to both fire fighters and occupants also increases. Numerous other factors, such as extreme temperatures, strong winds, blocked stairways, and locked passageways will further complicate operations. The danger is substantially decreased if the building has a properly installed, working sprinkler system.

Many high-rise residential buildings house elderly residents. This special population can be partially debilitated, making evacuation more labor intensive. Further, these elderly occupants are much more prone to injury and death from the products of combustion.

Scenario 3: Church Fire

Older, gothic-style churches are characterized by very high ceilings, with huge concealed spaces above the main body of the church. Newer churches come in a variety of styles, but both old and new churches have a common characteristic: a large, open area to accommodate the congregation and altar. This one factor means that structural collapse is probable once a fire reaches the structure supporting the roof. The open area requires an unusually high rate of flow because of the high ceiling. (Fire flow calculations will be explained in Chapter 8.) It will be necessary to use 2½″ (64-mm) or 3″ (76-mm) hand lines with solid streams to obtain the necessary reach.

Many people congregate during services, especially on religious holidays. During these times, life safety is a key tactical consideration. There are subsequent long periods of time when these buildings are unoccupied, allowing a fire to gain considerable headway before it is discovered. Many churches are not protected by fire suppression systems or automatic alarms.

In this scenario, a gothic-style church is involved in fire on a Monday afternoon **Figure 3-8**. The first report to the fire department is for a fire in the bell tower. The dispatcher relays a message to responding units from a pre-incident plan notation that there is a very large bell in the bell tower.

The first-arriving engine company confirms a large volume of fire and smoke from the bell tower. The fire appears to be confined to the bell tower at this time, and the doors to the church are locked.

Risk-Versus-Benefit Analysis
Expected Benefits
Saving property only. The church is not having services at this time, doors are locked and the fire has gained headway. There is a very low probability that anyone is in the church. This old gothic church has many stained glass windows and artwork that is considered priceless.

Expected Risk
Extreme. The heavy bell in the bell tower combined with a large volume of fire makes collapse of the tower inevitable. Fire extension to the interior of the church is unknown at this time.

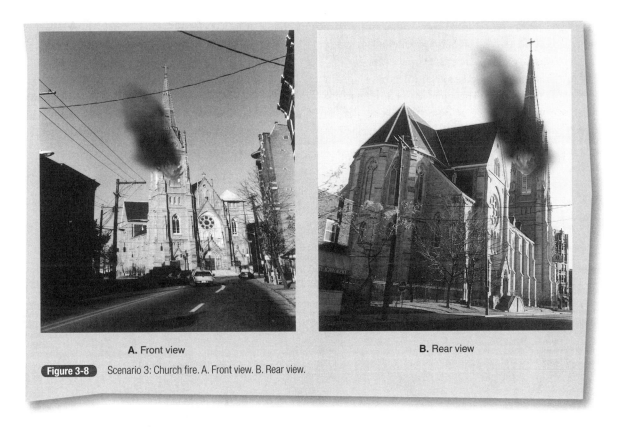

Figure 3-8 Scenario 3: Church fire. A. Front view. B. Rear view.

Structural Stability

Partial failure inevitable. As was stated above, the bell tower with bell should be expected to fail and collapse. The height of the bell tower would indicate a fairly large collapse zone that may be beyond the effective reach of exterior fire streams. If the fire is in the concealed space above the church interior, the church roof should be expected to collapse.

Offensive/Defensive Decision

Defensive. Structural collapse potential is great; no one should enter the bell tower area. It may be possible to force entry to the rear of the church to check for fire extension and make certain that everyone is out of the building. In checking for extension, evidence of fire in the concealed area may not be apparent. If the fire is evident, it is usually too late to save the roof. Therefore, it is preferable to make a defensive attack on the bell tower using streams placed beyond the collapse zone.

Resource Needs

A large water supply and two or three elevated master streams are needed.

Incident Action Plan

- Extinguish the fire in the bell tower.
- Check the interior of the church via a rear door.

Deployment

- *First truck, crew 1 (two fire fighters)*: Set up an elevated master stream.
- *First engine*: Supply water to the first truck's master stream.
- *First truck, crew 2 (two fire fighters)*: Force entry to the rear of the church; check for visible fire extension; make certain the church is unoccupied.
- *Second truck*: Set up a second elevated master stream, supplied by the second engine company.
- *Third truck*: Set up a third elevated master stream, supplied by the third engine company.
- Members of the second and third engine companies who are not involved in laying out the water supply would form the RIC.
- One company is staged ready to enter the church to conduct a thorough search and check for extension after the bell tower is extinguished, if structural conditions warrant.

The staffing requirements outlined above differ from those recommended in Chapter 2 for interior operations, because this scenario was an exterior, defensive operation.

The three scenarios used as examples here are for discussion purposes only. The incident action plan for each scenario would change as objectives were met. In the church fire example, the IC should be planning for a change from a defensive to an offensive operation if the bell tower can be extinguished before the fire enters the main church structure. However, there is a real possibility that the fire has already entered the concealed spaces above the church, thus requiring a larger defensive operation with little hope of saving the building. A limited degree of risk is acceptable as long as there is property to be saved.

If a defensive operation is initiated for the entire church, the only salvageable property will be external exposures. Therefore, the amount of risk should be reduced to as near zero as possible. This may mean using unstaffed master streams or pulling staffed master streams even farther back from the church.

Summary

The development of an incident action plan based on accurate information is critical to meeting the three priorities of life safety, extinguishment, and property conservation. Through the sound application of risk management techniques, safety can be addressed while meeting these three priorities.

Wrap-Up

Key Terms

incident action plan The objectives for the overall incident strategy, tactics, risk management, and member safety that are developed by the IC; these are updated throughout the incident.

risk-versus-benefit analysis The process of weighing predicted risks to fire fighters against potential benefits for owners/occupants and making decisions based on the outcome of that analysis.

Suggested Activities

1. Critique recent fires in your jurisdiction and evaluate the operations in the same way as the three examples in this chapter.

   A. Identify a reasonably expected benefit, and compare the benefit to the risks taken.

   B. Was the structure sound? Or was it in danger of collapse?

   C. Was the operation offensive or defensive?

   D. Was this the correct attack? Explain why.

   E. Were there sufficient resources to conduct a safe and effective offensive or defensive attack?

   F. List the objectives of the incident action plan in priority order.

2. Use a case study to evaluate the operation in the same way as explained above. Attempt to pick out one or two major problems in the implementation of the incident action plan. Case studies are available from the NFPA, NIOSH, NIST, several fire service magazines, and the U.S. Fire Administration. The following six fire investigation reports are particularly well suited to evaluating the risk-versus-benefit analysis described in this chapter:

   - *NIOSH Fire Fighter Fatality Investigation and Prevention Program Report F2003-12*, Cincinnati, OH, March 21, 2003. Available online at the NFPA website.
   - NFPA Fire Investigations Department. *Supermarket Fire, Fire Fighter Fatality, Phoenix AZ, March 14, 2001*. Available online at the NFPA website.
   - NFPA Fire Investigations Department. *Residential High Rise, Six Fatalities, North York, Ontario, Canada, January 6, 1995*, NFPA Fire Investigation Report. Quincy, MA: NFPA. Available online at the NFPA website.
   - NFPA Fire Investigations Department. *Fraternity House Fire, Chapel Hill, NC, May 12, 1995*, NFPA Fire Investigation Report. Quincy, MA: NFPA.
   - NFPA Fire Investigations Department. *Elderly Housing Fire, Johnson City, TN, December 24, 1989*, NFPA Fire Investigation Report. Quincy, MA: NFPA.
   - Gordon Routley. *Wood Truss Roof Collapse Claims Two Firefighters, December 26, 1992*, Report 069. Emmittsburg, MD: U.S. Fire Administration, 1992.

Chapter Highlights

- An incident action plan outlines major tactical objectives and provides the central focus for operations.
- Primary strategic considerations are always life safety, extinguishment, and property conservation.
- The key to a successful fire-ground operation is keeping incident action plans simple and clear.
- Life safety needs are determined by comparing the risks to fire fighters for a particular action to the benefit that action might have in saving lives.
- A structure in immediate danger of collapse should not be entered. Fire fighters may enter a building to rescue occupants even if the building may eventually collapse, but the IC must constantly review structural conditions.
- Knowledge of the structure's components and the extent of fire involvement are essential in assessing structural conditions.
- Lives and property are best saved by conducting an offensive (interior) attack.
- Resource capability and resource needs must be compared early in the incident.
- Actions taken during an attack must be coordinated as offensive (interior) or defensive (exterior). An offensive attack should be used whenever it is safe to do so; it may be changed to a defensive strategy if conditions change.
- A defensive attack should be used whenever the risk-versus-benefit analysis indicates that the risk associated with an offensive operation outweighs the possible benefits (e.g., interior conditions are judged to present an extreme danger to fire fighters and there is no reasonable chance to save lives).
- The IC sets the objectives, decides on the tactics, and assigns units to complete the tasks using the incident action plan.
- A written incident action plan is needed when resources from multiple agencies are being used, when several jurisdictions are involved, or when a continuing operation requires a transfer of command at shift change.
- The incident action plan must be followed up with status reports, as changing conditions may require changes to the plan.

Company Operations

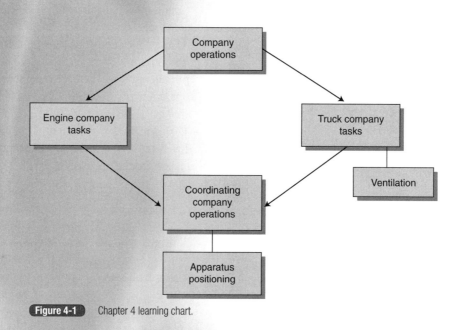

Figure 4-1 Chapter 4 learning chart.

Chapter 4

Learning Objectives

- Compare the division-of-labor concept as used in industry to company-level deployment.
- Describe structural firefighting functional assignments.
- Enumerate engine and ladder company fire-ground functions and tasks.
- Analyze tactics for the first-arriving fire company in relation to life safety.
- Apply engine and ladder company tasks to coordinating and controlling company-level deployment.
- Explain the importance of company unity to incident organization and accountability.
- List situations where splitting companies may be acceptable.
- Evaluate the positive and negative aspects of pre-assigning tasks, tools, and fire-ground positions.
- Assess proper and improper ventilation methods in regard to achieving the operational priorities of life safety, extinguishment, and property conservation.
- Given the fire and victim locations, determine the best vent location.
- Describe safe and efficient positioning of apparatus.
- Apply engine and ladder company tasks at a structure fire.
- Analyze company operations at a structure fire scenario.
- Develop an incident action plan utilizing engine and ladder company functional assignments.

Introduction

The primary responsibility of all companies working on the scene of an emergency is to work within the overall incident action plan. Freelancing cannot and should not be tolerated. Through the establishment of SOPs, first-in units can quickly go to work, allowing the incident commander (IC) time to evaluate options while maintaining control during those first few critical minutes on the fire ground. Assigning company-level operations should be included in the department's SOPs. This chapter focuses on identifying and coordinating engine and truck company tasks Figure 4-1 .

Companies and crews are supervised by a company officer, thus reducing the span of control by having one person report for the entire crew. In terms of both incident scene safety and operational efficiency, the importance of having a strong command system cannot be overstated.

The **division-of-labor principle** was the basis for the Industrial Revolution. It was found that developing job skills in a concentrated area allowed greater productivity, as compared to one individual developing a wide variety of skills. The problem implicit in the old saying, "jack of all trades, master of none," is overcome by developing, within an individual, a high level of expertise in a limited number of tasks. In the fire service the division-of-labor concept is applied by dividing tasks into the following functional areas:

- Command responsibilities
- Engine company tasks
- Truck company tasks
- Emergency medical services (EMS)
- Special operations
  - Hazardous materials teams
  - Technical rescue teams
  - Air operations
- Support services

The division-of-labor concept is not strictly adhered to in the fire service because fire fighters should be cross-trained to accomplish many different fire-ground tasks. The division-of-labor concept stresses the need for pre-assigning duties (functions) at an emergency to ensure that, before the incident occurs, everyone is familiar with what will be expected of them during an emergency.

Members of hazardous materials teams, paramedics, and heavy rescue squads are required to complete specialized training and certification requirements that ensure proficiency in their particular disciplines. However, all fire fighters should have a general understanding of these special duties, just as members of the special teams understand the duties of line fire fighters. As a minimum, line fire fighters ought to be able to function as part of either an engine or a truck company crew and be certified in basic emergency medical and hazardous materials mitigation skills.

The incident action plan governs company-level operations. Engine company tasks during an offensive attack focus on the interior and involve "rescue as you go" tactics. The main function of the engine company is to apply water directly on the burning material. During a defensive attack, the engine company sets up master streams and applies water to nearby buildings, deluges the main body of fire, or both.

Truck companies expend a great deal of effort ventilating, laddering, forcing entry, and conducting primary and secondary searches during an offensive attack. However, during defensive attacks, truck company duties are typically limited to setting up elevated master streams and support activities.

Simply put, engine company work involves applying water to the fire. Truck companies assist engine companies in gaining entry, laddering, controlling the fire spread through ventilation, and evacuating occupants. Heavy rescue companies fill in wherever needed, but are ideal rapid intervention crews (RICs).

Engine Company Tasks

An engine company is the basic firefighting unit. On close examination, engine company operations can be dissected into the following functions:

- Performing rescue operations
- Establishing a water supply
- Advancing and operating hose lines

In most cases, an engine company is first to arrive at the scene. The first-in officer initiates the attack and provides a means of moving water from a water supply to the fire.

Life safety is the first priority of everyone on the fire ground—not only the safety of the victims, but also that of the fire fighters. Engine companies usually accomplish life safety objectives by placing attack lines in position to protect victims and rescuers and by providing safe evacuation routes.

Many times it is possible to conduct a cursory search and rescue as fire lines are being advanced. Occupants may only be in need of direction and guidance, rather than actual physical assistance. The determination by an engine company officer to position hose lines or to remove occupants is a difficult decision. This tactical decision is based on the following concerns:

- Immediate danger to the occupants
- Available staffing and resources
- Time before additional resources arrive
- Extent of fire involvement
- Equipment available to perform the rescue

If there is a potential rescue that can be made, life safety will always be the top priority. The incident action plan should direct all resources (either directly or indirectly) toward the successful evacuation of all endangered occupants.

Truck Company Tasks

The terms "ladder company" and "truck company" are synonymous. It is not necessary to have an aerial ladder or elevated platform at every fire, or even in every fire department. However, it is necessary to have truck company *functions* assigned to a specific group of fire fighters on the fire ground. General truck company duties should be pre-assigned. Fire operations are more often unsuccessful because of poor or nonexistent truck company work than because of a lack of water.

The following tasks are normally assigned to truck companies:
- Conduct primary search
- Rescue trapped victims
- Ventilate
- Force entry
- Ladder the building
- Check for fire extension
- Access concealed spaces

As this list indicates, truck companies are responsible for a wide variety of tasks when an offensive attack is in progress. Therefore, it is crucial to properly staff companies that are charged with carrying out truck company duties. Other than providing ladder pipe and other elevated master streams, truck company activities are very limited during defensive attacks. In low-staffing situations, the truck company may need to assist the engine company in initial placement of hose lines during an offensive attack. However, the tendency is for the truck company to continue handling hose lines, even when adequate engine company personnel are available. This may not be the most effective use of truck personnel.

Coordinating Company Operations

For engine companies, rescue is typically an indirect activity. Getting hose lines between the victims and the fire makes rescue or self-evacuation possible. For truck companies, rescue may be a more direct activity. Truck companies will generally ventilate to control the spread of heat, smoke, and fire while conducting the primary search and removing any victims. Truck companies will also ladder the building to establish alternate evacuation routes. **Figure 4-2** illustrates how an engine company and a truck company coordinate a rescue.

Figure 4-2 Engine and truck company coordinated rescue.

The IC must coordinate all activities on the fire ground. Control and coordination are vital if the engine company is unable to control the fire or a decision to change from an offensive to a defensive attack is made. Truck companies will normally be working in areas above the fire while the engine company works at controlling the fire. If the fire is not controlled, truck companies working inside the structure need to be notified and, if at all possible, their retreat should be protected by the engine company.

Safety and control dictate that operating units work as groups (companies or crews). The first-in truck company at an offensive operation is the exception to that rule. Given the variety of tasks and the fact that many of these tasks can be performed by two-member teams, it is permissible for truck companies to split into separate crews. At a working structure fire, two members may go to the roof to ventilate, while a second two-member crew may conduct the primary search. In addition, once the first task is completed, these "truckies" may reposition to perform other ladder company tasks. For example, fire fighters who are assigned to the roof may also perform property conservation on the floor below the fire. Once these quick initial operations are accomplished, truck company members should be reunited as a company or full crew under the direction of the truck company officer. It is not freelancing when a company officer splits his or her crew into separate teams. Each crew should have a team leader who can communicate with the company officer. In some departments, the crew leader will be a higher ranking company member (such as a sergeant); in others, the crew leader will be a senior fire fighter.

Other exceptions to the general rule of maintaining company unity under the direct control of the company officer include the first-arriving company splitting into an inside and outside crew in compliance with two-in/two-out rules.[1,2] When a company is split into two or more crews, the NIMS organization chart should account for each crew. There are also times when the apparatus operator should remain at the apparatus to operate pumps or an aerial device. The company should remain together as a unit whenever possible. Maintaining company unity facilitates accountability because the officer can immediately verify the safety of the entire crew.

As has been pointed out, general truck company duties should be pre-assigned, usually to fire fighters arriving on aerial or elevated platform apparatus. This brings about a point of discussion: pre-assignment of specific tasks, tools, and positions. Many departments establish SOPs requiring truck company members to carry specific tools. For example, the officer carries a utility bar (K-tool, Hux Bar, etc.), another member an ax, and still another member a salvage cover or rope rescue equipment. Are these good SOPs?

In a high-rise office building, is the same equipment needed for a fire on the first floor during working hours as is needed for a fire on the top floor when the building is unoccupied and secured? Forcible entry tools will be needed in many cases, but will the same tools be needed at every fire?

Considering that most jurisdictions have a wide variety of occupancies, from single-family dwellings to high-rise structures, carrying the same tool into every situation, or even the same tool for different fires within the same structure, may not prove to be a productive policy. The best that can be said for the practice of pre-assigning tools and equipment is that establishing default SOPs for tools has merit. If there is no reason to bring a different tool (e.g., nothing-showing situations), bring the assigned tool. This pre-assignment of tools should be based on experience and should be limited to one or two members, rather than the entire crew having pre-assigned tools. Company officers must be prepared to change tool assignments according to circumstances.

The same can be said for pre-assigned tasks such as ventilation, forcible entry, and search and rescue. Ventilation may be required, but its priority changes depending on fire conditions, the building, the location of the fire, and the ventilation options. Primary search may require the entire truck crew or could be accomplished by the engine company as it advances.

Pre-assigning positions, unless done in a very general way with company officers having the latitude to use discretion, may result in the inefficient use of resources. Assigning fire fighters to the roof when the fire is several stories below the roof simply places them in a useless—and often dangerous—position when they could be accomplishing more effective truck company activities. Few fire departments have sufficient staffing to place fire fighters in unnecessary positions when a fire is in progress.

Fire officers and fire fighters should be highly trained, rational human beings. SOPs that limit their ability to make reasonable decisions are generally counterproductive. Fire fighters who are assigned to truck companies should bring tools into the structure, but the tools should be those needed for the specific operation. The roof assignment may be extremely important in many fires but useless in others. Some departments have discontinued the practice of roof ventilation on single-family dwelling fires. Forcible entry may be the single most important fire-ground task if victims are trapped in secured areas. Other times, windows and doors may be open, making forcible entry an unnecessary task. Given the fact that timing ventilation and forcible entry can make the difference between success and failure, the preferred approach is to have "truckies" attending to necessary truck work while engine crews are advancing lines.

A Word About Quint and Quad Companies

Some departments staff multi-function apparatus that are equipped to perform both engine and truck company operations. These are usually referred to as **Quint** or **Quad** companies. Some departments provide sufficient staffing on this single apparatus to achieve both engine and truck company objectives, while other departments do not provide the staffing needed to perform both engine and truck company tasks. It is essential that SOPs describe the actions the Quint/Quad company is to take when it is the first pumping or aerial apparatus on the scene.

Ventilation

Ventilation is one of the IC's most important tactical considerations. In the past, some fire departments initiated indirect attacks. This meant closing the structure (rather than ventilating) to gain maximum heat absorption. However, this is a dangerous tactic that simply does not work on most structure fires. Modern methods involve a coordinated effort in which ventilation is given a high priority.

Proper ventilation can have a positive effect on all three fire-ground priorities (life safety, extinguishment, and property conservation). However, ventilation is a double-edged sword. Improper ventilation can adversely affect all three priorities. If ventilation is inadequate, an offensive attack will rarely be successful. Even worse, improper placement of a fire stream into a vent opening or venting in the wrong location could place fire fighters and occupants in extreme danger and result in additional property damage. The effects of both proper and improper ventilation are shown in **Table 4-1**. Proper vent locations are shown in **Figure 4-3**.

TABLE 4-1	The Effects of Proper and Improper Ventilation	
	Proper Ventilation	**Improper Ventilation**
Life safety	• Pulls fire away from trapped occupants or their means of egress.	• Draws fire toward victims or can extend fire through their exit path.
Extinguishment	• Limits fire spread by channeling fire toward nearby openings and allows fire fighters to safely attack the fire.	• Spreads the fire into previously undamaged areas • Can cause a backdraft or let the fire gain headway while lines are being advanced.
Property conservation	• Limits smoke, heat, and water damage by allowing an interior attack and removing the products of combustion, which are often the primary cause of damage.	• Causes excessive and unnecessary damage. • Does not remove the damaging products of combustion, but rather spreads them throughout the building.

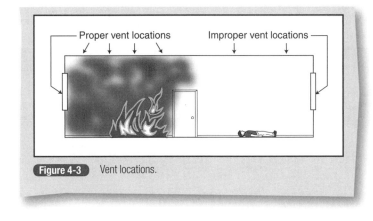

Figure 4-3 Vent locations.

Apparatus Positioning

Apparatus positioning and the fire company's assignment are directly related. The positioning of apparatus must be consistent with the company's objective. If aerial ladders or towers are needed, the first-arriving truck company will position for aerial or tower placement. All companies should make a conscious effort to keep the fire zone as accessible and safe as possible.

The company commander should always think about positioning. If the first-in engine company will be used to lay attack lines, position the apparatus as close to the curb line as possible with the side discharge gates aligned with the desired entry point. In most residential fires, this will be directly in front of the house. If the apparatus is set up to use rear discharges for pre-connected attack lines, then the pumper should be positioned close to the curb but forward of the desired entry point. Positioning of this first-arriving apparatus should be addressed in SOPs and, among other things, depends on the layout of the hose bed.

Department SOPs generally address actions to be taken by the second-in engine company. SOPs for this company may allow more discretion on the part of the company officer. Some departments specify that the first-in and second-in engine companies secure separate sources of water. Other departments permit an **attack pumper** configuration, in which the first engine company goes directly to the front of the building and the second engine company provides the water supply. Each department must decide what works best for its jurisdiction.

For purposes of discussion, when the second engine company is supplying the attack pumper from a hydrant, it should select the suction intake that will allow the apparatus operator to get close to the curb while keeping the intersection open.

If the first-in truck company will be using its aerial device, the truck must be positioned to obtain a safe operating angle and reach the desired point for roof access, rescue, ventilation, or other tasks. If the aerial device is not being used, the truck should be parked out of the way but close to the building's entry point. Getting tools, ladders, and other equipment to the building is a labor-intensive job if the truck is parked too far from the entry point. In positioning the truck, it is also important to remember that ladders usually feed off the back of the truck. Therefore, be certain that access to the rear of the truck is not blocked.

Figure 4-4 shows an example of apparatus positioning for an attack pumper operation with the aerial ladder being used to gain access to the roof.

Some common errors when positioning apparatus include the following:

- Placing aerials under wires where they cannot be safely raised
- Placing aerials and platforms in unsafe and unstable positions
- Not allowing enough room to extend the outriggers
- Apparatus/staff cars blocking access to the fire area or front of the building
- Pumpers placed in positions where pre-connected attack lines are difficult to lay
- Apparatus unnecessarily blocking streets or fire hydrants
- Unnecessarily blocking streets with large-diameter hose
- First-arriving companies not securing essential attack positions
- Positioning apparatus in dead ends or other locations where they cannot be quickly repositioned
- Companies assigned to staging responding through the fire area, causing unnecessary congestion

Staging operations must be defined in department SOPs.

Summary

The IC improves efficiency in coordinating activities at a structure fire using the fire company division-of-labor concept. The basic fire department unit is an engine

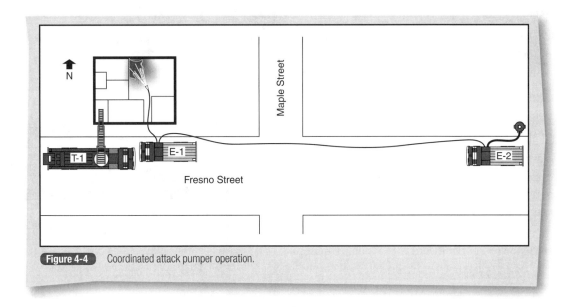

Figure 4-4 Coordinated attack pumper operation.

company whose primary duty is to apply water to extinguish the fire. This involves several tasks during an offensive attack including obtaining a water source, setting up and engaging pumps, advancing hose lines into the structure, and then advancing and operating the hose line to extinguish the fire. The engine company fulfills the life safety priority primarily by extinguishing the fire. When the fire is extinguished, the building becomes much safer for fire fighters as well as occupants. The engine crew also does a quick check of areas for victims as they approach the fire. Occupants in closest proximity to the fire are likely to be in greatest danger, and the engine company personnel will often assist in moving these occupants out of harm's way.

Truck company personnel are primarily responsible for gaining entry, laddering the building, controlling the fire spread through ventilation, and evacuating occupants. As a general rule, truck companies conduct the primary and secondary searches during an offensive attack. Truck company members are often assisted by engine personnel, with both groups working as a team, assisting and complementing the other's work. At first it would appear that engine company tasks are more important; however, as previously mentioned, fire operations are more often unsuccessful because of poor or nonexistent truck company work than due to a lack of water application. At times, less experienced ICs will be remiss in assigning truck company tasks simply because a truck apparatus is not yet on the scene. Remember, truck company tasks need not always be performed by fire fighters assigned to traditional truck apparatus. As long as the proper tools and equipment are available—and most engine companies are equipped with basic forcible entry, ventilation, and rescue tools—and the fire fighters are properly trained, it doesn't matter what type of apparatus is on the scene. What is necessary—in fact crucial—is that the truck company tasks discussed above be completed in a timely and safe manner.

Wrap-Up

Key Terms

attack pumper The first-arriving engine company goes directly to the fire building without securing a water supply.

division-of-labor principle Breaking down an incident or task into smaller, more manageable tasks and assigning personnel to complete those tasks (developing job skills in a concentrated area to allow for more productivity).

Quad Multi-function apparatus that are equipped to provide for the following four functions: water, pumps, hose, and ground ladders. In other words, this is a Quint minus the aerial ladder.

Quint Multi-function apparatus that are equipped to perform both engine and truck company operations. These are equipped to provide for the following five functions: water, pumps, hose, ground ladders, and an aerial ladder.

Suggested Activities

1. Using resources from your department or a department with which you are familiar, assign tasks to the first-arriving and subsequent arriving companies for the scenario described below:
   - Time: Sunday morning at 0700 hours
   - Weather: 60°F and sunny
   - Structure: Two apartments above a store, as shown in Figure 4-5.
   - Fire and smoke are visible from the first floor; no occupants can be seen from the exterior.
   - Hydrants with a flow of 1000 GPM (63 L/sec) each are located 300 ft (91.4 m) in each direction.

2. A fire is reported in the apartment building shown in Figure 4-6. See Chapter 2, Figures 2-7 and 2-8 for pre-plan information.
   - Time: Monday afternoon at 1400 hours
   - Weather: 60°F and clear
   - Report of smoke in the second and third floor hallways with multiple calls to dispatch
   - E-23 is the first-arriving engine company

   Engine 23 radios dispatch with the following message: "Engine 23, Memory Lane Command, verifying the address as 25 Memory Lane. We have light smoke visible at the second floor rear of a large, four-story apartment building with occupants exiting the building at this time. We

Figure 4-5 Business/apartment.

Figure 4-6 Apartment building. A. Front view. B. Rear view.

are going to attack the fire from the Memory Lane side with a 1¾" line."

Engine 23 secures a water supply using large-diameter hose from the hydrant on Memory Lane and positions their apparatus at the front of the building.

The officer and one fire fighter, with the assistance of the pump operator, advance a 1¾" (44-mm) hose line into the main entrance up the stairway to the second floor. They encounter heavy smoke in the hallway and advance down the hallway to Apartment 207. There is a heavy volume of fire in Apartment 207 and the door to the hallway is open, allowing smoke and fire to enter the hallway. They begin the fire attack using a 1¾" (44-mm) hose line.

The apparatus operator and hydrant fire fighter form the two-out team (initial RIC).

Truck 23 arrives shortly after Engine 23. Engine 23's officer orders them to the second floor fire area to begin forcible entry into apartments adjacent to the fire and to initiate the search-and-rescue effort on the fire floor.

Assume the role of District Chief 1 with the following call to dispatch: *"District Chief 1, Memory Lane Command, we have a working fire in a large apartment building."*

You call Engine 23 via portable radio and ask for a status report: *"Engine 23, this is District 1. I have assumed command. What is your status?"*

Engine 23 replies, *"We have a working fire on the second floor, but are making significant progress. Truck 23 is with us and they are assisting occupants out of the building. This place is full of elderly people. We could use some help here on the second floor."*

A. Evaluate the tasks being performed by Engine 23 and Truck 23. Are they consistent with engine and ladder company duties described in this chapter?

B. How much assistance would you call, e.g., second alarm?

C. Assign companies that you would summon to the scene using the pre-incident plan for the Memory Lane Apartments shown in Chapter 2, Figures 2-7 and 2-8.

Chapter Highlights

- All companies working at an emergency must work within the overall incident action plan.
- The division-of-labor concept stresses the need for pre-assigning duties (functions) at an emergency.
- Although specialized training is required for specific disciplines, all fire fighters should have a general understanding of special duties such as emergency medical and hazardous materials operations.
- An engine company is the basic firefighting unit. Engine company operations include rescue, establishing water supply, and operating hose lines.
- General truck company duties include search and rescue, ventilation, forced entry, laddering, utility control, and reconnaissance.
- Truck company activities are varied and many during offensive attacks but limited during defensive attacks.
- Control and coordination of truck and engine activities are vital to fire fighter safety.
- Safety and control dictate that operating units work as companies or crews.
- Maintaining company unity improves safety and accountability; however, it is permissible to split the first arriving company to comply with the two-in/two-out rule. Further, the first-arriving truck company is often split into two crews.
- Communication between company officers or crew leaders and the IC must be maintained.
- Pre-assigned tools, tasks, and operations should be adjusted to specific circumstances; strict adherence to tool and task SOPs can sometimes be counterproductive.
- Offensive firefighting involves a coordinated effort of hose placement and ventilation.
- Proper ventilation can pull fire away from victims and limit fire spread, but improper ventilation is counterproductive and dangerous.
- The fire zone should be kept as accessible and safe as possible.
- Proper positioning of apparatus promotes efficient use of resources.

References

1. National Fire Protection Association, *NFPA 1500: Standard for Fire Department Occupational Safety and Health Program.* Quincy, MA: NFPA, 2007.

2. *Code of Federal Regulations, 49 CFR 1910.* Washington, DC: U.S. Government, April 18, 1998.

Fire Fighter Safety

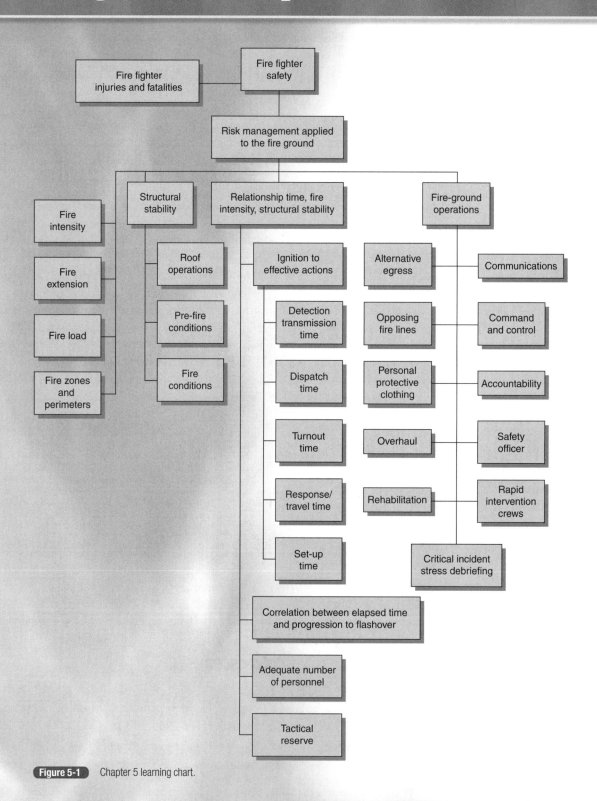

Figure 5-1   Chapter 5 learning chart.

Chapter 5

Learning Objectives

- Identify and analyze the major causes involved in on-duty fire fighter fatalities related to health, wellness, fitness, and vehicle operations.
- Analyze the trend in the number of fire fighter on-duty deaths over a 30-year period.
- Define frequency and severity as they relate to fire fighter injuries.
- Enumerate fire-ground safety issues addressed in *NFPA 1500*.
- Compare and contrast fire trends and fire fighter on-duty deaths.
- Describe the relative risk to fire fighters combating fires in different occupancy types.
- Analyze the trend in number of fire fighter injuries.
- Discuss risk management principles applied to the fire ground.
- Discuss and give an example of an imminent life-threatening situation.
- Use a probability analysis to assess the occupied status of a building based on time and occupancy.
- Estimate the collapse time based on burn time, fire intensity, content load, and construction type.
- Examine the difference between a managed retreat and an evacuation due to an imminent hazard.
- Evaluate the difference between lightweight and heavy structural components.
- Discuss and contrast pre-fire and fire conditions that contribute to structural collapse.
- Examine hazards presented by suspended ceilings.
- Compare construction methods in terms of structural stability, fire extension, and fuel contribution.
- Review the basics of building construction and how they relate to pre-fire planning.
- Estimate the collapse zone for a building in imminent danger of collapse.
- Describe exclusion zones other than collapse zones.
- Develop zones and perimeters around a structure fire.
- Define and explain the five time segments from ignition to effective action.
- Evaluate the survivability, structural stability, and flashover from ignition to effective action.
- Evaluate set-up time in regard to staffing on the first-arriving engine company.
- Compute the staffing necessary to achieve the tasks enumerated in *NFPA 1710*.
- Define and compare flashover and backdraft.
- Explain the relationship between NIMS and a fire fighter accountability system.
- List situations when a personal accountability report (PAR) should be initiated.
- Explain the importance of alternative egress for fire fighters conducting an offensive attack.
- Define rapid intervention crew (RIC).
- Explain the role of the RIC.
- Explain the importance of having a RIC immediately available from initial attack and throughout the operation.
- Determine the number of personnel to be assigned to the RIC based on the size and complexity of the building and incident.
- Describe safe interior operations.
- Construct an emergency message for a disoriented fire fighter needing assistance.
- Explain measures that can be taken to improve the chances of survival when fire fighters are lost and out of air in a large building.
- Describe methods used to supply air to a trapped fire fighter who has exhausted his or her air supply.
- List tools that should be available to a RIC.
- Compare the advantages and disadvantages of a mobile RIC versus a stationary RIC.
- Recognize hazards in operating opposing fire lines.
- Evaluate hazards to fire fighters during overhaul operations.
- Define immediately dangerous to life and health (IDLH) atmospheres and the relationship to SCBA usage.
- List factors the IC should consider when formulating an incident action plan to be used during overhaul.
- Describe informal rehabilitation at the fire scene.
- Describe hot weather rehabilitation.
- Describe cold weather rehabilitation.
- List the signs of critical incident stress.

Introduction

Fire departments are dedicated to saving lives and property from the perils of fire. Saving lives is the highest priority at the incident scene. The fire fighter's life is valued as being as important as any other life, and the fire fighter is the most valued resource of any fire department. Too often, fire fighters lose their lives in the process of saving the lives and property of others. Even more tragic are incidents where fire fighters lose their lives when there are no lives or property to be saved.

Fire fighter safety is closely related to the risk-versus-benefit analysis discussed in Chapter 3. This chapter describes how NFPA 1500 is applied to assessing fire fighter safety. It also relates fire fighter safety to pre-fire and fire conditions—time, fire intensity, and structural stability—and to the correlation between response time and progression to **flashover**.

This chapter also discusses some of the factors relating to fire fighter safety using statistics, as well as a series of case studies and established practices. The learning chart in **Figure 5-1** lists the topics covered in this chapter and points out how applying risk management principles relates to all facets of fire fighter safety.

Fire Fighter Injuries and Fatalities

Identifying and analyzing how fire fighters are killed and injured is critically important to reducing fire fighter injuries and deaths. The National Fire Protection Association (NFPA) has compiled fire fighter fatality statistics for 30 years. The annual fire fighter fatality report for the previous calendar year is published each year in the July/August edition of the *NFPA Journal*. The annual fire fighter injury report for the previous year is published in the November/December issue of the *NFPA Journal*. The charts and graphs used here are taken from data provided by the NFPA. The NFPA, National Institute for Occupational Safety and Health (NIOSH), U.S. Fire Administration (USFA), and others publish fire investigative reports analyzing individual fires that resulted in fire fighter fatalities. These reports provide information, which if properly used and analyzed, can prevent future on-duty fire fighter fatalities and reduce injury **frequency** and **severity**.

In examining the chart in **Figure 5-2**, it is obvious that the number of fire fighter fatalities is trending downward over this 30-year period.[1] Many measures have been taken to improve fire fighter safety during this period. Personal protective equipment and breathing apparatus used by fire fighters today are vastly improved compared to protective gear being used in 1977 when the NFPA began gathering data on fire fighter injuries and fatalities. In 1987, the first edition of *NFPA 1500: Standard on Fire Department Occupational Safety and Health Program* was adopted with subsequent revisions every five years.[2] Following the provisions outlined in NFPA 1500 can substantially reduce injury frequency and severity.

NFPA 1500: Standard on Fire Department Occupational Safety and Health Program provides the requirements for

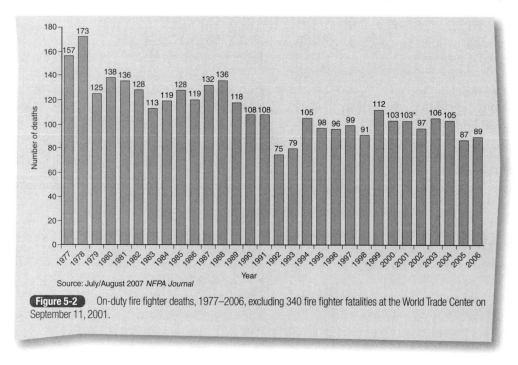

Figure 5-2 On-duty fire fighter deaths, 1977–2006, excluding 340 fire fighter fatalities at the World Trade Center on September 11, 2001.

safety measures to be taken at the incident scene. These measures represent the absolute minimum. Incident commanders (ICs) and safety officers must be familiar with this standard to be effective in protecting their most valued resource: the fire fighter. Several important fire-ground safety issues are identified in the *NFPA 1500* standard:

- Risk management principles must be applied to fire-ground operations.
- The IC is responsible for overall safety at the scene.
- An incident management system (IMS) must be used at all emergency scenes.
- The IC maintains command and control of all operating forces within a common strategy based on situation analysis. The situation analysis must be ongoing with changes in strategy consistent with the changing situation.
- An adequate number of personnel must be available to implement the strategy.
- Pre-established standard operating procedures (SOPs) must be implemented.
- An accountability system must be used at all incidents.
- RICs must be provided at all stages of the incident, beginning with the first-arriving unit utilizing a two-in/two-out process.
- Inexperienced members must be directly supervised by more experienced members.
- Minimum basic and continued fire training must be provided in accordance with SOPs.
- Medical treatment and rehabilitation must be available as needed.
- Full personal protective clothing must be worn.
- Self-contained breathing apparatus (SCBA) must be worn in hazardous atmospheres, atmospheres that are suspected of being hazardous, or atmospheres that may rapidly become hazardous.
- **Personal alert safety system (PASS)** devices must be worn and activated before entering a hazardous area.
- Critical incident stress debriefing teams must be available when needed.
- Post-incident analysis is necessary for significant incidents or those causing serious injury or death.

NFPA 1500: Standard on Fire Department Occupational Safety and Health Program contains a detailed description of these minimum requirements. OSHA standards also address measures to be taken to protect fire fighters.[3]

There is no doubt that attention to safety and the attitude toward safe operations has improved. Unfortunately, even with these improvements, the fire service is still experiencing a large number of on-duty deaths. Most troubling is the fact that we still experience a large number of fire fighter deaths on the fire ground. Figure 5-3 compares the number of fire fighter fatalities to the number of fires. This chart indicates that fire fighters are responding to fewer fires, but when operating at structure fires they dying at nearly the same rate in the late 1990s as in the late 1970s.

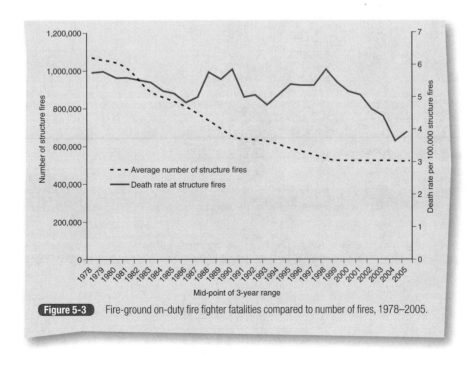

Figure 5-3 Fire-ground on-duty fire fighter fatalities compared to number of fires, 1978–2005.

Considering all that has been done to improve safety, this is a very disturbing statistic. Could this fatality rate be due to a lack of fire experience and training? With all of the duties being performed by fire fighters, and subsequent training requirements for the extra duties, some departments are not providing sufficient structural firefighting training and education to make up for the lack of fire-ground experience. Nothing completely takes the place of experience, but task- and tactic-oriented fire training and education can improve safety at fire-ground operations. The trend in fire fighter fatalities per 100,000 fires has improved since 1999. Hopefully, this trend will continue.

The fire ground is becoming more hazardous. Lightweight construction combined with heavy fuel loads (including large quantities of plastics) in very large buildings significantly increase the risk to fire fighters. Truss roofs, which are present in residential and larger commercial buildings, will be discussed later in this chapter. Large-span truss roofs create a substantial hazard to fire fighters. Modern personal protective equipment provides a higher level of protection, thus permitting fire fighters to make a closer approach to the fire. Advancing further inside the building and closer to the fire places fire fighters at greater risk. Fire fighters must understand the limitations of their personal protective equipment.

The everyday residential fire provides experience for larger fires, but strategy and tactics must be changed to meet the challenges posed when fire involves a large, complex building. Occupancy plays a role in fire fighter safety. No fire should ever be considered routine. Most fire fighters die in residential fires because most structure fires occur in residential occupancies. In 2005, 78% of all structure fires occurred in residential property (396,000 of 511,000 total structure fires).[4] However, once a fire occurs, the risk to fire fighters is much greater in some of the other occupancy types, as shown in **Figure 5-4**. The risk of a fire fighter being killed is twice as great in a manufacturing occupancy than in a residence. The comparable risk is in the range of 3:1 for store/office, public assembly and vacant/special. It is important to avoid a single-family residential mind-set. Escape routes are closer and easier to find in most single-family residential structures as compared to larger buildings, where more time is needed to escape and a higher probability of disorientation exists. Depleting the SCBA air supply or a SCBA failure often results in survivable smoke

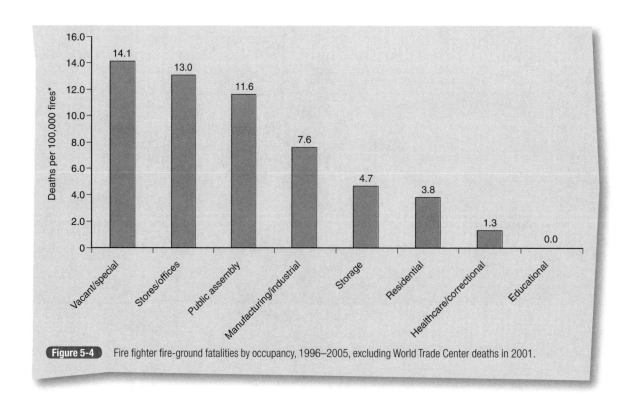

Figure 5-4 Fire fighter fire-ground fatalities by occupancy, 1996–2005, excluding World Trade Center deaths in 2001.

inhalation in a one- or two-family dwelling fire; however, the loss of air supply is more likely to be fatal in a larger, more complex building.

Figure 5-5 shows the percentage of deaths by type of duty. Given the lower number of fires and increased activity in non-fire emergencies such as EMS, the fire ground remains a very dangerous workplace.

In analyzing fire fighter on-duty deaths it is also important to consider the nature and cause of the injury leading to fire fighter fatalities, as shown in Figure 5-6 and Figure 5-7.

Sudden cardiac deaths are responsible for nearly half of all fire fighter on-duty deaths. Many of these deaths are preventable through proper medical evaluation and wellness programs.

When fire-ground statistics are analyzed, cardiac deaths remain the most common cause of fire fighter fatalities at structure fires, with asphyxiation second, followed by crushing injuries and burns, as shown in Figure 5-8.

In evaluating the most common non-cardiac on-duty deaths inside structure fires for the 30-year period that

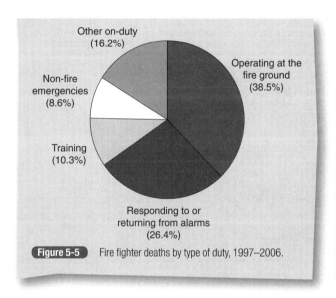

Figure 5-5 Fire fighter deaths by type of duty, 1997–2006.

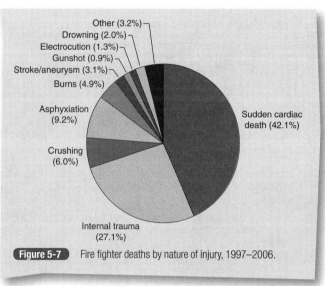

Figure 5-7 Fire fighter deaths by nature of injury, 1997–2006.

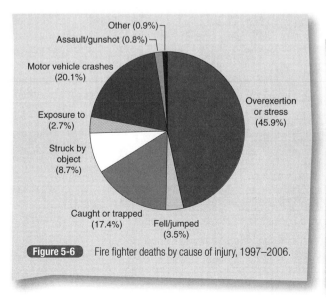

Figure 5-6 Fire fighter deaths by cause of injury, 1997–2006.

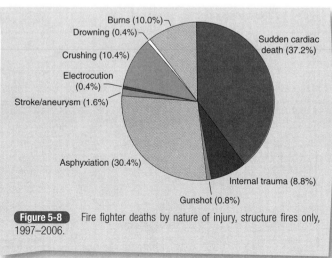

Figure 5-8 Fire fighter deaths by nature of injury, structure fires only, 1997–2006.

the NFPA has been keeping statistics Figure 5-9, there is a slight increase in the rate of deaths per 100,000 fires due to burns and crushing injuries. Prior to 2000, the trend was also upward for smoke inhalation, but this trend has been dramatically reduced.

Evaluating fire fighter deaths in regard to situations and causes is essential to a continued decrease in the number of fire fighter fatalities. The company officer and IC can also impact the non-life-threatening injuries that often lead to pain, suffering, and disability. Figure 5-10 shows a slight decrease in the number of fire-ground injuries per year for a 10-year period, but this decrease is less than the rate of decrease in the number of fire responses over the same period of time. Nearly 50% of all injuries occur on the fire ground, and nearly 50% of the injuries are strains and sprains.[5]

Fire administrators must be committed to fire fighter safety and provide the necessary procedures, training, and equipment. ICs have overall safety responsibility during an incident and must make the crucial offensive/defensive attack decision, then monitor, organize, coordinate, and provide safety measures described throughout this text. Company officers have a responsibility to properly supervise members of their company. Fire fighters must take personal responsibility

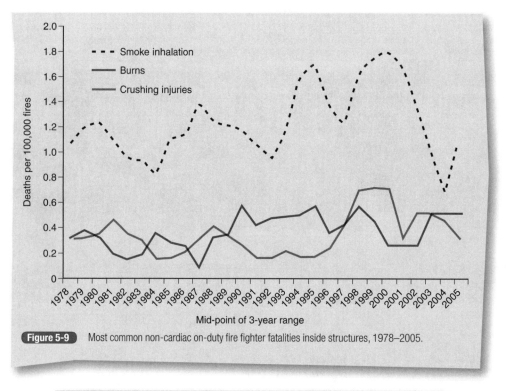

Figure 5-9 Most common non-cardiac on-duty fire fighter fatalities inside structures, 1978–2005.

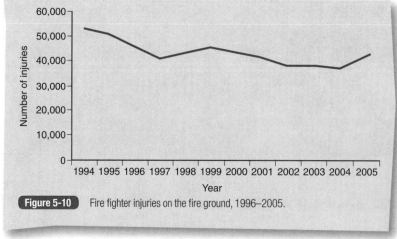

Figure 5-10 Fire fighter injuries on the fire ground, 1996–2005.

for their safety and the safety of members working with them by following procedures, maintaining firefighting skills, and properly using the equipment provided.

Risk Management Applied to the Fire Ground

Probably the most important element of the incident safety program is applying risk management to fire-ground operations. *NFPA 1500* specifically addresses risk management principles to be applied at the incident scene:

> **8.3.2** The concept of risk management shall be utilized on the basis of the following principles:
>
> **(1)** Activities that present a significant risk to the safety of members shall be limited to situations where there is a potential to save endangered lives.
>
> **(2)** Activities that are routinely employed to protect property shall be recognized as inherent risks to the safety of members, and actions shall be taken to reduce or avoid these risks.
>
> **(3)** No risk to the safety of members shall be acceptable when there is no possibility to save lives or property.
>
> **(4)** In situations where the risk to fire department members is excessive, activities shall be limited to defensive operations.

These important guidelines must be applied to every situation. Probably the single most important concept for the IC to learn is recognizing the point at which the risk to fire fighters' lives outweighs the possible benefits of saving lives and property. Too many fire fighters are injured or killed in body recoveries or wetting down the ruins.

The first statement in *NFPA 1500*-8.3.2 above regarding saving endangered lives can be somewhat ambiguous and has been the subject of much discussion. The controversy heightens with the provision of an exception to the two-in/two-out rule in the face of an imminent life-threatening situation. At the time of this writing, OSHA's exact definition of an imminent life-threatening situation meeting the exception to the two-in/two-out rule is not completely understood. The degree of risk associated with entering a burning building without the safety of two-in/two-out is higher than the risk taken with two fire fighters standing by outside the building. Therefore, the perceived benefit would have to be more absolute to justify the risk.

It is relatively easy to visualize a situation with an imminent life-threatening situation. Or is it? An argument could be made for urgency in the late night residential fire with an assumption that occupants are home asleep, possibly unaware of the fire, or already in serious trouble. This process could be continued with assumptions that the possibility of a threat to life always exists until

> Sometimes it is necessary to enter vacant structures to determine the life safety hazard. ICs must be extremely careful when confronted with a well-involved vacant structure.

a primary search verifies that no one is endangered. A warehouse or office building could also be occupied at night. Even supposedly vacant buildings can be occupied by the homeless. When a fire breaks out in a vacant structure, there is a possibility that the person who caused the fire is still in the building.

The concept of probability is important to the risk management process. The IC considers the possibility of people being in the building not as a yes-or-no proposition, but rather as a degree of probability: How does the IC evaluate the probability that people are in the building?

The probability that a nursing home is occupied is very high at any time of the day or night, as is the potential for a dwelling to be occupied at 2:00 AM. Finding someone in an office complex at 2:00 AM on a Saturday night or Christmas Eve would be less likely but possible. Thus, time of day, day of the week, and time of year are factors in determining the probability of the structure being occupied. Probability is important to the size-up process.

The *NFPA 1500* committee undoubtedly expects an IC to exercise his or her judgment on the issue of an imminent life-threatening situation.

Fire Intensity

Time and occupancy factors are not the only considerations. Fire intensity and building construction are equally important in determining what there is to save. The hazards to fire fighters entering a given area increase as the fire progresses toward flashover. The building is getting weaker throughout the process, to the point of complete

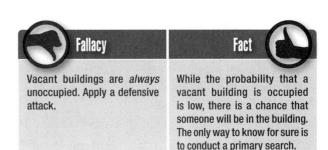

Fallacy	Fact
Vacant buildings are *always* unoccupied. Apply a defensive attack.	While the probability that a vacant building is occupied is low, there is a chance that someone will be in the building. The only way to know for sure is to conduct a primary search.

Case Summary

In Worcester, Massachusetts, six career fire fighters died in a vacant, six-floor, maze-like, cold storage and warehouse building while searching for two homeless people who accidentally started the fire and escaped prior to the arrival of the fire department. Approximately 30 minutes after the first alarm was struck, two fire fighters searching for victims and checking for extension sounded an emergency message. A personal accountability report (PAR) confirmed two missing fire fighters. The search-and-rescue operation was expanded to find the two missing fire fighters and the homeless people thought to be in the building. Four additional fire fighters became disoriented during this part of the operation. The fire was believed to have 30 to 90 minutes of pre-burn time. After more than an hour on the scene, a company on the interior notified command of structural problems and an arson investigator on the exterior reported the fire was venting from the roof. A defensive attack was ordered approximately 1 hour and 45 minutes after the initial alarm. The six missing fire fighters perished.

Figure CS5-1 The combination of a massive fire load and the maze of rooms in this Worcester, Massachusetts, warehouse contributed to the death of six fire fighters.

Source: National Institute for Occupational Safety and Health (NIOSH), Fire Fighter Fatality Investigation 99 F-47, Six Career Fire Fighters Killed in Cold Storage Warehouse Building Fire–Massachusetts.

---

failure. No one can determine exactly when flashover will occur or when the building will fail. However, there are useful approximations. In post-flashover fires, the chance of occupant survival is minimal within the flashover compartment.

In the past, the time from ignition to flashover was given as 10 minutes. The actual time can vary significantly, depending on a number of variables, including the following:

- Compartment size
- Ventilation
- Ignition source
- Fuel supply
- Fuel geometry
- Distance between fuel cells
- Location of the fuel
- Heat capacity of the fuel
- Geometry of the enclosure

There are several computer programs that can provide rough estimates of when flashover will be reached in a compartment. However, given the wide range of variables that must be entered into these computer programs, it is very difficult to provide a reliable time frame that the IC can use on the fire ground. Generally, one assumption that can be made is that the larger the volume of the enclosure where the fire is located, the longer the time required to reach flashover.

Building Design Loads

Loads imposed on a building are categorized as **live loads**, **dead loads**, and seismic, wind, snow, and ice loads. Building loads affect structural stability. Unusually high building loads can result in premature collapse. Loads placed on lightweight roof structures are particularly hazardous to fire fighters.

Fuel Load

The fire or **fuel load** consists of fuels provided by the contents and combustible building materials. Most of the construction materials used in wood frame buildings will burn, thus creating a large fuel load. Combustible building materials are very limited in fire-resistive and non-combustible construction. The primary fuel load for most structure fires is made up of the combustible contents.

Structural Stability

Structural failure can occur at any time. Always consider structural stability in sizing up a fire.

Fire intensity, burn time, content loads, and construction methods and materials all affect structural stability. An old rule of thumb called the 20-minute rule was based on all of these factors except the content load. The **20-minute rule** states that when a heavy volume of fire is burning out of control on two or more floors for 20 minutes or longer, structural collapse should be anticipated. This rule is based on ordinary construction. A frame structure may be completely consumed in less than 20 minutes, and unprotected steel can fail in less than 20 minutes when subjected to a heavy volume of fire. Fire entering a concealed truss area causes failure in a relatively short time, as has happened in several fires that have killed fire fighters. Heavy timber or fire-resistive structures, subjected to the same volume of fire, would be expected to withstand fire attack longer than a frame structure.

There is a subjective element in the 20-minute rule: heavy volume of fire. What is perceived as a heavy volume of fire by one IC is a moderate volume of fire to another. There is also reason to question the "two or more floors" part of the rule. A heavy volume of fire on a single floor could also result in collapse. Is 20 minutes an exact time? Absolutely not! Likewise, the length of burn time at heavy volume is subjective, as there is no way of knowing the exact time of ignition or the intensity of burn before arrival.

Time and intensity factors must be considered in determining whether it is safe to enter or be near a structure. The IC should "start the clock" when making a decision to conduct operations in the offensive mode when confronted with a well-involved fire. If, after a given period of time, the fire is still not under control, the incident action plan should be reviewed and the operation possibly changed to a defensive attack.

Given the amount of information that an IC must assimilate, it is easy to lose track of how much time has passed on the fire ground. *NFPA 1500: Standard on Fire Department Occupational Safety and Health Program* requires dispatch centers to notify command every 10 minutes until the fire is knocked down. This notification helps the IC track elapsed time. It is critical that the elapsed time take into account the time the fire was burning prior to notification and while units were responding, not just the time that has elapsed since arrival on the fire scene.

When changing from an offensive to a defensive operation, a managed retreat is best. Engine companies operating hose lines should provide protection while other companies, especially companies above the fire, evacuate the building. The planning section and safety officer can be of great assistance to the IC in making this decision and assisting in managing the retreat. Whenever signs of imminent collapse are observed, the offensive operation must immediately be abandoned. All units must be notified and immediately move to the exterior. Usually, an announcement is given over the radio followed by a pre-planned signal such as several long blasts on an air horn (ten three-second blasts, for example). Forces operating on the outside must be moved back to a safe distance from the building.

A wide variety of construction materials and methods have been used in buildings over the years. These range from heavy timber construction to lightweight wood trusses. Because of the different materials, the behavior of buildings under fire conditions will vary significantly.

Although some broad assumptions can be made, such as that a heavy timber building will survive longer than a lightweight wood truss structure, such assumptions are not universally true.

Fire spread inside of the building can occur through a variety of openings, either horizontally or vertically. Some buildings are constructed with large vertical shafts for utility lines or for other reasons. In the MGM Grand fire in Las Vegas in 1980, fire was able to spread vertically through seismic joints and stairways. In the NFPA report on this incident, a $\frac{5}{8}$"-(1.6 cm-)wide space in one of the stairways that had not been properly sealed off was cited as one of the conditions that contributed to the spread of smoke and fire into the stairway.

What the IC can learn from these incidents and others is that a building's performance under fire conditions can be unpredictable, and exterior appearances may not be accurate indicators of interior conditions.

Renovations to a building can dramatically affect how it performs under fire conditions. Openings made in floors to run utilities are not always properly sealed.

The potential conditions that can lead to structural collapse and fire spread cannot be reliably predicted. No building is completely immune to structural failure, but some buildings will withstand a very large and intense fire without a catastrophic (total) collapse. Other structures are known to experience early collapse under intense fire conditions. When deciding on an offensive or defensive strategy or when placing companies for rescue or fire attack, the IC must take structural stability into account.

Modern construction methods conserve materials by using lighter weight structural members providing the same load-bearing capabilities as earlier construction methods using massive structural members. The primary way this is accomplished is through truss construction. Lightweight trusses take the place of large wood beams or steel I-beams. Truss construction methods are structurally sound under normal conditions. As a matter of fact, the added strength allows builders to place trusses 24″ (61 cm) on center, whereas the earlier methods required 16″ (41 cm) on center supports.

The problem is that these lightweight structural members are adversely affected by the fire much sooner than the heavier building materials. Compounding the problem is the fact that once the truss loses its triangular configuration, it loses load-bearing capacity. Truss construction translates into failure with little warning to the fire fighter. A well-involved fire in a truss space may not be detected until it breaks out of containment. Several fire reports mention that fire fighters working on the interior stated that there were few signs of a serious fire prior to collapse.

Unprotected metal trusses used in non-combustible construction **Figure 5-11** fail quickly, and wood trusses **Figure 5-12** burn through sooner than structures that are built of larger structural members. Truss roofs above large, open areas are particularly dangerous. A truss roof failure

Figure 5-11 Metal truss construction.

Figure 5-12 Wood truss construction. A. Under construction. B. Completed building.

over a compartmented area may be partially suspended by walls separating the rooms below. A fire entering an undivided attic area can extend unimpeded throughout the entire attic. Truss roof collapse generally results in a large area of damage. More fire-resistive buildings can suffer partial interior collapses that are not often noticed by the IC or companies working in other areas of the building.

Structural connections can play a critical role in a building under fire attack. A large I-beam supported by a single bolt is only as strong as the single bolt. Wood truss construction typically uses gusset plates in place of nails Figure 5-13 . Nails form a stronger connection under fire conditions, owing to their depth. Gusset plates usually penetrate only a fraction of an inch and form a large surface area to collect heat. As soon as the gusset plate teeth lose their strength or the fire burns through the wood connecting surfaces, the wood truss loses its stability Figure 5-14 .

Figure 5-13 Gusset plates.

Figure 5-14 Fire damaged gusset plate.

Case Summary

In Hackensack, New Jersey, five fire fighters died in the collapse of a bow-string truss building. Fire conditions inside the building were relatively clear, but the concealed ceiling space was fully involved in fire.

Source: Thomas J. Klem, "The Hackensack Fatalities," *Fire Command*, October 1988, p. 26.

Figure CS5-2 This deadly fire at a Ford dealership killed five fire fighters on July 1, 1988, when the building collapsed.

Wooden I-beams are now in use, using the same logic as a steel I-beam. A 2 × 4″ (approximately 50 × 100 mm) board is placed on the top and bottom with a piece of plywood between, thus increasing the load-bearing strength. Again, this method is satisfactory under normal conditions but lacks reliability and stability when exposed to fire conditions.

Many older buildings are modernized by suspending ceilings or by adding walls or paneling and installing a facade. These renovations provide additional concealed spaces for the fire and present additional dangers to fire fighters.

Understanding the dynamics associated with structural collapse is critically important to the fire fighter. The IC must pay close attention to signs of impending structural failure. Many decisions to conduct the attack in the defensive rather than offensive mode are based on structural stability. When a safety officer is on scene, he or she must pay close attention to signs of collapse.

Roof Operations

Many departments automatically assign members to the roof for every fire. This can be a dangerous practice. A conscious decision should be made regarding roof safety. Lightweight truss roofs are particularly dangerous. When a fire has self-vented, the roof is likely to be unsafe, and the need for roof operations is questionable. The same kind of risk-versus-benefit analysis that is made before deciding on interior operations should be made before placing fire fighters on or under the roof.

If the roof is weakened and unsafe for fire fighters, working inside the building under the weakened roof is also too dangerous. If the IC or crews working on the roof determine that the roof is not safe, evacuate fire fighters working on and below the roof.

Some departments prohibit roof operations on truss roofs or at specific occupancies as part of their SOPs.

Pre-fire Conditions and Fire Conditions

Several factors must be taken into consideration in evaluating the collapse potential for a building that is under fire attack. These factors fall into two general categories: pre-fire conditions of the building and fire conditions.

Pre-fire Conditions

<u>Pre-fire conditions</u> include the type of construction, as was mentioned previously. In addition, the following pre-fire conditions can contribute to a collapse in a building that is heavily involved in fire:

1. **Weight.** Live and dead loads placed on floors, walls, and roofs can include the following **Figure 5-15**:
   - Air conditioning units
   - Tanks containing liquids
   - Large signs and marquees
   - False fronts (facades)
   - Cantilever appendages
   - Heavy machinery
2. **Fuel loads.** The type, location, and arrangement of combustible/flammable fuel loads inside of the building should be determined.
3. **Damage.** Consider damage to the building's structural support system from previous fires, weather, or collapse.

Figure 5-15 Roof loads.

4. Renovations. Older buildings, especially those that have had adjoining structures removed or have undergone renovations, must be evaluated.
5. Deterioration. Buildings or areas in poor repair, including vacant structures, may have deteriorated to dangerous conditions.
6. Support systems. Buildings with long spans, such as churches and warehouses, should be carefully evaluated for collapse potential.
7. Truss construction. Identify roofs and floors supported by trusses. This is best done during pre-planning as truss construction may not be recognized during fire operations.

Fire Conditions Leading to Structural Collapse

Fire conditions leading to structural collapse are sometimes difficult to read, and structural failures have occurred without warning. However, the proficient fire officer must recognize signs of imminent danger and maintain the span of control necessary to be able to react quickly when fire fighters are endangered **Figure 5-16**. As was mentioned previously, time and fire intensity are major factors. Signs of structural collapse include, but are not limited to the following:

1. Bulging, cracked, or unsupported walls
2. Walls leaking water or smoke
3. Falling bricks
4. Floors holding large volumes of water or stock soaked with water, thereby increasing the weight bearing on the floor (an indicator would be large quantities of water pouring into the building but little water draining out of the building)
5. Movement in floors or roof (if it feels as though it is unsafe to walk on, it probably is)
6. Any other signs of structural movement, including unusual noises
7. Vertical structural members that are out of plumb (columns, walls, etc.)

Figure 5-16 Fire conditions leading to structural collapse.

Fire Extension

Some buildings limit fire spread and contain the products of combustion better than others. As the fire enters concealed spaces, it can extend to remote locations **Figure 5-17**. This **extension** can result in a sudden increase in heat intensity when the fire breaks out of concealment or self-vents. A sudden increase in temperature can occur at a location remote from the original fire and result in fire breaking out at multiple locations. This is especially problematic when resources are limited and there is a need to attack the fire from several positions. Fire fighters performing search-and-rescue operations without a hose line are at particularly high risk, as are units working above the fire floor.

Fire extension can also cut off the primary means of egress for fire fighters. Fires that enter and are confined to concealed ceiling areas are particularly prone to getting behind the fire fighter. The pushing effect of hose streams, especially fog patterns, can accelerate the movement of fire in false spaces. Proper venting will direct the fire out of the building and protect fire fighters; improper venting pulls fire toward fire fighters or into areas behind fire fighters.

Fire fighters must check concealed spaces to determine whether the fire is in these areas. Thermal imaging cameras can be used to find hot spots in concealed spaces, but these areas should be opened up if there is any doubt regarding fire extension.

Suspended ceilings are used to hide open construction methods such as trusses, thus creating a concealed space. In most cases, suspended ceilings are supported by wires attached to lightweight metal grid-work that is known to prematurely collapse under fire conditions. The area between the suspended ceiling and actual ceiling or roof provides a convenient location for lighting fixtures, ventilation equipment, wiring, and other utilities. Heavy light fixtures, fans, and equipment are often located in this area, creating a serious falling-object hazard when the grid-work supporting the ceiling fails. Coaxial television cables, wires, conduit, and the supporting grid-work also present an entanglement hazard.

Truss floor assemblies **Figure 5-18** are also being used to reduce construction costs. The truss floor, like the truss roof, creates a concealed space and is less fire-resistive than heavier solid beam construction.

Table 5-1 compares building construction methods in terms of structural stability, fire extension, and fuel contribution. It is important to remember that the five building categories listed do not represent all possible construction methods. As an example, non-combustible buildings are further subdivided into protected and non-protected. Variations occur within all types of building categories. Buildings being renovated or remodeled may contain structures different than those prescribed by the original construction method. Additions and alterations sometimes result in what can best be described as a hybrid building (a combination of construction types).

The noncombustible building is sometimes mistaken for fire-resistive due to the non-burning characteristics of the structure. The differences are immense. The non-combustible building will normally be masonry or metal on the exterior with lightweight metal trusses as a roof

Figure 5-17 Fire entering truss space in an enclosed room.

Case Summary

In a 1995 fire in a warehouse in Seattle, Washington, most of the building was built of heavy timber structural members. However, one critical support for the floors had been modified by using 2 × 4″ (approximately 50 × 100 mm) wood studs. This support failed between 34 and 36 minutes after arrival on the scene, dropping four fire fighters into the fire below, killing them.

Source: Edward R. Comeau, *Fire Fighter Fatalities, Seattle, Washington, 1995*, NFPA Fire Investigation Report. Quincy, MA: NFPA.

Figure CS5-3 Seattle fire fighters pour water on flames at a warehouse fire on January 5, 1995. The fire trapped and killed four fire fighters.

Figure 5-18 Truss floor assembly.

structure. These buildings often have large open areas with long structural spans. Once the fire enters the truss space, expect imminent roof collapse. Unfortunately, this lesson is being repeated time and time again, and fire fighters are continuing to lose their lives in these buildings. Modern, big-box retail stores, such as Wal-Mart, Sam's Club, or Lowe's, are generally of non-combustible construction with large open areas, but most of these buildings are protected by an automatic sprinkler system. A properly installed and operating sprinkler system will usually control fires in these stores. However, it is important to remember that many code variances or "trade-ups" may have been allowed because the building was constructed with sprinkler protection. Therefore, if the sprinkler system is not controlling the fire in a building of this type, consider the hazards involved in entering this large-span truss space with a heavy fire load.

TABLE 5-1 Comparing Construction Types

Construction	Type	Structural Stability	Fire Extension Probability	Fuel Contribution
Fire resistive	I	Outstanding	Low	Low
Non-combustible	II	Poor	Average	Low
Ordinary	III	Average	Average	Average
Heavy timber	IV	Excellent	Low	Average
Frame	V	Poor	Average to high	High

Case Summary

At the tragic collapse of the Hotel Vendome in Boston, Massachusetts in 1972, nine fire fighters were killed during overhaul operations when five floors in one corner of the building collapsed. During the investigation it was found that a structural modification in a basement wall was a critical factor in the collapse that killed the fire fighters.

Source: "Collapse of the Hotel Vendome, Nine Fire Fighters Killed," *NFPA Journal*, January 1973, pages 34–41.

Figure CS5-4 A victim being removed from the rubble of the Hotel Vendome fire in Boston.

Case Summary

On June 18, 2007 a fire in a big-box furniture store resulted in nine on-duty fire fighter fatalities in Charleston, South Carolina. The furniture store occupied a former supermarket, which had been constructed with concrete block walls supporting a lightweight steel truss roof. The retail area had been expanded by erecting two pre-engineered steel structures, one on either side of the original building. An additional pre-engineered steel building had been constructed as a large rack-storage warehouse at the rear. A covered loading dock and furniture repair area, constructed of wood with sheet metal walls and roof covering, had been added in the space between the warehouse and the rear of the retail store. No automatic sprinklers were provided in any part of the complex.

The fire originated in the loading dock area and quickly extended into the rear of the furniture store, as well as the warehouse. Fire fighters entering through the front of the furniture store encountered light smoke and believed the fire was confined to the loading dock and a small area at the rear of the store. Hand lines were extended through the retail area in an attempt to stop the fire from extending further into the store. Fire fighters were unaware that the fire had entered the void space between the suspended ceiling and the truss roof.

Figure CS5-5 The deadly fire at the Sofa Super Store in Charleston, South Carolina, killed nine fire fighters.

The fire fighters were deep inside the store and became disoriented as it filled with smoke and their air supplies were exhausted. When fire broke through the suspended ceiling, conditions within the store instantaneously became untenable. Nine of the fire fighters inside the store died due to thermal injuries. Catastrophic collapse of the structure did occur, which is always a concern with truss construction; however, based on the investigation and coroner's reports, the nine fire fighters all perished prior to the collapse.

Sources: Charleston.net, accessed 8/28/07. Charleston Fire Department audiotapes of radio communications, Charleston Task Force.

Truss roofs and/or concealed spaces have played a major role in fire fighter fatalities at structure fires. Table 5-2 lists some fires where truss roofs and/or concealed spaces were contributing causes according to fire reports analyzing on-duty fire fighter fatalities.

A complete discussion of structural collapse is beyond the scope of this book. A number of textbooks discuss the different aspects of building design and how they can affect fire-ground operations. *Brannigan's Building Construction for the Fire Service* by Frank Brannigan and Glenn Corbett is considered a companion text to this book.[6] Fire officers should be familiar with the material contained in the Brannigan-Corbett book and how this information applies in their respective jurisdictions.

Fire Zones and Perimeters

Establishing a **collapse zone** is critical when structural stability is the reason for a defensive attack. Construction features combined with fire factors indicate the most probable type of structural failure. Given the fact that the IC is always working with incomplete and imperfect information, it is impossible to accurately predict the type of collapse and resultant collapse zone. The only safe collapse zone is one that is equal to the height of the building plus an allowance for scattering debris. A good rule of thumb for setting a collapse zone for most buildings is to establish an area 1½ times the height of the fire building. This sometimes presents a dilemma as the safe zone is beyond the street width, meaning that effective defensive positions are within the collapse zone.

A risk-versus-benefit analysis is essential. The crucial question that any IC must ask is, "What could I potentially save in relation to the risk being taken?" Obviously, no building is worth a fire fighter's life; therefore, imminent risk to a fire fighter's life to save a building is unacceptable. Also, remember that nothing should be risked to save what is already lost.

When a defensive stand represents a reasonable risk, positions at the corners of buildings are normally safer than those on the flat side of a wall. Consideration should also be given to using unstaffed ground monitors to reduce the risk of placing personnel in exposed positions.

When total collapse is imminent, collapse zones represent **exclusion zones** that no one is permitted to enter, regardless of the level of protective clothing. Exclusion zones can also exist within buildings, especially when roof structures are suspect. In addition, exclusion zones would include other areas containing imminent hazards such as falling glass, areas containing atmospheres within or near the flammable range, and any other area that the IC or safety officer deems too hazardous to enter.

Collapse and exclusion zones are not the only safety considerations regarding access. The concept of limiting

TABLE 5-2 Fire Fighter Fatalities with Truss Roof and/or Concealed Spaces as Contributing Factors

Location	Occupancy	Year	Number of Fire Fighter Fatalities
Hackensack, New Jersey*	Automobile dealership	1988	5
Memphis, Tennessee‡	Church	1992	2
Indianapolis, Indiana	Hotel/office	1992	2
Branford, Connecticut*	Carpet store	1996	1
Chesapeake, Virginia*	Auto parts store	1996	2
Lake Worth, Texas*	Church	1999	3
Texas†	Restaurant	2000	2
Arkansas†	Church	2000	4
Coos Bay, Oregon†	Auto parts store	2002	3
Tennessee†	Retail store	2003	2
Charleston, South Carolina	Furniture store	2007	9

*NFPA Investigative Report available (www.nfpa.org).
†NIOSH Fire Fighter Fatality Investigation available (www.cdc.gov/niosh/fire/).
‡USFA Technical Report Series report available.

access to the fire scene is defined in a variety of ways. It seems appropriate to extend the lessons learned from hazardous materials responses regarding zones, as similar zones are possible at structure fires. In this text, working areas are called zones, whereas a wide area beyond the working zones is known as the **fire perimeter**. The fire perimeter is usually staffed by police, who keep unauthorized people away from the scene, as in an isolation area at a hazardous materials incident. Incident conditions must be considered when determining the dimensions of the fire perimeter. Two blocks in all directions beyond the building on fire is a good rule-of-thumb approximation for the fire perimeter.

Within the **fire zone**, there could be several subdivisions. In most cases, there will be a **cold zone** where personal protective clothing is not required (similar to the cold zone at a hazardous materials incident). The command post, as well as other staff and command functions, would be based in this safe area. The cold zone would also include rehabilitation and medical treatment areas.

The **hot zone** would be an operating area, considered safe only when wearing appropriate levels of personal protective clothing (much like the hot zone at a hazardous materials incident). The IC and safety officer have a responsibility to establish and enforce the hot zone. Everyone has a responsibility to abide by their decision.

It is not always necessary to establish a **warm zone** during a structure fire. A warm zone is established when different levels of protective clothing are needed for various areas. The warm zone is an intermediate zone (between the hot and cold zones). As an example, fire fighters working close to a structure (warm zone) may need to be in full protective clothing, but not breathing air from their facepieces. Members entering the structure (hot zone) would be required to don their facepieces.

An **accountability system** must be established on the fire ground for two purposes. The first is to ensure that

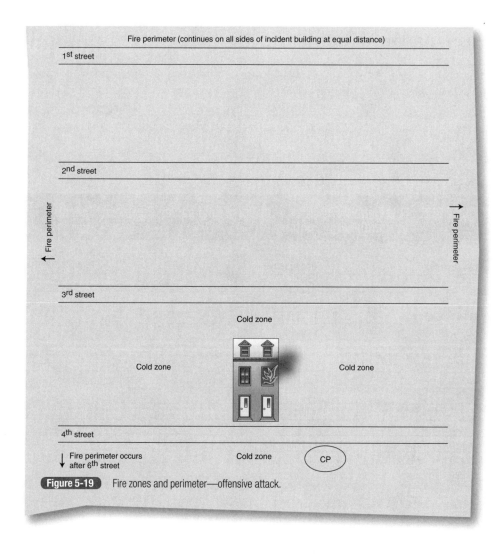

Figure 5-19 Fire zones and perimeter—offensive attack.

everyone entering the area has a specific assignment. This is done to eliminate freelancing on the fire ground, which has led to several fire fighter fatalities.

The second purpose of an accountability system is to be able to track all personnel at the scene and to identify the location of any missing personnel if a catastrophic event should occur, such as a collapse. It also serves to identify whether any personnel or crews are overdue so that search and rescue operations can be initiated.

Figure 5-19 illustrates fire zones with the building's interior being considered the hot zone. Fire fighters entering the building would be expected to be in full turnout gear, breathing from their SCBA. The area on all sides of the exterior would be the cold zone, and the fire perimeter would be maintained by police to keep unauthorized personnel away from the scene.

If this fire were in a larger building, donning the facepiece may be delayed until the fire fighter reaches an area that is or could be an **immediate danger to life and health (IDLH)** atmosphere. As mentioned earlier, this intermediate area is referred to as a warm zone.

If the fire is not contained and an exterior (defensive) attack becomes necessary, the hot zone is moved far enough away from the structure to place fire fighters outside the collapse zone. The collapse zone then becomes an exclusion zone, as shown in Figure 5-20.

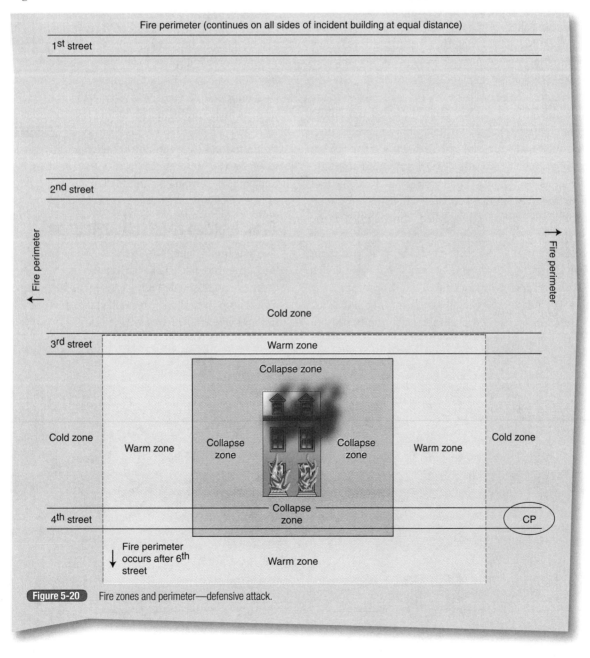

Figure 5-20 Fire zones and perimeter—defensive attack.

Relationship of Time, Fire Intensity, and Structural Stability

Figure 5-21 shows a relationship between time, structural stability, and survivability of occupants. At first, the fire progresses slowly, and the chance of safely exiting the building is very good. The structure is only minimally affected during the **ignition** and early **growth phases** of the fire. Structural stability is compromised as the fire continues to burn through the growth phase to the **fully developed phase**. The longer the fire burns, the greater the risk to fire fighters and occupants, until the point of flashover and/or structural failure. Ten minutes is used for flashover in Figure 5-21 with structural failure occurring later. The area shadowed under the curve represents the temperature rise to flashover, then a temperature decline due to less fuel being available during the **decay phase** of the fire. Structural collapse could occur before flashover, particularly in a large building where localized heating of the area above the fire could result in structural failure before the entire area reaches its flashover temperature. The line marked "structural stability" starts at the top left corner of the chart indicating 100% structural strength. Downward deflection of the line represents a weakening of the structure to the point of collapse where the line angles steeply downward. The line marked "survivability" is the probability of occupant survival. As the smoke and toxic gases reach the floor level, the probability of occupant survival is very low. Occupant survival is improbable in a post-flashover compartment, and even fully protected fire fighters will quickly succumb under flashover conditions. The "survivability" line will always reach the bottom of the chart prior to flashover, as temperatures necessary for flashover are lethal. Likewise, the probability of occupant survival is very low after a structural collapse. A different fuel configuration, a different ventilation profile, or a different structure could radically change the time value and curve configuration shown in Figure 5-21. The principle to keep in mind is that fire growth and time are critical factors.

Brannigan makes an important distinction between building fires and structural fires. He categorizes fires involving the contents of the building as building fires, whereas fires involving actual structural members are considered structure fires. This differentiation is not made here, but the concept is useful. According to Brannigan's terminology, ignition and early growth phase fires are generally building fires, but somewhere during the progression to flashover, the fire generally finds its way into concealed spaces and begins to involve the structure. As mentioned previously, fires entering concealed spaces present many problems, including the potential to harm fire fighters by cutting off their egress. **Table 5-3** lists the time variables from ignition to when effective actions are taken on the fire ground. Each of the variables is subject to change depending on conditions.

Time: Ignition to Effective Actions

The time from ignition to effective actions is critical. The goal is to arrive prior to flashover and intervene, thus interrupting the fire's progression to flashover. Depending on conditions, progression to flashover in small enclosures will be rather fast and will probably occur prior to the arrival of the fire

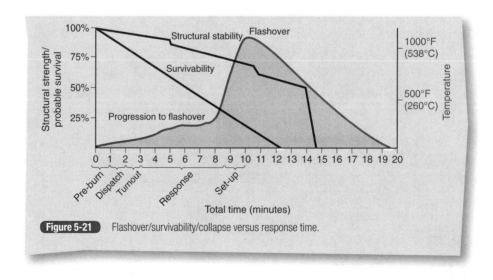

Figure 5-21 Flashover/survivability/collapse versus response time.

TABLE 5-3 Time Components: Ignition to Effective Operations

	Definition	Time Objective
Pre-burn time	From ignition until the fire is reported to the Public Safety Answering Point (PSAP)	Unknown Variable No time objective
Dispatch time	From receipt of the report from the public until fire units are notified	NFPA 1221 75 seconds to 105 seconds*
Turnout time	From when fire units are notified until apparatus leaves the station	NFPA 1710 1 minute
Response time	From when fire units leave the station until they arrive at the scene	NFPA 1710 4 minutes
Set-up time	From when fire units arrive at the scene until fire units take effective action	NFPA 1720 2 minutes

*Based on 15 seconds to answer call, 60 seconds to process, and 30 seconds for remote PSAP operation where the call is not received directly at the dispatch center.

department. The objective then is to contain the fire to the original flashover compartment. In analyzing the time from ignition to effective action it is necessary to consider the five components shown in Table 5-3. These times can vary depending on alarm systems, travel distance, proficiency of the attack team, and other factors.

Detection/Transmission

The first time segment shown in Figure 5-21 and Table 5-3 is the pre-burn time, which is the time from ignition until the fire is reported to the dispatch center. Unless a supervised detection/alarm system is in the fire area, this time component is dependent upon a person discovering the fire and then calling in the alarm. This time will vary greatly. A fire in an occupied area where everyone is awake will probably result in less detection/alarm time than one in an unoccupied building where the fire is not noticed until it shows on the exterior, which will probably be during the fully developed phase. When detection relies on a person detecting then transmitting an alarm, the time can be very long. There are many case histories where a fire burned undetected until it reached sufficient intensity to break through a window or roof. Likewise, fire investigations of large uncontrolled fires often note a delayed alarm, while occupants investigate the source of the fire or try to extinguish the fire before notifying the fire department. If the building is protected by a properly installed and operating supervised detection/alarm system, this time can be estimated based on experience. Otherwise this time component is unknown.

Dispatch Time

The second time segment is the dispatch time. This includes the time for the dispatcher to take the call, select the units for the assignment, and then dispatch the companies. *NFPA 1221: Installation, Maintenance, and Use of Emergency Services Communications Systems*[7] establishes time goals for dispatch centers. When the Public Safety Answering Point (PSAP) is at the communications center, the dispatcher is expected to answer the call from the public in 15 seconds, then dispatch responders within 60 seconds 95% of the time, for a total of 75 seconds (15 + 60) from call receipt to notifying units. *NFPA 1221* continues with a goal of 40 seconds to answer the call plus 90 seconds to dispatch units 99% of the time. When the Public Safety Answering Point (PSAP) is not at the communications center, 30 seconds is added to both goals to allow time to answer the call at the remote PSAP and then call the communications center.

Turnout Time

The third segment is the turnout time. This is the time from the receipt of the alarm until the apparatus crosses the front door sill of the station. Turnout time can differ greatly between fully staffed stations and on-call stations. *NFPA 1710: Standard for the Organization and Deployment of Fire Suppression Operations, Emergency Medical Operations, and Special Operations to the Public by Career Fire Departments*[8] defines turnout time as:

> **3.3.37.5** Turnout Time. The time beginning when units acknowledge notification of the emergency to the beginning point of response time.

NFPA 1710 sets a time objective of one minute for turn-out time:

> **4.1.2.1** The fire department shall establish the following time objectives:
>
> **(1)** One minute (60 seconds) for turnout time

NFPA 1720: Standard for the Organization and Deployment of Fire Suppression Operations, Emergency Medical Operations, and Special Operations to the Public by Volunteer Fire Departments[9] does not address turnout time.

Response/Travel Time

Response time depends on road conditions, terrain, distance, traffic, and other factors. Average response times can be established by using computer models. If computer models are not available, reasonably accurate estimations can be determined by using empirical methods, such as the ISO formula (1.7 × distance + 0.65 = travel time) or RAND average speed of 35 MPH.

NFPA 1710 establishes two response time goals:

> **4.1.2.1** The fire department shall establish the following time objectives:
>
> **(2)** Four minutes (240 seconds) or less for the arrival of the first arriving engine company at a fire suppression incident and/or 8 minutes (480 seconds) or less for the deployment of a full first alarm assignment at a fire suppression incident
>
> **4.1.2.2** The fire department shall establish a performance objective of not less than 90 percent for the achievement of each response time objective specified in 4.1.2.1.

Many fire fighters are killed and injured while responding to incidents. Response/travel time improvements are realized by reducing the distance from the fire station to the response area, not by increasing the speed of the apparatus or ignoring negative right-of-way situations.

Set-up Time

Set-up time is the time necessary to position the apparatus, advance the first hose line into the fire area of the building, and apply water. Staffing levels and training radically affect this part of the response time. The two-in/two-out rule is included in *NFPA 1500* and has been adopted by OSHA.[3] This, in effect, changes the setup time. Now four people must be on the scene, and two must be positioned outside the hazard area.

NFPA 1710 does not address set-up time. *NFPA 1720* states:

> Upon assembling the necessary resources at the emergency scene, the fire department should have the capability to safely commence an initial attack within 2 minutes 90 percent of the time.

NFPA 1410: Standard on Training for Initial Emergency Scene Operations[10] sets training goals ranging from 3 to 6 minutes to establish a water supply and discharge water through two hose lines. It is highly recommended that each fire department time their initial operations during training using their equipment and SOPs. In many cases, a significant amount of time can be saved through practice, providing more accessible equipment and revising SOPs.

These five time segments are valid for the initial response only. Pre-burn time does not figure into on-scene calls for assistance, and the dispatch time is generally less when fire units are calling for help, as the dispatcher does not have to ask questions of the caller. Turnout and response times can also be reduced for subsequent calls for assistance by placing units on alert status, moving them into vacated stations, or placing them in staging. Set-up time will change depending on the task assignment.

Adequate Number of Personnel

Set-up time is related to staffing. If the initial company is staffed with less than four personnel and an imminent life-threatening situation does not exist, the attack should be delayed until additional personnel arrive. As was discussed previously, *NFPA 1500* stipulates a minimum of four fire fighters as an initial crew at a working structure fire. The standard allows less than four in circumstances of an imminent life-threatening situation:

> **8.5.17** Initial attack operations shall be organized to ensure that, if on arrival at the emergency scene, initial attack personnel find an imminent life-threatening situation where immediate action could prevent the loss of life or serious injury, such action shall be permitted with less than four personnel.

From a commonsense point of view, a minimum crew size of four is logical, with three being an acceptable compromise in situations of imminent danger. This is based on one person at the pump panel maintaining a reliable but limited water supply. The pump operator also acts as the outside safety backup during initial operations with a single crew on the scene. The buddy system is required whenever fire fighters enter a burning building, so two fire fighters would be required to enter for fire suppression and/or search and rescue. Suppression is usually the best first action, and it takes at least two people to move and operate a standard attack line. The fourth person would be the hydrant or water supply person, thus providing a continuous water supply. In most cases, the fire fighter who is assigned to the hydrant is the second person outside using the two-in/two-out rule.

There are situations in which one or two fire fighters could start operations by setting up the pump, advancing a line toward the building, or securing a water supply. This would be acceptable under the *NFPA 1500* safety standard, provided that entry was delayed until a team of at least two could enter the building with two remaining outside.

There is also a possibility of a first crew of one or two fire fighters raising ladders to rescue people from the exterior. This would meet the standard, as they would not be making entry into the structure. However, this is a low-probability scenario. OSHA regulations require a two-in/two-out attack configuration for the initial attack when fire fighters enter a fire area but with an imminent life-threatening situation exception. The determination of imminent life-threatening situation quite often is somewhat subjective, thus inviting second-guessing.

Assembling a minimum four-person team is highly recommended when conducting an offensive attack. Exterior attacks may require fewer people, but they are generally not effective in saving occupants and quite often result in a total loss of the building or buildings.

The initial attack is designed to immediately save lives by extinguishing the fire or making a quick rescue. Many additional tasks are required to save lives and protect property. Some of these include the following (in no particular order):

- Additional attack lines to meet flow requirements
- Attack line above the fire
- Attack line to concealed spaces
- Backup for the initial attack line
- Exposure protection
- Forcing entry
- Laddering the building
- Opening up concealed spaces
- Salvage or property conservation
- Search and rescue of the area around the fire
- Search and rescue of the area immediately above the fire
- Search and rescue of other areas
- Utility control
- Ventilation

Not all of these tasks will be needed at every working structure fire, and not all need to be performed by a separate team of two people. The exact number of fire fighters needed to perform these tasks will vary depending on conditions as well as the size and complexity of the building. It is obvious that numerous fire fighters will be needed to attack the fire, perform the primary search, provide backup for inside crews, ventilate, force entry, perform salvage, and, where necessary, protect exposure buildings.

NFPA 1710: Standard for the Organization and Deployment of Fire Suppression Operations, Emergency Medical Operations, and Special Operations to the Public by Career Fire Departments, establishes minimum staffing levels in terms of tasks to be accomplished and the number of personnel needed to accomplish specific tasks.[8]

5.2.4.2.2 The initial full alarm assignment shall provide for the following:

(1) Establishment of incident command outside of the hazard area for the overall coordination and direction of the initial full alarm assignment. A minimum of one individual shall be dedicated to this task.

(2) Establishment of an uninterrupted water supply of a minimum 1520 L/min (400 gpm) for 30 minutes. Supply line(s) shall be maintained by an operator who shall ensure uninterrupted water flow application.

(3) Establishment of an effective water flow application rate of 1140 L/min (300 gpm) from two handlines, each of which shall have a minimum of 380 L/min (100 gpm). Each attack and backup line shall be operated by a minimum of two individuals to effectively and safely maintain the line.

(4) Provision of one support person for each attack and backup line deployed to provide hydrant hookup and to assist in line lays, utility control, and forcible entry.

(5) A minimum of one victim search and rescue team shall be part of the initial full alarm assignment. Each search and rescue team shall consist of a minimum of two individuals.

(6) A minimum of one ventilation team shall be part of the initial full alarm assignment. Each ventilation team shall consist of a minimum of two individuals.

(7) If an aerial device is used in operations, one person shall function as an aerial operator who shall maintain primary control of the aerial device at all times.

(8) Establishment of an IRIC that shall consist of a minimum of two properly equipped and trained individuals.

A minimum of 14 fire personnel would be required to staff the identified positions—15 if an aerial device is in use.

Tactical Reserve

Planning ahead to obtain required resources is crucial. The number of fire fighters needed to safely and effectively combat a working structure fire can be significant. Tactical efficiency can reduce the number of people necessary to fight the fire. For example, proper ventilation will assist the primary search team, allowing it to cover a much larger area with fewer people. This same venting also permits the attack team to extinguish the fire and move to another task more rapidly.

After the IC staffs all of the attack and support positions, there is a need to establish a tactical reserve. The size of the reserve force depends on the stage of the incident and the number of units working, as well as the type of incident. It can be accurately said that if all units are assigned, you need additional personnel.

Correlation Between Elapsed Time and Progression to Flashover

The main point to be made in studying the flashover/survival/collapse chart (Figure 5-21) is that time is a critical factor. The longer the fire has been burning, the less chance there will be to rescue occupants and the greater the chance of structural collapse. From a fire fighter safety viewpoint, the worst scenario is one in which the fire fighter arrives near the end of the buildup to flashover with occupants' lives at risk in a large, undivided area. Once flashover occurs, the chance for survival is near zero in the flashover compartment. The risk to fire fighters has increased because of the fire intensity and the potential for significant structural damage. This is a situation in which risk management is critical to fire fighter safety and survival.

In general, it can be said that the smaller the compartment, the sooner the progression to flashover or **backdraft**. Most of the fuel load in a structure fire will be in the form of solid materials. Ordinary solid materials must be pre-heated to a point where vapor fuel is released through a process called **pyrolysis**. The vapor fuel will ignite if there is an available and sufficient ignition source. However, as the fire progresses, it pre-heats solid fuels that release unignited pyrolytic vapors. If the fire area is enclosed, pyrolytic vapors and products of combustion such as smoke and carbon monoxide will accumulate. If there is sufficient oxygen available when the contained vapor fuels reach ignition temperature, a flashover occurs. If there is a lack of oxygen, the fire could go into the decay phase. However, if oxygen is introduced into this superheated and fuel-rich atmosphere, a backdraft explosion can occur. Fire fighters must use extreme caution when opening doors and windows when conditions indicate a potential backdraft situation.

Small, enclosed rooms quickly consume available fuel and/or oxygen, creating potential backdraft conditions. In a smaller room, the fuel packages tend to be closer to each other, allowing for heat radiation to raise their temperature toward ignition sooner. The volume of a smaller room allows the overall temperature to be raised more quickly than a larger room, decreasing the potential time to flashover or backdraft.

Observing smoke and temperature conditions is critical when entering small, enclosed areas. Sometimes these areas simply burn out before entry. Initially, flashover and backdraft tend to be the most critical problems in small areas, such as in the apartments shown in **Figure 5-22**. The chance of burning all available fuel or using up all the available oxygen is much less in larger areas.

Flashover and backdraft will usually occur much later, if at all, in large fire compartments. However, the concealed space created between suspended ceilings and truss roofs may be a relatively small-volume area owing to the height of the area. In fact, the authors of this text have

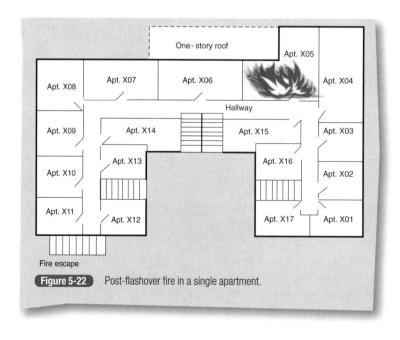

Figure 5-22 Post-flashover fire in a single apartment.

both witnessed backdrafts in these concealed spaces. One such incident, on December 17, 1960, at the Parkmoor Bowling Alley in Louisville, Kentucky, resulted in the deaths of three fire fighters. Two fire fighters were fatally injured in a 6670 ft² retail store area. In a 2004 incident investigated by NIOSH, a possible backdraft contributed to a roof collapse and entrapment when ceiling tiles were removed from below.[11]

Most buildings have more than one enclosed area or compartment. It may be possible to conduct search and rescue operations in other compartments after the compartment of origin has flashed over. As an example, suppose a fire occurs in the apartment building shown in Figure 5-22. The apartment of origin sustains flashover, but surrounding apartments are not yet involved. The fire in the apartment of origin may break out of containment or simply burn out. Survival chances for occupants remaining in the apartment of origin are near zero, but occupants in other areas near the fire can be rescued. If this is a multi-story building, there is a critical need to extinguish the fire and perform post-flashover rescue operations on the floor above the fire.

Fire-ground Operations

Communications

Communications are the lifeblood of any command organization. Without communications, the situation is at best perplexing, and at worst chaotic. At minimum, each crew working inside the structure or inner fire zone must be provided with radio communications to the outside. Some departments issue every member a radio; this practice has been a documented lifesaver on several occasions. Radio discipline can prove to be a significant challenge, but it is imperative when everyone on the scene is assigned a radio. If everyone is talking simultaneously, it will be impossible to communicate assignments or emergency messages. Proper

Case Summary

In Pittsburgh, Pennsylvania, in a residential fire, three fire fighters were missing and trapped inside of the building for over 45 minutes before their bodies were found by crews conducting firefighting operations. No one realized that they were even inside the building because they had not reported their activities to the IC.

Source: Michael Isner and L. Charles Smeby, *One Family Dwelling, Three Fire Fighter Fatalities, Pittsburgh, PA*, NFPA Fire Investigation Report. Quincy, MA: NFPA.

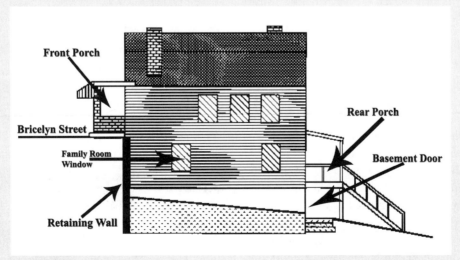

Figure CS5-6 East side of 8361 Bricelyn Street dwelling. This house looked deceivingly typical from the front, but had not only a basement, but also a subbasement, and was built on a sloping grade. This confusing layout was a contributing factor in the deaths of three fire fighters.

Case Summary

In the fire in Seattle, Washington, described earlier in this chapter, the crews who had entered the building were not aware of a subbasement and did not realize that the main body of fire was below them. They thought they were making good progress on the fire and were in the mode of knocking down spot fires. Although every fire fighter was issued a radio, they did not communicate their actions over their radios, but instead provided a progress report to the IC face-to-face when they changed out bottles. If they had reported on the radios, the crew in the basement—which was facing a large fire but not applying any water because of the potential for opposing hose streams—would have realized that there was a significant mismatch between the two situations. Although face-to-face is often the most effective way to communicate information, it is also the IC's responsibility to ensure that everyone involved in the operation is aware of critical information.

Source: Edward R. Comeau, *Fire Fighter Fatalities, Seattle, Washington, 1995*, NFPA Fire Investigation Report. Quincy, MA: NFPA.

---

use of the radio in exchanging important information in a clear, calm, concise manner is essential to good scene management and fire fighter safety.

Frequent progress reports are essential to the IC, who should have a good overall view of the incident. Interior crews and crews working in areas not visible to the IC are the eyes and ears of the IC.

Progress reports also provide everyone on the fire ground with information on other aspects of the fire that relate to their own particular operations.

Some departments use a Mayday code indicating that a fire fighter is in trouble. A special radio alert tone indicating a Mayday message is a worthwhile enhancement. It is important that SOPs spell out the response to a Mayday alert. The tendency is for all crews to stop what they are doing to go to the assistance of fire fighters who are calling the Mayday. If everyone responds to the Mayday call, however, there will be too many people in the area of the Mayday. In most cases, the RIC will be assigned to rescue the fire fighters who are in danger. However, there are some commonsense exceptions, such as when another fire crew is in the immediate area of the problem.

More important, critical functions such as fire suppression will not be accomplished if all personnel begin focusing on the rescue operation. Controlling or extinguishing the fire provides the time needed for the rescue operation. Fire officers must understand the importance of remaining within the command system, and the IC must remain in total control at all times. Specific assignments should be made for the rescue operations so that all tasks can be coordinated.

Departments should establish an emergency evacuation signal. One method is to use apparatus air horns to signal the retreat, such as 10 three-second blasts. Radio channels in use at the scene should also transmit an emergency evacuation message. Again, SOPs should be in place describing actions to be taken during an emergency evacuation. Just as the hose line controlling the fire buys critical time to accomplish a rescue, it can also provide additional time to evacuate and protect fire fighters above the fire. Only in rare cases is a "drop everything and run" evacuation warranted. The "drop everything and run" tactic is more often used during defensive operations. For offensive operations, an organized retreat is generally a better alternative.

Command and Control

Command and control was the subject of Chapter 1, but it is important to point out that the National Incident Management System (NIMS) is also a safety system. Taking measures to have everyone working toward a common goal in an organized fashion is the basis of a safe and effective operation. Conversely, freelancing leads to injuries and fatalities. NIMS is the foundation for accountability and rapid intervention procedures.

Accountability

A good organizational structure accounts for all personnel operating at the scene and assigns responsibility for each crew with a reasonable span of control via sectoring. Crew unity is an essential part of this process. Crew members should not be separated within the structure. The company officer should be able to account for all members of his or her crew working within the hazard area. Splitting crews inside the hazard area is inviting disaster. However, during offensive operations, it is acceptable to have an apparatus operator or other crew members working outside the hazard area while the officer and remaining crew members are inside the structure.

Many department SOPs allow the first-arriving truck company crew to split into ventilation and search and rescue teams. This should be a temporary arrangement, with someone supervising each crew. Each of the split crews should have a radio to enable them to provide progress reports, receive assignments, and call for emergency assistance if needed. Even then, the crews should reunite as a company as soon as their initial assignments are finished. Except in situations of imminent life-threatening circumstances, splitting a crew to provide a two-out team during initial operations is required by *NFPA 1500* and OSHA regulations when only one company or unit is on the scene and entry is made into the structure.

In addition to the accountability provided by NIMS, a separate system is needed to track and account for everyone at the scene Figure 5-23 . Several systems are in use, and most are effective if they are properly used. High-tech tracking systems are currently available and in use, primarily by the military. The current challenge, which is being addressed by the Metropolitan Fire Chiefs Association and similar organizations, is to transfer the available technology to the private sector at a cost the fire service could afford. Computerized card readers are being used by some departments. However, most accountability systems use roster sheets, personal tags, or labels. These accountability systems provide a means of gathering roster sheets or tags and establish a physical entry and exit location for the collection of accountability rosters or tags under the direction of an accountability officer(s).

If a situation deteriorates rapidly, the IC calls for a PAR. SOPs generally call for PARs in the following situations:
- When the IC thinks it is necessary
- When the safety officer requests one
- When the IC changes from an offensive to a defensive attack
- When sudden changes occur, such as backdraft, flashover, or collapse

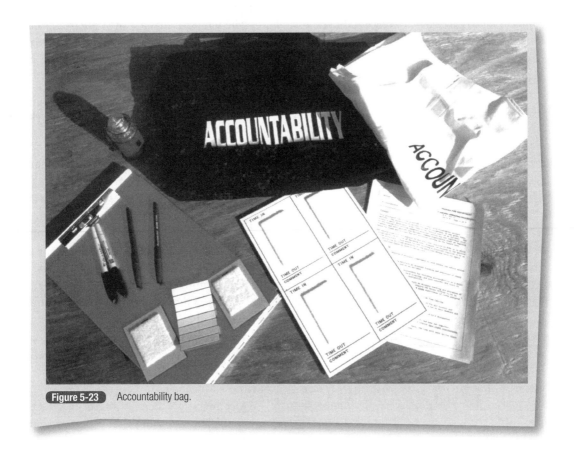

Figure 5-23 Accountability bag.

Case Summary

A fire in Washington, D.C., that killed an officer demonstrates the need to maintain control during these critical operations. The officer fell through the floor into the basement of a convenience store. As personnel were rapidly exiting the building because of the deteriorating conditions, it was realized that the officer had not come out with his fire fighter. Ad hoc emergency rescue operations were put together by another officer, and several entry attempts were made without any coordinated plan. The IC was not notified of the missing officer until after several unsuccessful rescue attempts had been made.

Source: District of Columbia Fire and Emergency Medical Services Department, Report from the Reconstruction Committee, *Fire at 400 Kennedy Street, NW, Washington, DC.*

- When the entire building has been searched
- When the fire is extinguished
- At prescribed times, such as every 10 minutes when dispatch notifies the IC of the elapsed time as prescribed in *NFPA 1500*

This routine request for a PAR either at a specific, designated time or whenever there is a change in tactics, reinforces the concept of personnel accountability and ensures that officers are aware of the location of the personnel assigned to them.

Because accountability procedures become more important as the incident increases in size and complexity, it is very likely that mutual aid resources will be at the scene. Therefore, a regional approach to an accountability system is the only logical choice.

The safety officer should not be the accountability officer. The safety officer position is mobile. The accountability officer should be at a fixed location. The safety officer must enter into the accountability process by keeping the accountability officer informed of his or her location. Likewise, the accountability officer is a valued informational resource for the safety officer, RIC, and IC.

Safety Officer

There is no exact time or situation that requires the handing off of the safety officer position at a structure fire. Each department should develop SOPs outlining when this staff position must be established. In general, the safety officer should be separately staffed whenever the IC believes that he or she can no longer effectively monitor safety at the scene. This situation usually happens fairly early, as the IC should be located at a fixed command post. The safety officer monitors all areas where fire fighters are working. The IC should be stationary, while the safety officer is mobile.

Alternative Egress

Most large buildings have multiple egress routes, as required by building and fire codes. When this is the case, the IC should send crews into each stairway for the purpose of managing the stairs. In high-rise fire operations, lobby control (see Chapter 12 for a complete explanation of lobby control) manages the stairs and elevators. This procedure will help to ensure that alternative routes that can be used during an emergency are identified and under control of the fire department. The lobby control function can also be useful at buildings that are not technically classified as high-rise structures.

The interior stairs are the preferred means of access and egress, but they are sometimes untenable or inadequate. When available, fire escapes provide additional means of escape and can also provide fire fighters with access to, and escape routes from, upper floors of the building **Figure 5-24**.

> Safety is everyone's responsibility at all times at the fire scene, whether or not someone has been specifically assigned the role of safety officer.

Fire escapes would be a second preferred choice of egress/access, because using fire escapes requires fewer fire fighters than positioning fire department ground ladders, and, if properly maintained, they also are safer. The building shown in **Figure 5-25** has two fire escapes and three stairways and thus provides alternative egress from every location within the building. However, if the use of a fire escape is included in the pre-incident plan, the fire escape should be thoroughly inspected on a regular basis and its capacity should be noted. A proper inspection oftentimes requires more than a visual inspection, because the support structure may be imbedded in the building or otherwise hidden.

A common practice in the past was for the first-due truck company to "claim the building" by placing a ladder

Figure 5-24 Building with exterior fire escape.

somewhere (anywhere) on the building. This practice had a fundamental purpose of providing an alternative means of egress for occupants and fire fighters. Unfortunately, the purpose was often forgotten and the practice was eliminated as staffing on truck companies was reduced.

Fortunately, this practice is making a comeback, with many departments now "laddering the building" to provide an alternative means of escape for fire fighters working inside a burning building. Properly laddering the building is an important task that should be accomplished early in the

Figure 5-25 Secondary means of egress from second floor.

operation. Placing a ladder to the one-story roof area shown in Figure 5-25 would provide an alternative egress from several apartments on the second floor of this building. SOPs should address laddering the building, possibly calling for ladders to be placed at predictable locations, such as the center window at the rear of the building or at the end of hallways. If ladders cannot be placed at a prescribed location, or if SOPs do not address where ladders are to be placed, it is important for exterior crews placing ladders to communicate the location of ladders to the crews working on the interior.

Just as codes require two separate and distinct means of egress, there should be provisions for a second way out for fire fighters. The alternative egress should be remote from the first. Access to the roof is generally by ladder, but there should also be a second means of egress from the roof, usually a second ladder placed remote from the first. Some departments assign laddering the building to the RIC.

Rapid Intervention Crews

Staffing is not sufficient until all safety and tactical positions are covered, and a reserve force is available for deployment. There is a critical need to provide rescuers for the fire fighters. It is important to point out that the RIC is not a substitute for safe and effective operations. Formulating an incident action plan based on a risk-versus-benefit analysis, combined with good tactics and company-level attention to safety, will reduce the need for emergency rescues. Tactics such as extinguishing the fire and venting the building can greatly reduce the possibility of fire fighters being trapped or disoriented.

Safe interior operations include:
- Maintaining crew integrity with at least two members working together and in contact with one another
- Providing hose line protection in areas where fire fighters are working
- Providing a means of communications to the exterior and among units working in an area
- Maintaining contact with the hose when operating a hose line
- Maintaining contact with a wall or rope when operating without a hose line
- Placing a fire fighter, signal light, or audible signal at the door leading to the room where fire fighters are working

Despite these measures, the unexpected can occur. Fire fighters must be taught self-survival techniques including how to operate the emergency features included on the SCBA and PASS.

When a fire fighter or group of fire fighters suspect they are in trouble and might need assistance, they should immediately transmit a call reporting their situation. When requesting assistance, fire fighters should give their exact location, if possible, or if they do not know their exact location, give the best estimate, e.g., we are located on the second floor near the rear. State the problem, such as out-of-air, disoriented, trapped in collapse, entanglement, or fell through floor or roof. Also state the number of fire fighters needing assistance and company number. Even after requesting assistance, it is sometimes possible to continue self-rescue efforts or take measures that could extend survival time, such as finding a source of outside air at a window or under a door. If the fire fighters requesting assistance are merely disoriented and not in imminent danger, it may be best for them to activate their PASS and conserve air while awaiting assistance.

The PASS device should be activated as soon as assistance is needed or thought to be needed. It is possible to increase the efficiency of the PASS by directing it away from the fire fighter's body. Trapped members can sometimes indicate their location by breaking a window, tapping on doors, directing a light around the room, or otherwise signaling their location. A team of fire fighters in Cincinnati depleted their air supply deep inside a smoke-filled warehouse/manufacturing building. They were able to extend their survival time by placing their breathing hose at a crack at the bottom of a metal overhead door that was locked on the exterior. They used a spanner wrench to bang on the door to indicate their location. They were eventually rescued and survived.

In the initial stages of an operation, with only one crew operating at the scene, the pump operator and one additional fire fighter are generally assigned as the RIC. This outside crew must know the position of the fire fighters working inside the structure. Later in the operation, a minimum of two (preferably four or more) fire fighters with full turnouts are assigned as the RIC. This crew must be available to immediately rescue fire fighters should the need arise. The company officer in charge of the crew must maintain the crew's readiness at all times.

As can be seen in Figure 5-26, becoming caught or trapped is the cause of injury in nearly half of all fire-ground fatalities, and many of these fire fighters could have been rescued by a properly staffed RIC.

The RIC should assemble necessary tools and equipment as prescribed in department SOPs. This tool list will be based on department operations, tools available, types of buildings protected, and other factors. The RIC must have SCBA with connections to supply air, as stated in *NFPA 1500*:

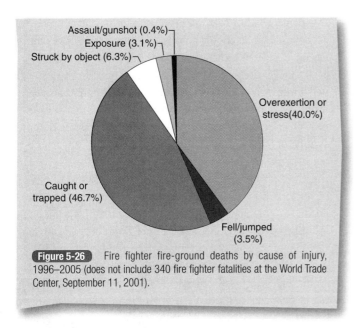

Figure 5-26 Fire fighter fire-ground deaths by cause of injury, 1996–2005 (does not include 340 fire fighter fatalities at the World Trade Center, September 11, 2001).

8.8.2.2 The rapid intervention crews/companies (RIC) at an incident where any SCBA being used are equipped with a RIC universal air connection (UAC) shall have the specialized rescue equipment including a fully charged breathing air cylinder with a NIOSH-certified rated service time of at least 30 minutes and compatible pressure and capacity with the SCBA being used at the incident, or a high-pressure air line of sufficient length to reach the location of the entrapped or downed fire fighter(s) and supplied by a pressurized breathing air source that can provide at least 100 L of air per minute at the RIC UAC female fitting and at a pressure compatible with the SCBA being used at the incident.

Tools commonly carried as part of the RIC cache include:
- Rescue ropes, search ropes, guideline ropes
- Thermal imaging camera
- Patient carrier, webbing, or harness
- Portable ladders for above- and below-grade rescues
- Forcible entry tools
- Wire cutters and other hand tools
- Lighting equipment

A thermal imaging camera is an indispensable tool when the RIC does not know the exact location of fire fighters needing assistance. The RIC officer should determine whether the situation calls for tools that may not be listed as part of standard RIC equipment. As an example, a building equipped with hardened security bars might require a special saw or cutting torch. The construction type, occupancy, fire location, and other factors will also indicate the type of rescue that may be needed as well as the tools necessary to perform the rescue. Additional tools and equipment needed because of building features could be known in advance and thus identified in the pre-incident plan.

A RIC that is assigned non-RIC tasks is at a severe disadvantage. The dedicated RIC can determine the best access locations, assemble tools and equipment, and generally pre-plan potential rescue operations according to conditions. They are also immediately available for assignment with full breathing apparatus. The dedicated RIC monitors conditions and provides assistance to the safety officer.

The availability status of the RIC poses a dilemma. A RIC stationed at a central location near the tool cache would be able to immediately respond with tools and equipment needed to rescue fire fighters calling for assistance. However, there is a benefit in having the RIC do a complete walk-around to determine the location of alternative means of ingress and egress and other building features that should be considered if a Mayday occurs. While doing a walk-around, the RIC can also improve egress by forcing doors (being careful that the means of doing so does not create an unwanted vent opening), placing ladders, and removing obstacles. A mobile RIC can recon the building and provide valuable information to the IC. Mobile and stationary RICs both have advantages and disadvantages. Some departments opt for a combination of the two by assigning some RIC members to recon with others remaining at the equipment cache location. When writing RIC procedures, it is important to consider the advantages and disadvantages of each based on RIC staffing and other factors in the response area. If a pre-incident plan is available for the building, the RIC should have access to that plan.

The U.S. Fire Administration conducted a study of fire department RIC operations.[12] Eighty-three (mostly career) departments responded to the study. The number of members assigned to the RIC (when a RIC is established) and RIC operations vary greatly. For most departments, formal RIC operations are relatively new. The fire service is still learning the best procedures to use when rescuing our own. Even the name RIC is not standardized. Crews assigned to rescue fire fighters are sometimes titled RIT (Rapid Intervention Team), RAT (Rapid Action Team), FAST (Fire fighter Assist and Search Teams), and Go Teams, among others.

Serious questions have been raised regarding the need for additional staffing for the RIC. Most departments now use the two-out team as an initial RIC but assign a full fire company as a RIC as the operation continues. Experience indicates that a two-person RIC is inadequate except in situations where the fire fighters are easily located and able to escape with minimal assistance. Even a four-person RIC may not be enough in a large building or in situations where fire fighters need to be extricated.

When notified that a fire fighter needs assistance, the RIC must locate, extricate, remove, assist, and provide medical attention for fire fighters trapped inside a building. The sooner the RIC reaches trapped fire fighters and is able to get them to a place of safety, the better the chances of survival. In a very large structure, this could require establishing multiple RICs located at various locations around the building.

Some departments categorize the RIC operation as a rapid removal or an extended operation. Rapid removal would normally consist of finding and assisting fire fighters to safety. In many cases, simply assisting a disoriented fire fighter to a safe area is all that is needed. Other times, a fire fighter may be unconscious or debilitated, but located very near a safe area or near the exterior. In these cases, it may be best to quickly remove the fire fighter rather than trying to establish an air supply and performing other special rescue techniques. Extended operations may require two or more RICs. The first RIC would make every effort to locate fire fighters needing assistance while a second team stands ready to extricate the fire fighters once they are found.

Fire departments may or may not require RICs to have a hose line as part of their standard procedure. Extending a hose line to the area where fire fighters need assistance can substantially delay search and extraction operations. However, it may be necessary to protect trapped fire fighters or to provide a path to the exit by extinguishing the fire. Fire Department New York assigns a truck or rescue company as the RIC and an engine company as the Engine-RIC. The Engine-RIC provides hose line protection. Other departments assign more members to the RIC team to provide sufficient personnel to find, extract, and protect the fire fighter rescue. If the RIC deploys a hose line or another unit is assigned to protect RIC operations, it is best to provide a separate water supply for the RIC hose line, as hose line failure may be the reason for the call for assistance.

In a high-rise operation on an upper floor, the RIC and RIC equipment should be located one or more floors below the fire floor so RIC members do not expend their air supply while on stand-by, but are near areas where fire fighters are working.

It is critically important to train and practice RIC procedures. Many times, training under simulated fire conditions will point out deficiencies in procedures that can be changed to improve RIC operations. During training sessions, it is essential to develop a variety of scenarios in different buildings and use full protective clothing as you would while performing a rescue.

Opposing Fire Lines

Hose streams or fire lines are important safety tools, but they can also do much harm when the fire is pushed toward fire fighters or victims. The IC assigns units so that everyone is working either offensively inside the building or in a defensive

Case Summary

In Phoenix, Arizona, a fire on the exterior of a large supermarket extended into a concealed space between the ceiling and roof in the storage area. A crew of fire fighters entering the store encountered worsening smoke conditions as they advanced. The low air supply warning sounded on two SCBAs, causing the crew to retreat. As they retreated, three members lost contact with the hose line that they had advanced into the store, and they became disoriented. One member was able to reestablish contact with the hose and find his way out. The other two fire fighters remained disorientated. One of the disoriented fire fighters heard another crew nearby and followed the sound to their location and ultimately to safety. The third member was found, but several crews were required to remove him, thus increasing his time in the hazard area. He was transported to the hospital, where he was pronounced dead.

Figure CS5-7 Debris fire fighters encountered while trying to find and rescue a missing fire fighter. Fire fighters needing rescue in large commercial structures may require more staffing and time to find and remove.

Source: Robert F. Duval and Stephen N. Foley, *Supermarket Fire, Phoenix Arizona, March 14, 2001, 1 Firefighter Fatality.* NFPA Fire Investigative Report. Quincy, MA: NFPA.

mode outside. Master streams improperly operated on the exterior will push fire into the building, endangering anyone inside. Interior lines improperly operated from one side of a compartment will push fire toward the other side. If fire fighters or victims are in the fire's path, they will experience a rapid increase in heat and smoke. It is imperative that the overall strategic plan avoid opposing hose lines at all costs because of the significant danger this improper tactic creates for personnel inside the building. Whenever possible, all interior hose lines should attack from the same access point. If this is not possible or practical, communications between units attacking the fire is essential.

 Figure 5-27 shows a fire in an enclosed space. Enclosed fires will naturally consume available oxygen and fuel, gaining in intensity and moving upward until a barrier is encountered. The fire will move horizontally when it cannot proceed upward and finally will move downward. As long as the fire is enclosed, it will continue to increase in size until it consumes the fuel or reduces the oxygen below the concentration needed for flaming combustion. In Figure 5-28 , a door is open, and a fire stream is applied through an open window. Thus, the fire is no longer enclosed, and the fire stream is affecting the movement of the fire. In the case shown in Figure 5-28 the door provides a natural pathway for the fire, and the fire stream accelerates the movement of the fire through the open door. If this door leads to a stairway or hallway that is being used for evacuation, the occupants trying to escape and the fire fighters who are assisting them are placed at great risk.

Personal Protective Clothing

The IC should establish the level of protective clothing necessary to enter fire zones (hot, warm, and cold). The safety officer should be consulted on this issue.

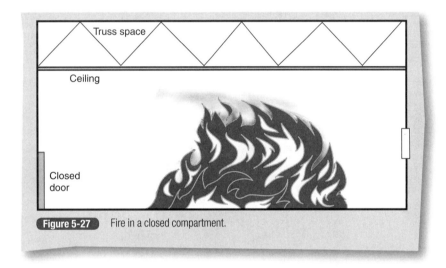

Figure 5-27   Fire in a closed compartment.

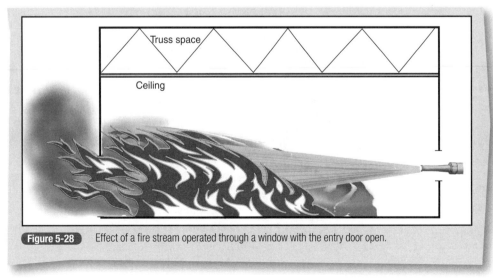

Figure 5-28   Effect of a fire stream operated through a window with the entry door open.

A tendency to dress down during the overhaul phase of an operation can lead to unnecessary injuries. Even after airing out the structure, removing the SCBA facepiece is questionable. Residual burning materials will produce toxic gases, and concentrations of fire gases will be present for a considerable time within dead air spaces.

In a study conducted by the National Institute of Occupational Safety and Health (NIOSH),[13] air monitoring was conducted at actual fire scenes immediately following fire suppression. The focus of the study was on the airborne health hazards to which fire investigators would be exposed. According to the study:

> Formaldehyde concentrations exceeding the NIOSH ceiling limit of 0.1 ppm [parts per million] and exposures to several PAHs [polycyclic aromatic hydrocarbons] (which are suspected of having carcinogenic potential in humans) were measured. This indicates that fire investigator exposures to irritants which cause acute effects and carcinogens which have chronic effects are of concern. Airborne contaminants, such as asbestos polycarbonated biphenals should also be suspected. The burning contents found in structure fires today include more plastics. Plastics produce deadly fire gases including cyanides and hydrogen chloride (HCl).

NFPA 1500 addresses the use of SCBA in stating:

> **7.9.7** When engaged in any operation where they could encounter atmospheres immediately dangerous to life or health (IDLH) or potentially IDLH or where the atmosphere is unknown, the fire department shall provide and require all members to use SCBA that has been certified as being compliant with NFPA 1981, *Standard on Open-Circuit Self-Contained Breathing Apparatus for the Fire Service*.

> **7.9.8** Members using SCBA shall not compromise the protective integrity of the SCBA for any reason when operating in IDLH, potentially IDLH, or unknown atmospheres by removing the facepiece or disconnecting any portion of the SCBA that would allow the ambient atmosphere to be breathed.

After a working structure fire and subsequent operations to gain access to concealed spaces, there are often many other hazards and dangers. Therefore, removing any part of the turnout ensemble inside the fire building is an unsafe practice.

Overhaul

The time between suppression and final <u>overhaul</u> is an ideal time to rotate personnel through rehabilitation while the building airs out. This recommended rehabilitation between active suppression and final overhaul is not meant to preclude the need for earlier rehabilitation when gaining control of the fire takes longer. The IC should use the time between active firefighting and overhaul to develop an overhaul plan (a non-emergency incident action plan). Safety considerations include:

- Structural damage/structural stability
- Smoke and airborne contaminants
- Cutting hazards (broken glass, jagged metal, protruding nails)
- Holes in floors
- Damaged stairways
- Utility hazards (gas, oil, electric)
- Overhead hazards (damaged ceiling, loose materials)
- Visibility (provide lighting or wait until there is adequate natural light)

When the pre-overhaul inspection reveals imminent hazards or risks that cannot be determined, overhaul should be delayed. On some occasions, overhaul is conducted using heavy equipment. The building shown in **Figure 5-29** was heavily damaged by fire, and the structural stability could not be determined. Master streams remained in place for several days with the area being secured. Eventually, heavy equipment was used to raze the building.

When conditions do not represent an imminent hazard, the identified hazards should be discussed with crews working in that area. If possible, mark the hazardous conditions using barrier tape, ladders, and other means. See Chapter 10 for more information on overhaul.

Figure 5-29 Delayed overhaul due to structural damage. This photo was taken approximately two months after the fire occurred.

Rehabilitation

NFPA 1584: Recommended Practice on the Rehabilitation of Members Operating at Incident Scene Operations and Training Exercises[14] provides guidelines for hot and cold weather **rehabilitation**.

Rehabilitation can be divided into three phases:

1. Pre-incident hydration and preparation
2. Incident rehabilitation
3. Post-incident recovery

Thirst is a sign of mild dehydration; members should not wait until they are thirsty to ingest liquids. While awaiting calls, members should pre-hydrate by drinking 6 to 8 ounces of water every six hours plus liquids that are ingested with meals. This is especially important during hot and humid weather conditions. However, water or sports drinks should not be ingested in extreme quantities. Members should avoid activities that cause extreme fatigue or exhaustion while waiting for a response.

On-scene rehabilitation can be formal or informal depending on weather conditions, length of time on the scene, and activity level. Working structure fires tend to require very high energy levels, thus they cause fatigue at a rapid rate. Informal rehabilitation usually takes place at the company apparatus. An apparatus used for rehabilitation should be in the cold zone so members can "dress down" while resting and rehydrating. Resting members should not be placed in areas where they will be breathing exhaust fumes. Formal rehabilitation involves setting up an area for rehabilitation. This area should provide shade and mechanical cooling equipment during hot weather and a warm location (possibly using portable heaters) during cold weather. The formal rehabilitation area must have water or sports drinks available. If members are on the scene for an extended period of time, healthy food should also be provided (see *NFPA 1584* for guidelines). **Figure 5-30** shows the layout of a typical rehabilitation/treatment area.

NFPA 1584 guidelines for hot weather rehabilitation are:

> **6.5.1** In extreme heat conditions, the rehabilitation process should provide for the following:
>
> **(1)** Members removing all protective clothing
>
> **(2)** Fluid and food to replace water, electrolytes, and calories lost during the incident
>
> **(3)** A shaded or misted area for initial cool-down of members, with fans to create air movement, if necessary
>
> **(4)** An air-conditioned area for extended rehabilitation to which members can be moved after their body temperatures have stabilized in the initial cool down area
>
> **(5)** Medical evaluation and treatment for heat emergencies per local EMS protocols
>
> Water or sports drink intake should be in the range of 12 to 32 ounces during a 20 minute rehabilitation.

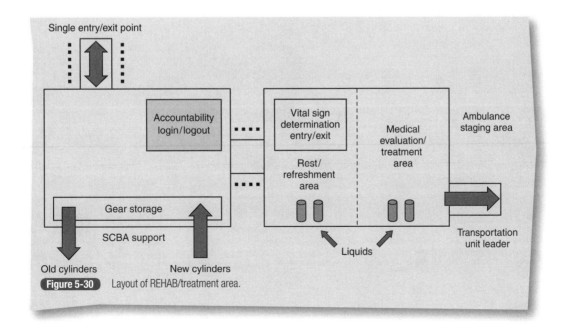

Figure 5-30 Layout of REHAB/treatment area.

Recent studies recommend active cooling techniques such as forearm emersion.[15]

For cold weather rehabilitation, *NFPA 1584* recommends:

> **6.5.2** In extreme cold conditions, the rehabilitation process should provide for the following:
>
> **(1)** A dry area shielded from the wind or other elements
>
> **(2)** Fluid and food to replace water, electrolytes, and calories lost during the incident
>
> **(3)** Members removing wet protective clothing and wet garments and donning dry clothing
>
> **(4)** A heated area for extended rehabilitation
>
> **(5)** Medical evaluation and treatment for cold emergencies per local EMS protocols. Departments should have an emergency medical system (EMS) unit on the scene of a working structure fire, preferably an advanced life support (ALS) unit.

As noted in the *NFPA 1584* guideline, an additional EMS unit, preferably an ALS unit, should be assigned to rehabilitation when a formal rehabilitation is implemented to evaluate fire fighters entering and exiting rehabilitation (REHAB). This unit's responsibility is to determine when members can return to action, as well as treating injuries and transporting personnel when necessary. Rehabilitation protocols should be spelled out in department SOPs. Many departments use a two-cylinder rule, which requires REHAB after fire fighters have exhausted a second 30-minute SCBA cylinder. The rest period after using two cylinders should be at least 20 minutes. During temperature extremes, REHAB becomes a more critical issue, and longer rest periods may be needed.

A sheltered REHAB area is preferred, and fluid replacement is required **Figure 5-31**. Before fire fighters are released from the REHAB area to return to active firefighting duties, their vital signs must be checked and be at safe levels. These can include pulse, respiration, and temperature. The department physician should be consulted when establishing re-entry guidelines.

Maintaining company unity and the accountability system is important as members are cycled through REHAB.

Figure 5-31 Rehabilitation area.

If a company member cannot be returned to duty, the accountability system board or ride list should be revised accordingly.

Critical Incident Stress Management

Critical incident stress management is related to REHAB, as it is best to take action to prevent or reduce stress at the incident scene. Much has been written about critical incident stress. Fires in which fire fighters are seriously injured or killed are obvious cases for critical incident stress management. Fires in which children are seriously injured, or a fire that results in one or more fatalities, could be reason to summon the critical incident stress debriefing team. ICs must be cognizant of symptoms of critical incident stress and take measures to mitigate the problems as soon as possible. Delayed signs of critical incident stress are many times masked as other illnesses. However, immediate assistance should be provided to emergency response personnel who display physical or mental abnormalities after an incident. Some signs of critical incident stress that may be observed at the scene of an emergency include the following:

- Shaking or trembling
- Loss of muscular control
- Blurred vision
- Respiratory difficulties
- Confusion and disorientation
- Chills
- Signs and symptoms of shock (weak rapid pulse, pale and clammy skin, sweating, nausea, falling blood pressure)

Some proactive measures that can be taken at the emergency scene, particularly at a stressful event, to diminish critical incident stress are as follows:

- Schedule breaks as often as possible at a designated rehabilitation area.
- When operations extend beyond one hour, rotate frontline personnel even if they do not want to be relieved.
- Check personnel in the rehabilitation area for signs and symptoms of critical incident stress.

Summary

This chapter has primarily discussed what needs to be done to keep the fire fighter from becoming a victim. It should come as no surprise that many of these same tactics are necessary to save civilians. The primary rescue technique in a structure fire is simply extinguishing the fire while venting the heat, smoke, and toxic gases. Once the fire is extinguished, the structure becomes safer for fire fighters and occupants. Improved fire fighter protective clothing, personal alert safety systems, rapid intervention crews, accountability systems, NIMS, rehabilitation, and other measures are now in place to reduce the number of fire fighter injuries and deaths.

Wrap-Up

Key Terms

20-minute rule When a heavy volume of fire is burning out of control on two or more floors for 20 minutes or longer in a building of ordinary construction, structural collapse should be anticipated.

accountability system Established on the fire ground to ensure that everyone entering the area has a specific assignment, to track all personnel at the scene, and to identify the location of any missing personnel if a catastrophic event should occur.

backdraft Occurs when oxygen (air) is introduced into a superheated, oxygen deficient, compartment charged with smoke and pyrolytic emissions, resulting in an explosive ignition.

cold zone An area where personal protective clothing is not required, where the command post, rehabilitation, and medical treatment are based.

collapse zone The area endangered by a potential building collapse, generally considered to be an area 1½ times the height of the involved building.

dead load The weight of a building; consists of the weight of all materials of construction incorporated into a building, including but not limited to walls, floors, roofs, ceilings, stairways, built-in partitions, finishes, cladding, and other similarly incorporated architectural and structural items, as well as fixed service equipment.

decay phase The phase of fire development in which the fire has consumed either the available fuel or oxygen and is starting to die down.

exclusion zone An area that is unsafe regardless of the level of personal protective equipment and which must be cleared of all personnel, including emergency response personnel.

extension Fire that moves into areas not originally involved, including walls, ceilings, and attic spaces; also, the movement of fire into uninvolved areas of a structure.

fire perimeter A wide area beyond the working zones, usually staffed by police to keep unauthorized people away from the scene.

fire zone The area where emergency responders are working and which can be subdivided into exclusion, hot, warm, and cold zones.

flashover An oxygen-sufficient condition where room temperatures reach the ignition temperature of the suspended pyrolytic emissions, causing all combustible contents to suddenly ignite.

frequency As related to fire fighter injuries, a measure of how often an injury occurs; e.g., sprains and strains are the most frequent fire fighter injuries.

fuel load Fuels provided by a building's contents and combustible building materials; also called fire load.

fully developed phase The phase of fire development where the fire is free-burning and consuming much of the fuel.

growth phase The phase of fire development where the fire is spreading beyond the point of origin and beginning to involve other fuels in the immediate fire area.

hot zone An operating area considered safe only when wearing appropriate levels of personal protective clothing; established by the IC and safety officer.

ignition phase The phase of fire development where the fire is limited to the immediate point of origin.

immediate danger to life and health (IDLH) Describes any atmosphere or condition that poses an immediate hazard to life or is capable of producing immediate, irreversible, debilitating effects on health.

live load The weight of the building contents, people, or anything that is not part of or permanently attached to the structure.

overhaul Examination of all areas of the building and contents involved in a fire to ensure that the fire is completely extinguished.

personal alert safety system (PASS) A device that continually senses a lack of movement of the wearer and automatically activates the alarm signal indicating that the wearer is in need of assistance. The device can also be manually activated to trigger the alarm signal.

pre-fire conditions Factors that can be contributing factors to a collapse in a building that is heavily involved in fire, and which include construction types, weight, fuel loads, damage, renovations, deterioration, support systems, and related factors such as lightweight truss ceilings and floors.

pyrolysis The chemical decomposition of a compound into one or more other substances by heat alone; the process of heating solid materials until combustible vapors are emitted.

rehabilitation The process of providing rest, rehydration, nourishment, and medical evaluation to members involved in extended incident scene operations and/or extreme weather conditions.

severity The extent of an injury's consequence, usually categorized as death, permanent disability, temporary disability, and minor.

warm zone Intermediate area between the hot and cold zone where personal protective equipment is required, but at a lower level than in the hot zone.

Suggested Activities

1. Use the most recent NFPA fire fighter fatality report to determine the type of duty as well as the nature and cause of injury most likely to result in an on-duty fire fighter fatality.

2. Use the most recent NFPA fire fighter injury report to determine the most common cause of injury. Using a 1 to 10 scale where 10 is most severe, estimate the probable severity for various injury types (e.g., heart attack on the fireground = 10).

3. Use a fire report from NFPA, NIOSH, USFA, or other source to evaluate the risk to fire fighters conducting an offensive attack. Classify the incident using the four risk/benefit categories from *NFPA 1500-8.3.2*.

4. Develop a fire scenario for a pre-planned building in your jurisdiction where fire fighters become disoriented. Construct a simulated emergency message to be transmitted by the disoriented fire fighters.

5. List and describe examples of imminent life-threatening situations at a structure fire that would justify entering the building prior to establishing a two-out team.

6. Use the scenarios below to analyze and discuss the probability of the building described being occupied at the time of the alarm. Express the probability of occupancy as a percentage where 0% means there is no chance anyone is inside the building and 100% indicates that the building is definitely occupied by at least one person. Explain your answer.

    A. Single-family residential building at 2:00 AM, Sunday, July 10, clear and mild weather conditions.

    B. Abandoned center city five-story building at 2:00 AM, Sunday, December 10, very cold and windy weather conditions.

    C. High-rise hotel in vacation resort area at 2:00 PM, Saturday, August 1, sunny and warm weather conditions.

    D. Nursing home at 1:00 PM, Wednesday, July 13, sunny and warm weather conditions.

    E. High school, 1:00 PM, Monday, October 2, mild and rainy weather conditions.

    F. High school, 1:00 PM, Monday, July 11, hot and rainy weather conditions.

7. Use the scenarios in Question 6 to compute the staffing needed to safely and effectively conduct an offensive attack.

Wrap-Up, continued

Chapter Highlights

- There is a correlation between response time and progression to flashover.
- NFPA, NIOSH, and USFA reports analyze on-duty deaths and injuries to improve fire fighter safety by sharing experiences.
- Minimum safety measures described in *NFPA 1500* should be familiar to the IC and safety officers.
- Advancements in safety standards and equipment have improved overall safety, but trends in fire fighter fatalities are only slightly improved.
- Fire-ground conditions are more hazardous due to changes in building contents, materials, and construction methods.
- Fire-ground safety is every member's responsibility.
- Risk to fire fighters must be weighed against the benefit of an action; it is unacceptable to risk fire fighters' lives if there is nothing to be gained. The likelihood that a building is occupied is an important consideration in determining risk-versus-benefit.
- Fire intensity and building construction are key factors in assessing the risk to fire fighters as the longer and more intense the fire, the greater the likelihood of structural failure.
- Time from ignition to flashover can vary significantly, depending on a number of variables related to the compartment size, ventilation, and fuel load.
- Live and dead loads affect structural stability, particularly on lightweight roof structures.
- A building's performance under fire conditions can be unpredictable, and exterior appearances may not indicate interior conditions.
- Fire fighters must understand the dynamics associated with structural collapse and recognize the signs of impending structural failure.
- Pre-fire and fire conditions are key elements in evaluating a building's structural integrity during a fire.
- Fuel loads, prior damage, renovations, deterioration, support systems, and construction features (e.g., truss construction) all must be assessed to determine structural soundness.
- Fire entering concealed spaces can extend to remote locations and break out with a sudden increase in intensity.
- Fire extension can cut off the primary means of egress for fire fighters.
- The area between a suspended ceiling and the roof/ceiling is a common concealed space allowing fire to extend unnoticed.
- Fire zones can be subdivided into a hot zone (highest level of protective equipment required), a warm zone (lower level of protective equipment required), and a cold zone (no protective equipment needed).
- The fire perimeter is usually staffed by police to keep unauthorized people away from the scene.
- A safe collapse zone is equal to the height of the building plus an allowance for debris scatter usually calculated as the distance equal to 1½ times the height of the fire building.
- When total collapse is imminent, no one is permitted to enter the collapse zone, regardless of the level of protective clothing.
- Accountability for all persons entering the fire zone is essential.
- Structural stability decreases as time and fire intensity increase.
- Ignition-to-flashover time varies.
- Time components include time from detection to dispatch, time from dispatch to notifying fire units, turnout time, response (travel time), and setup time.
- The time from ignition until the fire is reported to the dispatch center will vary greatly.
- Presence of an alarm system or alert occupants can reduce the time from ignition until the fire is reported to the dispatch center.
- The turnout time objective is 1 minute.
- Response time goals are 4 minutes or less for the first arriving unit, and 8 minutes or less for arrival of the complete initial alarm assignment.
- Set-up time components include the time necessary to position the apparatus, advance the first hose line into the fire area of the building, and apply water.
- Set-up time can be significantly reduced through training, providing more accessible equipment, and revising SOPs.
- Risk to fire fighters and occupants generally increases the longer a building is on fire.

- Smaller compartments progress to flashover faster than larger compartments.
- Safe fire-ground operations require attention to many areas, such as communications, command and control, accountability, protective clothing, rehabilitation, the placement of fire lines, ventilation, and the use of rapid intervention crews.
- For the IC to maintain a safe and efficient operation, communications must be properly managed and everyone must work toward a common goal in an organized fashion.
- Response to a "mayday" by the RIC must be orderly and organized. An emergency evacuation signal is crucial to warn of imminent collapse.
- The company officer should be able to account for all members of his or her crew working within the hazard area.
- A personnel accountability report (PAR) at a specific, designated time, or whenever there is a change in tactics, ensures that officers are aware of the location of the personnel assigned to their command.
- Accountability should not be assigned to a safety officer; the safety officer must be mobile while accountability is a stationary position.
- The safety officer monitors all areas where fire fighters are working.
- Safety is everyone's responsibility at all times at the fire scene, whether or not someone has been specifically assigned the role of safety officer.
- Interior stairways are the preferred means of egress, but fire escapes and ladders are important alternatives.
- Fire fighters assigned to RIC duties must be trained and equipped for RIC operations.
- There are advantages and disadvantages to mobile versus stationary RICs.
- Whenever possible, all interior hose lines should attack from the same access point.
- Communications among units attacking the fire is essential to avoid opposing hose lines.
- Removing any part of the turnout ensemble inside the fire building is an unsafe practice.
- During prolonged operations, rehabilitate personnel at set intervals. Always rehabilitate personnel during the post-suppression/pre-overhaul interval.
- Develop an overhaul plan for handling various aspects of the damaged structure.
- Mark or protect identified hazards prior to commencing overhaul operations.
- Rehabilitation can be divided into three phases: pre-incident hydration and preparation, incident rehabilitation, and post-incident recovery.
- On-scene rehabilitation is especially important in extremes of temperature, inclement weather, or during extended incidents.
- Rehabilitation activities should include checking for signs of critical incident stress.

References

1. Rita F. Fahy, Paul R. LeBlanc, and Joseph L. Molis. Fire fighter fatalities in the United States, 2006. *NFPA Journal,* July/August 2007.
2. National Fire Protection Association. *NFPA 1500: Standard on Fire Department Occupational Safety and Health Program.* Quincy, MA: NFPA, 2007.
3. Code of Federal Regulations, 49 CFR 1910. Washington, DC: U.S. Government, April 18, 1998.
4. Michael J. Karter, Jr. U.S. fire loss for 2005. *NFPA Journal* Sep/Oct:46–51, 2006.
5. Michael J. Karter, Jr., and Joseph Molis. Fire fighter injuries for 2005. *NFPA Journal* Nov/Dec:76–85, 2006.
6. Francis Brannigan, Glenn Corbett, and National Fire Protection Association. *Brannigan's Building Construction for the Fire Service.* Sudbury, MA: Jones and Bartlett Publishers, 2007.
7. National Fire Protection Association. *NFPA 1221: Installation, Maintenance, and Use of Emergency Services Communications Systems.* Quincy, MA: NFPA, 2007.
8. National Fire Protection Association. *NFPA 1710: Standard for the Organization and Deployment of Fire Suppression Operations, Emergency Medical Operations, and Special Operations to the Public by Career Fire Departments.* Quincy, MA: NFPA, 2004.
9. National Fire Protection Association. *NFPA 1720: Standard for the Organization and Deployment of Fire Suppression Operations, Emergency Medical Operations, and Special Operations to the Public by Volunteer Fire Departments.* Quincy, MA: NFPA, 2004.

10. National Fire Protection Association. *NFPA 1410: Standard on Training for Initial Emergency Scene Operations*. Quincy, MA: NFPA, 2005.
11. National Institute for Occupational Safety and Health. Partial roof collapse in commercial structure fire claims the lives of two career fire fighters–Tennessee. NIOSH fire fighter fatality investigation F2003-18, July 26, 2004. Cincinnati, OH: NIOSH. Accessed July 10, 2007 at www.cdc.gov/niosh/fire/reports/face200318.html.
12. James Williams and Hollis Stambaugh. *Rapid Intervention Teams and How to Avoid Needing Them: Special Report*. USFA-TR-123. Washington, DC: Homeland Security/U.S. Fire Administration, March 2003. Accessed July 10, 2007 at www.in.gov/dhs/training/firefighter/rapidintervention.pdf.
13. Gregory Kinnes and Gregg Hine. *Health Hazard Evaluation Report 96-0171-2692, Bureau of Alcohol, Tobacco, and Firearms*. Cincinnati, Ohio: NIOSH, May 1998. Accessed July 19, 2007 at www.cdc.gov/niosh/hhe/reports/pdfs/1996-0171-2692.pdf.
14. National Fire Protection Association. *NFPA 1584: Recommended Practice on the Rehabilitation of Members Operating at Incident Scene Operations and Training Exercises*. Quincy, MA: NFPA, 2003.
15. G.A Selkirk, T.M. McLellan, and J. Wong. Active versus passive cooling during work in warm environments while wearing firefighting protective clothing. *Journal of Occupational and Environmental Hygiene* 1:521-531, 2004. Defence Research and Development Canada, 8/2004.

Life Safety

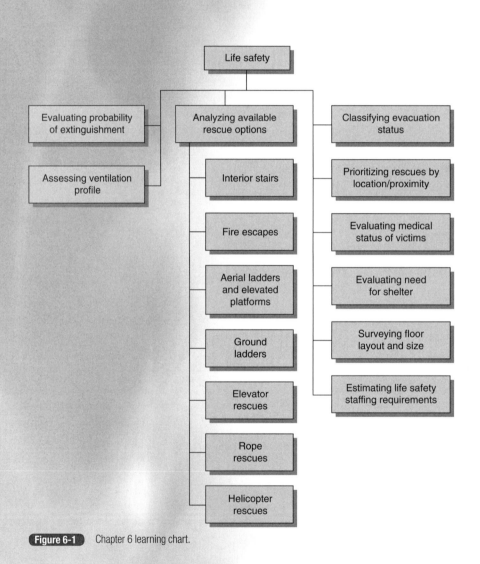

Figure 6-1   Chapter 6 learning chart.

Chapter 6

Learning Objectives

- Explain the relationship between life safety and extinguishment.
- Discuss the positive and negative aspects of ventilation in regard to life safety.
- Evaluate ventilation options as they relate to fire location and select the best option given a scenario with several vent options.
- List and evaluate rescue options.
- List rescue priorities in terms of occupant proximity to fire.
- Define a mass-casualty incident.
- Explain the medical (EMS) function at a large structure fire with multiple casualties.
- Describe conditions that affect life safety staffing requirements.
- Evaluate tactics at a fire scenario where a large number of occupants need to be rescued.
- Compare and contrast the positive and negative effects of entering an enclosed fire area.
- Use a scenario to select and describe proper ventilation techniques.
- Use a scenario to describe and apply rescue options.
- Use a scenario to evaluate priorities as they relate to occupant proximity to the fire.
- Use a scenario to estimate staffing requirements at a structure fire occupied by a large number of people.

Introduction

In this chapter, the most important of all fire-ground functions—life safety—is discussed. Life safety presents the ultimate challenge to the incident commander (IC). Decisions of whether to control the fire, remove the victims, or both might seem obvious. However, during the heat of battle, the IC is faced with numerous high-priority challenges. Deciding how best to assign initial limited resources at the incident scene is difficult. Initial strategic decisions at the incident scene must be made quickly and on the basis of the situation and the resources available to the IC. Understanding the interrelationship of various life safety tactics discussed in this chapter is essential for an IC. The learning chart in Figure 6-1 shows that the actual rescue of occupants is not the only way to save lives in a structure fire. Fire suppression and ventilation are also used to accomplish this most important objective. Figure 6-1 lists rescue options from the best method (using the interior stairs) to more hazardous and more staff-intensive methods.

> In the vast majority of cases, the best life safety tactic is suppressing the fire.

Although fire extinguishment is generally considered the second priority, it is an essential part of most rescue operations. Fire fighter safety is a critical component of the life safety priority. The protection provided by fire streams, combined with proper ventilation, is an important part of protecting fire fighters as well as the occupants.

Evaluating the Probability of Extinguishment

Determining the probability of extinguishment is a major factor in life safety decisions. If the fire can be quickly extinguished, the top life safety priority is applying the required rate of flow. Only on rare occasions is it best to rescue visible occupants rather than controlling or extinguishing the fire.

A number of factors must be taken into account by the IC upon arrival at the fire scene. A common tactical error is prioritizing victims according to visibility. Occupants who are still inside the building may be in grave danger, and it may be best to rescue people who are at windows via the interior stairs *after* the fire is under control or extinguished.

In some cases it is possible to apply a **defend-in-place** tactic if the fire can be quickly controlled. Electing to leave people in a burning building is a calculated risk that the IC must be prepared to make on the basis of fire conditions, available resources, and the extent of danger to victims.

A fire may require a long lead time to suppress or even be impossible to contain when large-volume flows are needed. (Rate-of-flow calculations will be explained in Chapter 8.) Large interior flow requirements from handheld lines translate into larger staffing needs, which, in turn, may mean more time to assemble the required resources.

If the fire cannot be extinguished immediately, building features such as fire doors and fire walls can aid in extending the time available for evacuation by providing a barrier between the victims and the fire. Hose lines can also be used to push the fire or otherwise protect egress routes. It is crucial to ensure that sufficient personnel and resources are available to provide the required water flow. Opening a fire door to apply water below the required rate of flow could have severe consequences. The opening that is made for the fire hose will allow the fire and smoke to extend into the stairway at an accelerated rate.

It is essential to get a hose line into position to extinguish the fire, which is usually accomplished by the first-arriving crew. The crew advancing the hose line will assist victims they come upon but cannot be expected to perform a complete primary search of the area surrounding the fire. Their primary responsibility is extinguishment.

The floor or floors above the fire are critical areas that must also be searched. The number of crews needed for search and rescue depends on the number of floors to be searched, the size of the building, fire intensity, smoke conditions, and occupant status.

A single crew can search and evacuate an entire floor in a single-family detached dwelling where the occupants are able to assist themselves. If occupants are missing or unable to make their own way to safety, it may well take an entire crew to rescue one individual plus an EMS crew to provide treatment and transportation once the victim is removed from the building.

A high-rise building may have many floors above and below the fire to be searched. Floor areas in these buildings are generally large. More crews will be needed when confronted with large areas and/or when several floors are to be searched. There is sometimes a tendency to assign multiple

Fallacy	Fact
Occupants who are visible at windows are in the greatest danger.	People who are visible at windows might or might not be in grave danger. Occupants who are unable to reach a window are likely to be in more danger.

> Vent to control the fire. Apply water to extinguish the fire.

companies to search a floor before getting at least one company in each exposed area above the fire to determine fire and smoke conditions and evacuation needs. It is best to get at least one company into all areas on the fire floor and above as soon as possible. Once all critical areas have been assigned, additional units should then be assigned to each area as needed to complete search and rescue activities.

Only in the most extreme circumstances should the first-in engine crew do anything other than advance an attack line. Generally, more lives will be saved by controlling the fire than by performing ladder rescues or knocking on doors to evacuate. Often, the only hope for occupants in the immediate fire area is fire control, especially if the fire is between them and their way out. What does all of this mean? The first-in engine crew should *not* be included in the staffing needed for rescue and evacuation. If multiple lines are needed to confine the fire, add the staffing needed for additional fire lines to the total staffing requirement.

Assessing the Ventilation Profile

Ventilation has a positive component and a negative component. If fire fighters are working in an enclosed area when a window breaks, the fire will gain in intensity, but the products of combustion will be vented out of the area. For this reason, ventilation must be coordinated with the placement of attack lines to control the fire. In most cases, ventilation makes the job of finding the fire and victims much easier. However, under certain conditions, improper venting could produce a backdraft, and the fire fighter or occupants could be seriously injured or killed by the resultant fire and explosion. Further, a vent opening made between fire fighters or victims and their path to egress could prove fatal.

Ventilation is done to relieve the products of combustion, allowing fire fighters to advance on the fire. When venting to support suppression, it is important to coordinate hose line placement with ventilation. Close coordination of venting means that the hose line is ready to quickly overcome the increase in combustion that will likely occur.

Venting can be a very effective life safety procedure. When venting for life safety purposes, the principle is to pull the fire, heat, smoke, and toxic gases away from victims, stairs, and other egress routes.

A common life safety venting tactic involves opening a scuttle at the top of a stairway. This can be a very effective means of venting the stairs because the smoke will generally rise, due to the natural stack effect. However, there is a negative component to this method of vertical venting: If a door to the fire area is opened, the vent opening will pull the fire and smoke from the fire area into the stairs (see Figures 2-12, 2-13, and 2-14 in Chapter 2). Likewise, a vent opening made in the wrong location may pull the fire into areas containing fire fighters and occupants. **Figure 6-2** is a plan view of a one-story warehouse building with overhead doors identified as A, B, C, D, and E. Roof vents are labeled 1, 2, 3, and 4. Given the location of the fire, the best possible

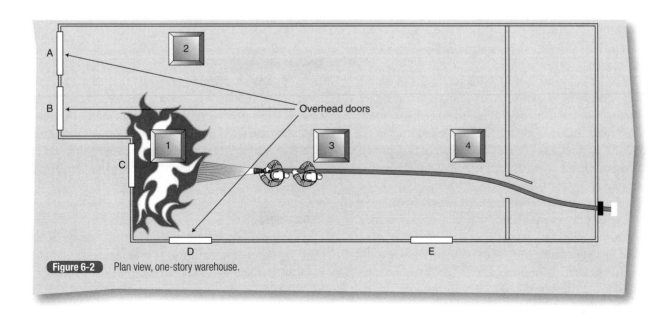

Figure 6-2 Plan view, one-story warehouse.

> Regardless of the intended results, the fire is bound by the rules of physics and chemistry.

vertical vent would be the roof vent directly over the fire (roof vent 1). However, the roof could be weakened in this area, or the fire intensity below could cause injury when the roof is opened. Vent 2 would be an acceptable alternative, although it would spread the fire into unburned areas. Cutting open the roof in the area near the fire would also be acceptable, but cutting through the roof of this commercial building could be difficult. Opening roof vents 3 or 4 would be unacceptable as they would pull the fire toward fire fighters operating the hose line. Likewise, horizontal ventilation through door C would be best if horizontal venting was chosen with the understanding that opening this door might be hazardous due to the sudden release of fire at the opening. Door D would be the next best horizontal opening. Doors A and B could be used, but opening either one would not be an ideal choice as venting through these doors would spread the fire. Opening Door E could have disastrous effects, as it would pull the fire toward fire fighters advancing on the fire and through a large portion of the warehouse.

Strategically placed hose lines can offset the negative effects of improper venting, but pulling the fire toward occupants and fire crews is *never* a good tactic. At times, particularly when a large rate of flow is required, the hose line will be unable to stop the forward progress of the improperly vented fire.

When venting to save lives, it may be necessary to vent before having a hose line ready. However, this is a risky tactic that should be used only as a last resort. Building features can retard fire growth and prevent extension into occupied areas if they remain closed and in place. For this reason, it is usually best to avoid opening doors and windows in the fire area before charged hose lines are in position.

Fire fighters may break windows as they conduct their primary search. This venting tactic is often necessary; however, when possible, it is better to open the windows rather than breaking them. Reversible venting is the preferred method, as once a window or door is shattered, the vent opening cannot be easily closed. If the windows or doors are opened so that they can again be tightly closed, negative, unintended consequences caused by the vent opening can be reversed simply by closing the window or door.

In making openings for ventilation, it is important to consider the fact that the fire and heated gases will naturally move upward, laterally, and then downward.

Positive-pressure ventilation has become popular and is a useful tool. The key to successful positive-pressure venting is to control the outlet openings. If too many doors and windows are opened, positive-pressure venting will prove ineffective. As with all other ventilation methods, there is a negative side to positive-pressure venting. The fire can be directed toward victims, toward their escape routes, or into unburned areas.

Analyzing the Available Rescue Options

Occupants should be rescued by the shortest, easiest, and most direct route. Victims who are in danger above or below grade level should be evacuated by using the interior stairways whenever possible.

A defend-in-place concept is now being used in occupancies such as high-rise buildings and healthcare facilities where occupants are moved away from the fire area but remain in the structure. Many large buildings are constructed in a manner that would allow a defend-in-place strategy. The following question arises: Would the occupants be safer in their rooms, which have not been invaded by the products of combustion, or would it be better to bring them out through smoke-filled corridors and stairwells in an attempted evacuation? This decision is particularly perplexing on floors above the fire that are not yet threatened, especially if stairwell capacity is needed for evacuation of more endangered occupants or for moving fire control forces and equipment to the fire area.

On the next few pages, some of the most common means of building evacuation from upper floors are discussed. They are listed in preferential order; for example, fire escapes are generally preferred over ground ladders.

Interior Stairs

Interior stairs are preferred over all other means of egress from upper floors when they are tenable. The safest and easiest way to move occupants is by using the stairs, which are intended for that purpose. Any other method is considered a less desirable alternative, and the use of alternative methods should be justified by special circumstances.

Fire Escapes

Fire escapes are poor substitutes for interior stairs, and their structural integrity is sometimes questionable, especially when large numbers of people are using the fire escape. However, when properly maintained, they are

Case Summary

Figure CS6-1A The door to the stairway where victims were found in the North York, Ontario fire.

Figure CS6-1B The MGM Grand fire caused by an electrical short killed 83 people and injured hundreds more.

In several fatal fires, victims were found in the corridors and stairways, but people who remained in their rooms were not injured. In a fatal high-rise fire in North York, Ontario, six victims were found in the stairways in the upper stories of a 29-story building. The fire occurred on the fifth floor. In the MGM Grand fire in Las Vegas, almost 60% of the fatalities in the high-rise tower were found in corridors, stairs, and elevators.

Sources: Residential High Rise, North York, Ontario, NFPA Fire Investigation Report. Quincy, MA: NFPA; R. Best and D. Demers, *MGM Grand Hotel Fire,* NFPA Fire Investigation Report. Quincy, MA: NFPA.

the second-best option. If the fire escape is structurally sound, it is preferred over other alternative methods.

Aerial Ladders and Elevated Platforms

Following interior stairs and fire escapes, the next rescue preference is the aerial ladder or elevated platform. There are two schools of thought when using these apparatus for rescue: Aerials, as compared to elevated platforms, can move larger numbers of people from a single area in a shorter period of time. However, elevated platforms provide a more secure, less stressful rescue for the occupant, and articulating platforms can access areas that an aerial would have difficulty reaching. Fire departments should survey hazard areas before purchasing apparatus, matching apparatus to identified needs.

Ground Ladders

Ground ladders follow aerials and elevated platforms on the rescue preference list. Ground ladders are generally less stable, do not have the reach of most aerial ladders, and require more personnel to position. However, in certain circumstances, such as when buildings are set back from the street or when two or three fire fighters can quickly access the second or third floor, ground ladders are preferred over aerial devices. When fire and/or smoke are between the victim and the interior stairs, aerial ladders, elevated platforms, and ground ladders are the best alternatives when a quick knockdown is not possible or when fire extinguishment tactics further endanger occupants. Again, remember that if the stairways and corridors leading to the stairways are relatively clear of smoke, use the stairs.

When the fire threat seems real to the occupants, they will sometimes go to a window or other visible position to

> Rescues and evacuations via interior stairs, fire escapes, aerial apparatus, and ground ladders are considered acceptable, and are even the preferred means of egress under certain circumstances. Elevators, scaling ladders, ropes, helicopters, and air bags are all questionable rescue methods that are potentially more hazardous.

await rescue. They may or may not have checked the stairs before deciding to wait at the window. Do not attempt a ladder rescue just because people are at the window. Reassure them that you are there to help, but determine first whether they must be moved, and second whether stairways are available. Raising and positioning ground ladders may require several fire fighters. Likewise, the rescue will involve at least one fire fighter assisting the occupant and probably another trying to assist the person onto the ladder with a third fire fighter at the foot of the ladder.

Elevator Rescues

An interesting question arises regarding the use of elevators for rescue. Fire safety professionals warn against occupants using the elevators as a means of escape. However, they could be safely used when under the fire department's control. Under certain conditions, the use of elevators for evacuation is justified in buildings that are subdivided with good, fire-resistive construction. If an elevator is remote and separated from the fire area with an auxiliary power supply, using it for rescue purposes, especially to evacuate immobile occupants, may be the best tactic. The use of elevators in the immediate fire area is hazardous to everyone, including fire fighters. The safe use of elevators is discussed further in Chapter 12.

Rope Rescues

The use of ropes for rescue purposes can be justified only in extreme cases in which the victims are beyond the reach of aerials and elevated platforms. Before resorting to this last-option rescue technique, every effort should be made to perform an interior rescue. Rope rescues are extremely slow and dangerous and require specialized equipment and expertise. There is an added hazard to the rescuer and the victim because of the inherent danger of technical rescue operations. This rescue method is a tactic of last resort.

Helicopter Rescues

Before deciding on rooftop rescues, the IC should ask whether the occupants who have reached the roof are safe where they are. Many times it is possible for occupants to wait out the fire on the roof. Most roofs provide a difficult operating platform for helicopters, owing to obstructions that include antennas, satellite dishes, and ventilation equipment **Figure 6-3**. These obstructions make it impossible for the helicopter to land on these roofs. Even if the roof is clear of obstructions, it may not be a good landing platform because of the hot air currents caused by an intense fire.

Because of the limited number of people who can be safely carried in a helicopter, the choice of who is going to be rescued and who will stay behind will also need to be made. Victims may think they are in immediate danger and believe that they cannot survive long enough to wait for another trip. They may attempt to climb into the helicopter, overloading it beyond its safe operating capacity. The helicopter has been used to advantage in some rescue situations, but its usefulness has been overstated. At many fires, dramatic helicopter rescues were completely unnecessary.

Some departments use helicopters to place fire fighters on the roof. These fire fighters then assist and reassure occupants on the roof or descend down the stairway to perform rescues. They could also use the standpipe for extinguishment. Helicopters can provide quick and less fatiguing access to upper floors in a high-rise building. However, fire fighters typically use the path of ingress as their egress. A helicopter that drops fire fighters on the roof may not be able to return and remove them. If your department anticipates the use of helicopters, continuous training of both fire personnel and helicopter crews is absolutely essential. Most private helicopter owners will resist getting involved in rescue work because they lack rescue equipment, training, and insurance coverage for performing rescues.

Classifying Evacuation Status

During the early stages of an operation the IC is forced to make staffing estimates based on incomplete or even inaccurate information. As companies are deployed, crews will give status reports, request assistance, or possibly release fire fighters for other assignments. Information becomes more complete and reliable as fire crews provide reconnaissance by entering endangered areas.

Fires with a significant life safety challenge reinforce the need to build a command structure from the very start of the incident so that a systematic search-and-rescue operation can be made without duplication of effort. A marking system that indicates what areas have been searched can be very helpful in managing a systematic search. Many departments use plastic door hangers or other marking procedures to indicate rooms that have been searched. Even a simple chalk-marking system, in which an "X" or company number is placed on a door after the area is searched, can be effective. Adopting a marking system as an SOP is highly recommended.

Occupants who escape on their own may need direction, emergency medical treatment, a place of safe refuge, or other assistance. The care of these victims is usually a non-

Figure 6-3 Rooftop obstructions.

emergency function that can be delayed if necessary until personnel have verified that no one is in immediate danger.

Other victims may not be aware of the fire or of egress routes. They may need direction to safety, or fire fighters may need to provide a safe egress route for them. These victims will then be able to escape with little assistance. It may not be necessary to escort them all the way to the outside or, for that matter, to a place of safe refuge. When there is a limited number of emergency personnel available, these victims should be permitted to evacuate on their own whenever possible, allowing available fire fighters to perform the more crucial task of search and rescue.

Sometimes it is possible to designate specific stairways as evacuation routes and other stairways for fire operations. This is a good tactic because fire fighters advancing hose lines through the stairs to the fire area will allow smoke to enter the stairs through the partially open door and impede egress because of the hose laying on the stairs. Six occupants of the Cook County Administration Building perished while attempting to exit via the stairway being used by fire fighters to attack the fire. The Cook County Administration Building fire is discussed further in Chapter 12.

A practical problem occurs in attempting to channel the flow of occupants into a designated stairway. In most cases, fire forces will not be in total control of the evacuation until they reach each floor to direct escaping occupants to the stairway of choice. Simply designating the stairway is not enough; specific instructions must be given to both fire fighters and occupants. Some large buildings have voice announcement systems or public address systems that permit the IC to relay messages to evacuees. Unfortunately, occupancies equipped with these systems are the exception, rather than the rule.

Immobile or unconscious victims will place additional requirements on the rescuers. In most cases, these victims

Case Summary

A fire occurred in a high-rise building with elderly occupants in St. Louis, Missouri. An attempt was made to attack the fire from both stairwells, but because of the number of elderly people coming down one of the stairwells, this was not possible. Whenever the door to the fire floor was opened, the smoke and heat entered the stairway, endangering the occupants. Once the flow of residents was stopped, it was then possible to enter the floor from the second stairway to fight the fire and rescue an injured fire fighter. However, this operation was delayed because of the danger in which the occupants were placed whenever the door was opened.

Source: Edward R. Comeau, personal correspondence following the investigation.

---

must be physically removed to the outside or to a place of safe refuge. This effort usually requires a minimum of two fire fighters (often more) to rescue one victim.

A commonly held fallacy is that the occupants will be able to provide rescuers with reliable information concerning occupant status and that the IC will be able to account for the building's occupants.

 Fallacy
It is possible to account for all the occupants of the building without actually conducting a thorough and complete search.

 Fact
The only way to be sure the building is completely evacuated is to perform a primary and secondary search throughout the entire structure.

The *only* reliable way to verify evacuation status and fire conditions is for fire fighters to enter the structure and systematically check every room. Occupancies such as primary schools and health care facilities conduct regular fire drills, and staff should have control over students and patients. But even in these occupancies, the staff may not account for visitors in the building. Some industrial settings have evacuation plans that include accountability. However, the mobility of employees within the facility may make these plans less effective than would be possible in a school or healthcare facility, where the location of occupants is known. In the typical apartment building, some of the occupants may not be home when the fire occurs, and others may have escaped the building and then left the scene.

Case Summary

An early morning fire in a large apartment complex in Bremerton, Washington is an example of the difficulty in determining how many victims remain inside the burning structure. The fire broke out at approximately 6:00 AM, and a number of people had already left for work. During the initial stages, it was feared that possibly 20 people were unaccounted for and still trapped in their apartments. During the course of the day, a more complete accounting was made, and it was ultimately determined that there were four people missing who had died in the fire.

Source: Edward R. Comeau, *Apartment Fire, Bremerton, Washington,* NFPA Fire Investigation Report. Quincy, MA: NFPA.

Figure CS6-2 The apartment complex was four stories high, with balconies on the outside. A number of people had to be rescued from these balconies by ladders because their primary escape route had been blocked by the fire.

Case Summary

The fire chief in Chapel Hill, North Carolina, faced problems accounting for occupants when a fraternity building caught fire. The fire occurred on a Sunday morning that was also graduation day. A number of the residents had already moved out for the semester, but no one was sure who or how many, or whether any visitors were in the building. It was originally feared that as many as 25 people were inside the building. Ultimately, it was determined that five people had been trapped and killed by the fire.

Source: Michael Isner, *Fraternity Fire, Chapel Hill, North Carolina,* NFPA Fire Investigation Report. Quincy, MA: NFPA.

Figure CS6-3 Fire fighters entering the fraternity building in Chapel Hill, North Carolina.

Whenever possible, the IC should drive around the fire building to see all sides before setting up a command post. Fire crews should view the exterior, looking for occupants in need of assistance. However, this outside reconnaissance should not be done at the expense of the fire attack or interior search. Reports from occupants may be unreliable, but they should not be ignored. In many cases, occupants will be able to specifically direct rescuers to the location of a victim.

Flashover is the critical landmark: before flashover, rescue is possible; after flashover, rescue is highly improbable within the flashover compartment. Recognizing the fire stage is important to the life safety effort. In the very early stages of a fire (ignition and early growth phases), occupants tend to rescue themselves once they are aware of the fire. Later, as the fire becomes larger and produces greater quantities of smoke and toxic gases, occupants may be conscious but unable to exit until the fire is extinguished. The most difficult victim to rescue is one who has been felled by the smoke, heat, or gases in an unknown location.

Fire conditions are related to many factors, such as the time that elapsed before the fire was detected, the time it takes to notify the fire department, and the response time of the units. If a fire is extending through a structure, the threat to occupants is directly related to their awareness of the fire, their ability to escape, the construction of the building, and the provisions that have been made for egress. Once a fire is in progress, the IC has little control over these factors.

<u>Critical time</u> (the time available until the structure becomes untenable) will vary depending on numerous building factors, as well as the building's fuel load. No exact figures can be given, but experience shows that fires spread rapidly upward in balloon-frame construction or, for that matter, in any construction that allows vertical fire spread to occur unimpeded. The time from ignition to full involvement in a balloon-frame structure will be very short in comparison to a fire in a fire-resistive building. The fire-resistive building allows fire forces or occupants to confine the fire by using built-in features such as fire doors. An enclosed fire-resistive area can quickly become untenable because the products of combustion, including heat, are not relieved, as they would be in a less compartmentalized structure.

If the IC determines that time does not permit the rescue of all the occupants, victim triage should be established

Case Summary

In the apartment fire in Bremerton, Washington that was summarized earlier in the chapter, the first-arriving officer was faced with a large number of victims who were on their balconies, needing to be rescued. Their only means of egress was blocked by the fire. One person went so far as to lower a rope and slide down four stories, severely burning his hands.

The officer knew that he could not possibly rescue all of the people who were in danger by removing them from the fire. Therefore, he elected to advance a hose line into the courtyard to fight the fire. Thus, the fire spread was slowed until other units arrived on the scene and were able to begin rescuing the occupants using fire department ladders.

Source: Edward R. Comeau, Apartment Fire, Bremerton, Washington, NFPA Fire Investigation Report. Quincy, MA: NFPA.

Figure CS6-4 The fire at the Bremerton, Washington, apartment complex when fire fighters arrived.

to rescue as many victims as possible. The importance of fire control to life safety cannot be overstated. Buying time with fire control tactics is a critical strategy that must be implemented whenever possible. If the fire is extinguished, or at least in the knockdown stage, the fire fighters and victims within the building have a much better chance to evacuate safely.

Estimating the Number of People Needing Assistance

Once the pre-incident plan, visual, and reconnaissance information have been considered, the IC should be able to evaluate life safety requirements in terms of staffing and equipment.

Fire fighters assigned to search-and-rescue tasks should be equipped with forcible entry tools. Heavy metal doors with locks will extend the time and effort required for the primary search. In forcibly entering an interior area, it is important to consider the value of the opening being forced. A fire door is designed to resist the fire. It can contain a fire for a significant period of time or keep fire from extending to an exposed area. Whenever possible, fire doors should be closed after the search is complete. Some department SOPs call for a rope, hose strap, or similar device to be attached to the door to assist in closing it if necessary.

The fire attack crew should have hose lines in place and charged before forcing the door to the room or area of origin. In buildings with strong compartmentalization, conditions beyond the door are unknown, and extreme caution must be exercised when making entry. If the door is hot to the touch and evacuation is not complete, it may be best for the crews to stand ready at the door and allow the door to contain the fire while other crews complete the evacuation.

By making return trips to the fire area, fire fighters can rescue more than one person even in situations in which carry-and-drag rescues are needed. However, this is very demanding work and will quickly exhaust the available rescuers. In sizing up the personnel requirements, estimate the physical condition of the occupants, the number of occupants on the fire floor, and the travel distance to safety, as well as fire growth.

After estimating the number of victims needing to be rescued, the method of rescue and victim condition must be considered. Immobile or unconscious victims will require at least one rescuer for each victim or, more typically, two rescuers per victim.

Few, if any, fire departments could successfully evacuate a high-rise building with 1000 or more occupants in need of assistance. Further, if all of the occupants of a high-rise building are placed in the exit facilities at the same time, more people might be placed in danger. Trade-offs must be made. An evacuation must be prioritized by removing the most endangered occupants first. Meanwhile, fire fighters can plan ahead to move potential victims in less immediate danger before they are felled by the products of combustion, allowing rescuers to conserve strength by directing or assisting victims to a safe area while they are still mobile.

Surveying Floor Layout and Size

The size of the area to be searched and evacuated, as well as the location of stairways, halls, and fire escapes, has much to do with organizing the search effort. Quite often, the primary and secondary searches are done by assigning floors to each search team. Large buildings may be further subdivided by wing or building side.

Many buildings have mazelike designs, like the one for the Beverly Hills Supper Club, shown in **Figure 6-4**. This is particularly true of buildings with multiple additions or office buildings that use cubicles for workspaces. In some cases, building additions result in confusing egress paths.

Fire fighters working in multi-story buildings should, whenever possible, survey an uninvolved floor before ascending to the fire floor or floors above. The layout may be different from floor to floor, but some building features tend to remain constant, such as stairs, elevators, and standpipes.

A few precautions are in order. The general layout of the space can be different on each floor. One floor may be wide open, while the next is subdivided into small offices or suites. Even the stair, elevator, and standpipe locations can change, particularly on the lower levels, where elevators are in banks serving different areas or in tower-type high rises.

Another problem is consistency in floor numbering. Buildings that are constructed on a grade may have unusual floor numbers. Sometimes what appears to be the first floor of a building from one side is actually identified as a higher or lower floor. For example, the first floor main entrance to Good Samaritan Hospital in Cincinnati is the sixth floor. The terrain slopes away from the front of this hospital, and a section that was added to the building has a grade-level entrance six floors below the main entrance. To avoid mixing floor numbers at the same level, the entire hospital floor numbering system was changed to be consistent with the new addition. This same situation can exist even in single-family detached dwellings that are built on a grade. In high-rise buildings, there may be more than one set of windows on the first floor. The mezzanine may not be counted as a floor, and these buildings sometimes do not have a 13th floor. Each of these factors adds to the confusion. The grade-level sixth floor at Good Samaritan Hospital or the absence of a 13th floor can be known in advance through pre-incident planning.

In smaller buildings, it is often possible to get a good idea of the interior layout from the exterior of the building. Large windows indicate areas that are most likely common areas (living or family rooms), whereas bedrooms tend to

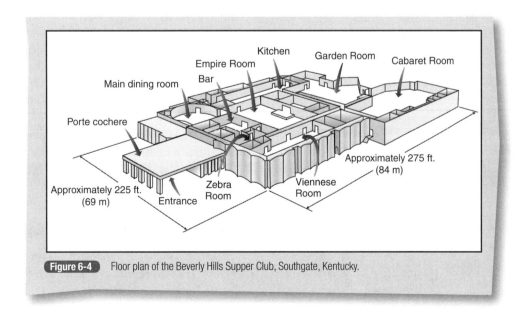

Figure 6-4 Floor plan of the Beverly Hills Supper Club, Southgate, Kentucky.

Case Summary

At two fires—one in Seattle, Washington, and another in Pittsburgh, Pennsylvania—a total of seven fire fighters were killed in buildings that were built on sloping grades. Because of the confusing topography, it was not clear to everyone on the fire ground how many floors the buildings had or the exact location of the crews operating inside of the buildings in relation to the fire.

Sources: Edward R. Comeau, *Warehouse Fire, Seattle, Washington*, NFPA Fire Investigation Report. Quincy, MA: NFPA; Michael Isner, *Residential Fire, Pittsburgh, Pennsylvania*. NFPA Investigation Report. Quincy, MA: NFPA.

Figure CS6-5 Sloping topography can make it difficult to determine the exact location of the fire and how many stories are in a building.

have smaller windows. Buildings of similar construction in an area will typically have similar floor layouts. For instance, the locations of stairs to the second floor of a single-family detached dwelling tend to be near the main entry.

Even a small building that has been renovated or one that is at various grade levels on different sides can be difficult to search. These buildings can be deadly to the fire fighters who go inside to conduct offensive operations.

Larger buildings, particularly large, open buildings, present an additional hazard to fire fighters and occupants when fire conditions seriously reduce visibility. Most fire fighters have experience in residential fires, because residences are where most fires occur. The same tactics and procedures used in small, compartmented buildings could be fatal when used in a large, open structure with low visibility. Searching a small room using a right- or left-hand search can be safe and effective. A larger open area would require the use of a rope or other guideline and different search techniques. A thermal imaging camera can be used to quickly scan a small room prior to entry. Larger rooms may require the fire fighters conducting the search to advance with the thermal imaging camera as they check around obstacles and advance throughout the compartment. A very large room may require substantial escape time—more than the time available when the low pressure alarm on the SCBA sounds. The person in charge of the search may need to assign someone as timekeeper. As one crew depletes their air supply, a second crew will need to move forward to continue the search. Large compartments can also produce large, overwhelming fires that can overrun a hose crew.

Prioritizing Rescues by Location/Proximity

After determining the total number of occupants and developing a strategy for a full or partial evacuation, the search-and-rescue priority should be established. This must be determined by

> The key to a successful search is to be systematic.

deciding who is in the greatest danger from the fire. Search and rescue is then prioritized on the basis of rescuing those in the greatest danger first. The priority list is as follows:

1. People on the fire floor nearest to the immediate fire area
2. People in proximity to the fire area on the same level as the fire
3. People on the floor above the fire, especially immediately over the fire area
4. People on the top floor (unless fire conditions result in smoke stratification, as discussed in Chapter 12)
5. People on the floors between the floor above the fire and the top floor
6. People on the floors below the fire
7. People in nearby buildings
8. People outside (in the collapse or falling glass zones)

Item 8 could be placed anywhere in the priority list, depending on the structural integrity of the building, falling glass, etc.

Rescue priorities are shown in **Figure 6-5**. Those in the greatest danger are rescued first. One exception is when it is not possible to save everyone. In these situations the IC should opt to save the largest number of people possible. This will require committing resources to areas where the largest number of occupants can be rescued. An important point to remember is that most victims either will escape on their own or, if given direction, will be able to evacuate without assistance.

The key to successful search operations is to be *systematic*. The IC should follow the priorities listed above in assigning primary search areas. Crews should use a consistent method of marking and recording areas that have been searched so that all threatened areas are checked before the secondary search begins. The primary search is a quick but thorough search of the area. The secondary search ensures that no one was missed the first time through. If conditions and resources permit, a secondary search should be conducted as soon as the primary search is complete.

To ensure complete coverage during the secondary search, good practice dictates using a different crew. For example, if the first ladder crew conducted the primary search of the fire floor and the second truck crew checked the floor above during the primary search, the secondary search assignments should be reversed (second ladder crew assigned to the fire floor and the first ladder crew to the floor above). Some fires require multiple search assignments, making this reassignment more complex. The officer in charge of the search-and-rescue group would manage the reassignment

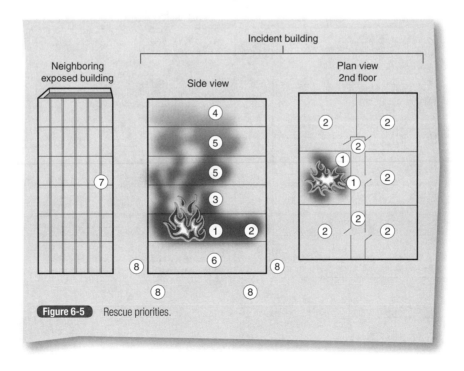

Figure 6-5 Rescue priorities.

> One system of prioritizing patients uses the categories of 1: life-threatening priority; 2: serious, but not life-threatening priority; 3: walking wounded; and 4: deceased. Color-coded tags are also used to categorize victims.

from primary to secondary search. Figure 6-6 provides an example reassignment for the secondary search where five companies are assigned to primary searches in different areas of the building and in an exposure building.

The time and effort expended conducting the primary search dictates whether these same companies should be reassigned to a secondary search location or whether additional companies should be assigned to perform the secondary search. When necessary, additional secondary search companies can be summoned to the scene, but having a tactical reserve available in staging would be preferred.

The person managing the search-and-rescue effort should assign specific areas of responsibility, and then track these assignments and record progress reports on a status board. When a large evacuation effort is required, the units involved in search and rescue will rely on communications to ensure that all areas are searched and to avoid duplication of effort. If at all possible, the search group should be assigned a separate radio channel or use telephones specifically designated for this purpose. It may also be possible to use the existing telephone system within the building as long as it is not affected by the fire.

Evaluating the Medical Status of Victims

The IC must have wide discretion in calling medical assistance to the scene. Medical assistance can be requested to treat and transport injured fire fighters or occupants. The relative hazard, number of potential victims, and type of incident will dictate the overall need for emergency medical services (EMS) at the scene. A general rule for calling assistance applies here: If you think you need help, you do. It is much better to have medical assistance standing by at the scene than to have injured fire fighters and occupants waiting for treatment while EMS is responding.

Medical units should be set up at one or more designated locations within the cold zone or beyond. Medical personnel should remain with their ambulance, ready to receive the injured. It may be a good idea to set up the medical triage/treatment area near the rehabilitation area. However,

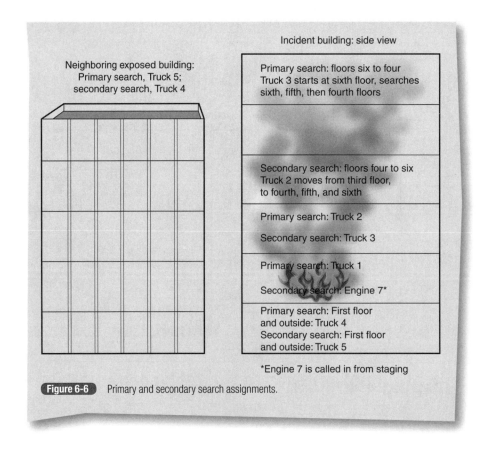

Figure 6-6 Primary and secondary search assignments.

sufficient room should be available to physically separate the two areas.

Evaluating Victims in Mass-Casualty Incidents

Mass-casualty situations are the result of specific incidents, such as transportation accidents, hazardous materials releases, or natural disasters. Fires in buildings with large numbers of occupants have the potential to become mass-casualty incidents. Providing a system that includes triage, treatment, and transportation is a proven method of managing the medical component when large numbers of victims are encountered.

Mass casualty is defined in terms of department and community capabilities. A mass-casualty incident occurs when one or more of the following situations exist:

- The number of victims and the nature of their injuries make the normal level of stabilization and care unattainable.
- The number of immediately available trained personnel and transportation vehicles are insufficient.
- Hospital capabilities are insufficient to handle all of the victims requiring care.

Triage, Prioritizing, and Transport

Triage is the first medical priority in managing a medical disaster. The first-arriving EMS personnel should not leave the scene until they are relieved of triage responsibilities. Emergency medical teams are accustomed to treating one or more individuals, followed by immediate transport. Categorizing victims without treatment and transportation runs contrary to their normal role and must be overcome by rigid enforcement of triage procedures.

When staffing permits, it is best to physically separate the dead from living patients. The walking wounded should be directed to another location for treatment. The walking wounded could be located in a separate area within the shelter being used for evacuated occupants.

Once triage teams have prioritized the victims on the basis of their injuries, treatment teams will follow. Treatment teams first treat those needing immediate care (priority 1) and arrange for their transportation. After this, priority 2 and 3 patients' needs can be addressed. If sufficient personnel are on the scene, multiple teams can be formed to treat as many patients as possible. Paramedics should be assigned to treat the most seriously injured; EMTs and first responders can be assigned to take care of the remaining patients.

Providing transportation for victims is an important function of the medical transportation officer. A drive-through arrangement for ambulances will keep traffic lanes open.

At times it is beneficial to use helicopters to transport patients. In using helicopters, the landing zones must be in safe areas, far enough away from treatment and triage areas that they will not interfere with those activities. Helicopters that are used to transport victims should not be flown over the fire area or command post. The noise and effect of the downdraft from the helicopter rotors can be extremely disruptive and can create a hazardous situation.

Staging and categorizing transport capabilities of medical units are essential in managing transportation needs. Incoming EMS transportation vehicles should be directed to an established staging area during mass-casualty operations and requested as needed. It may be possible to transport several priority 2 or 3 victims in the same vehicle. Priority 1 victims often require advanced life support (ALS) personnel on board the transport vehicle. Minor care (priority 2 or 3) victims are usually staged near the site for later treatment. Depending on the nature of the injuries, it may be possible to transport priority 2 and 3 patients in vehicles other than ambulances, such as buses.

Communication with hospital facilities is necessary to determine how many patients and what types of injuries can be treated at each facility. Hospitals determine their limits on the basis of personnel, staff expertise, and space. Hospitals must be prepared to exceed their normal patient capacity; however, they must also be trained to recognize when they have reached their emergency capacity.

Patients should be directed to hospitals and trauma care centers according to hospital capabilities. A medical planning officer, who performs functions similar to those assigned to the Planning Section, can be assigned to the Medical Branch or Medical Group as the medical planning officer. The medical planning officer should document where and when each patient is transported. The mass-casualty incident may require that only priority 1 patients be transported. Field hospitals could be used to treat priority 2 and 3 patients. Even with transportation available, it is generally not a good idea to flood hospitals with the walking wounded.

During normal operations, EMS personnel can treat individual fire fighters and occupants. When the number of potential victims exceeds normal capacity for individual treatment, the IC is well advised to staff a medical branch or medical group to coordinate medical activities. When the number and/or condition of the victims approach mass-casualty status, separate staffing of a medical branch is essential. Figure 6-7 is a flowchart for a mass-casualty incident showing victims needing various levels of care.

The mass-casualty flowchart should be read from the top down. The objective is providing the best possible medical assistance within the limitations of available resources.

Evaluating the Need for Shelter

During weather extremes, there is a need to rehabilitate fire fighters and provide shelter for occupants. Occupants of nursing homes, hospitals, and other places where special needs exist will require shelter even during normal weather conditions.

In many cases, evacuees will find their own shelter by leaving the scene to go home or to the home of a friend. When shelter is required, a nearby building—well outside the fire zone—should be used. Agencies such as the Red Cross generally have personnel available to assist in setting up and managing shelters. However, their response times may be fairly long, making it necessary to assign fire, police, or EMS crews to this task until other agencies arrive.

Estimating Life Safety Staffing Requirements

The IC must have sufficient staffing to extinguish or at least contain the fire, conduct search-and-rescue operations, treat and transport the injured, and remove victims to a place of safety while preventing re-entry. The number of fire fighters needed depends on several factors:

- Number of victims
- Rescue methods used

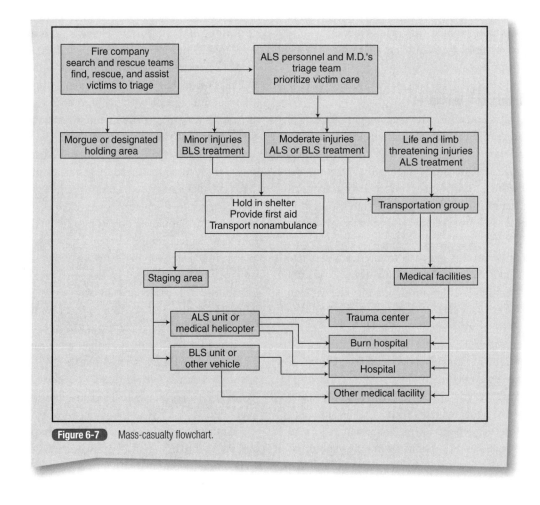

Figure 6-7 Mass-casualty flowchart.

- Condition of victims
- Fire conditions
- Smoke conditions
- Victim mobility
- Weather conditions
- Accessibility (need to force entry)

Additional staffing will be needed in the following situations:

- Victims are close to the immediate fire area.
- Victims have existing or fire-related physical impairments.
- The fire occurs during weather extremes, especially freezing weather.
- Evacuation routes other than the interior stairs must be used.
- It is necessary to force entry to rooms or hallways.

Evacuation status is determined by the following:

- Pre-incident planning information
- Occupant information
- Visual observation
- Reconnaissance
- Alarm information

Each fire department must develop SOPs based on local conditions and resources. Throughout most of the United States, crews, units, or companies are assembled according to apparatus type and function. Many different tasks are required on the fire ground; therefore, a division-of-labor approach is common practice. Fire department duties normally include the following:

- Water supply and application (engine company duties)
- Ventilation, entry, search and rescue, and property conservation (ladder company duties)
- Search, rescue, and rapid intervention (rescue company duties)
- Triage, treatment, transportation, and rehabilitation (emergency medical duties)
- Planning, organizing, coordinating, and establishing command (duties of the IC)

This division-of-labor concept is important. Without the pre-assignment of general duties, actions are delayed and/or duplicated. In a rescue scenario with occupants showing at windows, there would be a tendency to assist the visible victims at the expense of addressing more critical tasks. In other cases, fire fighters might neglect ventilation, entry, and search in favor of water application.

Many references are available that complement the information presented here or describe actual tasks necessary to fulfill the objectives of life safety. Suggested activities at the end of this chapter are designed to extend the learning experience by providing an opportunity to apply the material presented.

Summary

The most important fire-ground activity is saving lives. To accomplish this, the IC must evaluate a number of factors on arrival to ensure that rescues and evacuation are performed in the safest, most effective manner possible. Sound risk-management principles must be applied throughout the incident as conditions change to ensure that fire fighter safety is addressed while every reasonable effort is made to rescue those who are in danger.

The following is a sample of the information that must be considered and the numerous decisions the IC must make:

- Consider department SOPs and pre-incident plan information.
- Consider size-up factors.
- Determine the number and location of the victims.
- Determine the number of personnel that are needed to effect the rescues.
- Determine the resources needed to deliver the required rate of flow to control the fire.

Much of the information needed to develop an incident action plan should be included in the department's SOPs and the pre-incident plan for the property. Other decisions must be made at the scene, depending on information obtained through the IC's size-up. However, all of this information is interrelated and will require rapid decision making to ensure the safety of both the victims and the fire fighters.

Wrap-Up

Key Terms

critical time The time available until the structure becomes untenable.

defend-in-place A tactic utilized during a structure fire when it is very difficult to remove occupants from the building. Occupants are either protected at their present location or moved to a safe location within the building.

Suggested Activities

1. Evaluate a fire scenario in which a large number of occupants were rescued. The objective of this activity is to apply and evaluate rescue priorities and tactics. Answer the following questions related to the rescue operation:

   A. Was the fire attack adequate?

   B. Classify the occupancy type, and explain some of the specific problems encountered in this type of occupancy.

   C. Could the IC reasonably expect an accurate accounting of building occupants in the first 15 minutes of the operation?

   D. What rescue methods were used (interior stairs, fire escapes, ladders, helicopters, etc.)?

   E. Were the proper rescue options used? For example, were ground ladders used to rescue occupants? If so, was this the safest and most effective option?

   F. How many people would be expected to occupy this building at peak times?

   G. How many people occupied the building at the time of the fire?

   H. How many occupants were rescued or assisted to safety?

   I. Were any occupants found through search-and-rescue procedures?

   J. Were any victims physically carried or dragged from the fire area?

   K. How large an area and how many floors were searched?

   L. Were there separate primary and secondary searches? If so, were different teams used for each search?

   M. Were task assignments consistent with the rescue priorities established according to victim locations?

   N. Was the search-and-rescue operation systematic?

   O. Was a marking system used to indicate areas searched?

   P. What effect did ventilation have on the rescue operation?

   Q. Was the ventilation provided by fire forces adequate and necessary?

   R. Was forcible entry adequate?

   S. Were ventilation and forcible entry methods reversible?

   T. Were adequate medical resources called to the scene in advance?

   U. Was adequate shelter provided for displaced occupants?

2. It is 2:00 AM. An alarm is transmitted for 1000 Rialto Blvd. Weather conditions are fair, calm, and 0°F (−18°C). Upon arrival at the four-story apartment building shown in **Figure 6-8**, the building manager reports a fire in Apartment 207 on the second floor. No fire or smoke is visible from the exterior, but occupants are evacuating the building. Residents can also be seen on several second, third, and fourth floor balconies. Although there is no pre-incident plan for this building, your past experience relative to EMS responses indicates that the building is primarily occupied by elderly occupants, many of whom are disabled or only partially mobile. There is an elevator in the center section of the building with open stairways near the elevator and in each of the two end sections of the building.

   As officer of the first-arriving engine company (Engine 1), you and a fire fighter advance a 1¾" (44-mm) hose line to the second floor. The pump operator and hydrant fire fighter remain outside as the two-out team. Smoke conditions are light in the hallway, which runs the entire length of the

Figure 6-8 Four-story apartment building.

building with apartments off both sides and ends of the hallway. As you advance the 1¾″ (44-mm) line down the hallway, you encounter several elderly occupants who are self-evacuating. You check the metal fire doors leading to apartments for heat as you advance toward Apartment 207. All of the doors leading to apartments are ambient temperature. The door to Apartment 207 is very hot. You pass by Apartment 207 to check Apartment 209. The elevator and center stairs are located just beyond Apartment 209. The door to Apartment 209 is ambient temperature and the stairway has very light smoke.

A. What do you expect to happen if you do not open the door to Apartment 207?

B. What are the advantages and disadvantages in leaving the door in a closed position?

C. What do you expect to happen if the door to Apartment 207 is opened?

D. What are the advantages and disadvantages of opening the door vs. leaving the door in a closed position?

3. Use the fire scenario in Question 2 to answer the following question: You are the officer on the first-arriving truck company (Truck 1); the officer of Engine 1 has assumed a fast attack command and orders you to split your company with one crew assigned to vent the building and the other to assist them on the second floor.

There is access to three sides of the building with a large paved parking lot to the rear and streets at the front and on one side of the building. Apartment 207 is located to the rear of the building with one window facing the rear.

Your 100′ (30.5 m) straight stick aerial apparatus is positioned at the rear of the building. Select the best vent option (assume all of the following are possibilities):

A. Vent the roof by cutting a large hole at the center of the roof.

B. Vent the center stairway by removing the scuttle cover at the top of the stairs.

C. Vent an end stairway by removing the scuttle cover at the top of the stairs.

D. Vent the elevator shaft by opening the penthouse door.

E. Vent Apartment 207 via the rear window.

167

Wrap-Up, continued

4. You are the first-arriving chief officer and assume command of the scenario described in Questions 2 and 3. What would be the best method of rescuing the occupants on the third and fourth floor balconies?

   A. Remove them via aerial ladders.

   B. Remove them via ground ladders.

   C. Evacuate them using the stairs at each end of the building.

   D. Control the elevator and escort each person on a balcony to the elevator for removal.

   E. Remove them using a rescue rope.

5. Reports from Engine 1 indicate that they are attacking the fire in Apartment 207, but conditions in the hallway are becoming untenable. What would be the best method of rescuing second floor occupants visible at windows or on the balconies?

   A. Remove them via aerial ladders.

   B. Remove them via ground ladders.

   C. Evacuate them using the stairs on each end of the building.

   D. Control the elevator and escort each person on a balcony to the elevator for removal.

   E. Remove them using a rescue rope.

6. Using the same scenario and conditions explained in Question 5, categorize occupants in greatest danger to least danger, where 1 equals greatest danger and 5 equals least danger.

   A. First floor

   B. Second floor, center section

   C. Second floor, end sections

   D. Third floor

   E. Fourth floor

7. Based on status reports, you determine that there are approximately 20 apartments per floor. As IC, you decide to rescue remaining occupants on the second floor via ladders from the exterior windows and balconies. You do not know how many occupants remain in the building; therefore, you order companies to force every door on the second, third, and fourth floors and remove the occupants by means of the stairways on each end of the building. The first floor is also to be evacuated.

   Estimate the total number of companies (officer, driver, and two fire fighters per company) that will be needed to carry out the incident action plan, which includes not only rescue/evacuation, but also extinguishment, interior exposure protection, ventilation, and medical care.

Chapter Highlights

- Life safety is the first incident priority.
- Suppressing or extinguishing the fire is often the best way to protect lives.
- If the fire cannot be extinguished immediately, sometimes evacuees can be protected by using fire doors/walls and hose streams.
- The fire floor and floor(s) above the fire are critical areas and searching them must be a high priority.
- Ventilation must be coordinated with the placement of attack lines to control the fire.
- Improper venting could produce a backdraft; vents opened between fire fighters or victims and their path to egress could be fatal.
- Ventilation strategy must consider the fact that the fire and heated gases will naturally move upward first, then laterally, and finally downward.
- Whenever possible occupants should be rescued by the shortest, easiest, and most direct route using the interior stairways for above- or below-grade rescues.
- A defend-in-place strategy may be a better option when occupants are located far from the fire zone or are not able to evacuate rapidly.
- Interior stairs are the best means of egress from above grade, but fire escapes, aerial ladders, elevated platforms, ground ladders and other less desirable alternatives may be used as circumstances require.
- If aerial ladders can be properly positioned, they can move large numbers of people from a single area in a short period of time.
- Elevated platforms provide a more secure rescue and can access areas that an aerial ladder would have difficulty reaching.
- Ground ladders are generally less desirable than aerial ladders and elevated platforms, but can provide quicker access in some situations.
- When feasible, occupants at the windows should be directed to use safer alternative exits before a ladder rescue is initiated.
- Elevators should not be used for rescue except when they are verified as safe to use or when they are remote and separated from the fire area with an auxiliary power supply.
- Elevators confirmed as being safe may be helpful in rescuing immobile occupants.
- Rope rescues are among the least desirable rescue options, as this tactic is often slow and dangerous.
- When considering helicopter rescues remember that landing zones may be obstructed and air currents above a burning structure may be hazardous to helicopter operations.
- Adopting a marking system for rooms that have been searched is a highly recommended SOP.
- Occupants may need direction to safety, or fire fighters may need to provide a safe egress route for them.
- The *only* reliable way to verify evacuation status is for fire fighters to enter the structure and systematically check every room.
- Before flashover, rescue is possible; after flashover, rescue is highly improbable within the flashover area.
- Fire fighters working in multi-story buildings should survey an uninvolved floor before ascending to the fire floor or floors above.
- Floor plans have features that tend to be constant on upper stories, such as stair, elevator, and standpipe locations.
- Search-and-rescue priority should be established by deciding who is in the greatest danger from the fire.
- When it is not possible to save everyone, the IC should opt to save the largest number of people possible.
- The key to successful search operations is to be systematic.
- Medical units should be set up at one or more designated locations within the cold zone or beyond so that injured fire fighters and occupants may be treated quickly.
- A system to expedite triage, treatment, and transportation is essential in mass-casualty situations.
- During weather extremes, shelter is required for rehabilitating fire fighters and for evacuated occupants.
- Agencies such as the Red Cross generally assist in setting up and managing shelters, but fire fighters sometimes must begin the process until they arrive.
- The staff needed for life safety operations depends on the number of victims, available rescue options, condition of victims, and fire, smoke, and weather conditions.

Fire Protection Systems

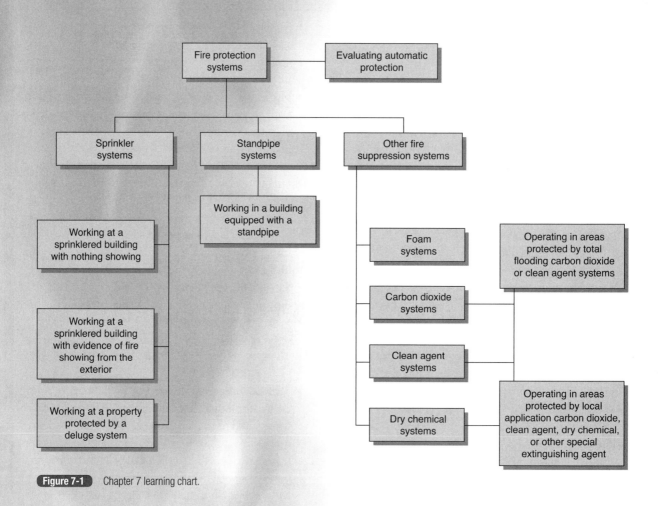

Figure 7-1 Chapter 7 learning chart.

Chapter 7

Learning Objectives

- List information related to fire protection systems that should be included in a pre-incident plan for a protected building.
- Compare residential sprinkler systems to commercial sprinkler systems.
- Explain why it is important to "Let the System Do Its Job" when conducting operations in a building protected by an automatic fire suppression system.
- Recognize the differences between wet pipe, dry pipe, and deluge sprinkler systems.
- Compare the reliability of a wet pipe sprinkler system to dry pipe and deluge sprinkler systems.
- Compare and contrast operations at a sprinkler protected building with and without obvious signs of a fire or system operation.
- Describe fire department operations at a building protected by a deluge system.
- Identify, classify, and describe different types of standpipe systems.
- Describe fire department operations at a building equipped with a standpipe system.
- Explain discharge pressure differences in standpipe systems and how these differences affect operations.
- Describe the advantages and disadvantages of solid bore and automatic nozzles when operating from a standpipe.
- Compute the pump discharge pressure needed to supply a fire line in a high-rise building equipped with a standpipe.
- Develop a list of standard standpipe equipment.
- List and describe fire protection systems other than sprinkler or standpipe systems.
- Explain fire department operations at a facility protected by a Class B foam system.
- Describe fire department operations at a property protected by a total flooding carbon dioxide system.
- Describe fire department operations at a property protected by a total flooding clean agent system.
- Define the term interlock and provide an example of an interlock on a carbon dioxide system.
- As it relates to company responses, explain the possible problems with habitual false alarm system activations.
- Pre-plan a building protected by a sprinkler system.
- Pre-plan a building protected by a standpipe system.
- Pre-plan a building protected by a foam system.
- Pre-plan a building protected by a non-water-based extinguishing system.
- Evaluate operations at a fire in a building protected by a fire protection system.

Introduction

When fire protection systems are available, the offensive attack takes on an entirely different character. By properly using an installed fire protection system, the incident commander's (IC's) job can be made much easier, and the risk to fire fighters can be significantly reduced. This chapter covers operations at properties protected by the various types of fire protection systems, as shown in **Figure 7-1**.

Fire protection systems can greatly simplify firefighting efforts. When automatic fire protection systems are present, a considerable part of the offensive attack strategy involves properly supporting or using the system. If the building or area that is on fire is equipped with an automatic fire suppression system, the primary tactic is to support the system and let it do its job.

It is much safer to let the system handle the fire than to expose fire fighters to positions deep inside the building. Efforts should be directed toward maintaining the system in a fully operational status, while laying backup lines for final extinguishment. Pre-incident planning is essential for any building that is protected by an automatic fire suppression system.

During pre-incident planning, fire fighters need to familiarize themselves with the general layout of the building. Telephone numbers of owners and building managers should be available to responders. A lock box is an ideal place to keep this information. If a lock box is not available, emergency contact information should be kept on first-alarm fire apparatus and at the dispatch center. The people included on the notification lists can then be contacted for assistance in gaining access to all parts of the building, restoring the system to service, and having employees assist in salvage operations. The location and operation of various water supply components should be known in advance, including:

- Main control valves
- Divisional control valves
- Fire pump (electric, diesel, other)
- Fire department connections
- Water supply (gravity tank, pump, public water system)
- Hydrant water supply (same as sprinkler supply, independent supply)

System limitations and peculiarities should also be addressed during pre-incident planning.

Sprinkler Systems

As of this writing, residential sprinkler system installations, as described in *NFPA 13R: Standard for the Installation of Sprinkler Systems in Residential Occupancies up to and Including Four Stories in Height,*[1] and *NFPA 13D: Standard for the Installation of Sprinkler Systems in One- and Two-Family Dwellings and Manufactured Homes,*[2] are still fairly rare. However, the available statistical data[3] indicate that these systems provide a high degree of reliability. The primary purpose of these life safety systems is to allow additional escape time. The hardware requirements for life safety systems are different from those of a sprinkler system in an industrial or office occupancy. For example, fire pumps are not generally included in the *NFPA 13R* and *13D* systems. The tactics described on the next few pages apply to all buildings that are protected by sprinkler systems; however, some modifications are necessary because of the lack of fire pumps as well as the differences in the size and construction of residential structures.

Commercial-type automatic sprinkler systems meeting the requirements of *NFPA 13: Standard for Sprinkler Systems*[4] have an exceptional record in controlling fires. Large losses of life are practically nonexistent in buildings that are equipped with a properly designed, maintained, and operating sprinkler system. When large-loss fires do occur in these properties, some degree of human error is generally involved. *NFPA 13E: Recommended Practices for Fire Department Operations in Properties Protected by Sprinkler and Standpipe Systems* lists some frequent causes of sprinkler system failure that fire fighters should prepare to resolve.[5]

> **4.1.1** Fire department personnel should be knowledgeable of and prepared to deal with the following three principal causes of unsatisfactory sprinkler performance:
> 1. A closed valve in the water supply line
> 2. The delivery of an inadequate water supply to the sprinkler system
> 3. Occupancy changes that render the installed system unsuitable

Although rare, large losses do occur in sprinkler protected buildings. There are times when an explosion or extremely

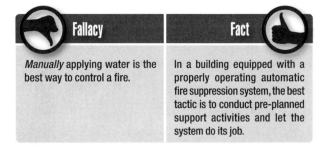

Fallacy: *Manually* applying water is the best way to control a fire.

Fact: In a building equipped with a properly operating automatic fire suppression system, the best tactic is to conduct pre-planned support activities and let the system do its job.

fast-moving fire overwhelms a sprinkler system. Sometimes improper design, such as a building feature constructed below the sprinkler head, may shield the fire from sprinkler water. Losses also occur when the sprinkler system is shut down or otherwise out of service. When the system or part of the system is out of service, the system should be tagged as specified in *NFPA 25: Standard for the Inspection, Testing, and Maintenance of Water-Based Fire Protection Systems*:[6]

> **14.3.1** A tag shall be used to indicate that a system, or part thereof has been removed from service.
> **14.3.2** The tag shall be posted at each fire department connection and system control valve indicating which system, or part thereof, has been removed from service.
> **14.3.3** The authority having jurisdiction shall specify where the tag is to be placed.

Proficient ICs are familiar with the general operation of fire protection systems. Department SOPs should address working in structures that are protected by fire protection systems.

The system that is most often encountered is the **wet pipe sprinkler system**, which is very effective and one of the most reliable fire suppression systems. The wet pipe sprinkler system is connected to a reliable water supply such as a water main or tank (or both). The water is distributed throughout the protected structure and applied to the fire through sprinkler heads. Individual sprinkler heads are self-contained detection/application devices, which account for their reliability. Valves control water distribution to sprinkler piping, and fire pumps may be needed to provide the necessary water pressure and volume to the system.

Dry pipe sprinkler systems are used in areas that might be subject to freezing temperatures. The difference between wet and dry pipe sprinkler systems is that the piping in a dry system is filled with air that is under pressure instead of water. When a sprinkler head opens because of a fire, the air bleeds out of the system. This reduction in air pressure causes the main valve to open, flooding the system with water that ultimately discharges out of the open sprinkler head. Because of this design, water may take a longer time to reach the fire than in a wet pipe system.

Pre-action sprinkler systems are also filled with air that may or may not be under pressure. The one difference is that a sensing device, such as a smoke or heat detector, opens a valve, flooding the piping with water. If a sprinkler head has also fused, then water will come out of the sprinkler onto the fire. In some systems the air in the sprinkler piping will be under pressure. If there is a loss of air pressure, as when a sprinkler head fuses, the valve will open, flooding the system with water.

In both systems, if the capacity of the piping exceeds 500 gallons (1893 L), then *NFPA 13* requires that the system be equipped with an accelerator or exhauster. These devices rapidly remove the air from the system, reducing the time necessary for the water to fill the piping.

Pre-action systems can be used when a warning is desired before actual water discharge, such as in computer rooms. In some installations, both the sprinkler head and the pre-action sensing device must actuate for the system to operate. In other words, the system will not discharge water when a sprinkler head is damaged or fused unless the secondary sensing device controlling the valve also detects a fire. This safeguard is used where incidents of vandalism require a backup alarm, such as in parking garages.

Another type of sprinkler system is the **deluge system**. With this system, there is no water in the sprinkler piping, and all sprinkler heads (or applicators) are open. When it is activated, a detector-operated control valve (which is normally closed) opens, releasing water that fills the piping and then discharges through the open sprinkler heads (or applicators) into the protected area. Deluge systems protect areas with high-challenge fires, such as flammable liquids, conveyors moving combustible commodities, and transformers. They may also be installed in aircraft hangars as combination water/foam systems. Deluge systems can also be used in many special applications, such as fixed water spray systems **Figure 7-2**. Such applications are used to protect aboveground liquefied petroleum gas tanks from fires,

Figure 7-2 Fixed water spray system piping.

dilute flammable liquids, disperse flammable gases, and protect tanks from exposure fires. Deluge systems are also equipped with manually operated override control valves that can be used if the detection devices fail to operate Figure 7-3 . Many times the manual override is located at the deluge control valve, but it could be located elsewhere. Fire fighters should know where these control valves are located and how to manually activate the system. An exterior sprinkler system, designed as a water curtain to protect a building from exposure fires, may also be available. Some of these systems will require manual actuation, and it may also be necessary to supply or augment the water supply to the exposure protection system. This is yet another reason for pre-incident planning. Many deluge and exterior sprinkler systems will have operating instructions printed near the manual control valve.

The most common operational error at properties protected by automatic sprinkler systems is shutting down the system prematurely. Fire fighters working at the scene must be sure the fire is totally under control before shutting down the system. When the system is shut down, a fire fighter should be assigned to stand by the valve in case it is necessary to reopen it quickly. This fire fighter should be equipped with a radio.

In many cases the sprinkler system will control the fire, but it may not completely extinguish fires that are shielded from direct water contact Figure 7-4 . In a warehouse, commodities may be stored in shelves that block the spray pattern and do not allow the sprinkler system to apply water directly on the burning stock. Hand lines must be in place, ready to extinguish any remaining fires, and the fire fighter who is assigned to the sprinkler valve must stand by, ready to reactivate the system should it become necessary.

> If the sprinkler system is not equipped with a back-flow valve or the back-flow valve is not operating properly, water supplied via drafting could pollute the potable water supply.

Fire department operations at a property protected by an automatic sprinkler system should not deprive the sprinkler system of water. A properly designed, installed, and maintained sprinkler system should have a calculated water requirement that includes enough water to support hose streams. However, the IC must be extremely careful to avoid depleting the sprinkler system of the water pressure and volume necessary to properly support operating sprinkler heads. The best practice is to connect attack lines to a water supply separate from the one supplying the sprinkler system. If a decision must be made between the use of hose lines and properly supplying the sprinkler system, it is usually best to supply the sprinkler system. If it becomes obvious that the sprinkler system is not achieving the desired result because of damaged piping or related problems, then the IC may elect to redirect water from the sprinkler system to support fire department hose lines.

Efforts should be made to obtain additional water supplies from sources that will not affect the sprinkler system's operations. This can be done by identifying nearby water mains that are not part of the system supplying the building. Water can then be pumped from the remote water mains to the incident to either augment the sprinkler system or supply apparatus on the scene. The local water department may also be able to bring additional water pumps on-line and provide further volume and pressure to the area.

In rare cases, usually where sprinkler piping is damaged, it may be necessary to shut the system down so that enough water is available for manual suppression. If a sprinkler system is having difficulty controlling the fire, it is much better to have hose lines at strategic locations and secure additional off-site water supplies than to shut down the system if the sprinkler piping is still in place.

Following are general guidelines that could be incorporated into department SOPs for operations at protected properties.

Figure 7-3   Manual control.

Figure 7-4 A fire shielded by obstructions.

Working at a Sprinkler-protected Building with No Signs of Fire

Listed here are tasks that should be performed at buildings that are equipped with a sprinkler system when no signs of fire or system operation are evident from the outside. It is assumed that there was some report of an unusual condition or a fire alarm system was activated that caused the fire department to be notified.

Gaining Entry

If a key to the building is contained in a lock box, it may not be necessary to force entry. Likewise, if someone who can provide access is responding and will arrive within a few minutes of the fire department's arrival, it may be advisable to wait for assistance. Conversely, if there will be a considerable delay in entering the building, it may not be appropriate to wait for assistance. The potential fire or water damage generally outweighs the damage done by forcible entry. When it is necessary to force entry, crews should do so with consideration to property damage. Many times, ladders can be used to gain entry through upper-story windows. Guidance concerning forcible entry should be outlined in SOPs. Pre-incident plans should identify the location of lock boxes and identify if keys are available, and include information concerning potential entry locations.

Checking the Main Control Valve

A fire fighter equipped with a radio should be sent to the sprinkler system riser (main shutoff). Depending on available staffing, the fire fighter may be assigned to remain at the valve throughout the entire operation. This fire fighter determines whether the system is flowing and checks the valve to ensure that it is in the fully open position. One method of checking for water flow, if the riser is not equipped with a device such as a water motor gong, is to place an ear against the riser and listen for water flowing through the pipe. If this area has the potential to become a hazard area, then the buddy system would apply, with two fire fighters assigned to the control valve.

Large buildings or complexes could have multiple fire department connections and multiple sprinkler or standpipe systems. Multiple fire department connections are often interconnected; supplying any of the fire department connections therefore provides supplemental water to the entire building or complex. However, there are buildings or complexes where the fire department connection supplies only one system. Pre-plans must address this issue. If the intakes are not interconnected, the area covered by each intake should be specified on the pre-plan and a marking system used to identify intakes.

Several types of main valves are used to open and close sprinkler systems. The two most commonly used are the outside stem and yolk valve (OS & Y) and the post indicator valve (PIV) Figure 7-5 . Again, in preparing pre-incident plans, the sprinkler valve type, location, and operation should be noted. In most situations the sprinkler valve will be locked in the open position. It may be necessary to cut or break the lock to shut down the system.

Checking the Fire Pumps

A fire fighter (or two fire fighters if conditions warrant) equipped with a radio should be assigned to check the fire pumps. If the main pump is operating, there is a good chance that the system is discharging water, either accidentally or onto a fire. If a fire is detected but the pumps are not operating, the fire fighter can manually start the pumps at the direction of the IC. It is a poor practice to rely on remote annunciator panels to determine whether the fire pumps are running. An actual physical check of the pumps is needed.

The fire fighter who is assigned to the main control valve may also be in a position to monitor the fire pump. In many cases the fire pumps are located near the main shutoff valve. This is an important position that should not be relegated to a part-time task, requiring the fire fighter to walk a considerable distance between the riser and pumps. A good rule of thumb is to separate the tasks if the pumps cannot be seen or heard by someone standing at the main control valve.

Checking the Building for Fire and/or Sprinkler Operation

Fire fighters should be assigned to conduct a systematic check of the entire building. If there was a reason to notify the fire department because of an unusual condition, then it must be verified that nothing is amiss.

Figure 7-5 Sprinkler valves. A. Outside stem and yolk valve at riser. B. Post indicator valve.

Supplying the Fire Department Connection

A pumper connected to an adequate off-site water supply should connect two 2½″ (64-mm) or 3″ (76-mm) hose lines (large-diameter hose if the system is so equipped) to the fire department connection. The water supply for this pumper should be large-diameter hose, two 2½″ (64-mm) or 3″ (76-mm) hose lines, or a direct hydrant connection. A single 2½″ (64-mm) or 3″ (76-mm) supply line is inadequate. During pre-incident planning, off-site water sources should be identified. In department SOPs, engine companies should be assigned the task of supplying water to the fire department connection.

Working at a Sprinkler-protected Building with Evidence of Fire Showing from the Exterior

If there are signs of a fire and/or the sprinkler system is operating, the main objective is to support the system while ensuring that occupants are safe.

Gaining Entry

Again, gain entry to the building, using the minimum force necessary. However, with a fire in progress or a system operating, the time spent gaining entry will increase the risk to the occupants and will cause additional fire, water, and smoke damage, thereby justifying a more forceful entry.

Checking the Main Control Valve and Fire Pump

Fire fighters should be assigned to the main valve and pump as when there is no visible sign of fire. However, when there is a confirmed, working fire, their primary responsibilities are to ensure continued operation of the system and to provide rapid shutdown when appropriate. These positions are critical and should be staffed throughout the operation.

Large systems will be equipped with control valves on portions of the system. These valves are not as prone to accidental closing as main control valves are, but they should be checked as soon as possible.

Supplying the Fire Department Connection

When there is a confirmed, working fire, an engine company must supply water by pumping into the fire department connections, as shown in **Figure 7-6**. The department SOPs should identify minimum water supplies and pump pressures. Some publications, including *NFPA 13*, recommend a pressure of 150 psi (1034 kPa) at the fire department connections. As was previously noted, the pre-incident plan modifies SOPs whenever necessary according to the characteristics of the specific occupancy.

Letting the System Do Its Job

It is better to shut down a sprinkler system too late rather than too early. The sprinkler system should be permitted to operate until the IC is certain that the fire is under control. When the sprinkler system is shut down, the only fire remaining should be small spot fires. Hose lines should then be used to complete extinguishment.

Backing Up the System

Prepare for an offensive attack and overhaul by positioning hose lines. The hose lines should be staffed by fire fighters in full protective clothing, including SCBA. Hose lines should not be operated except to perform rescue operations and limit fire spread or for overhaul operations after the sprinkler system has been shut down. However, if the sprinkler system is ineffective due to damaged piping, malfunction, or inadequacy, hose lines may take priority.

Figure 7-6 Pumper supplying water to a fire department connection at a sprinkler-protected building.

> Positive-pressure ventilation increases pressure within a structure (or part of a structure) by using a fan or blower to force air into the building while limiting vent openings. Negative-pressure ventilation places the blower or fan at an opening inside the building to "pull" or exhaust smoke out of the structure.

Ventilating

When it is safe to do so, crews should make ventilation openings above the fire. Proper ventilation will channel the fire and limit its extension. The cooling effect of sprinkler water can inhibit upward smoke movement, possibly making it more difficult to ventilate. Recommended ventilation tactics should be part of the pre-incident plan.

Some large-area, sprinkler-protected buildings are equipped with automatic or manual roof vents and **draft curtains** that are designed to limit fire spread. Overhaul will most certainly require ventilation, usually positive- or negative-pressure mechanical ventilation. Fire fighters should not shut down the system to locate the fire; they should ventilate.

Performing Property Conservation Tasks

When possible, property conservation tasks should be accomplished while extinguishment is in progress. Extinguishment takes priority, but water damage often dictates that the IC summon enough assistance to start property conservation simultaneously.

Placing the System Back in Service

If fire crews are properly trained to do so, they should place the sprinkler system back into service by replacing fused sprinkler heads and reopening valves. Most codes require that a supply of spare sprinkler heads, of the type used in the system, be kept on the premises for this purpose. If the system is too complex or the sprinkler heads are not available, the owner or manager should restore the system. If the system is equipped with division control valves, restoring the unaffected part of the system may be possible. Some department SOPs prohibit reactivating the system, owing to possible legal implications; others believe that it is important to place the system back in service as soon as possible. The property owner should contact a licensed sprinkler contractor to inspect the system after it has operated during a fire. There is a possibility that sprinkler heads or piping could be damaged. Department policy and guidance on this subject should be included in the department's SOPs.

> When replacing sprinkler heads, it is crucial that the proper sprinkler head be used. In some buildings, the same type of sprinkler head will be used throughout the building. In other buildings, pendant, upright, sidewall, and other types of heads may be used. Sprinkler heads designed to operate at different temperatures may also be in use. The sprinkler head being replaced must be of the same type and temperature rating.

Working at a Property Protected by a Deluge System

The tasks required at a sprinkler-protected building are basically the same for wet, dry, or pre-action systems. The deluge system presents at least one additional consideration: manual operation of the deluge valve. Many of these systems are located outside of buildings. Usually, the hazard protected by deluge systems can create extreme risks for fire fighters who are attempting to manually suppress the fire. System operation will be obvious, thus negating the need for the thorough investigation required in sprinkler-protected buildings. Following are general steps to be taken at these properties. However, pre-incident planning is the key to a successful operation.

Checking the Control Valve and Fire Pump

Just as with the wet and dry pipe systems, it is important to ensure that valves are open and that the pumps are operating properly.

Operating the Deluge Valve

It is possible, though improbable, that a fire would be in progress in an area protected by a deluge system that failed to operate properly. If the system is needed for fire control, the deluge valve should be activated manually. However, it is more likely that an exposure fire would threaten an area protected by a deluge system. It may be possible to cover these protected exposures with the deluge system by operating the deluge valve if operation of the system does not create a safety hazard or cause additional damage. Consideration must be given to the water supply requirements for these systems when they are being used for exposure protection. A deluge system may deplete a private water supply system. In the case of a transformer fire, it is usually best to stay out of secured areas and allow the system to do its job.

Checking Interlocks

Deluge systems often trip interlocking devices when activated. For example, system operation may de-energize electric transformers, shut down conveyor belts, or shut off a fuel supply. In most cases, there is a means of manually activating the interlock. If these interlocks have failed to operate and it is possible to safely shut down fuel supplies, conveyor belts, or other processes that contribute to the spread of fire, this should be done. Usually, the deluge system will control the fire even if the interlocks do not function. Sometimes, it may be advisable to wait for plant personnel to shut off fuel supplies or perform other related activities.

Letting the System Do Its Job

As with the wet and dry pipe systems, it is better to shut down the deluge system too late rather than too early. A determination must be made that the fire is completely under control before shutting down. This system will be flowing large quantities of water, increasing the temptation to shut down prematurely. Remember, even with hose lines in place, it may be impossible to effectively apply the quantity of water necessary to control the fire if the system is shut down too soon. Many facilities that are protected by deluge systems are such that total loss of the facility has already occurred in the area of operation. The deluge system is merely preventing extension. For example, a transformer that catches fire, setting off a deluge system, has probably already sustained the maximum fire loss. The deluge system is designed to protect exposures.

Backing Up the System

Hose lines staffed by fully protected fire fighters are required at strategic locations for fires such as those involving conveyors. However, manually operated hose lines usually create a substantial safety hazard if the deluge system is protecting high-voltage transformers. The IC must evaluate this situation carefully, knowing that charged hose lines offer a dangerous temptation. Most of the time, the protected transformer could completely burn out without endangering lives or additional property.

Working in a Building Equipped with a Standpipe System

Standpipe systems are not automatic fire suppression systems. Unlike sprinkler systems, standpipe systems cannot operate without human intervention. A properly operating and maintained building standpipe system is helpful in conducting offensive attacks. In high-rise structures, it may be impossible to conduct a safe and effective interior operation on upper stories if the standpipe system is inoperative.

NFPA 14: Standard for the Installation of Standpipe and Hose Systems,[7] lists five major types of standpipe systems:

- *Automatic Dry:* filled with pressurized air that automatically admits water into the system when a discharge is opened
- *Automatic Wet:* filled with water with an adequate water supply that provides water when the discharge is opened

> The automatic wet standpipe system is the most common and most reliable standpipe system.

- *Semiautomatic Dry:* dry standpipe that admits water into the system piping upon activation of a remote control device located at a hose connection
- *Manual Dry:* does not have a water supply; the system relies exclusively on a supply provided via the fire department connection to supply the system demand
- *Manual Wet:* system filled with water connected to a water supply that maintains water in the system, but is not capable of providing water for firefighting purposes unless it is supplied by a fire department pumper

NFPA 14 also defines three classes of standpipes (Class I, Class II, and Class III). In addition to these variances, the standpipe can be independent from or connected to the sprinkler system. As mentioned previously, it is important to evaluate the effect of diverting water from the sprinkler system to manually operated hose streams. Expect high-rise buildings to be equipped with standpipes, but other buildings may also have standpipes. Most new "big box" mercantile properties will have a drop from the sprinkler system with a hose connection. Given the different types and classes of systems, as well as variations in pressure and design, it is imperative that pre-plan information be available for all buildings equipped with a standpipe system.

When the standpipe system is connected to a fire pump, it is important to determine whether the building's fire pump is providing sufficient pressure and volume to support firefighting operations on upper floors. If not, it may be necessary for the fire department to either supplement the water supply or provide all of the water needs through the fire department connections. In very large high-rise buildings, this may create some difficulties.

The laws of hydraulics tell us that there is 0.434-psi **back-pressure** for every 1 ft of elevation (3-kPa back-pressure for every 0.3 m of elevation). A 60-story high-rise building in which each floor is 10′ (3 m) high will yield a back-pressure of 260 psi (1800 kPa).* *NFPA 1962: Standard for the Inspection, Care, and Use of Fire Hose Couplings and*

---

*10 ft × 60 stories = 600 ft × 0.434 psi/ft = 260.4 psi; 3 m × 60 stories = 180 m × 3 kPa/0.3 m = 1800 kPa.

Nozzles, and the Service Tracking of Fire Hose[8] requires that 2½″ (64-mm) or 3″ (76-mm) hose be tested to a minimum of 300 psi (2069 kPa) and large-diameter hose be tested at a lower pressure unless it is attack-grade hose. Just overcoming the back-pressure in a high-rise building may cause hose failure. When the additional required nozzle pressure and friction loss are included, hose failure becomes likely **Figure 7-7**.

It is poor practice to use the hose lines that are preconnected to the building's standpipe system. A preconnected hose supplied in a hose cabinet is seldom tested or properly maintained. Fire fighters should bring their own hose, nozzles, and adapters into the building.

Manual wet and dry standpipes will require water to be supplied through the fire department connection. Manual dry standpipe systems are sometimes unserviceable due to valves being left open, hose thread damage due to caps missing, pipe damage, and other problems. With a manual dry system, these problems might not be known until the fire department attempts to use the system. An automatic or wet manual standpipe system will flow water if damaged or otherwise compromised, bringing immediate attention to the problem, unless the main control valve has been shut down.

Checking Fire Pumps and Main Control Valves

Just as with a sprinkler system, it is important to check to be sure fire pumps are operating and the main control valve is open, as a properly operating system is crucial when a fire occurs in the upper stories of a high-rise building.

Supplying Fire Department Connections

The water supply requirements for standpipe operations are much the same as those for a sprinkler-protected building: two 2½″ (64-mm) or 3″ (76-mm) hose lines connected to

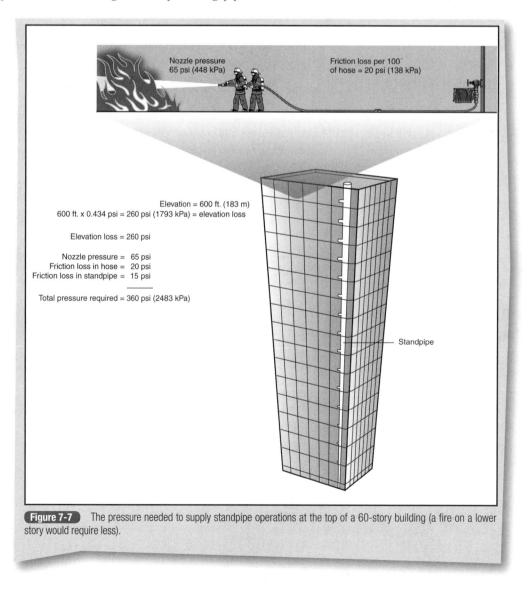

Figure 7-7 The pressure needed to supply standpipe operations at the top of a 60-story building (a fire on a lower story would require less).

the fire department connection (large-diameter hose when the connection is so equipped), the pumper being supplied from a hydrant by large-diameter hose, or the direct, soft suction connection.

As noted previously, some standpipe systems will not have a fixed water supply and will rely entirely on fire department pumpers to provide a water supply. As shown in Figure 7-7, the volume supplied must be hydraulically calculated, allowing for elevation loss, friction loss in the hose, friction loss in the standpipe piping system, and nozzle pressure. Standard fire-ground hydraulic calculations should be adequate with a slight allowance for system piping. In most operations an allowance of 10- to 15-psi (69- to 103-kPa) friction loss is adequate for the standpipe system piping. It is not necessary to determine the exact friction loss in the standpipe system when a fire occurs. However, if additional information regarding friction loss in the system piping is available through pre-planning, it should be noted, especially if it is appreciably more than 15 psi (103 kPa).

Providing Standpipe Equipment

As a minimum, standpipe equipment should consist of the following:

1. First-arriving engine company:
   - Two 50′ (15-m) lengths of 1¾″ (45-mm) or larger diameter hose
   - Smooth-bore or automatic variable-stream nozzle capable of providing sufficient flow at low pressures
   - A set of adapters, including a 2½″- to 1½″ (64- to 38-mm) reducer

2. Second-arriving engine company:
   - Three 50′ (15-m) lengths of 2½″ (64-mm) hose
   - Variable stream or smooth-bore nozzle. If a variable stream nozzle is used, it should preferably be one that is capable of being converted to a smooth bore.

3. Truck company:
   - Forcible entry, ventilation, and salvage equipment as required

This list of equipment will provide an attack line that is capable of containing most fires and a larger backup line. However, pre-incident planning could identify the need for additional, specialized equipment. Rate-of-flow determinations, as described in Chapter 8, are as essential in standpipe operations as they are in any other manual fire suppression operation.

Figure 7-8 shows standpipe equipment that would be placed in two quick-opening carrying cases with each fire fighter carrying a hose pack. This arrangement allows fire fighters to share the equipment load and provides an efficient means of carrying the standpipe equipment into the building and deploying it at the standpipe outlet.

Some buildings have pressure- or flow-reducing valves in the standpipe system. These valves can cause problems for fire fighters using the system, as was experienced at the Meridian Plaza fire in Philadelphia that killed three fire fighters.[9] In this case, pressure-reducing valves were improperly set to reduce the pressure to less than 60 psi (414 kPa) at the valve. Steps have been taken to eliminate this problem, but good pre-incident planning and routine inspections will reduce the possibility of this occurring. Many pressure- or flow-reducing valves are field adjustable.

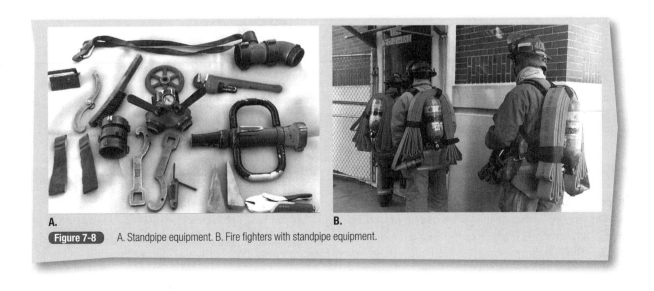

Figure 7-8 A. Standpipe equipment. B. Fire fighters with standpipe equipment.

Some field adjustments are as simple as removing a fitting that reduces the size of the discharge opening. Other valves are more difficult to change and may require special tools to regulate the pressure. Instructions for increasing the flow and/or pressure should be included in pre-plans, and members must be shown how to adjust the pressure setting during training sessions. The pre-planning survey should include arrangements to obtain special standpipe adjustment tools. Buildings equipped with complex pressure-reducing valves (PRVs) will likely be buildings with elevators; therefore, it might be desirable to place standpipe adjustment tools with fire service elevator keys. The pressure-regulating tool should be added to the standpipe equipment before ascending to the fire floor. PRVs must be installed and maintained properly to ensure that they can provide the required volume and pressure for firefighting operations. With proper pre-planning and maintenance, it should not be necessary to field-adjust PRVs.

Automatic nozzles do not provide good flows at pressures below their design parameters. Some automatic nozzles require a nozzle pressure of 100 psi (690 kPa) and will not provide any flow at lower pressures. These nozzles should not be used as standpipe nozzles. Conversely, smooth-bore nozzles can produce an adequate stream at low discharge pressures and are often the better choice for standpipe operations. A 2½″ (64-mm) line with 50-psi (345-kPa) nozzle pressure and 1⅛″ (29-mm) tip will flow approximately 265 GPM (approximately 17 L/sec). At 30-psi (207-kPa) nozzle pressure, a 1⅛″ (29-mm) smooth-bore nozzle will flow over 200 GPM (13 L/sec).

At the other extreme, it is possible to find standpipe discharge pressures above 150 psi (1034 kPa). A 2½″ (64-mm) line with an automatic nozzle will produce a large flow at this pressure. The 1⅛″ (29-mm) smooth-bore nozzle is not designed to operate at such high pressures. After discounting a friction loss of 40 psi (276 kPa), the 1⅛″ (29-mm) tip will be flowing nearly 400 GPM (25 L/sec) with a nozzle reaction force of nearly 220 psi (1517 kPa). The solution is having the right tools at the right place, as well as knowing how and when to use them.

Some departments have automatic 1¾″ (45-mm) nozzles with a 1½″ (38-mm) adapter above the shutoff, allowing the fire fighter at the nozzle to change to a smooth-bore nozzle when appropriate **Figure 7-9**. When this is the case, the nozzle tip that is not attached should be included with the standpipe equipment. When this configuration is available, some officers carry the smooth-bore tip in the pocket of their turnout coats for easy conversion to a solid stream. Nozzle manufacturers are now designing nozzles for standpipe use that can be changed from variable to solid bore without changing the tip. When deciding on standpipe hose and nozzles, it is important to determine the actual flow capabilities at the pressures available at the standpipe discharge. Some nozzles are significantly more efficient than others. When changing from a variable-stream to a smooth-bore nozzle, it is important to recognize that there may be a significant increase in flow with a resultant increase in nozzle reaction (see Chapter 8 for more information regarding rate-of-flow and nozzle reaction). Department training should include the use of the various types of nozzles at low and high pressures. Departments should also conduct flow tests to determine the actual flows discharged using various hose layouts at expected pressures. The time to safely experiment and learn how to properly use firefighting equipment is during fire department training sessions. Trial and error on the fire ground can produce deadly results for both civilians and fire fighters.

Connecting to the Standpipe Discharge

Some difference of opinion exists regarding where to connect to the standpipe outlet. Many departments have standing orders requiring connection to the standpipe one floor below the fire. Other departments allow the connection at the fire floor, provided that the standpipe valve is in a stairway or other protected area. In either case, excess hose may prove to be a problem.

In the book *Firefighting Principles and Practices*,[10] William E. Clark presents an excellent method of dealing with the excess hose. Hose is laid up the stairway above the fire floor before fire fighters enter the fire floor. This allows them to enter the fire floor pulling the hose *down* the steps rather than up, while keeping excess hose out of

Figure 7-9 Interchangeable smooth-bore and automatic nozzle.

the way. In using this tactic, it is important to lay the hose up the steps before making entry to avoid being above the fire when the door is opened. Recognize that fire hose in the stairway will impede occupant egress and that smoke conditions in the stairway may place occupants at greater risk. Whenever possible, it is best to dedicate a stairway for fire operations and other stairs for occupant evacuation. See Chapter 12 for a more in-depth discussion regarding protecting occupant egress in high-rise buildings.

In conducting pre-incident planning, the standpipe outlets must be located and special operational requirements identified. As with sprinkler operations, the pre-incident plan may identify operational needs that are different from the department's SOPs.

Non-water-based Extinguishing Systems

In addition to water-based systems, there are many other fixed extinguishing systems that use different agents. Following is a list of fire suppression agents and types of systems that use them:

- Foam
  - Surface application
  - Subsurface application
  - Deluge
- Halon (and other clean agents)
  - Total flooding
  - Local application
- Carbon dioxide
  - Total flooding
  - Local application
  - Extended discharge
  - Hand lines
- Dry chemical
  - Local application
  - Hand lines
- Other inerting systems (using inert gases to extinguish or contain a fire)

There is an extensive variety of systems and system components that can be used, depending on the hazard being protected. Each of these specialized systems will influence the development of a strategic plan. Therefore, ICs should have a good working knowledge of all systems. In most cases, the systems will probably have activated before the fire department's arrival. In other situations, the IC may need to activate the systems manually.

The proper use of in-place suppression devices is yet another reason for pre-incident planning. All properties that are protected by a fire-suppression system should be pre-planned. The pre-incident plan should include a drawing showing the location of system components (risers, shutoffs, pumps, and agent supply containers).

Some systems can create hazards for the occupants or fire fighters. For instance, carbon dioxide and high concentrations of halon can suffocate anyone inside the enclosure. Dry chemical can cause physical harm and, at the very least, obscure vision.

Foam Systems

High-expansion foam systems are designed to protect buildings; however, these systems are rare. They are designed to fill an area, such as a basement, with foam, thereby smothering the fire.

Low-expansion foam systems are usually found at properties storing large quantities of flammable and combustible liquids. Some low-expansion foam systems are automatic; others require fire department support. Even fully automatic foam systems have provisions for manual operation.

Refineries and petroleum storage depots normally protect aboveground storage tanks with a foam system. A **foam house** will be located on the property or nearby, containing additional quantities of foam agent and a means of manually operating the system. System operation will be different at each facility, and the responding fire department should be familiar with the hazards being protected and the operation of the foam system. During the pre-planning survey, consider the proximity of the foam house to the flammable or combustible liquid hazard being protected. If the foam house is located too close to the hazard, it may be unsafe to enter the area under some fire conditions. A department protecting such a facility that fails to prepare a pre-incident plan is rightfully open to public criticism.

Fire departments must train on how to operate the foam system, but should also insist that written operating instructions be posted within the foam house. Foam is the number one defense against flammable liquid fires but is nearly useless on pressurized liquids or gases. Remember, for the automatic system, let the system do its job and support it with additional water supply and foam as needed. For the manual system, gain access to the foam house and operate the system as required.

At bulk storage facilities the quantity of foam required to suppress a fire may be very large. The quantity should be determined either by the property owner or during pre-incident planning. If the required volume is not immediately available on-site, then part of the pre-incident plan should be directed toward protecting exposures until additional

Case Summary

A fire occurred in the tank farm at the Denver airport on November 25, 1990. It took two days to extinguish the fire. Fire suppression efforts were hampered by several factors, including fire impingement on adjacent tanks (which ultimately failed and added more fuel), lack of control valves necessary to isolate the various tanks, severe weather, and insufficient supplies of aqueous film-forming foam (AFFF).

Source: Michael Isner. *Tank Farm Fire, Denver, CO,* Fire Investigation Report. Quincy, MA: NFPA.

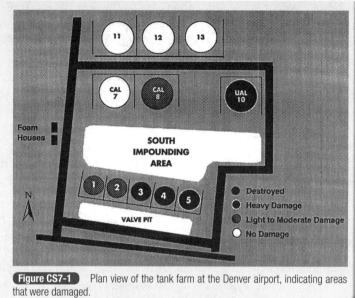

Figure CS7-1 Plan view of the tank farm at the Denver airport, indicating areas that were damaged.

quantities of foam can be obtained. Without the required quantity of foam on-site, it may be counterproductive to launch a fire attack that will not succeed. Furthermore, foam will not suppress a three-dimensional fire, such as may occur with leaking fuel.

To the authors' knowledge, automatic **Class A foam** systems are not in widespread use at the time of this writing. However, Class A foam increases the effectiveness of water and adheres well to exposed structures. Therefore, at incidents where these are important operational objectives, the use of Class A foam should be considered.

Carbon Dioxide Systems

There are two types of automatic carbon dioxide systems: total flooding and local application. In addition, some occupancies will provide carbon dioxide hose reels for manual application of the agent. Carbon dioxide is used in areas where preventing water damage is a prime objective or where this extinguishing agent is more effective than water or dry chemical. Carbon dioxide systems rely on detectors for their activation. Storage of carbon dioxide is limited; therefore, the system has a limited discharge time.

Total flooding systems depend on agent containment for a period of time (soak time) to be effective. If ventilation systems are not shut down or the compartment is opened, the carbon dioxide will quickly dissipate, and the fire may rekindle. Carbon dioxide extinguishes by depleting the oxygen supply, and a carbon dioxide system generally requires carbon dioxide concentrations ranging from 34% to 75%. A room flooded with carbon dioxide may appear normal while actually being oxygen deficient. Carbon dioxide is very cold when discharged, as well as being heavier than air, and may accumulate in low or remote locations. Fire fighters entering a room where a carbon dioxide system has discharged must wear SCBA.

Halon and Other Clean Agents

Many large computer installations are protected by sprinklers because of the reliability factor. Clean agent systems are also used because of their ability to react quickly and to suppress a fire in its beginning stages, without damaging sensitive equipment. The quick extinguishing capabilities of Halon have long been recognized in explosion-suppression systems where deflagrations are actually suppressed before pressure builds up.

Halogenated agents were greeted by many as the extinguishing agent of the future. They were considered the ultimate agents for all situations, being nontoxic, non-damaging, easily applied, and leaving the compartment safe for human habitation during discharge. The euphoria that surrounded the use of halogenated agents in the early days has disappeared. We now know that halogenated agents can damage equipment, and although some do not cause immediate death, a health risk may be associated with staying in the compartment after the agent has discharged. Halon is not considered toxic, but the products of decomposition (hydrogen fluoride, hydrogen chloride, and hydrogen bromide) are harmful to humans and might damage some electronic components.

Further, Halon and other clean agents are expensive, and studies now indicate that Halon causes environmental harm by destroying the ozone layer. What was once thought to be a miracle agent is on its way to oblivion. Other clean agents are now being used, and halogenated agents are being phased out.

Fixed clean agent systems rely on smoke detectors for activation. They have a limited supply of agent that must be discharged into a confined room or area. Most clean agents do not require suffocating quantities to extinguish the fire; a low concentration is usually sufficient. However, to be effective, the room must remain closed, and the ventilation system must automatically shut down on discharge.

Clean agents are generally considered nontoxic to humans in the concentrations that are found in computer rooms and other installations. However, personnel entering a room where a discharge has occurred must wear SCBA until the room has been completely ventilated.

Dry and Wet Chemical Systems

Dry chemical systems are used in a number of different applications. One of the most common is in kitchen hoods in restaurants. Because of the ability of dry chemical to suppress fires in cooking appliances, ductwork, and other related areas, it is widely used for this purpose. Dry chemical systems are used in many other applications, such as dip tanks and gasoline-dispensing facilities. However, kitchen hood systems are the most predominant application.

A kitchen hood system can be activated either automatically or manually by a pull station. For automatic operation, fusible links are located over the area being protected or in the ductwork. In the event of a fire, the link will fuse, releasing the tension on a cable that, in turn, will cause the dry chemical agent to be discharged through a series of nozzles.

Wet chemical systems, similar in design to dry chemical systems, are also found in kitchen hood applications. Wet chemicals react with hot grease or oil to form a foam blanket that suppresses the release of combustible vapors. Property owners often prefer wet agent systems because cleaning up after the system discharges is much easier.

As with the other specialized systems, these systems will probably have already discharged before the fire department's arrival.

Operating in Areas Protected by Total Flooding Carbon Dioxide or Clean Agent Systems

Letting the System Do Its Job

If the system is controlling the fire, maintain the chemical concentration by keeping the doors closed. Unless occupants failed to escape, there is no need to enter the area if the fire is being controlled. Entering the area will allow the carbon dioxide or clean agent to dissipate, thereby reducing its effectiveness. Unlike the sprinkler and standpipe systems, carbon dioxide and clean agent systems have a very limited supply of extinguishing agent.

Final Extinguishment and Rescue

If it becomes necessary to enter the room to perform a rescue or for final extinguishment, members must wear full protective clothing and must don their facepieces before entering. After a carbon dioxide release, the area may appear to be completely clear yet pose a serious hazard due to a lack of oxygen. Clean agents may pose a threat because of corrosive decomposition gases. Overhaul operations must be completed, especially where Class A materials are involved. Otherwise, rekindles may occur as the agent concentration is diluted over a period of time.

Manual Activation

These systems will generally be equipped with a manual actuation device that can be operated in the event of fire where automatic detection/activation sensors fail. Some systems are equipped with an abort switch, which can be held to prevent agent discharge. Sometimes the abort switch is located outside the protected area with no provisions to determine what is going on inside the protected area. Employees may decide to keep the agent from discharging for a number of reasons. The IC must determine whether preventing system discharge is justified.

Checking Interlocks

Fire suppression systems often trip interlocking devices when they are activated. At minimum, there will be provisions for shutting down the ventilation system when using carbon dioxide and clean agents. Carbon dioxide systems and many clean agent systems have pre-action alarms interlocked to the system, allowing occupants time to escape. In most cases there is a means of manually activating the interlock. If ventilation systems are operating, it is imperative that they be closed, especially with carbon dioxide total flood application. However, if the ventilation system has been running for a period of time after discharge of either carbon dioxide or clean agents, then the concentration of agent within the room may be insufficient to suppress a fire, at which time manual hose streams may be necessary.

Checking Agent Supply
On occasion a cylinder or other container holding clean agents or carbon dioxide may be accidentally shut off. There may be extra containers of agent that can be connected to the system once the original supply has been depleted. A fire fighter should be assigned the task of checking supply valves, or if extra supplies of agent are available, an entire company may be needed to connect the additional supply. However, this may require specialized knowledge, and by the time an assessment is made, the fire may have grown too large for the system to control. Manual suppression may be required, and the IC should be prepared for this eventuality.

System Restoration
System restoration will, by necessity, be left to the property owner and/or a contractor who is capable of recharging and resetting the system.

Operating in Areas Protected by Local Application Carbon Dioxide, Clean Agents, Dry Chemical, or Other Special Extinguishing Agents

Letting the System Do Its Job
Make sure valves are fully open and do not interfere with system operation.

Checking the Interlocks
Interlocks may shut off fuel supplies or de-energize equipment. Manual operation of the interlocks may be possible, or employees may be able to shut down equipment as needed.

Manual Activation
Support the system by activating manual devices when necessary. These systems are generally protecting Class B or Class C hazards and do not depend on an enclosure.

Backing Up the System
Be prepared with backup equipment, such as hose lines, foam lines, or portable extinguishers as required to augment the system and/or complete overhaul.

Case Summary

An early morning fire occurred in a restaurant in Boston, Massachusetts. The cleaning crew was attempting to remove grease from the burners by covering them with aluminum foil and then turning the burners on high. A fire occurred and spread into the ductwork above the stove. The dry chemical system did not automatically discharge, and the fire spread upward, inside the ductwork, until it reached a mechanical room on the fifth (top) floor. From that point, the fire was able to ignite the roof structure. The roof on this historic building was completely burned off as a result of a kitchen fire five stories below it.

Source: Edward R. Comeau, unpublished NFPA fire investigation.

In kitchen hood systems, it is common for dry chemical systems to also protect the ductwork leading from the cooking appliances to where it exhausts outside of the building. It is important to inspect the entire ductwork and exhaust system to verify that the fire did not spread beyond the cooking appliance.

System Restoration

System restoration should be left to the property owner or a contractor who is capable of recharging and resetting the system.

A Word About Responses to Building Fire Alarm Systems

False alarms transmitted by fire suppression or fire alarm systems are common. In some areas, they have become so frequent that special responses are sent to fire alarm activations. Although the number of fire alarms can be a nuisance, it is important to remember that a percentage of these alarms will be for actual fires. The high number of false alarms causes apathy, and apathy lulls fire forces into complacency. Then, when least expected, the inevitable occurs and lives and property are lost—lives and property that could have been saved had the alarm been heeded.

False alarms are often the result of poor system maintenance or improper installation. Unfortunately, no one has all of the answers to this problem. Jurisdictions have assessed penalties on property owners who have an excessive number of false alarms, but this may cause the property owner not to call when the alarm sounds or to shut down the system. For good reason, departments insist on a rapid call when there is an emergency. If the property owner investigates the alarm before calling the fire department, the call for assistance will be delayed.

Particularly difficult are automatic alarms sounding inside a tightly secured building. These situations often present the IC with a difficult dilemma. Forcing entry can cause significant damage; however, failing to enter the building may place people or property at risk.

To assist fire department personnel in gaining entry after business hours, the building owner may be willing to provide a lockbox with emergency access keys or make similar arrangements. However, if a decision is made to force entry, it should be done in a manner that results in as little damage as possible.

Direction should be provided to the owner through code adoption, local ordinance, or official notification from the fire chief or fire marshal. Further, the fire department should work with the building owner to identify entry options in advance and include them in the pre-incident plan.

An April, 1996 *Fire Engineering* magazine article describes several methods of investigating a secure building with an alarm sounding, including gaining access to the roof to look for signs of fire.[11]

Summary

It is essential that the IC take advantage of a working fire suppression system. If necessary, the IC should support and back up the system, being aware that manual fire suppression activity could reduce the effectiveness of the more efficient fixed extinguishing system.

The key to successful operations in buildings protected by fire protection systems is having SOPs that address these operations in a general way and having pre-incident plans that describe important features of the fire protection systems in specific properties within your jurisdiction.

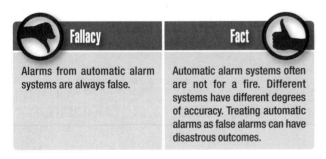

Fallacy	Fact
Alarms from automatic alarm systems are always false.	Automatic alarm systems often are not for a fire. Different systems have different degrees of accuracy. Treating automatic alarms as false alarms can have disastrous outcomes.

Wrap-Up

Key Terms

back-pressure Also known as elevation pressure, this is the pressure required to overcome the weight of water in a piping or hose system. Each vertical foot of water in a pipe, hose, or tank exerts a pressure of 0.434 psi (3 kPa) at the base.

Class A foam Foam for use on fires involving Class A fuels such as vegetation, wood, cloth, paper, rubber, and some plastics.

deluge system A sprinkler system in which all sprinkler heads or applicators are open. When an initiating device, such as a smoke detector or heat detector, is activated, the deluge valve opens and water discharges from all of the open sprinkler heads simultaneously.

draft curtain A wall designed to limit horizontal spread of the fire and which extends partially down (usually no more than 20% of the height of the compartment) from the underside of the roof.

dry pipe sprinkler system A system in which the pipes are normally filled with compressed air. When a sprinkler head is activated, it releases the air from the system, which opens a valve so the pipes can fill with water.

foam house A fixed facility consisting of an enclosure housing a foam concentrate supply tank, a foam solution proportioning system, a pump, and sometimes an extra supply of foam concentrate that can be added to the proportioning system.

pre-action sprinkler A dry sprinkler system that uses a deluge valve instead of a dry-pipe valve and requires activation of a secondary device before the pipes fill with water.

standpipe system A piping system with discharge outlets at various locations; in high-rise buildings an outlet will normally be located in the stairway on each floor level; most are connected to a water source and the pressure is boosted by a fire pump.

wet pipe sprinkler system A sprinkler system in which the pipes are normally filled with water.

Suggested Activities

1. Using the information provided in this chapter and in Chapter 2, develop a pre-incident plan for a sprinkler-protected property.

2. Using the information provided in this chapter and in Chapter 2, develop a pre-incident plan for a building with a standpipe system.

3. Compute the pump discharge pressure needed on the top floor of the tallest building in your response district including areas covered by mutual/automatic aid agreements.

4. Using the information provided in this chapter and in Chapter 2, develop a pre-incident plan for a facility equipped with a Class B foam system.

5. Using the information provided in this chapter and in Chapter 2, develop a pre-incident plan for a building protected by a non-water-based extinguishment system.

6. Evaluate a fire report from the NFPA, USFA, or other source for a fire in a building protected by an automatic fire protection system. Determine why the system failed to control the fire and list possible ways to improve the operation of the fire protection system. Some reports that could be used for this activity include:
   - K-Mart Fire—Falls Creek, PA (Sprinkler System)[12]
   - Warehouse Fire—New Orleans, LA (Sprinkler System)[13]
   - Storage Warehouse—Phoenix, AZ (Sprinkler System)[14]
   - Restaurant—Boston, MA (Dry Chemical System)[15]

Chapter Highlights

- If the building is equipped with an automatic fire suppression system, the primary tactic is to support the system and let it do its job.
- Pre-incident planning is essential for any building that is protected by an automatic fire suppression system.
- Even though automatic sprinkler systems are the most effective way to save lives and property, these systems are not as common in residential occupancies as in commercial occupancies.
- There are different sprinkler installation standards for different occupancy types (e.g., *NFPA 13D* covers one- and two-family dwellings and manufactured homes, *NFPA 13R* covers residential occupancies up to and including four stories in height, and *NFPA 13* covers other installations).
- With few exceptions, a properly designed, maintained, and operated sprinkler system will prevent large loss of life and property.
- An explosion or extremely fast-moving fire can overwhelm a sprinkler system, particularly if the explosion damaged the system or if the building has been renovated or has changed usage in ways that affect the fuel load.
- Different types of sprinkler systems include the wet pipe system, dry pipe system, pre-action system, and deluge system.
- To avoid depriving the sprinkler system of an adequate water supply, fire department personnel should avoid taking water from the same source to supply hose lines.
- The main objective when working in a building protected by an automatic fire protection system is to support the system while ensuring that occupants are safe.
- When working in a building protected by a sprinkler system, back up the system with hose lines and ventilate.
- When working in a property protected by a sprinkler system it may be possible to begin property conservation tasks simultaneously with extinguishment.
- When working at a property protected by a deluge system it may be necessary to manually operate the deluge valve.
- Hazards protected by deluge systems may pose additional dangers to fire fighters.
- A properly operating and maintained building standpipe system may be helpful in conducting offensive attacks, but it is not an automatic system—it must be manually operated.
- Five major types of standpipe systems exist: automatic wet and dry, semiautomatic dry, and manual wet and dry systems. Automatic wet standpipes are most common and most reliable.
- When working from a standpipe system check for proper operation of the main control valve and fire pumps, and supply the fire department connection.
- Fire departments should provide equipment needed for standpipe operations along with SOPs for standpipe use.
- Foam, clean agents, carbon dioxide, dry chemicals, and inert gases may be used in some situations to protect sensitive electronics and other special hazards.
- Pre-planning of buildings with non-water fire-suppression systems is essential, as improper use of and/or exposure to some agents can be harmful to fire fighters.
- Unless there are victims inside, do not enter an area where a carbon dioxide or clean agent system has discharged if the fire is being controlled.
- If it becomes necessary to enter a room protected by a carbon dioxide or clean agent system to perform a rescue or for final extinguishment, members must wear full protective clothing and must don their facepieces before entering.
- False alarms are often the result of poor system maintenance or improper installation.
- The fire department should not assume that any alarm is false, even if multiple false alarms have occurred from the same system.

References

1. National Fire Protection Association, *NFPA 13R: Standard for the Installation of Sprinkler Systems in Residential Occupancies up to and Including Four Stories in Height.* Quincy, MA: NFPA, 2007.

2. National Fire Protection Association, *NFPA 13D: Standard for the Installation of Sprinkler Systems in One- and Two-Family Dwellings and Manufactured Homes.* Quincy, MA: NFPA, 2007.

3. Jim Ford, *Automatic Sprinklers: A 10 Year Study.* Unpublished report from Scottsdale, Arizona, fire department, undated.

4. National Fire Protection Association, *NFPA 13: Standard for Sprinkler Systems.* Quincy, MA: NFPA, 2007.

5. National Fire Protection Association, *NFPA 13E: Recommended Practices for Fire Department Operations in Properties Protected by Sprinkler and Standpipe Systems.* Quincy, MA: NFPA, 2005.

6. National Fire Protection Association, *NFPA 25: Standard for the Inspection, Testing, and Maintenance of Water-Based Fire Protection Systems.* Quincy, MA: NFPA, 2007.

7. National Fire Protection Association, *NFPA 14: Standard for the Installation of Standpipe and Hose Systems.* Quincy, MA: NFPA, 2003.

8. National Fire Protection Association, *NFPA 1962: Standard for the Inspection, Care, and Use of Fire Hose Couplings and Nozzles, and the Service Tracking of Fire Hose.* Quincy, MA: NFPA, 2003.

9. Thomas J. Klem, *One Meridian Plaza, Philadelphia, Three Fire Fighter Fatalities, 2/23/91, Fire Investigation Report*. Quincy, MA: NFPA Fire Investigations Department, 1991.

10. William E. Clark, *Firefighting Principles and Practices*, 2nd edition. Saddle Brook, NJ: Fire Engineering, Penwell Press, 1991.

11. William Shouldis, The 'REALITY' of Alarm Activation Response, *Fire Engineering*, April: 98–99, 1996.

12. Richard L. Best, *K-Mart Corporation Distribution Center Fire, 6/1/82, Fire Investigation Report*. Quincy, MA: NFPA Fire Investigations Department, 1983.

13. Edward R. Comeau and Milosh Puchovsky, *Warehouse Fire, New Orleans, LA, 3/21/96, Fire Investigation Report*. Quincy, MA: NFPA Fire Investigations Department, 1996.

14. Robert F. Duval, *Storage Warehouse, Phoenix, AZ, 8/2/2000, Fire Investigation Report*. Quincy, MA: NFPA Fire Investigations Department, 2000.

15. Edward R. Comeau, *Restaurant Fire, Boston, MA, 8/31/95, Fire Investigation Report*. Quincy, MA: NFPA Fire Investigations Department, 1995.

Offensive Operations

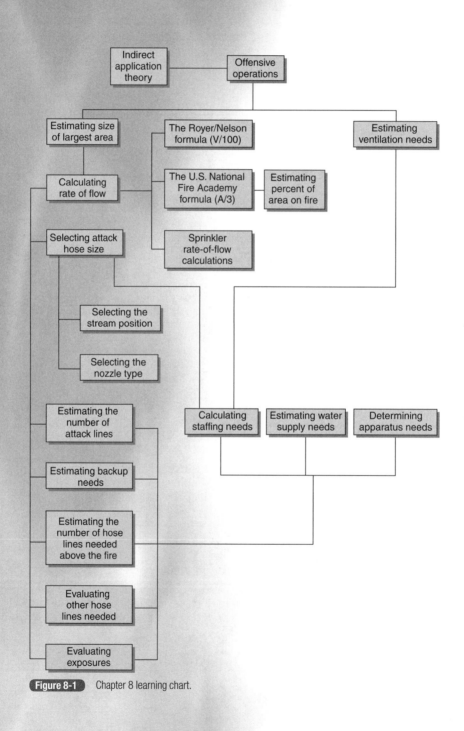

Figure 8-1 Chapter 8 learning chart.

Chapter 8

Learning Objectives

- Compare an offensive fire attack to a defensive fire attack, explaining the basics of each type of attack and identifying the rationale for each strategy.
- Describe trial-and-error methods of calculating rate of flow.
- Explain the theory of indirect extinguishment.
- Compare conditions within a fire compartment after pulsing versus after an indirect attack.
- Analyze rate-of-flow requirements using V/100, A/3, and sprinkler calculations.
- Define a ventilation-controlled fire.
- Define a fuel-controlled fire.
- Describe "area of involvement" and how it applies to rate-of-flow calculations.
- Write a brief position paper outlining the advantages of using the Royer/Nelson (V/100) rate-of-flow formula.
- Explain why a fire attack meeting or exceeding the calculated rate of flow could fail to extinguish the fire.
- Explain the relationship between nozzle type, rate of flow, and nozzle reaction force.
- Discuss the advantages and disadvantages of using an aerial device as a portable standpipe.
- Describe extinguishment of ordinary combustibles by inhibiting pyrolysis.
- Discuss the dangers of opposing fire streams and ways to avoid opposing fire streams.
- Define external exposure.
- Define internal exposures.
- List factors to consider when evaluating external exposures.
- Describe the purpose of a backup line and how it can be used to protect fire fighters attacking the fire.
- Evaluate water supply requirements based on rate of flow and other factors.
- Examine the relationship and proper use of ventilation during offensive extinguishment operations.
- Describe the factors that determine the number of apparatus needed at an offensive operation. Discuss apparatus management at a medium- to large-scale incident.
- Develop a list of advantages and disadvantages when using Class A foam during structural firefighting.
- Compute and compare the rate of flow for various areas using A/3 and V/100.
- Evaluate the available flow from standard pre-connected hose lines and determine when the rate of flow for a structure should be pre-incident planned.
- Estimate the number and size of hose lines needed to apply a calculated rate of flow.
- Assess staffing requirements for an offensive attack based on rate-of-flow and life safety factors.
- Assess the probability of an imminent life-threatening situation.
- Compare staffing available to staffing requirements.
- Using a fire scenario, assess the total water supply available and apparatus needs in terms of required fire flow.
- Given fire conditions and location, determine the ventilation possibilities and choose the best ventilation method(s).
- Evaluate the flow available from a standpipe system and standard fire department standpipe equipment based on a calculated rate of flow.
- Examine and evaluate various attack positions in a multi-story building.
- Discuss factors involved in choosing an offensive strategy.

Introduction

The objective of an offensive fire attack is to apply enough water directly to the burning fuel to achieve extinguishment. The essential question to be answered is how many gallons per minute (or liters per minute) are required to extinguish a given fire with properly placed hose lines. Calculating the rate of flow allows the incident commander (IC) to match the number and size of fire lines to flow requirements. In most cases, applying water directly to the burning fuel completes the extinguishment process. However, large offensive fire operations require the IC to consider many variables to successfully extinguish the fire, as can be seen in **Figure 8-1**.

An offensive fire attack is the preferred strategy whenever conditions and resources permit an interior attack. A defensive decision limits operations to the exterior, generally resulting in a larger property loss and limiting rescue options. The offensive versus defensive decision is based on staffing available to conduct an interior attack, water supply, ventilation, and a risk-versus-benefit analysis. Fire fighters should not enter a building in imminent danger of collapse or where fire conditions do not permit safe entry. Likewise, resource capabilities in terms of staffing, apparatus, and water must be able to meet incident requirements for a safe and effective operation. Rate of flow is a major factor in determining if resources are adequate.

> An offensive attack is preferred when conditions and resources permit interior operations.

Calculating Rate of Flow

Notable authors of fire tactics textbooks have addressed rate of flow in different ways. In his book *Firefighting Principles and Practices*, William E. Clark uses a derivation of the fire compartment volume in cubic feet divided by 100 (V/100) to determine the rate of flow in gallons per minute.[1] In *Fire Command*, Alan Brunacini states, "[w]hen the IC is able to apply more water than the fire can match with heat, we win. Until the IC reaches this level, the fire will continue to burn and eventually win (if the IC cannot overpower it)".[2] Others in the fire service take a trial-and-error approach.

> Rate-of-flow calculations as discussed in this text apply only to offensive attacks.

John Coleman[3] notes that a hose stream's extinguishing capability will generally be determined in about 30 seconds. He further recommends a trial-and-error approach in stating, "If you are at the top of a stairway and have a line directing water at a well-involved second floor or attic and you don't darken down the fire within 30 seconds or so, get more water."

This book takes an approach similar to Clark's methodology. However, trial-and-error methods are also recognized as being useful. With trial-and-error methods fire fighters start with their favorite pre-connect hose line, advance the hose line into the building, and, in many cases, make progress on the fire. Successes can reinforce bad habits, warding off any discussion of a more scientific approach. However, when trial-and-error fire fighters fail to make progress with their favorite hose line, they simply add more lines of the same kind. When they are forced out of a building after taking a severe beating during an unsuccessful effort, there is always an excuse. Seldom do they recognize their failure to flow enough water to overpower the fire. The continued use of initial attack lines when they are clearly overpowered by the fire is a sign of poor training and a lack of fire-ground discipline.

The success rate of the trial-and-error method is highly dependent on the flow rate of the standard pre-connected hose line. The larger the pre-connect line, the higher the success rate. With the advent of 1¾" and 2" (44- and 51-mm) pre-connected attack lines, success rates have improved dramatically from the days of 1" (25-mm) or booster line preference.

The trial-and-error method is probably the most commonly used method of determining flow rate. Unfortunately, many ICs fail to recognize the need for larger flows when the situation arises.

All of the methods of calculating rate-of-flow requirements have some imprecision built into them, because it is not possible to take into account every single variable that will be encountered at a fire scene. Simply stated, the question is how much water is needed to overwhelm the fire. Should operations begin with 2½" (64-mm) hose lines? Should the second hose line be a 2½" (64-mm) hose? Applying rate-of-flow principles should answer these questions.

> Placing a third small diameter hose line in the same compartment is generally a tactical error.

Fallacy	Fact
Rate-of-flow calculations are used to determine the exact amount of water needed and then to apply a prescribed flow in gallons per minute.	Rate-of-flow calculations yield a rough approximation of the flow. They assist the IC in deciding the number and size of fire lines needed for an offensive operation.

Three rate-of-flow calculation methods are in general use, and all have certain advantages and disadvantages. They are described and compared on the next few pages.

(1) Royer/Nelson: $\dfrac{\text{Volume in ft}^3}{100} = \text{GPM}$

$\dfrac{\text{Volume in m}^3}{0.748} = \text{L/min}$

(2) National Fire Academy: $\dfrac{\text{Area in ft}^2}{3} = \text{GPM}$

$\dfrac{\text{Area in m}^2}{0.074} = \text{L/min}$

(3) Sprinkler calculations: Fuel load specific calculations

Indirect Application Theory

Lloyd Layman's indirect attack theory[4] is based on the very efficient cooling effect of water being converted to steam. Indirect attacks and resultant heat absorption result in the ability to control fairly large fires with relatively little water. Fire tests conducted while Layman was in the Coast Guard are the basis of his theory. Theoretically, the indirect attack is valid. In practice, it places occupants' lives in jeopardy and is very ineffective under most structural fire conditions. This proves two important points about research:

1. A hypothesis can be theoretically sound, or even seem logical, yet be empirically incorrect.
2. Good test methods can lead to erroneous results when extended to conditions beyond the scope of the research.

Quite often, practitioners attempt to use a favorite tactic for every situation. This has often caused problems in the fire service. In the case of indirect application, fire departments attempted to apply the principle to all fires, even situations that obviously did not resemble the tightly closed shipboard fires from which the theory evolved. The prescribed rate of flow for indirect applications, as described by Lloyd Layman, would be much less than any of the flow calculations in common use today. Advocates of this system point to the reduced water usage and resultant reduction in property loss. In fact, the loss due to proper use of water in a direct attack is overstated, and failure to extinguish the fire will inevitably result in far more damage. It is important to note that indirect application disrupts the heat balance within the room, and the combination of high heat and humidity will greatly decrease the chance of survival for anyone occupying the compartment.

> The purpose of rate-of-flow calculations is to determine the size and number of hose lines needed when a fire progresses beyond the normal room and content residential fire, but when an interior, offensive attack is still practical.

There has been a recent resurgence in using a modified, indirect attack tactic (often referred to as pulsing) to reduce the flashover potential. Pulsing involves directing a short blast of water toward the ceiling to reduce the potential for flashover. There is some dispute as to whether pulsing should actually be categorized as an indirect attack, because it is not meant to extinguish the fire but rather to knock down the fire overhead allowing entry into the compartment to perform a direct fire attack.

The Royer/Nelson Formula

Keith Royer and Bill Nelson of Iowa State University developed fire-flow calculations for structural firefighting based on test fires. They hypothesized that most fuels release 535 BTU (564 kJ) when combined with 1 ft³ (0.0283 m³) of oxygen. Thus, the fuel load and fire type are not as important as the room volume and can be ignored because the fire is oxygen controlled. This proposition is somewhat in conflict with standards developed by fire-protection engineers in determining sprinkler flow calculations. Sprinkler flows are adjusted by the size and type of fuel.

The **Royer/Nelson formula** is based on the premise that the best rate of application is one in which the fire is controlled in 30 seconds of effective application (80% efficiency). Many test fires were conducted at Iowa State University to evaluate fire behavior; the Royer/Nelson volume formula was a result of observations and measurements taken at actual test fires. Royer and Nelson calculated the rate of flow in gallons per minute as the volume of the fire area in cubic feet divided by 100 (V/100), as illustrated in **Figure 8-2**. This is the formula adapted in Clark's book *Firefighting Principles and Practices*.

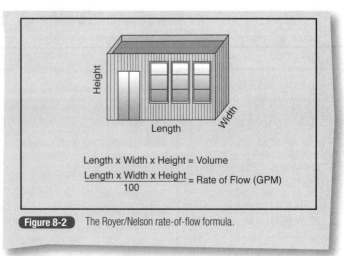

Figure 8-2 The Royer/Nelson rate-of-flow formula.

The major assumption of Royer and Nelson—that structural fires are primarily **oxygen (ventilation) controlled**—is true for many fires. However, as the area gets larger and ventilation occurs, the fire may become **fuel controlled**, thus negating the major assumption behind their theory. These large-area fires are the most likely to require larger hose lines and the application of rate-of-flow principles. Oxygen- (ventilation-) controlled fires require less water than a free-burning fire. Therefore, the Royer/Nelson formula may understate the needs in a well-ventilated, large-area fire.

Gross underestimation is most likely to occur when fires reach a point beyond the capability of an interior attack and other factors, such as the reach of fire streams, water supply, pump capacity, and staffing, become more important. Therefore, the Royer/Nelson formula (V/100) will be valid for most fires where an interior attack is advisable.

Although Clark advocates the V/100 formula, he acknowledges that the progression to flashover is based primarily on the ratio of the surface area of the fuel to the size of the enclosure. FPETOOL[5] (see Appendix A) graphically shows this relationship. If other factors such as fuel type and fuel load are held constant and the volume of the enclosure is increased, the time to flashover will also increase.

The process leading to flashover and backdraft are described in basic fire fighter training manuals, such as *Fundamentals of Fire Fighter Skills*.[6] Although the conditions necessary for backdraft (an oxygen-deficient atmosphere) and resultant explosive effect are much different than flashover, the process leading to each event is similar. The major difference is the oxygen-starved condition necessary for backdraft and the oxygen-sufficient condition required for flashover. The process begins with a fuel being heated until the vapors being generated (via pyrolysis) ignite. The fire continues to burn and grow to a point of oxygen depletion. This results in a ventilation-controlled fire, followed by a vented or completely involved fire. If the fire is allowed to smolder at high heat levels rather than venting, the potential for backdraft exists. Too many variables occur in actual fires to accurately predict the progress of an ignition-phase fire to the fully developed phase. However, the effect of volume is fairly predictable. A large fire area has more oxygen and therefore remains a fuel-controlled fire for a longer period of time. The heating of all combustibles is delayed as heat rises and reaches equilibrium in large-area fires. All of this leads to the conclusion that no rate-of-flow calculation is completely accurate.

The U.S. National Fire Academy Formula

The use of rate-of-flow calculations is simply to determine whether the fire is beyond the capacity of the standard attack. Most departments begin an interior attack using one 1¾″ (44-mm) hose line with a required backup line at least as large as the initial attack hose line. The backup line can be used to augment flow if necessary. Therefore, the normal attack for most fire departments would be equal to the flow from two standard pre-connected hose lines. Thus, the flow delivered using SOPs will vary depending on the size of the backup line and the flow produced by the standard pre-connected hose line. With this in mind, the more conservative—and probably less accurate—rate-of-flow calculation introduced by the U.S. National Fire Academy[7] cannot be fully discounted. To these authors' knowledge, test fires, such as those used to develop the Royer/Nelson formula, were not used as a basis for the **U.S. National Fire Academy formula**. Therefore, the validity of the formula is yet to be proven.

The U.S. National Fire Academy developed the A/3 formula for use in their fire tactics courses **Figure 8-3**. Many experts believe that calculating the volume at the scene of a structure fire is not practical, thus justifying the simpler area calculations. An article by Royer and Nelson

> ASETBX is an application within a computer program called FPETOOL, produced by the NIST Building and Fire Research Laboratories. ASETBX estimates the rise in temperature and descent of the smoke layer based on a single compartment's size.
>
> Such programs are especially useful in reconstructing a fire.

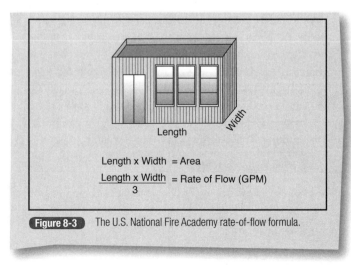

Length × Width = Area

$$\frac{\text{Length} \times \text{Width}}{3} = \text{Rate of Flow (GPM)}$$

Figure 8-3 The U.S. National Fire Academy rate-of-flow formula.

recognizes that the V/100 rate-of-flow formula is primarily a planning tool.[8]

The A/3 and V/100 formulas yield very different numbers. Typically, the A/3 formula requires more water than the V/100 formula does. The A/3 formula generally overstates the need, but as the ceiling height increases, the A/3 formula becomes less conservative.

Large-volume fires are the fires that require the application of a rate-of-flow formula. Any mathematical calculation on the fire ground is difficult, but experience can be gained through practice by estimating volume or area during training. These estimates can then be compared to the actual area and volume.

Given the difficulty in calculating the rate of flow at the fire scene, a strong recommendation is made to include rate of flow as part of the pre-incident plan information when the flow exceeds the capabilities of two standard initial attack hose lines used by your department. For example, many departments attach an automatic nozzle to 200' (60 m) of 1¾" (44-mm) hose with a pump discharge pressure of 150 psi (1034 kPa) as their initial attack line. Depending on the hose and nozzle configuration, this one 1¾" (44-mm) initial attack configuration yields a flow of approximately 125 GPM (473 L/min). Therefore, this department would pre-incident plan any building with a compartment where the rate of flow exceeds 250 GPM (946 L/min). The rate of flow would then be listed for applicable compartments within the structure. If one 2" (51-mm) attack line with a 1" (25-mm) smoothbore nozzle (approximate flow of 240 GPM [908 L/min]) were used as the initial attack hose line with a matching backup line, then a pre-incident plan would be suggested for any building with a compartment requiring more than a 480-GPM (1817-L/min) rate of flow.

Sprinkler Rate-of-Flow Calculations

Both the Royer/Nelson and National Fire Academy formulas ignore the fuel load and fuel type; this is a substantial weakness in these formulas. For this reason, **sprinkler system calculations** may prove useful in pre-incident planning. There are several sources of sprinkler calculations, including NFPA documents and Factory Mutual Data Sheets.

Fires go through a series of stages from ignition phase to flashover. These stages are characterized by being fuel dependent in the early stages and then oxygen dependent in later stages. A well-involved fire in a vented area will be controlled by both fuel and oxygen. Therefore, the IC must consider the weaknesses of the rate-of-flow calculations being applied.

A fire in an area containing a small fuel load will generally require less water than one in a similar area with a large fuel load unless the fire is totally oxygen controlled in a tightly enclosed space. Once this space is vented, either by the fire or by fire fighters, the fire will react to the amount and type of available fuel. Remember, rate of flow is being used to determine the size and number of hose lines needed and whether the fire is beyond the capabilities of available resources. The IC decides when it is time to use larger, more powerful streams. The IC also uses rate-of-flow calculations to determine whether there are extenuating circumstances—reasons other than rate of flow—when operations are unsuccessful. If two 1¾" (44-mm) lines are not controlling a fire with a calculated rate of flow of 100 GPM (379 L/min), the problem is probably a failure to apply water directly on the burning fuel. Proper ventilation often will make the fire more visible and allow fire fighters to get closer to the fire, thus improving the chances of successful direct application.

Estimating the Size of the Largest Area

Rate-of-flow calculations are based on the area or volume of the compartment(s) on fire. There is some difference of opinion among practitioners of rate of flow as to whether to make allowances for compartmentation. Some advocate applying rate of flow to the entire area or volume on fire; others see each individual compartment as a separate fire area.

This text recommends calculating each room as a separate fire area. This has been a point of discussion over the years, but Royer and Nelson's research clearly indicates that each enclosure should be calculated separately.

In a discussion with Keith Royer, he emphasized that flows should be determined considering each room capable of holding heat as a separate fire area. Royer elaborated by stating that he was talking not just about areas enclosed by fire walls and fire doors, but about any enclosure, even if the door is missing. This is an extremely important distinction. Few large areas are undivided, and in fact it is easier to fight fires in a series of smaller compartments than to do battle in one large, undivided area. Fires completely involving a large, undivided area often require a defensive attack.

> Rate of flow should be included in pre-incident plans when the size of the largest compartment exceeds the rate of flow established by the department's SOPs for the initial attack, including provisions for a backup line.

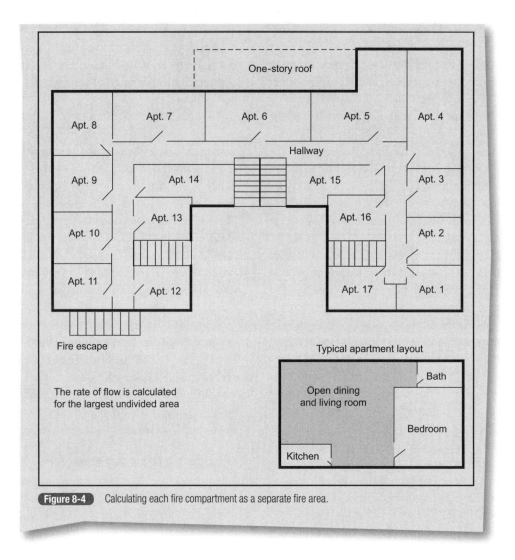

Figure 8-4 Calculating each fire compartment as a separate fire area.

In looking at the plan view for the apartment building in Figure 8-4, those advocating calculating the entire area would apply the rate-of-flow formula to the entire space. Some would recommend adding each floor to the total. The more rational approach would be to calculate each individual space as a separate fire area. If each individual space is calculated separately, it becomes possible to move down the hallway, then into each apartment, extinguishing the fire in one room at a time. Instead of calculating the entire area, only the flow needed for the largest area (the open dining and living room in this case) is required.

Practical limitations apply to this logic, as it probably would not be possible to enter the floor if the entire floor were on fire. This would lead to a defensive attack, in which the rate of flow does not apply. If several rooms are on fire, it may be advisable to use multiple lines to expedite extinguishment by attacking more than one room simultaneously.

Estimating the size of the largest involved area while on the fire scene is extremely difficult. The IC does not always have an exact location or a detailed floor plan. Pre-incident planning the rate of flow where it exceeds the capabilities of two standard attack lines is recommended. If the IC is confronted with a fire exceeding the rate of flow for two standard pre-connects with no pre-incident planned rate-of-flow data available, he or she will be forced to estimate the size of the involved compartment to determine

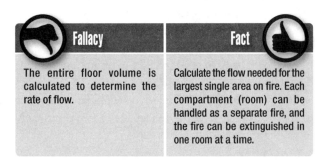

Fallacy: The entire floor volume is calculated to determine the rate of flow.

Fact: Calculate the flow needed for the largest single area on fire. Each compartment (room) can be handled as a separate fire, and the fire can be extinguished in one room at a time.

the rate-of-flow requirement, or else revert to trial-and-error methods. Fire fighters making the initial attack may be able to give the IC an estimated compartment size. Also, for a large, undivided building, the IC can estimate the length, width, and height by comparing them to lengths of fire hose or to the height of a pumper **Figure 8-5**. However, these methods are much more difficult and less reliable than consulting a pre-incident plan that includes rate-of-flow calculations. The department that fails to pre-incident plan buildings requiring a large rate of flow places their ICs at a considerable disadvantage.

Either of these calculations (A/3 or V/100) is possible at the fire scene, and each provides only a rough approximation. Both calculations assume that the entire area is undivided. If the area is subdivided, the rate of flow will be less, possibly within the capabilities of 1¾" (44-mm) hose lines if the compartments are small. If the building is not subdivided, the large rate of flow calculated by either method would indicate that once a fire involves a considerable portion of the building, an interior attack would probably be unsuccessful even if fire fighters could safely enter. This undivided structure most certainly would not be a situation in which additional 1¾" (44-mm) hose lines should be used. If the building is undivided and a considerable area is involved in fire, interior 2½" (64-mm) hose lines or interior master stream appliances will be required.

Given that calculating rate of flow is not an exact science, most ICs would estimate, round off, and shortcut the math. For example, a 32′ × 32′ (9.8 × 9.8 m) building would become 30′ × 30′ (9 × 9 m). Dividing the first 30′ by 3 yields 10′ × 30′ = 300 GPM (10 × 30 for the A/3 rate-of-flow calculation). The metric calculation yields 1136 L/min. This A/3 calculation results in a rate of flow exceeding the standard 1¾" (44-mm) pre-connect. Multiple 1¾" (44-mm) hose lines may be sufficient. Advancing a 2½" (64-mm) hose line could exceed the 300 GPM (1136 L/min) requirement for this fire.

By taking this concept of matching the hose line to the rate of flow one step further, it is possible to determine in advance the number and the size of hose lines needed to combat fires in one-family dwellings, apartment buildings, and small businesses. This is a form of pre-incident planning: pre-incident planning by general occupancy type. Examining a typical dwelling or apartment building will provide the IC with general rate-of-flow information that is very useful in initiating an offensive attack in these properties. For example, calculating the rate of flow for a 20′ × 15′ (6.1 × 4.6 m) living room with a 10′ (3-m) ceiling yields:

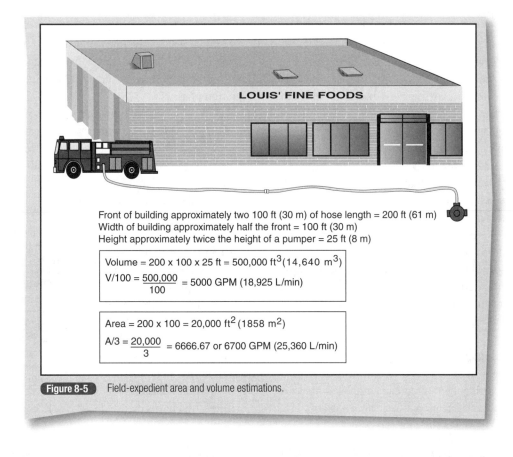

Figure 8-5 Field-expedient area and volume estimations.

$A/3 = 20' \times 15' = 300 \text{ ft}^2$ divided by $3 = 100$ GPM (379 L/min)

$$V/100 = \frac{20' \times 15' \times 10'}{100} = 30 \text{ GPM (117 L/min)}$$

For this situation, 1½" (38-mm) or 1¾" (44-mm) lines are acceptable.

Calculating the rate of flow for a 20' × 25' (6 × 8 m) basement recreation area with a 10' (3-m) ceiling yields:

$A/3 = 20' \times 25' = 500 \text{ ft}^2$ divided by $3 = 166$ GPM (643 L/min)

$$V/100 = \frac{20' \times 25' \times 10'}{100} = 50 \text{ GPM (189 L/min)}$$

A 1¾" (44-mm) hose line is acceptable only when using V/100.

These smaller areas within residential properties tend to reaffirm the lower V/100 rate-of-flow calculations. When discussing rate of flow in terms of the advantages of 1¾" (44-mm) and larger hose lines, fire fighters who prefer using 1" (25-mm) booster hose ("red line") often challenge the need for 1¾" (44-mm) hose, stating that they successfully extinguished many fires with the "red line." Most fires occur in dwellings and many in very small rooms within a dwelling. Many bedrooms are 10' × 12' × 8' or 960 ft³. Dividing the volume (960 ft³) by 100 yields a rate of flow of 9.6 GPM. The "red line" easily produces this flow. Calculating the same room using A/3 computes to 120 ft² divided by 3 or 40 GPM. The maximum flow from the low-volume nozzles typically attached to the "red line" is 40 GPM (182 L/min). Thus the successful encounters with the "red line" tend to verify the V/100 formula. However, we do not advocate using anything less than a 1¾" (44-mm) line for a structure fire. The larger the flow from the initial hose line, the greater chance of immediate success.

The basement recreation area example above represents a fairly large, single area for a dwelling. Given that the U.S. National Fire Academy figures tend to overstate the case and there should always be a backup line, it can be concluded with a fair degree of confidence that fires within one- and two-family dwellings are normally within the flow capabilities of a 1¾" (44-mm) or 2" (51-mm) hose line with backup line. Only extraordinarily large, open dwellings would require rate-of-flow calculations and subsequent pre-incident planning. Apartment buildings can also be included in this category, as they normally have smaller undivided areas, with two exceptions. Apartment buildings with common attics or basements can, and do, present large-area fires requiring additional water. Larger apartment buildings and complexes may also have common areas (recreation rooms, dining rooms, etc.) requiring a large rate of flow. Large apartment complexes may be pre-planned for reasons other than rate of flow. As an example, maps showing locations of buildings and streets may be needed to find specific addresses within a complex.

Estimating the Percent of Area on Fire

The Royer/Nelson V/100 formula is based on fuel consumption being oxygen controlled by the volume of the enclosure and *does not* reduce the rate of flow by the percentage of volume of the enclosure involved in fire. The U.S. National Fire Academy formula for rate of flow is based on the area of involvement rather than the volume of the enclosure; therefore, a reduction by the area of involvement is recommended. This difference has to do with an assumption by Royer/Nelson that the fire is oxygen controlled.

The U.S. National Fire Academy recommends a percentage-of-involvement modifier in applying the A/3 rate of flow. For example, if one-fourth (25%) of the floor area is involved, divide the total flow by 4 (or multiply by 0.25). This modifier appears to be reasonable. It takes less water to extinguish 100 ft² (9 m²) of fire than to extinguish 400 ft² (37 m²) of fire.

It is important to point out that the total V/100 flow will not always be required even though a percentage of involvement is not calculated. During the early stages of a fire in a large area, the fire will be fuel controlled, and less than the total calculated flow will be needed. Most fires can be controlled with less than 1 GPM (4 L/min) during the ignition phase.

In most cases the first company advancing to the fire area will provide the reconnaissance needed to determine the fire area in terms of percentage of involvement and compartmentation. Advancing small-diameter pre-connected hose lines (1½", 1¾", or 2" [38-mm, 44-mm, or 51-mm]) is faster than advancing 2½" (64-mm) hose. Unless the size of the fire is obvious, the first company advancing to the fire area may decide to use the standard pre-connect to size up the situation from the interior. It is important to get companies into all areas exposed to the fire as soon as possible to determine the life safety and extinguishment needs.

> When using A/3 to estimate the rate of flow, a percentage-of-involvement modifier can be used to reduce the fire area being calculated. For example, if only 25% of a 100-GPM area is burning, the required rate of flow would be 25% of 100, or 25 GPM. When using V/100, calculate the flow for the entire compartment. The percentage-of-involvement modifier does not apply to V/100.

The differences between the V/100 formula and A/3 are less significant if the percentage involvement is considered. This text recommends the V/100 for pre-incident planning, but if the A/3 calculation is used at the incident scene (because of a lack of pre-incident planning), a percentage-of-involvement modifier should be used. Since the purpose of calculating rate of flow is to determine the size and number of hose lines needed, it is reasonable to apply either formula. However, V/100 is based on a substantial body of research and is preferred for that reason.

Comparing Rate-of-Flow Calculations

A three-story building that is occupied as an auto parts store will be used as an example. The office and sales area are on the first floor, as shown in **Figure 8-6**. The second floor is a storage area for metal automobile parts in cardboard boxes. The third floor is used to store automobile and truck tires, on end, in metal racks. The building is 40′ × 60′ (12 × 18 m) with 15′ (5-m) ceilings on the first floor, and 20′ (6-m) ceilings on the second and third floors.

Royer/Nelson (V/100) rate-of-flow calculations would be as follows:

$$\text{First floor} = \frac{40' \times 60' \times 15'}{100} = 360 \text{ GPM (1362 L/min)}$$

This GPM is reduced by half for the first floor, as the first floor is subdivided into two 30′ × 40′ (9 × 12-m) areas, resulting in a flow of 180 GPM (681 L/min) for the first floor.

$$\text{Second and third floors} = \frac{40' \times 60' \times 20'}{100} = 480 \text{ GPM (1817 L/min)}$$

No reduction is made, as the second and third floors are not subdivided into smaller compartments.

The A/3 formula would calculate the same floors as follows:

$$\text{First floor} = \frac{40' \times 30'}{3} = 400 \text{ GPM (1514 L/min)}$$

$$\text{Second and third floors} = \frac{40' \times 60'}{3} = 800 \text{ GPM (3028 L/min)}$$

It is also possible to determine the required rate of flow by using sprinkler calculations. A number of variables would apply in making these calculations, such as the building type, the number of floors, the occupancy type, the commodity inside the structure, and the storage configuration of the commodity. These sprinkler requirements are addressed in *NFPA 13: Standard for the Installation of Sprinkler Systems.*[9]

Properties that are already equipped with a sprinkler system should have a plate on the riser with all of the required water-flow requirements stamped into the plate. This information could be very useful for comparing the two methods outlined above with the engineered water requirements.

> In hydraulically designed sprinkler systems, the fuel load, fuel type, and estimated area of involvement are computed when determining the required flow from the system.

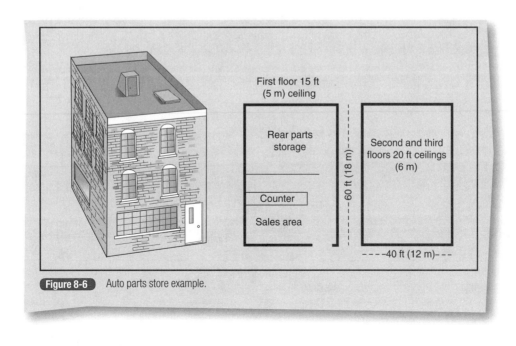

Figure 8-6 Auto parts store example.

At first, sprinkler calculations may seem complicated, but with the proper training, they are fairly easy to apply. However, sprinkler calculations will require additional time and a degree of judgment regarding the commodity's classification. Sprinkler rate-of-flow calculations are most accurate, as they are based on actual fire experience and consider the important factors of area of involvement and fuel load.

NFPA 1142: Standard on Water Supplies for Suburban and Rural Firefighting,[10] identifies occupancies where extra water supplies may be needed. This list could be used as a reference for deciding when to consult the *NFPA 13* commodity storage series of standards for extra hazard rates of flow.

Figure 8-7 and **Figure 8-8** compare the rate-of-flow calculations discussed in this chapter. Figure 8-7 uses the actual ceiling height of the auto parts store example. Figure 8-8 shows how A/3 and V/100 calculations are nearly the same as the ceiling height increases to 30′. Both figures calculate the sprinkler rate of flow for the rubber tire storage on the third floor using extra hazard sprinkler tables found in the *NFPA Fire Protection Handbook*.[11] Sprinkler calculations require a degree of engineering judgment, but the much higher calculated flow shows how high-hazard fuel storage can affect the rate of flow.

In the 30′ ceiling example, the calculated flow using A/3 is still slightly higher, but V/100 and A/3 calculations are much closer. Mathematically, the rates of flow would be equal for a 33⅓′ (10-m) ceiling. Ceilings of this height would be common in warehouse or mercantile occupancies but extremely rare on the upper floors of a high-rise.

Sprinkler rate-of-flow calculations are not suitable for on-scene use unless the pre-incident plan identifies flow requirements. With a pre-incident plan, no field calculation is necessary.

Sprinkler calculations apply to rate of flow, but the sprinkler system is far more efficient than manual operations. Sprinklers are more effective in "getting the wet stuff directly on the red stuff," and they pre-wet the exposed fuel much as fire fighters with hose lines wet exposures. Engineers designing sprinklers allow a margin of safety.

A close look at the Royer/Nelson and U.S. National Fire Academy rate-of-flow calculation systems shows that the U.S. National Fire Academy system yields a higher rate

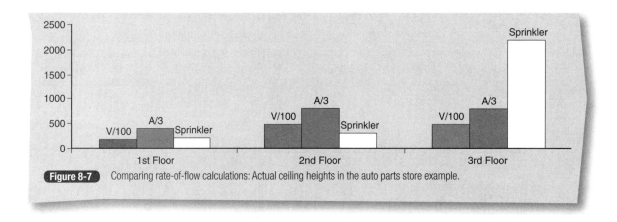

Figure 8-7 Comparing rate-of-flow calculations: Actual ceiling heights in the auto parts store example.

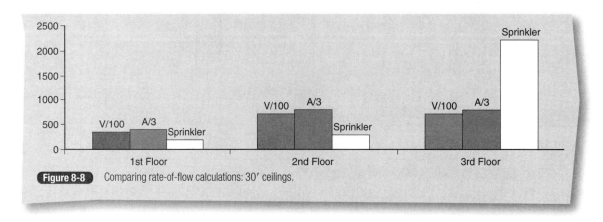

Figure 8-8 Comparing rate-of-flow calculations: 30′ ceilings.

of flow in most situations. Underestimating the flow will allow the fire to continue to make progress while fire fighters engage in a hazardous and futile battle. By contrast, if the need is overestimated, the fire is extinguished in less time with a reduction in the hazard to the fire fighters. Because all field-estimated rate-of-flow predictions are prone to error, overestimation is best. However, on closer examination it can be seen that gross overestimations could cause a delay in starting an aggressive interior attack or possibly lead to a defensive or non-attack decision because of a perceived lack of resources.

An article by C. Bruce Edwards in *Fire Engineering* points out that a rate of flow exceeding the minimum will extinguish the fire in less time and with less total water.[12] In calculating rate of flow at the incident scene, if in doubt, err on the side of higher rates of flow.

Remember that the U.S. National Fire Academy recommends a percentage-of-involvement modifier. If 25% of the auto parts store were involved in fire, the calculations would be as follows:

$$A/3 = \frac{40' \times 60'}{3} = 800 \text{ GPM} \times 0.25 = 200 \text{ GPM (757 L/min)}$$

V/100 remains the same at:

$$\frac{40' \times 60' \times 20'}{100} = 480 \text{ GPM (1817 L/min)}$$

The V/100 formula is modified only by the size of the compartment, not by the percentage of involvement, due to the assumption that fuels will produce a given heat quantity per volume of air regardless of fuel size and type. Sprinkler flow densities should be applied to the entire compartment.

Which Rate-of-Flow Calculation Is Best?

As the preceding discussion shows, there is merit in each of the rate-of-flow calculation methods. There are also assumptions and limitations for each. The best approach would build on the relative strengths and weaknesses of the various methods. With this in mind, this text recommends that all buildings requiring a rate of flow exceeding the flow of two standard pre-connects should be pre-incident planned by applying the V/100 formula.

If the fuel load is heavy in terms of total volume and/or fuel type, sprinkler calculations would be preferred, the larger of the two calculations (V/100 or sprinkler) being used to determine rate of flow. With the rate of flow calculated in advance, the IC can more accurately determine the size and number of fire lines needed.

For purposes of discussion, let's assume that the standard pre-connect hose line is four lengths of 1¾″ (44-mm) hose with an automatic nozzle and a predetermined pump discharge pressure of 150 psi (1034 kPa). Further, assume the flow rate for this standard pre-connect line is 125 GPM (473 L/min). SOPs should require that a backup hose line at least as large as, or larger than, the initial hose line be placed in service whenever there is a working structure fire. Therefore, the pre-incident plan rate-of-flow threshold for this department would be 250 GPM (946 L/min). Applying V/100 in reverse order would indicate that the compartment volume to be pre-incident planned in this case would be 25,000 ft³ (708 m³) or larger.

Returning to the auto parts store example, the second and third floors are 48,000 ft³ (1360 m³), because of being undivided and having higher ceilings. The first floor is 18,000 ft³ (510 m³).

- This building would not require a pre-incident planned rate of flow for the first floor.
- V/100 would be used to calculate the ordinary hazard on the undivided second floor.
- Sprinkler calculations would be recommended for the extra-hazard tire storage on the third floor.
- If there is no pre-incident plan for this building, after arrival on the scene, the IC may elect to use the A/3 or V/100 formula.

As discussed in Chapter 2, rate of flow is but one of many reasons to pre-incident plan a building.

When confronted with a fire in an area that is beyond the control of two standard pre-connect hose lines and where the fire flow has not been pre-planned, the A/3 formula may be easier to apply. However, this places the IC at a distinct disadvantage, as the building size must be estimated, and calculations performed under stressful fire-ground conditions might not be accurate.

Whether rate of flow is calculated in advance or on the fire ground, the IC should modify the formula on the basis of trial and error. If the fire fighters on the fire floor report that they have enough lines, they probably do (trial and error). Of course, there are times when they fail to see hidden or extended fires, but this is not where rate-of-flow calculations are useful. Likewise, if fire fighters report that the calculated 500 GPM (1893 L/min) rate of flow is not getting the job done, the IC should consider the possibility that water is not being effectively applied to the burning materials. In such cases, a change of tactics or stream placement may be needed. If this part

> An old axiom states: Failing to plan is planning to fail.[13]

of the operation is not being supervised by a division or group leader, the use of a tactical level management unit may improve extinguishment operations. This also could be a case where a lack of "truck work" is hampering the operation. Properly venting the area may allow the attack teams to better apply water directly to the seat of the fire.

Royer/Nelson assume an 80% efficiency in getting water on the burning fuel. If the nozzle team is failing to get water to the burning material, the 80% efficiency assumption is not valid. Royer/Nelson also assume the flow to be sufficient to knock down the fire in 30 seconds. This supports John Coleman's assertion that progress should be evident within 30 seconds and that the attack teams should be making steady progress on the fire if the rate of flow is sufficient and being properly applied.

Once again, using the auto parts store as an example, assume that the responding department uses 125 GPM (473 L/min) initial attack lines (1¾" [44-mm] hose line with an automatic nozzle). Further, assume that the first attack hose line will be backed up with another attack hose line of the same size or larger. Looking at each floor, we can apply the rate of flow in terms of the number and size of fire lines needed to control a fire.

First Floor

The first-arriving engine company advancing a 1¾" (44-mm) hose line into the front or rear of the building would be well positioned to extinguish the fire with the single 1¾" (44-mm) hose line, as the rate of flow of 180 GPM (681 L/min) is nearly met.

The second- or third-arriving engine company would back up the initial attack, with another 1¾" (44-mm) hose line, providing enough water to easily overpower the fire (approximately 250 GPM [946 L/min] for a 180-GPM [681-L/min] fire).

Second Floor

The first-arriving engine company would advance a 1¾" (44-mm) hose line into the building and have little chance of success if the fire involved a substantial portion of the floor area. The recommended rate of flow of 480 GPM (1816 L/min) is far from being met with the 125 GPM (473 L/min) attack line.

If the first-arriving engine company reported anything other than extinguishment with the 1¾" (44-mm) hose line, the IC should order a 2½" (64-mm) hose line as the second line. Such a line can generate a flow of up to 350 GPM (1325 L/min).[14] The 125 GPM (473 L/min) from the first hose line combined with a maximum 350 GPM (1325 L/min) from the 2½" (64-mm) hose line yields a total flow of 475 GPM (1798 L/min). This rate of flow nearly meets the 480 GPM (1817 L/min) requirement and has a good chance of success. If more water is needed, the IC could order an additional hose line or, in consultation with the first-arriving crew, increase the pressure in the 1¾" (44-mm) line, which would also increase the flow. The IC could also opt to replace the automatic nozzle with a ⅞" (22-mm) solid stream and reduce the pressure.

Third Floor

The initial 1¾" (44-mm) attack hose line would be effective only if the fire were contained to a relatively small area. Two-and-a-half inch (64-mm) lines or a master stream would be needed for extinguishment on the third floor. Pre-incident plan information could prove crucial in this situation. Given the limited access to the third floor and the flow requirements for a fully involved fire, the department may decide that fires beyond a specific level of progression (e.g., beyond the capabilities of two 1¾" [44-mm] lines) would result in a quick and limited primary search and a move to a defensive attack. The pre-incident plan may recommend a defensive attack if half or more of the third floor is involved on arrival.

When conditions permit, there is merit to initiating an attack on the fire using the standard pre-connect hose lines, thus applying a trial-and-error method. Quite often, it is impossible to determine the area involved in fire or the size of the fire from the exterior. A crew advancing a 1¾" (44-mm) hose line may be in position to begin their initial attack while performing an interior size-up. However, when the fire is beyond the resources of the available units, a defensive attack is the only option. If the rate of flow is pre-incident planned, it is much easier to determine when the fire is beyond the department's capabilities to initiate a successful offensive attack.

Selecting the Attack Hose Size

Many of the V/100 calculated flows will be less than the flow capacity of a 1" (25-mm) hose line, but given the variables involved in calculating rate of flow and the unknown factors in evaluating the fire from the exterior, booster hose is clearly inappropriate for structure fires. And, contrary to what many believe, the energy expended in stretching a booster line may be greater than that for a 1½" (38-mm) or 1¾" (44-mm) line.

> It is possible to have a small fire in a large building, but it is not possible to have a large fire inside a small building.

> There is absolutely no justification for using anything smaller than a 1½" (38-mm) line as an interior attack line. Further, a 1¾" (44-mm) hose line requires little additional effort and provides more flow, thus it is preferred over the 1½" (38-mm) hose line.

These authors recommend the use of 1¾" (44-mm) attack hose lines as a minimum. The flow rate for this size hose line will be adequate to extinguish most fires when the prescribed backup hose line follows the attack line. The work effort for 1½" (38-mm) and 1¾" (44-mm) hose is very similar, so why not go for the extra flow?

Pre-connected hose length, pump discharge pressures, and hose diameter are highly dependent on the area being protected. If the department's jurisdiction consists of single-family dwellings, anything more than a 1¾" (44-mm) hose line is probably unnecessary in most situations. If these dwellings sit well back from the street, a 250' (76-m) or longer hose line may be needed. When the average hose lay exceeds 250' (76 m), it would be wise to use 2" (51-mm) or larger pre-connects to compensate for the additional friction loss. Some departments connect one or two lengths of 2½" (64-mm) hose to the discharge and then connect four or five lengths of 1¾" (44-mm) to the 2½" (64-mm) hose to reduce friction loss in long hose lays.

As the hose size increases to 2½" (64-mm), mobility decreases. Using a larger hose size is a trade-off of mobility for an increase in flow. Handling a 2½" (64-mm) hose line in a small area is often unnecessary and extremely difficult. If the fire requires entering small areas to achieve extinguishment, 2½" (64-mm) hose lines will be a liability. Conversely, for fires in large, open areas the 2½" (64-mm) hose line is often the line of choice. However, the mobility of smaller hoses (1½" to 2") (38 to 51 mm) is of value in entering the structure to evaluate life safety and extinguishment conditions.

It is possible to use building features, such as fire doors, to hold the fire. However, this is not an ideal tactic in initiating an offensive fire attack. Either you advance on the fire, or it advances on you. Remember that the V/100 rate of flow is based on extinguishment in 30 seconds. Some fires will take longer, but steady progress should be possible if the rate of flow meets or exceeds the requirement. If water is applied at less than the required rate of flow, extinguishment will be due to burning all available fuel or to a lack of oxygen. In most cases, either of these situations is less desirable than applying the amount of water necessary to reduce the fuel temperature.

Charts from manufacturers sometimes show flow rates for variable-stream nozzles based on pump discharge pressure and the length of hose attached (e.g., 150 GPM (568 L/min) at 200 psi (1379 kPa) pump discharge pressure through 200' of hose). Other manufacturers of variable-stream nozzles base flow rates on nozzle pressure. Smooth-bore nozzles are rated using nozzle pressure; therefore, we have based our comparisons on nozzle pressure as well. Nozzle pressure is the pump discharge pressure, or in the case of standpipes the outlet pressure, minus friction loss in the hose and minus pressure loss due to elevation. Pressure loss due to elevation is 0.434 psi (3 kPa) per foot of elevation. In **Figure 8-9**, the

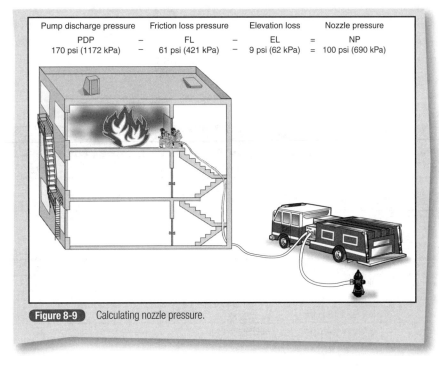

Figure 8-9 Calculating nozzle pressure.

friction loss for the hose is an estimate. The nozzle is operating on the third floor; therefore, if each floor is 10′ in height, the pressure loss due to elevation is 0.434 multiplied by 20 = 8.68 psi (60 kPa). Many departments simply use a rule-of-thumb calculation for elevation loss such as 5 psi per floor. As can be seen in Figure 8-9, the difference between nozzle pressure and pump discharge pressure can be significant.

Most of the loss in the Figure 8-9 example is due to friction loss in the hose. Calculating friction loss is a bit more challenging than elevation, because friction loss in the hose varies by size, hose type, manufacturer, age, and condition. It is best to use calibrated pressure gauges to determine the actual friction loss in the hose carried on department apparatus. Using manufacturer's friction loss information is a second alternative. Information provided on general friction loss tables, such as in **Table 8-1**, is a less desirable method and could be significantly different than the actual friction loss. Notice the difference in friction loss between 1½″ (38-mm) hose and other hose sizes at 150 GPM (568 L/min). **Table 8-2** shows the flow rate for various types of nozzles based primarily on data provided by manufacturers. The flows for automatic nozzles were derived from several manufacturers' data and are provided as examples. Prior to 1998, combination nozzles were rated at 100-psi (690 kPa) nozzle pressure. Today, fire departments choose from a wide variety of combination nozzles. Nozzles are available that offer adjustable flow rates while others limit the options to either low or high pressure. Other options include combination variable-stream and solid-stream nozzles, as well as variable-stream nozzles that are designed to operate at very low pressures. Many older style nozzles will shut down at low pressure. In addition to variable-stream nozzles, there are smooth-bore and Vindicator® nozzles.

In choosing a nozzle, it is important to identify features that will improve fire-ground operations. Problems associated with nozzle malfunctions increase with more complex nozzle designs. This is especially true if there are a wide variety of nozzles in use by the department. Do you really need to be able to adjust the flow at the nozzle? Is it important to change from a high-to-low pressure setting?

One of the advantages of a pre-connected hose line is establishing a standard pump discharge pressure, which eliminates the need to calculate hydraulics at the incident scene. However, it will be necessary to add or subtract elevation pressure to maintain a pre-determined nozzle pressure. In looking at Figure 8-9, a standard pump discharge pressure might be 150 psi (1034 kPa). With the hose line operating on the third floor, it would be necessary to add nine or ten pounds to the pump discharge pressure.

We also gathered data from fire departments that tested nozzles. It is important to note that actual test results varied, with some nozzles of the same type and manufacture providing greater or lesser flows. For new nozzles, the flow rates were generally within 10% of the manufacturers' data. However, through actual field testing, departments discovered that improperly operating nozzles produced lower-than-expected flows, and hose of the same size yielded higher or lower friction loss. Both of these factors affected flow rates. Flow rates will vary according to nozzle pressure, pump discharge pressure, and length of the hose lay. Flow rates will also vary between manufacturers and even for different nozzles produced by the same manufacturer. Table 8-2 provides general comparisons for a variety of nozzles.

TABLE 8-1 Approximate Friction Loss Per 100′ of Hose

Flow in GPM	1½″ hose (psi)	1¾″ hose (psi)	2″ hose (psi)	2½″ hose (psi)
40	4.5	3	1	Negligible
60	10	5	2.5	Negligible
100	25	12	6	3
125	37	21	10	4
150	54	26	13.5	6
175	Not Recommended	34	18	8
200	Not Recommended	45	24	10
250	Not Recommended	70	37.5	15
300	Not Recommended	95	54	21
350	Not Recommended	Not Recommended	78	28

TABLE 8-2 Nozzle Flows

Nozzle Type	Nozzle Pressure (psi)	Flow (gallons per minute)	Approximate Friction Loss (150' of 1¾" hose in psi)	Required Pump Discharge Pressure for 1¾" Hose in psi	Approximate Friction Loss (150' of 2½" hose in psi)	Required Pump Discharge Pressure for 2½" Hose in psi
Standard automatic nozzle	30	Not rated	N/A	N/A	N/A	N/A
	40	Not rated	N/A	N/A	N/A	N/A
	50	50	6	56	Negligible	50
	75	93	18	93	4	79
	100	256	110	210	23	123
Low-pressure automatic nozzle	30	110	23	53	5	35
	40	172	50	90	12	52
	50	193	63	113	14	64
	75	237	95	170	20	95
	100	273	122	222	26	126
Smooth-bore ⅞"	30	123	32	62	5	35
	40	142	37	67	9	49
	50	159	44	94	10	60
	75	195	65	140	16	91
	100	225	86	186	18	118
Smooth-bore 15/16"	30	142	37	67	9	39
	40	165	47	87	11	51
	50	184	59	109	14	64
	75	225	86	161	18	93
	100	260	113	213	23	123
Smooth-bore 1¼"	20	206	N/A	N/A	15	35
	30	253	N/A	N/A	23	53
	40	292	N/A	N/A	28	68
	50	326	N/A	N/A	37	87
	75	400*	N/A	N/A	54	129
Vindicator® heavy attack	25	175	51	75	12	37
	40	210	75	115	16	56
	50	250	105	155	23	73
	75	345	N/A	N/A	42	117
	100	440*	N/A	N/A	66	166

* Exceeds recommended maximum flow of 350 GPM for a handheld hose line.

Probably the most important point to be made is that each department should flow test their nozzles on a regular basis to be sure they are operating properly and to confirm the flow that can be expected from the hose/nozzle configurations available on their apparatus. If any part of the apparatus, hose, or nozzle configuration is changed, flow testing should be conducted. We recommend that every department perform flow tests on their hose and nozzle combinations using a calibrated flow meter. To do this, first place a flow meter in a hose line between a hydrant and the intake side of the pumper. The flow meter should not be connected directly to the hydrant. After making sure all drains are closed and the tank-to-pump line is shut, advance the standard attack line. Place a pressure gauge in the hose line behind the nozzle and another 100' (30 m) back from the nozzle. Increase the pump discharge pressure to the pressure specified in the department's SOPs. The pressure gauges will measure the actual friction loss in the 100' (30 m) of hose, and the flow meter will provide the actual flow for your hose and nozzle combination (Figure 8-10). This is also a good time to try various hose layouts. Most manufacturers do not recommend 1¾″ (44-mm) hose layouts longer than 250′ (76 m) because of excessive friction loss in 1¾″ (44-mm) hose at high flows. If longer hose lines are needed, larger diameter hose should be used to reduce the friction loss and increase the flow. Try different nozzles with different lengths and sizes of hose to determine the best combinations.

Training is essential, regardless of the hose/nozzle combination your department chooses. If the nozzle has flow or low-pressure selection devices, be sure fire fighters are completely familiar with how to operate these devices. If the nozzle has a flush mode, it is important for fire fighters to be trained in its proper operation. After flushing the nozzle, always check to make certain that it is returned to the normal operating mode prior to being placed back in service. Most important, fire fighters must train in full turnout gear with all of the nozzles available to them under conditions that simulate those found on the fire ground, such as lying on the floor, advancing the hose line through doorways, and directing the stream at different points in an enclosed area. It is essential that fire fighters experience the nozzle reaction force during training. The fire ground is not a good place to experiment. If the nozzle reaction force is too great, reduce the flow, or replace the nozzle with one that is easier to control. Remember that handling a nozzle from a standing position with the hose in a straight line behind the nozzle is very different than controlling the nozzle from a crawling or kneeling position with the hose at an angle. When nozzle reaction is near the upper limit for handheld lines, a surge in the hose line could cause fire fighters to lose control. Always allow a safety margin. Nozzle reaction force usually increases as the flow and nozzle pressure increases. However, some nozzles are designed to reduce nozzle reaction force or are equipped with devices, such as pistol grips, that assist in controlling the hose line. When purchasing nozzles, compare flow rates and nozzle reaction force in order to select a nozzle that produces the desired flow and can be safely handled under actual fire-ground conditions.

As noted previously, a handheld hose line can generate a flow of up to 350 GPM (1325 L/min). It is possible to control a hose line flowing considerably more than 350 GPM (1325 L/min) by forming a hose loop. However, forming hose loops to counteract the nozzle reaction force is not practical inside most buildings.

Pressures available at standpipes can be significantly lower than those provided by a fire department pumper and the pressure can be different depending on the building's elevation. Some high-rise buildings will have pressure or flow-reducing valves on the standpipe (see Chapter 7 for details). These valves are designed to maintain predetermined pressures throughout the building. However, there have been occasions when these valves were improperly set, resulting in very low pressures at the standpipe outlet. At the Meridian Plaza fire in Philadelphia, Pennsylvania (see Chapter 12), improperly set pressure-reducing valves created a major problem.

NFPA 14: Standard for the Installation of Standpipe and Hose Systems[15] requires that a sign be posted, usually at the main control valve, stating the location of the two most hydraulically remote hose connections with the designed flow and pressure at these connections. It is important to note the standpipe pressure and flow on the building pre-plan. If pressure- or flow-reducing valves are used, note the pressure and flow

> Pump discharge pressure is generally set at 150 psi (1034 kPa) to 200 (1379 kPa) psi for automatic nozzles. Pump discharge pressure can be directly read at the apparatus pump panel, which simplifies pump operations. Elevation loss as well as friction loss in the apparatus piping and hose reduces the pressure reaching the nozzle. The friction loss in a pumper and given length of hose vary, thus the flow will be different when operating from different apparatus and when using different hose.

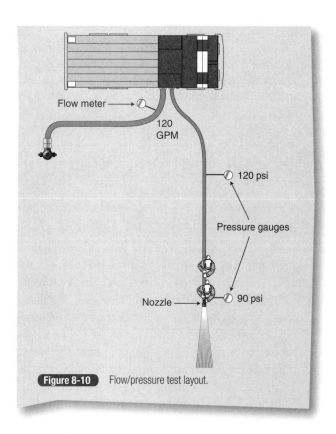

Figure 8-10 Flow/pressure test layout.

Case Summary

A fire began in the fully sprinklered K-Mart warehouse in Falls Township, Pennsylvania, where aerosol cans containing petroleum-based liquid fell from a pallet after a forklift entered the aisle. A flammable liquid vapor released from the damaged aerosol cans ignited. The fire quickly spread to involve large areas of the warehouse. The aerosol cans became flaming missiles penetrating through conveyor openings in the fire wall. The fire overwhelmed the sprinkler system, making manual extinguishment necessary. The fire department had pre-incident planned the building and decided that the best tactic would be to confine the fire to a single quadrant.

Calculating the rate of flow shows that a large fire in any quadrant would be beyond the capabilities of the fire department; therefore, using the building's construction features to contain the fire was a good plan. The height of the structure is 30′ (9 m). Quadrant A is the smallest fire area at 215,600 ft² (20,029 m²). If the fire flow were calculated for this building, it would be as follows:

$$A/3 = \frac{215,600 \text{ ft}^2}{3} = 71,867 \text{ GPM } (272,017 \text{ L/min})$$

$$V/100 = \frac{6,468,000 \text{ ft}^3}{100} = 64,680 \text{ GPM } (244,814 \text{ L/min})$$

This is for the smallest area. The fire actually began in the larger Quadrant B. Could any fire department apply 65,000 or 70,000 GPM (246,025 to 264,950 L/min) in an offensive mode? Could fire streams operated from entry points reach the far wall of any compartment which is a minimum of 440′ (134 m)? Adding to the problem was an unprotected steel-truss roof structure. Fire departments must carefully compare the risk to the benefit before committing to an interior operation that will ultimately prove futile. Even if the percentage modifier allowed in A/3 rate-of-flow calculations were used and only 1% of the area was heavily involved in fire, the value of property saving operations would be questionable.

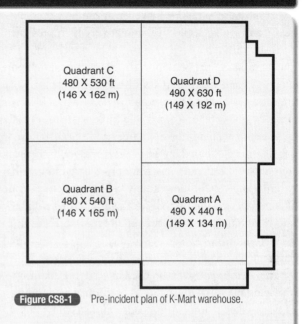

Figure CS8-1 Pre-incident plan of K-Mart warehouse.

Source: Richard Best, "$100 Million Fire in KMart Distribution Center," *NFPA Fire Journal*, March 1983, pages 36–42; rate-of-flow comparisons by authors.

---

at locations so equipped. Without pressure- or flow-reducing valves, the most hydraulically remote outlets should present the worst case pressure and flow scenario, with other outlets in the building providing higher pressure and flow capability. Standpipe systems must be inspected and tested on a regular basis. During pre-incident planning it is important to note the most recent residual test pressures.

Standpipe outlets with a residual pressure exceeding 100 psi (690 kPa) or a static pressure exceeding 175 psi (1207 kPa) are required to be equipped with pressure-reducing valves. If a pressure higher than this is needed for fire department operations, the pressure-reducing valve should be adjusted. For most situations, the maximum outlet pressure to a flowing hose line will be 100 psi (690 kPa). Some standpipe systems are only required to provide 65 psi (448 kPa) residual pressure. There may be older standpipe systems with even lower pressures due either to different code requirements at the time of installation and/or reduction in interior pipe diameter due to corrosion or other factors.

Some automatic nozzles are designed to operate at 100 psi (690 kPa) nozzle pressure (nozzle pressure will be less than the outlet pressure due to friction loss). A nozzle pressure below 100 psi (690 kPa) will result in a reduced—and possibly ineffective—fire flow. If the pressure is too low, some automatic nozzles will shut down. The low pressure automatic nozzles listed in Table 8-2 were designed to operate at lower pressures. The Vindicator® nozzle listed in Table 8-2 is designed to work at pressures as low as 25 psi (172 kPa). Solid-stream nozzles will provide some flow at most any pressure; however, at very low pressures, the stream will be ineffective, and at very high pressures, it will be difficult to safely handle the attack line. The flow increases in proportion to the nozzle pressure. Some automatic nozzles can be fitted with a smooth-bore tip, while others have a built-in solid-bore nozzle. When selecting a nozzle/hose combination for a standpipe system, consider the pressure at standpipe discharges and select standpipe hose and nozzles based on the lowest pressure. If there are large pressure differences in standpipes or buildings that require a large rate of flow, it may be necessary to provide different nozzles and/or hose for these buildings. In many cases, changing hose line from 1¾" (44 mm) to 2" (51 mm) or 2½" (64 mm) will result in a greater nozzle pressure and a larger flow.

Aerial apparatus can be used as portable standpipes by placing the aerial apparatus in position at a window on the floor where a hose line is needed and attaching hose to fittings at the elevating platform or top of the aerial. This method of providing an improvised standpipe is especially effective when the aerial device has a pre-piped waterway. This tactic does not place hose through the doorway from the stairs, thus it reduces smoke infiltration from the hallway into the stairway. Apparatus pumps will supply the pressure, thereby eliminating potential standpipe pressure problems. This tactic can also be used to advance hose inside a building that is not standpipe equipped. Of course, even when the aerial ladder is positioned at an ideal location close to the building, the portable standpipe operation will be limited to the eighth floor, which is the optimum reach of a 100′ (30 m) aerial ladder. One major disadvantage in using the aerial device for hose lines is that it is no longer available for removing occupants from the building. Another option is using a rope to advance a fire line to an upper floor. This is a more time-consuming and complex tactic, as it typically requires several fire fighters to advance the hose line, and the hose may need to be supported at intermediate levels. Also, when using a rope to advance a hose line because the stairways are unsafe, the attack team members are at a disadvantage, because they do not have the stairway or aerial device available as a means of escape. A rope can be used as a means of escape; however, fire fighters may not have the equipment, training, or experience necessary to perform a safe and effective rope escape.

When it is possible to conduct an offensive attack, the normal first action is to place the standard attack hose line in position, applying water directly on the burning material.

Pre-incident planning or the on-scene size-up may indicate the initial use of 2½" (64-mm) hose line.

Assuming that the first-arriving company will place a 1¾" (44-mm) or larger standard attack hose line in position, the trial-and-error method is generally applied. If this initial hose line does not allow the attack team to make immediate progress, more water is needed. The second crew may also lay the standard attack hose line as a backup line. Many times, department SOPs specify that the first-arriving engine company attack the fire and the second crew lay a backup line. Quite often, the IC is still in the process of developing an action plan as these pre-established procedures are implemented.

A mistake is made when the IC calls for another 1¾" (44-mm) standard attack hose line to combat a fire that is not being controlled by two 1¾" (44-mm) lines. When more than two 1¾" (44-mm) hose lines are needed to control the main body of fire, there is an obvious need for 2½" (64-mm) hose lines if an offensive attack is to continue.

Selecting the Nozzle Type

A discussion over the best type of nozzle is sure to result in a lively dialogue. Obviously, those from the indirect application school prefer finely divided fog (vapor) streams and the variable-stream nozzle that is capable of providing this pattern. The reality is that there is little need for fog streams during offensive structural firefighting. There are times when a fog stream may be appropriate, but these are mostly when the fire stream is being used to stop the forward extension of the fire, for exposure protection, or for Class B fires.

As Table 8-2 indicates, the automatic nozzle operated in the straight-stream position is capable of delivering as much water as solid-stream nozzles typically attached to 1¾" (44-mm) hose. From an empirical point of view, the solid core of water produced by the solid-stream nozzle would appear to deliver more water with better penetrating power than

> This text recommends the use of 2½" (64-mm) lines when the flow from two standard pre-connects does not equal or exceed the calculated rate of flow. The reason for using 2½" (64-mm) hose line during an offensive operation is to apply large quantities of water to the fire. The nozzle pressure and tip size will determine the actual flow from a 2½" (64-mm) hose stream. Fire departments should establish SOPs based on the nozzle types available and the flow requirement challenges in the community. The use of 2½" (64-mm) hose lines inside a building is an arduous task and requires the hose crew to gain experience through training under simulated fire conditions. As the flow increases, the nozzle reaction force will also increase.

> Keep in mind that it may be necessary to adjust pump pressures when changing nozzle tips from variable to solid stream.

a straight stream or large droplet stream. However, manufacturers have conducted technical tests measuring stream force at various distances that have disputed some of the claims made by smooth-bore advocates. Actual data from independent testing sources is sparse, so the controversy continues.

Stream force is an issue, as it affects the distance the stream will carry. In large, undivided areas, stream reach can mean the difference between success and failure. Likewise, force allows the nozzle crew to access hidden fires behind lath, plaster, and other lightweight building materials by penetrating the ceiling or wall with a powerful, compact stream.

In addition, there is sprinkler research that proves that large-orifice sprinkler heads have better extinguishing qualities. The spray pattern from these heads creates larger water droplets that are better at penetrating the fire plume and reaching the burning fuel. This research appears to favor the solid, large-drop or straight-stream nozzle.

It is possible to have the best of both worlds by attaching an automatic nozzle to the standard pre-connected hose lines, but with a 1½″ (38-mm) fitting above the shutoff. These nozzles can be changed to a solid-bore stream by removing the variable-stream tip. This allows the attack crew to quickly change the nozzle without shutting down the hose line at the pump. Nozzles are now designed with a built-in smooth-bore feature just above the shutoff. Be sure to measure the flow in each flow mode before deciding on a nozzle capable of a variable and solid stream. Fire fighters should be trained in the proper changing of nozzle tips and understand the potential difference in nozzle reaction force.

Selecting the Stream Position

Offensive fire attacks are normally described as __indirect__, __direct__, or __combination__. The indirect attack has very little application to structural firefighting. However, there are a few possibilities for the indirect attack. Because the steam produced by this attack will be dangerous to fire fighters and occupants, it is obvious that this is a poor choice in occupied areas, including areas occupied by fire fighters. Tightly enclosed areas aboard ships are the best application of this attack, but shipboard fires are beyond the scope of this book. However, picturing the tightly enclosed, unoccupied hold of a ship provides some indication of when this attack may be appropriate.

In conducting an indirect attack, the idea is to keep ventilation to a minimum while introducing a fog stream directed at the ceiling. This will result in maximum steam production. This tactic may prove useful in unoccupied basements, attics, or storage areas. A similar tactic of directing piercing nozzles into small, unvented attics and subterranean spaces has been used with success. Although the indirect attack has limited application, every tool should be known and understood.

The direct attack is preferred in most situations **Figure 8-11**. This tactic applies water directly on the burning material, thereby extinguishing the fire by reducing the temperature of the burning and superheated fuel. Class A fuels must be vaporized by heating the solid materials until combustible vapors are emitted; this is known as __pyrolysis__. By applying water directly to the burning fuel, the temperature of the fuel is reduced, thus reducing or eliminating pyrolysis. Cooling ordinary combustibles to shut down the pyrolysis process is much like shutting off the gas supply at a flammable gas fire.

Figure 8-11 A direct attack.

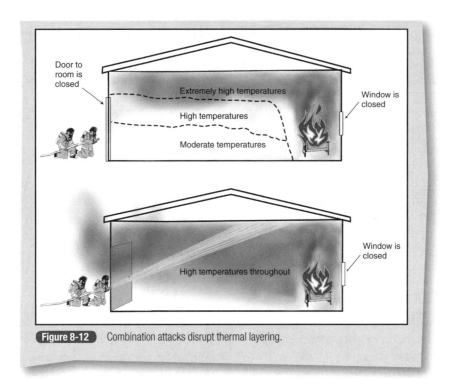

Figure 8-12   Combination attacks disrupt thermal layering.

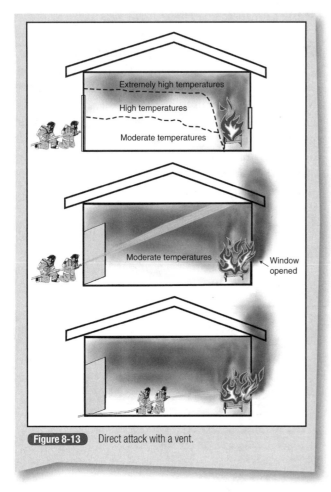

Figure 8-13   Direct attack with a vent.

Some would recommend the combination attack, using both the direct and indirect application. However, disturbing the heat balance and generating intense steam make little sense for most interior operations.

There may be times when an indirect attack is made from a point at the entrance of an enclosure and then the hose line is advanced to complete extinguishment. This combination attack would be the exception rather than the rule and should not be used in occupied areas. When fire fighters encounter very high heat conditions, indicative of an impending flashover, directing a stream overhead is advised, but this should be done in a pulsing fashion before actually entering the enclosure where the attack is taking place Figure 8-12. Using a solid or straight stream to reduce the heat overhead will be less efficient in terms of heat reduction but will cause less steam production and disruption of the heat balance. This can be particularly valuable in progressing from room to room in an attempt to find the fire or reach the room of origin. Of course, ventilating the enclosure is the safest, most effective means of preventing flashover.

Under these conditions, proper ventilation is critical to prevent flashover and make interior conditions tenable for victims and fire fighters. Proper ventilation Figure 8-13 may eliminate the need for an indirect or combination attack, thus allowing the preferred direct attack.

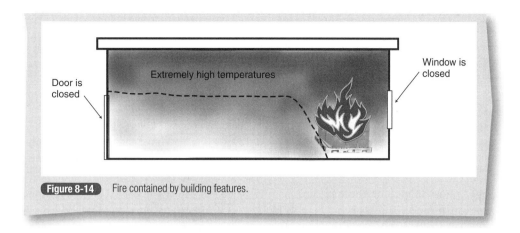

Figure 8-14 Fire contained by building features.

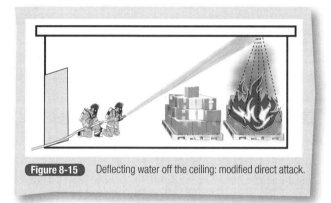

Figure 8-15 Deflecting water off the ceiling: modified direct attack.

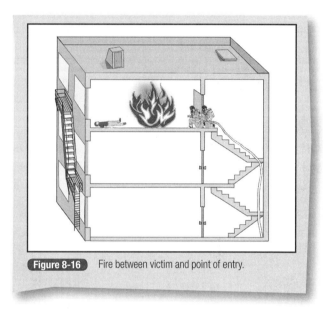

Figure 8-16 Fire between victim and point of entry.

From a safety standpoint, the officer should conduct a risk-versus-benefit analysis before entering an area of extreme heat. If there is a high potential for saving lives, the continued advancement may be justified. If the goal is keeping the fire out of the common hallway or stairs, it may be better to use the door to the apartment to contain the fire while protecting the stairs Figure 8-14. Each situation is different, and no one approach is correct for every incident.

There are times when the fire is shielded by non-burning materials, such as in a warehouse with rows of high-stacked commodities. If possible, the hose should be moved to a position that allows a direct attack. If this is not possible, then deflecting a solid or straight stream off of the ceiling and allowing the droplets to fall on the fuel is an alternative Figure 8-15. The objective is to get the water directly on the fuel. This tactic should be considered a derivation of a direct attack.

Whenever possible, the hose line must be placed between the victim and the fire, with entry and ventilation team activities coordinated by the IC. Pre-incident plan information, reconnaissance, and information from occupants who have escaped can assist the IC in placing attack crews and vent teams in the best possible location. In most instances the exact location of the victim or victims will not be known until the attack team enters the occupied compartment. When an occupant is located in a position beyond the fire and the attack team's entry point, as shown in Figure 8-16, there is a danger that the attack team will push the fire toward the trapped occupant.

If a window is opened beyond the person and the attack team, the fire will generally move toward the vent opening. This is a difficult situation. An attempt to quickly remove the victim from the window is one possibility, but the vent opening is problematic. Fire fighters entering

> Whenever possible, place the attack line between the victim and the fire.

the window opposite the attack team will, by opening the window, pull the fire toward this horizontal vent opening. For this reason, it is imperative that these fire fighters have a charged hose line in place. These fire fighters are now positioned to knock down the fire and rescue the victim. However, coordination and communication are critical to prevent an opposing hose streams situation.

Sending fire fighters inside, opposite a hose stream, may place them in extreme danger. Whenever possible, attack teams should be repositioned to get the hose line between the victim and the escape route. Again, to avoid injuries, communication and coordination are critical.

When conditions will not permit an alternative access point to attack the fire, it is best to have the attack team continue to extinguish the fire, using a solid or straight stream to minimize the pushing effect of the fire stream.

When deck-mounted master streams and elevated master stream apparatus (water towers) came into common use, fire fighters were anxious to put these new tools to work. One tactic that resulted was a so-called blitz attack, in which an outside stream is used by the first-arriving unit to knock down a heavy volume of fire that is visible from the exterior. There were reported successes in using this tactic. However, the potential for pushing the fire into other areas, including common hallways and stairs, greatly outweighs any potential benefit. If people are in the room where this exterior attack is directed, they will be placed in even greater danger. When conditions permit an offensive attack, the blitz attack is generally a poor tactic. These easily operated master streams are well suited to exposure protection during an offensive attack, but are not generally considered offensive tools. However, there are a few exceptions:

- Inside a large open building, large master streams could be used to contain a fire to one section of the building to facilitate evacuation in other areas.
- The pushing effect created by fog patterns can be used to advantage in pushing the fire away from interior or exterior victims, egress areas, and exposures, as shown in Figure 8-17.

Estimating the Number of Attack Lines

Previous examples showed that a single 1¾" (44-mm) hose line could probably extinguish most fires within a dwelling. It is good practice to lay a backup hose line to protect egress

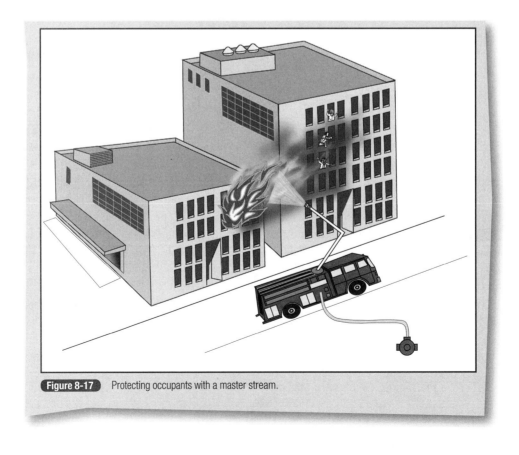

Figure 8-17 Protecting occupants with a master stream.

routes and bolster the attack. Even trial-and-error methods become more successful when provisions are made to double the size of the attack. The number of lines needed will be based on flow requirements, but there must always be a reserve. Simply stated, enough lines must be placed in service to apply the necessary rate of flow and protect egress routes for interior fire fighters, should conditions deteriorate.

In addition, the IC is responsible for the planning section function of predicting where the fire is going. Construction features may hold a fire within a given area or allow it to travel horizontally and/or vertically throughout the structure. In the case of an apartment building's common attic, it is necessary to have truck companies gain access to and check for extension in this hidden space while an engine company provides a hose line to limit fire extension. Just as it is necessary to perform search-and-rescue operations above the fire, it is also necessary to place hose lines above the fire.

Evaluating Exposures

Estimating the number of hose lines above the fire addresses the need to protect interior exposures. Hose lines are needed to prevent the fire from moving through natural pathways, such as concealed spaces, stairs, chutes, and shafts. Fire can also extend vertically up the exterior of the building from windows or other openings on the fire floor(s). All of these would be classified as **internal exposures** and need to be checked and protected with a hose line as required.

Fire can extend to adjacent buildings (**external exposures**) as well. Ventilation crews must be aware of this potential, as improper ventilation can spread the fire to exposed buildings—just as proper ventilation will pull the fire away from exposures. Nearby buildings may need to be protected once the fire breaks out of the containment of the building of origin. Evaluating the need to defend nearby buildings with additional hose streams involves many factors, including the following:

- Proximity to the fire building
- Wind direction
- Height of exposure compared to the level of fire involvement in the fire building
- Life hazard in the exposure
- Hazard presented by the exposed occupancy (e.g., explosive or chemical storage)

When evaluating exposures, be sure to consider the location of fire apparatus and equipment parked near the fire building **Figure 8-18**.

Figure 8-18 External exposures.

Protecting exposures can involve hose lines on the exterior or interior of the exposed building. In working on the exterior, master streams are often the streams of choice.

Estimating Backup Needs

Backup lines are needed to protect the crew on the initial attack line and to provide additional flow if needed. Given that a single 1¾″ (44-mm) hose line is sufficient for most residential fires (in terms of rate of flow), the backup hose line is additional insurance. In larger area fires, the backup hose line again provides insurance against rate-of-flow miscalculation and provides additional protection for retreating fire fighters. Backup lines should be at least as large as the initial attack line. If you started the attack with a 2½″ (64-mm) hose line, the backup hose line should be a 2½″ (64-mm) hose line. If the attack began with a 1¾″ (44-mm) hose line, the backup hose line could be 1¾″ (44 mm), 2″ (51 mm), or 2½″ (64 mm).

Although backup hose lines can be used to augment the initial attack, the defined purpose of a backup line is to provide flow in addition to what is needed to extinguish the fire, thus providing a measure of protection for fire fighters combating the fire.

Estimating the Number of Hose Lines Needed Above the Fire

It is a good practice to get a hose line on the floor immediately above the fire. Where there is a risk of extension to concealed spaces and attics, additional precautionary lines are needed at each of these areas. Backup lines may also be needed in other areas above the fire **Figure 8-19**.

Evaluating Other Hose Lines Needed

On occasion, hose lines are needed below the fire to protect egress routes and to protect external exposures. The total number of lines needed is as follows:
- Hose line(s) required to meet the rate of flow in the immediate fire area
- Backup hose line(s) for immediate fire area
- Hose line(s) to protect egress routes
- Hose line(s) to protect internal exposures (attic where vertical travel is likely, upper floors, concealed spaces, etc.)
- Other backup hose lines as needed
- Hose line(s) to protect external exposures

Figure 8-19 Attack and additional hose streams.

Estimating Water Supply Needs

Prior to the widespread use of large-diameter hose, the officer assigned to the first-arriving unit in a hydranted area had to decide whether to rely on a marginal water supply by forward laying a single 2½″ (64 mm) or 3″ (76 mm) supply line or use a reverse lay with a direct hydrant connection, which could result in placing the pumper a considerable distance from the fire building. Large-diameter hose eliminates this dilemma and has proven to be an effective tool in providing an adequate water supply at the fire scene **Figure 8-20**. In an area with hydrants, providing at least two water supplies for a working structure fire is considered good practice. In areas with a marginal or nonexistent water supply, estimating and providing an adequate

> Each hose line requires a crew of at least two fire fighters, but assigning a full company per line is preferred.

Figure 8-20 Large-diameter hose supply.

water supply can be a substantial challenge. Fortunately, many of the fires requiring large fire flows occur in areas with reasonably good water supplies. When a significant water supply challenge exists in an area with a limited water supply, operational options should be outlined in SOPs and the pre-incident plan.

Attacking a fire with anything less than a water supply from a source of water that is capable of supplying flow requirements over an extended period of time is a gamble. Attack pumper tactics (operating with water from the apparatus tank) work well in many situations, especially in fighting a fire in a small dwelling. However, relying solely on water from the apparatus water tank can be dangerous. Many things can happen that will cause fire fighters to lose their water supply. When 1¾" (44-mm) lines are used, even a 1000-gallon (3785-L) water tank can supply water for less than 10 minutes.

> A 1000-gallon (3785-L) booster tank, compared to a 500-gallon (1893-L) tank, will add over two tons to the weight of the pumper, which could be problematic especially in hilly terrain.

Often, this water discharge time is sufficient. However, if fire fighters need to flow water while advancing the hose line, they may deplete their water supply before exhausting their SCBA air supply.

If an attack pumper concept is implemented, its use must be well understood by all members, and it should be limited to fires in relatively small structures. Where the water supply is limited or nonexistent, consider using two pumpers or other means, such as a tanker, to augment the water supply to the pumper.

Some large urban areas with excellent water supplies will have areas containing small ancillary buildings, such as a shed on a golf course located a considerable distance from a hydrant. In many cases, a long, complex relay is planned to deliver water to these properties. Rather than relying on relay operations, it might be better to use multiple attack pumpers to deliver the needed water to the scene.

It is important to remember that a large flow of water for a short period of time will require less water and be more effective than a lesser flow over a longer period of time.

As an example, consider a fully involved 25′ × 16′ (7.6 × 4.9-m) room with 10′ (3-m) ceilings in a single-family dwelling. The required rate of flow is as follows:

$$\frac{25' \times 16' \times 10'}{100} = 40 \text{ GPM (151 L/min)}$$

If the apparatus booster tank contains 500 gallons (1893 liters), a 1″ (25-mm) booster line could discharge 20 GPM (78 L/min) for 25 minutes, or longer than the air supply of a standard SCBA air cylinder. However, the entire 500 gallons (1893 liters) would likely be exhausted, and the fire would not be extinguished because the rate of flow was insufficient to suppress the fire.

If 150 GPM (568 L/min) were discharged, the fire would be extinguished in less than a minute, with less than 150 gallons (568 L) of water. In the article by Edwards from *Fire Engineering*,[12] his calculations would indicate extinguishment in about two seconds with 5 gallons (19 L) of water at nearly four times the required flow. Seldom is water applied at an ideal rate, but the fact remains that the higher-flowing 1¾″ (44-mm) hose line would extinguish the fire quickly.

The need for a continuous supply of water having been established, standard guidelines should be followed. The use of a single, unpumped 2½″ (64-mm) supply line is discouraged. From a water-supply standpoint, this single line is much better than total reliance on the apparatus water tank, used with the attack pumper, and may be acceptable in residential fires. However, double 2½″ (64-mm) or 3″ (76-mm) lines are far superior to the single unpumped 2½″ (64-mm) or 3″ (76-mm) supply line. Furthermore, in using 2½″ (64-mm) or 3″ (76-mm) supply lines, both the volume and pressure can be improved by assigning a second pumper to relay the water from the hydrant. However, pumping into large-diameter supply hose (4″ to 5″ [102 to 127 mm]) during offensive operations is rarely necessary.

More often, it would be better to position the extra relay pumper in a position to provide a separate water supply or assign it to staging. Only in situations in which flows in excess of 1000 GPM (3785 L/min) are anticipated from a single pumper or for extremely long relay lines (over 700′ [213 m]) is it sometimes necessary to place another pumper in a large-diameter hose water supply layout. Further, once the pumper is placed in operation, the apparatus operator must remain at the pump panel. Thus, unnecessary use of apparatus also results in a reduction of available staffing.

Just as a backup hose line is needed inside structures, obtaining a second source of water is a good practice. This second source of water is essential at working fires, especially when operating off dead-end mains or limited flow hydrants. Water supply is not enhanced by securing a second source of water off the same dead-end main. Remember, for offensive operations, a minimum water supply layout must be capable of supplying the calculated rate of flow plus backup lines and other lines used to protect exposures.

Water supply should be addressed in department SOPs. The initial water supply is crucial. Companies responding must be given clear instructions regarding the water supply. See Chapter 2 for additional information regarding pre-planned water supplies.

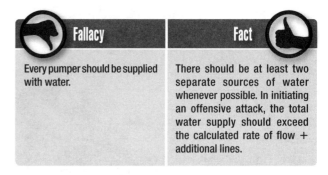

Fallacy	Fact
Every pumper should be supplied with water.	There should be at least two separate sources of water whenever possible. In initiating an offensive attack, the total water supply should exceed the calculated rate of flow + additional lines.

Estimating Ventilation Needs

In Chapter 6, venting was discussed as a life safety tactic. Venting is most often used to assist in extinguishment efforts by controlling fire spread and making conditions tenable for fire fighters and victims **Figure 8-21**. Timing is crucial when the venting tactic is intended to support extinguishment.

The ideal scenario calls for making the ventilation opening just before attack teams enter the fire area. If the vent openings can be controlled, positive-pressure venting virtually ensures the vent timing, as opening the door to the fire compartment allows the pressure to enter the area and exit through an existing or created vent opening. Natural horizontal and vertical venting should be timed and reversible (open the window rather than breaking it) whenever possible. Ventilation allows fire fighters to approach the fire and apply water directly to the burning materials. It also provides oxygen, which intensifies the burning rate if the fire is ventilation-controlled.

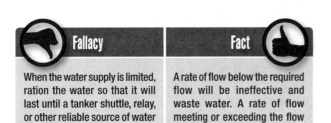

Fallacy	Fact
When the water supply is limited, ration the water so that it will last until a tanker shuttle, relay, or other reliable source of water can be supplied.	A rate of flow below the required flow will be ineffective and waste water. A rate of flow meeting or exceeding the flow requirement will be effective and use less water.

Figure 8-21 Venting controls fire spread; water extinguishes the fire.

Therefore, when venting is being carried out to support extinguishment, attack crews with charged lines should be positioned before opening up the building.

Improper vent placement can channel the fire into concealed spaces or otherwise extend the fire. Remember, regardless of our best intentions, fire and the products of combustion are bound by the laws of physics. When a door leading from the hallway to the room on fire is opened, smoke and heat will vent into the hallway. If vertical ventilation is provided via the stairway, expect the fire to enter the stairs unless extinguishment efforts are successful. Ventilation is an important tactic. The IC makes decisions as to what type of ventilation is best and coordinates the activities of hose teams and vent crews. Fire officers should be completely familiar with the various ventilation methods.

Venting is normally considered an offensive tactic; however, the **trench cut** is sometimes referred to as a defensive tactic. The trench cut normally involves interior operations, which is the definition of offensive operations used in this text. According to another definition, defensive can mean establishing a fire break and stopping the fire spread at that point. In this sense, the trench cut is defensive, as the objective is to protect interior exposures. The trench cut provides a fire break separating the burning area from the uninvolved area. To be successful, it must be made well ahead of the fire and effectively separate the burned from the unburned. It takes time to cut a trench completely across a roof. It is imperative that this tactic be planned with sufficient lead time.

Trench cuts can be used in any large structure and are particularly suited to U-shaped apartment buildings and garden apartments. The largest trench cut operation on record was at Tinker Air Force Base in Oklahoma,[16] where several long trenches were cut in a roof covering 2,531,965 ft^2 (235,219.5 m^2) for a fire operation spanning three days.

Calculating Staffing Needs

Staffing requirements are many and varied. Hose and ladder crews, rescue groups, rapid intervention teams, IMS positions, and a tactical reserve are examples of assignments that must be adequately staffed at the scene of a working structure fire. In addition, the IC needs to consider the total extinguishment staffing requirements based on rate of flow, backup lines, placement of lines above the fire, and the establishment of a secondary water supply. *NFPA 1710: Standard for the Organization and Deployment of Fire Suppression Operations, Emergency Medical Operations, and Special Operations to the Public by Career Fire Departments*,[17] defines tasks and minimum staffing for the initial response. These tasks and staffing include:

1. Establish command (minimum one person).
2. Establish a minimum uninterrupted water supply of 400 GPM (1514 L/min) (minimum one person).
3. Establish an attack and backup line flowing at least 300 GPM (1136 L/min) (minimum four people).
4. One support person per each line (minimum 2 people)
5. At least one search and rescue team (minimum 2 people)
6. At least one vent team (minimum 2 people)
7. Establish an initial rapid intervention crew (RIC; minimum 2 people).
8. If an aerial ladder is in use, a person operating the aerial ladder (minimum 1 person).

On some occasions the first-arriving unit(s) handle the fire and the total response defined in *NFPA 1710* is not required. As an example, the first arriving engine and truck companies conduct the initial attack on a mattress fire inside a residential property. The engine extinguishes the fire and provides the uninterrupted water supply as well as the two-out RIC. Truck company members conduct a quick search and rescue with one crew while the other crew vents the building.

On other occasions the staffing needs are greater than the initial response outlined in *NFPA 1710*. In these cases departments must call for mutual aid or additional alarms. Consider a fire on the second floor of the auto parts store example that was shown in Figure 8-6. In this scenario the building is believed to be unoccupied and has external exposures on three sides. Each fire company has an officer, a driver, and two fire fighters on duty. The standard pre-connect is a 200′ (61-m), 1¾″ (44-mm) hose line with a ⅞″ (22-mm) smooth-bore tip. The department's SOP establishes a 65-psi (448-kPa) nozzle pressure, resulting in a flow of 184 GPM (696 L/min).

Engine 1 (First-Arriving Engine Company)

Assume that the building has been pre-incident planned with an estimated rate of flow of 480 GPM (1817 L/min) based on V/100 calculations (sprinkler calculations are also 480 GPM [1817 L/min] in this case). The first-arriving officer takes command and gives an initial report indicating a working fire on the second floor while working with another member of the crew in advancing a standard pre-connect hose line to the second floor **Table 8-3**.

TABLE 8-3	Staffing with One Hose Line in Operation				
Company	Total Staffing	Staffing at Apparatus	Assignment	Staff at Assignment	Flow
Engine 1	4	2	2nd floor	2	184 GPM (696 L/min)
Totals	4	2	(N/A)	2	184 GPM (696 L/min)

As per SOPs, a forward lay of 5″ (127-mm) hose supplies the initial operation. The apparatus operator remains at the pumper, while the fire fighter assigned to the hydrant hooks up and then joins the apparatus operator as part of the initial RIC. This assumes that the fire fighter hooking up to the hydrant will be able to join the apparatus operator as a member of the initial RIC before the attack team enters the structure.

Engine 2 (Second-Arriving Engine Company)

In the absence of orders, SOPs for this department call for the second-arriving engine company to advance an additional hose line to the fire floor to increase the rate of flow or provide a backup hose line for the first-in attack crew. Otherwise, they are to advance a hose line to the floor above the fire. Realizing that there is a working fire in a pre-incident planned building that requires a large rate of flow, the officer of Engine 2 plans to back up Engine 1 by advancing a 2½″ (64-mm) hose line with a 1⅛″ (29-mm) smooth-bore nozzle into the building.

Engine 2 does not secure a source of water for itself. The Engine 2 officer radios command to advise that a second source of water has not been secured. One fire fighter from Engine 2 joins the fire fighter from Engine 1 to form a dedicated RIC **Table 8-4**. The remaining three crew members from Engine 2 advance a 2½″ (64-mm) hose line off Engine Company 1's pumper. Engine 2 is parked out of the way of other arriving apparatus, but in a position to be ready for response to protect exterior exposures, should it become necessary.

There is a good chance that water applied from these two hose lines will be adequate to control the fire, as the flow being applied is slightly above the 480 GPM (1817 L/min) required rate of flow. Had the department opted for an automatic nozzle with a flow rate of 125 GPM (473 L/min) as the standard pre-connect, the flow rate would be slightly below the needed rate of flow. This is where trial and error plays a role. If the fire is being extinguished with the present attack, the flow is sufficient. Otherwise, an increase in pump discharge pressure to 200 psi (1379 kPa) for the 1¾″ (44-mm) hose line with automatic nozzle would increase the flow by approximately 50 GPM (189 L/min), thus meeting the required rate of flow with the 2½″ (64-mm) line.

Reconnaissance is the best way to determine the need for more lines to control the fire. In any case there is still a critical need to back up the hose lines operating on the fire floor, secure additional water supplies, and get a hose line to the third floor.

Truck 1 (First-Arriving Truck Company)

If Truck 1 arrives early there may be a need for its crew to force entry into the building. Otherwise, the engine company crews will be responsible for this task. The truck company's crew will split up—two fire fighters providing truck duties such as horizontal ventilation for the engine companies working on the second floor, laddering the second floor, and controlling utilities. The other two members of the truck company will work above the fire, conducting the primary search and checking for fire extension on the third floor **Table 8-5**.

TABLE 8-4 Staffing with Two Hose Lines in Operation

Company	Total Staffing	Staffing at Apparatus	Assignment	Staff at Assignment	Rapid Intervention Crew	Flow
Engine 1	4	1	2nd floor	2	1	184 GPM (696 L/min)
Engine 2	4	0	2nd floor	3	1	305 GPM (1154 L/min)
Totals	8	1	(N/A)	5	2	489 GPM (1850 L/min)

TABLE 8-5 Staffing with Two Engine Companies and a Truck Company Working

Company	Total Staffing	Staffing at Apparatus	Assignment	Staff at Assignment	Rapid Intervention Crew	Flow
Engine 1	4	1	2nd floor	2	1	184 GPM (696 L/min)
Engine 2	4	0	2nd floor	3	1	305 GPM (1154 L/min)
Truck 1	4	0	2nd floor	2	0	0
Truck 1	(see above)	0	3rd floor	2	0	0
Totals	12	1	(N/A)	9	2	489 GPM (1850 L/min)

Chief Officer

The first-arriving chief officer establishes an exterior, stationary command post and confers with the company commander of Engine 1 who is serving as the IC **Table 8-6**. Engine 1's officer advises the chief officer that slow progress is being made on the fire. After a careful size-up, the chief officer assumes the position of IC and develops the following incident action plan:

- Conduct a primary search of all floors. (Companies currently on the second and third floors will be responsible for searching these floors.)
- Ladder the building to provide alternative egress for fire fighters.
- Advance a backup hose line to the second floor.
- Advance a hose line to the third floor.
- Secure a second water supply.
- Stage an engine and truck company as a tactical reserve.
- Plan for a possible defensive operation.
- Set up REHAB.
- Provide for medical treatment and transportation.

This would be a full first-alarm response for some departments, and in many cases mutual aid would be necessary to muster 13 fire fighters. Most urban departments would send a larger response to a fire in a commercial building. Comparing the task being performed and the staffing at the scene, the requirements outlined in *NFPA 1710* are being met, if the aerial ladder is not in use. However, the IC's incident action plan includes several additional tasks that have not been staffed and there is no tactical reserve. The two-person minimum is assigned as a RIC. Continuing operations will thus require additional resources.

Engine 3 (Third-Arriving Engine Company)

Engine 3 is instructed to supply its apparatus with a 5″ (127-mm) supply line and to advance a 1¾″ (44-mm) hose line with a ⅞″ (22-mm) smooth-bore tip to the third floor **Table 8-7**.

Engine 4

Engine 4 will back up the two lines operating on the fire floor with a 2½″ (64-mm) hose line laid from Engine 3's apparatus **Table 8-8**.

At this stage, the members of Truck 1 should be free to complete a primary search and check for extension on the first floor.

TABLE 8-6 Staffing with Two Engine Companies, One Truck Company, and a Chief Officer

Company	Total Staffing	Staffing at Apparatus	Assignment	Staff at Assignment	Rapid Intervention Crew	Flow
Engine 1	4	1	2nd floor	2	1	184 GPM (696 L/min)
Engine 2	4	0	2nd floor	3	1	305 GPM (1154 L/min)
Truck 1	4	0	2nd floor	2	0	0
Truck 1	(see above)	0	3rd floor	2	0	0
Chief officer	1	0	Command post	1	0	0
Totals	**13**	**1**	**(N/A)**	**10**	**2**	**489 GPM (1850 L/min)**

TABLE 8-7 Staffing with Three Engine Companies, One Truck Company, and a Chief Officer

Company	Total Staffing	Staffing at Apparatus	Assignment	Staff at Assignment	Rapid Intervention Crew	Flow
Engine 1	4	1	2nd floor	2	1	184 GPM (696 L/min)
Engine 2	4	0	2nd floor	3	1	305 GPM (1154 L/min)
Truck 1	4	0	2nd floor	2	0	0
Truck 1	(see above)	0	3rd floor	2	0	0
Chief officer	1	0	Command post	1	0	0
Engine 3	4	1	3rd floor	3	0	184 GPM* (696 L/min)
Totals	**17**	**2**	**(N/A)**	**13**	**2**	**489 GPM (1850 L/min)**

*The flow column reflects the rate of flow required to extinguish the main body of the fire. Engine 3's hose line is on the floor above; it is not added to the total flow calculation. The two water supplies (Engines 1 and 3) can easily handle the total potential flow of 673 GPM (2547 L/min) if the hose line on the third floor is operated.

TABLE 8-8 Staffing with Four Engine Companies, One Truck Company, and a Chief Officer

Company	Total Staffing	Staffing at Apparatus	Assignment	Staff at Assignment	Rapid Intervention Crew	Flow
Engine 1	4	1	2nd floor	2	1	184 GPM (696 L/min)
Engine 2	4	0	2nd floor	3	1	305 GPM (1154 L/min)
Truck 1	4	0	2nd floor	2	0	0
Truck 1	(see above)	0	3rd floor	2	0	0
Chief officer	1	0	Command post	1	0	0
Engine 3	4	1	3rd floor	3	0	184 GPM* (696 L/min)
Engine 4	4	0	2nd floor	4	0	305 GPM (1154 L/min)
Totals	21	2	(N/A)	17	2	794* GPM (3004 L/min)

*Although the backup hose line is not assigned to attack the fire, the additional flow is immediately available should the need arise.

Heavy Rescue, Truck 2, and Engine 5

Heavy Rescue is assigned as the dedicated RIC with one member assigned as the accountability officer. The two members assigned to rapid intervention from Engines 1 and 2 can now rejoin their companies. Truck Company 3 and Engine Company 5 are staged two blocks from the scene.

Two medic units are assigned to REHAB, treatment, and transport. Additional staff officers are assigned to fill the positions of safety officer, planning section chief, and Division 2 supervisor on the second floor **Table 8-9**.

While it may be possible to double up on some assignments shown on the various deployment charts, fire fighter safety must never be compromised. In a large jurisdiction this fire could be handled by on-duty forces. In smaller jurisdictions, fires such as this would require assistance from mutual aid companies. In some cases, mutual aid assistance would respond automatically to a reported structure fire.

Without an exposure or rescue problem, and with all companies taking appropriate action, this fire required what would be an extra alarm assignment for most departments. Note that three or four fire fighters are assigned to each hose line, with the exception of the first fire line, where only two fire fighters are available for interior assignments. Operating a hose line with two fire fighters is an absolute minimum and a very arduous task. Two fire fighters can safely operate a 1¾″ (44-mm) hose stream during practice sessions. During an actual fire, it may be necessary to advance the hose line up stairways, around obstacles, and through multiple doorways, then operate the hose from a

TABLE 8-9 Staffing with a Full Complement

Company	Total Staffing	Staffing at Apparatus	Assignment	Staff at Assignment	Rapid Intervention Crew	Flow
Engine 1	4	1	2nd floor	3	0	184 GPM (696 L/min)
Engine 2	4	0	2nd floor	4	0	305 GPM (1154 L/min)
Truck 1	4	0	2nd floor	2	0	0
Truck 1	(see above)	0	3rd floor	2	0	0
Chief officer	1	0	Command post	1	0	0
Engine 3	4	1	3rd floor	3	0	184 GPM* (696 L/min)
Engine 4	4	0	2nd floor	4	0	305 GPM (1154 L/min)
Rescue 1	4	0	RIC/Acc*	0	4	0
Engine 5	4	4	Staged	0	0	0
Truck 2	4	4	Staged	0	0	0
Medic unit	2	0	Treatment/transport	0	0	0
Medic unit	2	0	REHAB	2	0	0
Staff	3	0	Staff	3	0	0
Totals	40	10	(N/A)	26	4	794 GPM† (3004 L/min)

*Accountability officer
†Although the backup hose line is not assigned to attack the fire, the additional flow is immediately available should the need arise.

crawling or kneeling position, all of which greatly increase the effort needed to successfully place a hose line in position. Consider that Engines 2 and 4 in our scenario advanced 2½" (64-mm) hose lines to the second floor. Assigning all available personnel on a single company to a hose line not only provides the staffing necessary to properly position and operate the attack line, it also maintains company unity and accountability.

On occasion, there may be a need to assign more than one company to a hose line even in departments that assign four people to a company, as was done in the previous auto parts store example. As an example, a company is assigned to advance a hose line to the fifth floor of a large warehouse without a standpipe. If the company is supplying its own water, one member must remain at the pumps, several fire fighters will be needed to advance the hose line through the entry door and up the stairway (many warehouse buildings have 20′ (6 m) or higher floors), and the hose line must then be advanced through a door at the fifth floor level and possibly around storage racks and through aisles. In this scenario multiple companies might be needed to advance the hose line and operate the nozzle that could be flowing as much as 350 GPM (1325 L/min).

Determining Apparatus Needs

The apparatus that is used to deliver staffing to the incident scene is generally more than sufficient to support an offensive operation. Typically, the greater challenge is to properly position on-scene apparatus to best utilize their capabilities. Apparatus that are not needed at the incident scene should be parked out of the way, preferably in positions that allow access to water supplies and are available to support a possible defensive attack. During large-scale incidents, apparatus are often assigned to a staging area. The IC should designate a staging officer to manage and coordinate all companies assigned to the staging area. A staged apparatus has the staffing necessary to function as a unit, such as an engine company, a truck company, or a medic unit. Apparatus without adequate staffing are classified as out of service or parked, rather than staged.

There is a temptation to use on-scene apparatus whether they are needed or not. In the auto parts store example, there is some justification for a third source of water, and a ladder truck could be used to accomplish additional venting or to ladder the building. However, more often than not, during offensive operations there is a need

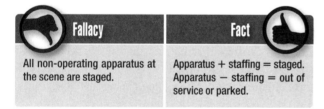

Fallacy: All non-operating apparatus at the scene are staged.

Fact: Apparatus + staffing = staged. Apparatus − staffing = out of service or parked.

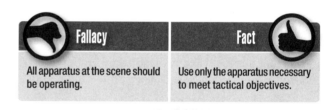

Fallacy: All apparatus at the scene should be operating.

Fact: Use only the apparatus necessary to meet tactical objectives.

for additional staffing, not additional apparatus. Placing unnecessary apparatus in the immediate fire zone tends to block access that may be needed should a change in tactics be required.

A Word About Class A Foam

Apparatus equipped with Class A foam capability may prove useful in areas with a marginal water supply, as buildings in these areas may require more water than is carried in apparatus water tanks. In these situations the use of Class A foam could make a significant difference.

The second expected advantage of Class A foam is that there is less water damage because less water is used. A hose crew properly applying untreated water at or above the required rate of flow will probably use more water than a crew applying Class A foam. However, are the additional gallons of water saved enough to justify the expense of the foam? Also, once materials are wet, are they further damaged by being wetter? If so, since one characteristic of Class A foam is to act as a wetting agent resulting in greater penetration, does its use result in additional content damage?

Finally, pre-wetting fuels and providing a foam layer on an exposure seem to be the most useful applications of Class A foam in an urban or suburban setting. More information is needed to better quantify these characteristics of Class A foam, but its use in these situations is promising.

The scope of this book is limited to structural firefighting. At this time, Class A foam has been used primarily in wildland and wildland/urban interface fire operations. Proponents of Class A foam claim that the product is superior to untreated water when used as a suppression agent or to protect exposures from radiant heat. Early tests and field experience indicate that these claims are valid.[18] However, we surveyed several fire departments throughout the United States and Canada and found that most fire departments were not using Class A foam to fight structure fires even when it was available on the apparatus. The jury is still out on the cost versus benefit of Class A foam.

Summary

To conduct a safe and effective offensive operation, the IC must consider a number of different factors. One of the most important considerations is to determine whether there are sufficient personnel and resources on the incident scene to deliver the required rate of flow needed to suppress the fire and to protect the fire fighters who are working in or on the building. This chapter discusses the advantages and disadvantages of three different formulas for calculating rate of flow. An experienced IC will be familiar with each of these formulas. Further, remember that rate-of-flow calculations are most valuable when they are included in a pre-incident plan. The authors strongly recommend using V/100 or sprinkler charts when calculating rate of flow during pre-incident planning.

Wrap-Up

Key Terms

combination attack A type of attack employing both direct and indirect attack methods.

direct attack Firefighting operations involving the application of extinguishing agents directly onto the burning fuel.

external exposures Buildings, vehicles, or other property threatened by fire that are external to the building, vehicle, or property where the fire originated.

fuel-controlled fire A fire in which the heat release rate and growth rate are controlled by the characteristics of the fuel, such as quantity and geometry, and in which adequate air for combustion is available.

indirect attack Firefighting operations involving the application of extinguishing agents to reduce the buildup of heat released from a fire without applying the agent directly onto the burning fuel; ventilation is kept to a minimum while a fog stream is directed at the ceiling; most useful in unoccupied tightly enclosed spaces.

internal exposures Areas within the structure where the fire originates or within buildings directly connected to the building of origin that were not involved in the initial fire ignition.

oxygen- (ventilation-) controlled fire A fire in which the heat release rate and/or growth of the fire are controlled by the amount of air available to the fire.

pyrolysis The chemical decomposition of a compound into one or more other substances by heat alone; the process of heating solid materials until combustible vapors are emitted.

Royer/Nelson formula A rate-of-flow calculation that calculates the rate of flow as the volume in the fire area in cubic feet divided by 100 (V/100). This formula is based on the assumption that structural fires are primarily oxygen-controlled.

sprinkler system calculations Specific rate-of-flow calculations for sprinkler systems based on the fuel load; can be found in various publications including NFPA documents and Factory Mutual Data Sheets.

trench cut A cut made from bearing wall to bearing wall to prevent horizontal fire spread in a building.

U.S. National Fire Academy formula A rate-of-flow calculation that calculates the rate of flow as the area in ft^2 divided by 3 (A/3).

Suggested Activities

1. Use A/3 to calculate the rate of flow for each room in **Figure 8-22**. The ceiling height in this two-story residence is 8′ (2.4 m) for all rooms. The first-floor layout should be considered an open layout due to large openings between rooms with decorative half-walls (shown as broken lines) between some areas. The only true enclosure on the first floor is the half bath (approximately 25 ft^2 [2.3 m^2]). The footprint (dimensions of the overall structure at the base) of this two-family house is 25′ × 40′ (7.6 m × 12 m). Rooms on the second floor should be calculated

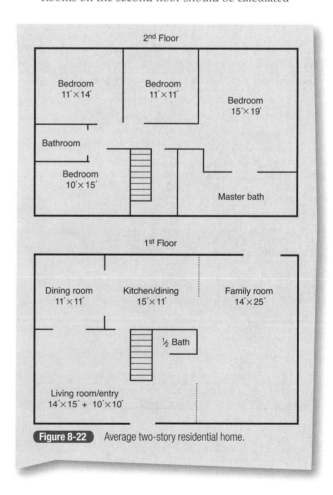

Figure 8-22 Average two-story residential home.

individually. There are no nearby buildings. Is there a need to pre-plan the rate of flow for this structure?

2. Use V/100 to calculate the rate of flow for each room in Figure 8-22. The ceiling height in this two-story residence is 8′ (2.4 m) for all rooms. The first-floor layout should be considered an open layout due to large openings between rooms with decorative half-walls (shown as broken lines) between some areas. The only true enclosure on the first floor is the half bath. Rooms on the second floor should be calculated individually. Is there a need to pre-plan the rate of flow for this structure?

3. Using Figure 8-22, compare the calculated A/3 rate of flow versus V/100 for a fire involving the family room.

4. **A.** Based on the V/100 calculations in #2 above, would the flow from your department's standard pre-connected hose line be sufficient to extinguish any fire that could be fought offensively in this house?

   B. Assuming a backup hose line will be advanced to the fire area, would the flow from two standard pre-connected hose lines be sufficient to extinguish any fire that could be fought offensively in this house?

5. **A.** Based on V/100, list the total staffing requirements for a fire involving the family room, entry, and living room of the house shown in Figure 8-22. The fire occurs late at night and the occupied status is unknown.

   B. Would you consider this an imminent life-threatening situation?

   C. What is the total water supply needed for this fire (include backup hose lines and hose lines protecting internal exposures)?

   D. List the vent methods and locations that would be best for this fire.

   E. Would the typical response to residential properties in your community provide sufficient personnel and apparatus to safely and effectively conduct an offensive operation at this fire?

6. **A.** Use V/100 to compute the required rate of flow for the open layout kitchen/family room/dining area in Figure 8-23. This home is situated near the center of a two-acre lot.

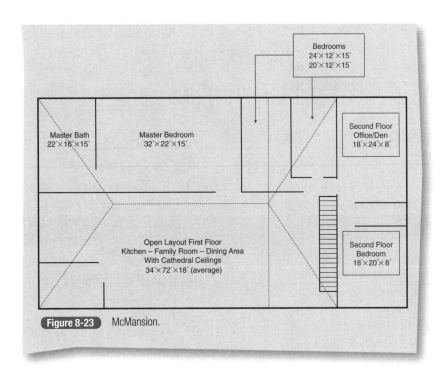

Figure 8-23 McMansion.

B. Based on V/100, would the flow from your department's standard pre-connected hose line be sufficient to extinguish any fire that could be fought offensively in this house?

C. Assuming a backup hose line will be advanced to the fire area, would the flow from two standard pre-connected hose lines be sufficient to extinguish any fire that could be fought offensively in this house?

D. Should the rate of flow for this house be pre-planned?

7. A. Based on the rate-of-flow calculation in #6 above, list the total staffing requirements. The fire occurs during the day with no occupants or cars visible.

B. Would you consider this an imminent life-threatening situation?

C. What is the total water supply needed for this fire (include backup hose lines and hose lines protecting internal exposures)?

D. List the vent methods and locations that would be best for this fire.

E. Would the typical response to residential properties in your community provide sufficient personnel and apparatus to safely and effectively conduct an offensive operation at this fire?

8. A. Use V/100 to compute the required rate of flow for the warehouse fire shown in **Figure 8-24**. This fire is on the fourth floor of a seven-story building of ordinary construction. A fire wall with double fire doors separates the two halves of the warehouse. The ceiling height is 25′ (7.6 m) with commodities stored on skids and in racks as high as 20′ (6 m). The storage is ordinary commodities with no known hazardous materials. There is no sprinkler system in the building, but there is a wet pipe automatic standpipe system with outlets on each level in each stairway. A standpipe test was conducted two months ago showing a pressure of 65 psi (448 kPa) on the seventh floor with a total flow of 1000 GPM (3785 L/min) from the two interconnected standpipes.

B. Based on V/100, would the flow from your standpipe hose line be sufficient to extinguish any fire that could be fought offensively in this warehouse?

C. Assuming a backup hose line will be advanced to the fire area, would the flow from two standard standpipe hose lines be sufficient to extinguish any fire that could be fought offensively in this building? Or, should the warehouse be pre-planned?

D. If more than two standpipe hoses are needed, how many and what size hose lines are needed to fight this fire offensively?

9. The warehouse shown in Figure 8-24 has a pre-incident plan and work shifts are scheduled 24 hours a day, Monday through Saturday. It is now 3 AM, lights are on in the building, several trailers are backed into the loading docks, and workers can be seen evacuating the building at ground level. Assume building and fire conditions permit an interior attack.

A. Based on the rate-of-flow calculation and subsequent analysis of hose line requirements in #8 above, estimate the total staffing requirements for an offensive operation.

B. Would you consider this an imminent life-threatening situation?

C. Including backup hose lines and hose lines protecting internal exposures, what is the total water supply needed for this fire?

D. List the vent types and locations that would be best for this fire.

E. Would the typical response to commercial properties in your community provide sufficient personnel and apparatus to safely and effectively conduct an offensive operation at this fire?

10. Analyze various attack positions and explain the advantages and disadvantages of each position.

11. A. Use V/100 to calculate the rate of flow for a large-area building in your response area (or a public building with a large open area). Large area, as used here, is defined as a building requiring a

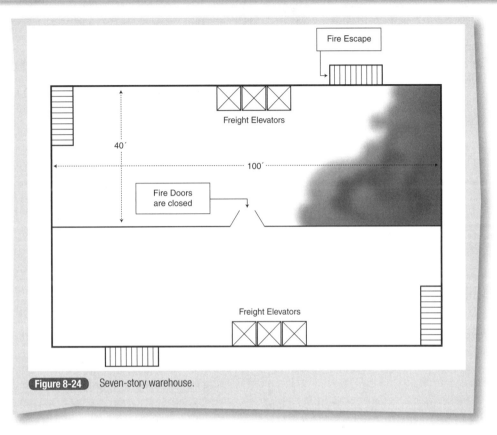

Figure 8-24 Seven-story warehouse.

flow greater than the combined flow of two of the standard pre-connected attack lines used by your department. Use the selected building from your jurisdiction or a scenario from an actual fire in which the dimensions of the fire area are shown. If you are not using an actual fire scenario, place the fire on a pre-incident plan drawing. Locate entry and exit locations that could be used to attack the fire. Do not select a building that is protected by a sprinkler system or other fire-suppression system for this exercise.

B. What is the structural collapse potential?

C. How long do you think the building will remain stable under the conditions described in the scenario?

D. Assume that the first-arriving company attacks the fire with the standard pre-connected hose line. What size hose line would you order from the second-arriving company?

E. How many and what size hose lines would be needed to extinguish the fire?

F. Compare attack positions from the various entry locations to the fire area. Which position would be best to avoid pushing the fire into uninvolved areas?

G. Consider distances from the entry location to the seat of the fire. Is it possible for the hose stream to reach the burning materials from the entry position?

H. Is a direct attack possible? Explain your answer.

I. What type of nozzle would be best for attacking the fire in this scenario?

J. What size hose lines should be used as backup lines?

K. What additional lines are needed to protect exposed areas?

L. What is the total water supply requirement?

M. What is the best location and type of venting to both control the fire spread and assist the nozzle crew in attacking the fire?

N. Would forcible entry be a problem? Explain your answer.

O. Are concealed spaces problematic? Explain your answer.

P. What is the most likely fire extension pathway?

Q. List the companies and staffing needed to conduct a safe and effective attack at a time when the building is known to be unoccupied.

R. Consider the effect a sprinkler system would have on the total operation.

Chapter Highlights

- Resource capabilities in terms of staffing, apparatus, and water must meet incident requirements for a safe and effective operation.
- Rate of flow is a major factor in determining if resources are adequate.
- Methods for determining rate of flow include trial-and-error, the Royer/Nelson formula, the U.S. National Fire Academy formula, and sprinkler system calculations.
- Each rate-of-flow approach has strengths and weaknesses.
- No rate-of-flow calculation is completely accurate, as variables such as fuel load, fuel type, ventilation, and the size of the compartment can affect fire parameters.
- Large-volume fires require the application of a rate-of-flow formula, which should be included in the pre-incident plan information.
- Sprinkler rate-of-flow calculations account for fuel type and load.
- It is better to err on the side of overestimating the required rate of flow, but gross overcalculation can result in an inappropriate strategy.
- Buildings requiring a rate of flow exceeding the flow of two standard pre-connect hose lines should be pre-incident planned by applying the V/100 formula with the exception of areas containing an extra-hazard fuel load.
- If the fuel load is heavy in terms of total volume and/or fuel type, sprinkler rate-of-flow calculations are preferable.
- A 1¾" (44-mm) hose line is the minimum recommended size hose line for an offensive attack at a structure fire.
- Using a hose size larger than 1¾" (44 mm) is a trade-off of mobility for an increase in flow.
- Hose size, elevation, nozzle size, and nozzle type are factors in calculating the pump discharge pressure.
- Nozzle and hose configurations should be flow tested regularly to confirm the flow and proper operation.
- Variable-stream nozzles provide fog or straight streams.
- Direct attack applies water directly on the burning material and is preferable to indirect or combination attacks in most situations.
- Hose streams should be positioned to avoid pushing fire toward fire fighters or victims.
- The number and size of attack hose lines will be based on flow requirements.
- Additional hoses should be laid to meet the rate of flow, to provide backup, to protect means of egress, and to protect internal, and external exposures.
- Fire can extend within a building to involve other areas or attached structures (internal exposures) or to adjacent structures (external exposures).
- Backup lines protect fire fighters on the initial attack line and provide additional flow if needed.
- Backup lines should be at least as large as the initial attack line.
- It is good practice to place a hose line on the floor immediately above the fire.
- Each hose line requires at least two fire fighters, but a full company per line is preferred.

- It is good practice to provide at least two water supplies for a working structure fire.
- When a significant water supply challenge exists, options should be outlined in SOPs and the pre-incident plan.
- Use of an attack pumper should be limited to fires in relatively small structures.
- A flow of water meeting or exceeding the rate-of-flow requirements applied for a short period of time will require less total water and be more effective than a lesser flow over a longer period of time.
- Proper ventilation aids in the extinguishment effort and provides for a safer working environment for fire fighters in the building.
- Staffing needs vary depending on tasks to be accomplished, but certain minimum staffing requirements can be predetermined.
- Apparatus used to deliver staff to the scene is usually sufficient to support an offensive operation.
- Safe and efficient positioning of on-scene apparatus is a key consideration.
- Class A foam may be useful where there is a limited water supply or to cover exterior exposures.

References

1. William E. Clark, *Firefighting Principles and Practices*, 2nd edition. Saddle Brook, NJ: Fire Engineering, Penwell Press, 1991.
2. Alan Brunacini, *Fire Command,* 2nd Edition. Quincy, MA: NFPA, 2002.
3. John Coleman, *Incident Management for the Street Smart Fire Officer.* Saddle Brook, NJ: Fire Engineering, Penwell Press, 1997.
4. Lloyd Layman, *Attacking and Extinguishing Interior Fires.* Quincy, MA: NFPA, 1955.
5. National Institute of Standards and Technology, *FPETOOL*, version 3.20. Gaithersburg, MD: U.S. National Institute of Standards and Technology, 1993.
6. IAFC, NFPA, *Fundamentals of Fire Fighter Skills.* Sudbury, MA: Jones and Bartlett Publishers, Inc., 2004.
7. U.S. National Fire Academy courses. Emmitsburg, MD: U.S. National Fire Academy.
8. Keith Royer and Floyd Nelson, *Water for Fire Fighting, "Rate of Flow" Formula, Extension Services Bulletin #18.* Ames, IA: Iowa State University, 1959.
9. National Fire Protection Association, *NFPA 13: Standard for the Installation of Sprinkler Systems.* Quincy, MA: NFPA, 2007.
10. National Fire Protection Association, *NFPA 1142: Standard on Water Supplies for Suburban and Rural Firefighting.* Quincy, MA: NFPA, 2007.
11. National Fire Protection Association, *NFPA Fire Protection Handbook,* 18th edition. Quincy, MA: NFPA, 1997.
12. C. Bruce Edwards, Critical fire rate. *Fire Engineering,* September: 97–99, 1992.
13. Warren E. Isman and Gene P. Carlson, *Hazardous Materials.* Encino, CA: Glencoe, 1980.
14. IAFC, NFPA. *Fundamentals of Fire Fighter Skills.* Sudbury, MA: Jones and Bartlett Publishers, Inc., 2004: p. 477.
15. National Fire Protection Association, *NFPA 14: Standard for the Installation of Standpipe and Hose Systems.* Quincy, MA: NFPA, 2003.
16. James C. Goodbread, *Fire in Building 3001.* Quincy, MA: NFPA, 1985.
17. National Fire Protection Association, *NFPA 1710: Standard for the Organization and Deployment of Fire Suppression Operations, Emergency Medical Operations, and Special Operations to the Public by Career Fire Departments.* Quincy, MA: NFPA, 2004.
18. J. Gordon Routley, *Compressed Air Foam for Structural Fire Fighting: A Field Test.* U.S. Fire Administration, 1993.

Defensive Operations

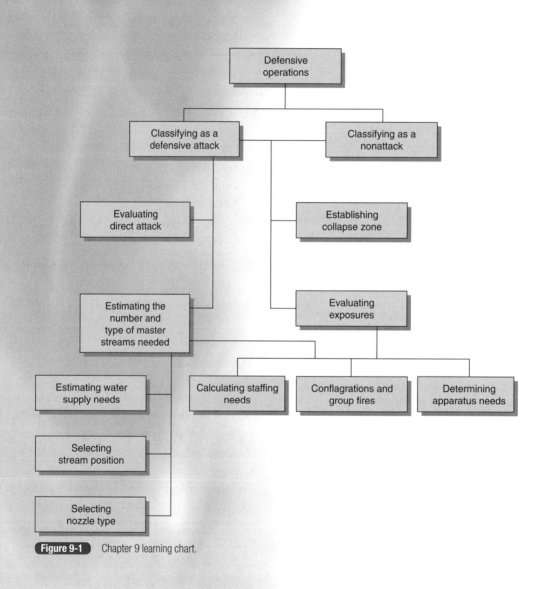

Figure 9-1 Chapter 9 learning chart.

Chapter 9

Learning Objectives

- Compare and contrast a defensive versus offensive fire attack, explaining the key differences.
- Explain why an offensive attack is preferred over a defensive attack.
- Enumerate conditions that would lead to a defensive attack.
- Describe how collapse zone dimensions are determined.
- Evaluate the effectiveness of master streams operated from distances required to maintain a safe collapse zone.
- Discuss the positive and negative effects of operating a hose stream into a window or roof opening.
- Describe conditions when a direct defensive attack is preferred as compared to an indirect defensive attack (i.e., covering exposures).
- Describe how water should be applied when protecting an exposure from radiant heat.
- Compare and contrast the use of handheld hose streams versus master stream appliances during defensive operations.
- List two ways the water utility may be able to increase the total water supply at the incident scene.
- Compare and contrast the use of fog versus solid streams during a defensive attack.
- Estimate staffing and apparatus needs when operating master streams.
- Define conflagrations and group fires.
- List common problems leading to conflagrations.
- Explain tactics used to control a conflagration.
- Discuss why conflagrations are likely to occur immediately after a natural disaster.
- List reasons for a non-attack strategy.
- Given a scenario, calculate the dimensions of the collapse zone.
- Prioritize exposures based on fire conditions, occupancy, and weather factors.
- Develop an incident action plan for a defensive fire.
- Develop an incident action plan for a conflagration.
- Apply defensive tactics to a defensive fire attack.
- Apply defensive tactics to a conflagration.
- Evaluate staffing, water supply, and apparatus needs for a large-scale defensive fire.
- Apply NIMS to a defensive fire scenario.
- Determine the probability of a conflagration for a specified response area.

Introduction

An <u>offensive</u> fire attack is the preferred strategy whenever conditions and resources permit an interior attack. A <u>defensive</u> decision limits operations to the exterior, generally resulting in a larger property loss and limiting rescue options. The offensive/defensive decision is based on staffing available to conduct an interior attack, water supply, ventilation, and a risk-versus-benefit analysis. A defensive operation is used whenever the risk-versus-benefit analysis determines that the risk to fire fighters' lives and safety outweighs any possible benefit that might be achieved through an offensive attack. Fighting a fire in a building that is structurally unsound and where there are no lives to be saved is a clear example of a defensive operation. However, situations are not always so clear cut, and the incident commander (IC) must rely on solid fire-ground information, coupled with training and experience, to determine whether the operation should be offensive, defensive, or non-attack.

Classifying as a Defensive Attack

The preferred fire attack mode is offensive, and it should be used whenever it is safe to do so. Defensive attacks are generally conducted in the following situations:
- Structural integrity concerns, fire conditions, or other hazards prohibit entry.
- Resource needs outweigh resource capabilities.
- A risk-versus-benefit analysis indicates that the risk is too great in terms of what can be saved.

When a defensive fire attack is initiated, the objective is to save property that has not already been destroyed and/or to protect the environment. Many times this means sacrificing the building of origin in favor of saving surrounding external exposures. Other times a defensive attack can knock down a heavy volume of fire that prevents entry, limiting or stopping the fire spread, thus saving the remainder of the building. The overhaul portion of a defensive attack, where part of the building is still standing, is extremely dangerous.

Defensive fires tend to be more spectacular, but in reality they are easier to handle and pose fewer risks to fire fighters if the proper precautions are taken. The IC must keep in mind the factors that lead to a defensive decision. Fire fighters should never be needlessly placed in a dangerous environment. When a fire is of such magnitude that the building's structural support system is threatened, the building should be evacuated and a defensive operation initiated.

Figure 9-1 is a learning chart for a defensive strategy. The remainder of this chapter is devoted to explaining defensive strategies using the chart as an outline.

Establishing a Collapse Zone

When a fire seriously threatens the structural integrity of a building, fire fighters should be moved outside the collapse zone to avoid life-threatening emergency retreats like the one shown in **Figure 9-2**.

Many theories exist regarding collapse zone distances. Some theorists believe that a building will fall within one-third of its height. This and other theories are generally based on sound principles but with questionable assumptions being made about how the collapse will occur. Buildings sometimes collapse within themselves; at other times the walls will fall away from the supporting structure as a unit. This type of structural failure results in the collapse zone being equal to the height of the building plus an allowance for debris to scatter **Figure 9-3**.

Any collapse zone that is less than the building's height plus an allowance for debris scatter is a calculated risk. The IC must ask whether the expected benefit is worth the risk.

> The rate-of-flow formulas described in the discussion of offensive attacks in Chapter 8 do not apply to defensive operations.

Figure 9-2 Fire fighters escaping a wall collapse.

> A good rule-of-thumb method of estimating the collapse zone is to use a distance equal to 1½ times the height of the building.

The inability to apply water from a distance greater than the height of the building may lead to a non-attack strategy for tall buildings that are in danger of collapse. Fire departments should measure the effective water application distance for master streams carried on their apparatus. Collapse zones can be pre-planned. The actual height of the building can easily be determined during pre-incident planning. At the time of an incident rough approximations must be made. For older multi-story residential and office buildings, using a factor of 12′ (4 m) per story is reasonably accurate. Modern buildings normally have lower ceilings and are often less than 12′ (4 m) per story, whereas warehouse buildings or special occupancies could have much higher ceilings. Using the 12′ (4 m) per story estimate, a five-story building would be 60′ (18 m) in height requiring a collapse zone of approximately 90′ (1.5 × 60) (27 m). The question becomes: Can available master streams effectively apply water from a distance of 90′ (27 m)? Wind conditions must

Figure 9-3 Wall collapse.

Case Summary

A fire occurred at a condominium complex in Florence, Kentucky, on August 11, 1999. Although the building was fully sprinkler protected, the fire was not controlled by the sprinkler system, as the fire originated in a concealed space created by the truss floor structure between the second and third floors. The fire eventually broke out of the concealed spaces to destroy internal exposures on both sides of the point of origin and threatened an external exposure 20′ (6 m) away. The original alarm was transmitted as an explosion with people trapped. Fire fighters began with an offensive attack supporting the primary search. As soon as it was verified that the building was cleared of occupants, fire fighters moved to a defensive attack and were successful in extinguishing the main body of fire with a direct attack. A fire stream was used to protect the nearby exposure building with an additional master stream placed in a position as a precautionary measure. The 24-unit condominium

Figure CS9-1 Condominium fire, Florence, Kentucky.

where the fire originated was severely damaged by fire, but the remainder of the complex was saved.

Source: Florence, Kentucky, Fire Department.

also be considered when estimating the effective distance for a nozzle. Solid-stream nozzles, as compared to straight-stream and fog nozzles, generally maintain better stream continuity for long distances, particularly when the stream is windward or there is a crosswind.

Another limiting factor for apparatus and fire stream placement is the width of the street. If a 90′ collapse zone is required and streets are only 40′ wide, personnel, apparatus, and equipment placed in the street would be within the collapse zone. Before committing people and apparatus to a potential collapse zone, the IC must seriously consider the expected benefits. If a decision is made to position personnel and equipment within the potential collapse zone, evaluate the building to determine the safest possible location. For example, positioning companies at the corners of the building is usually safer than a frontal attack. Again, anyone who is seriously considering being an IC is strongly encouraged to read texts such as *Brannigan's Building Construction for the Fire Service* to become familiar with a building's structural components.[1] Another excellent reference is *NFPA 1670: Standard on Operations and Training for Technical Rescue Incidents*.[2]

Evaluating Exposures

Exposures are generally divided into internal exposures and external exposures. Internal exposures deal with fire extending from one area of a building to another within a structure, as was described in the discussion of offensive fire attacks in Chapter 8. Defensive fire streams may spread the fire inside the building to uninvolved areas because improperly placed fire streams tend to push the fire. Fog streams have a much greater tendency to push the fire but may also effectively fill an enclosed area with steam, helping the suppression effort. Exterior streams may push fire into concealed spaces or common stairs, resulting in the total loss of a building. Elevated streams, improperly positioned, are notorious for pushing vented roof fires back into the building. Protecting the interior exposure is only marginally effective in a defensive mode. However, if the building is tightly constructed with closed doors separating areas, there is a good chance of confining the fire to a specific area.

Many times in a defensive operation, the building of origin is recognized as a total loss. Therefore, emphasis is placed on protecting external exposures. The IC should evaluate external exposures in terms of life safety, extinguishment, and property conservation.

The distance between exposed structures and the volume and location of the fire have much to do with prioritizing exposure protection. Radiant heat increases as the size of the flame front increases. Knocking down the main body of fire can reduce the flame front, thus reducing radiant heat energy. Therefore, the most effective way to protect exposures is to extinguish the main volume of fire. When inadequate resources, collapse probability, or fire volume make a direct attack on the fire ineffective, applying water directly on the exposed structures is the best tactic.

The energy levels for radiant heat are inversely proportional to the square of the distance between the heat source and the exposure. The closer the buildings are located to each other, the more radiant heat will be transferred from the fire building to an adjacent structure. Exposure buildings that are higher than the fire building are also at a greater risk.

The IC must determine the best method of protecting potential exposures. Available staffing, as well as matching apparatus resources to incident requirements, will greatly affect the IC's options. Wetting the exposure is generally the most effective way to prevent ignition. Radiant heat will travel through transparent materials such as water. Therefore, directing a water stream between buildings is less effective than applying water directly to the exposed building.

As was previously mentioned, Class A foam appears to present possibilities in terms of protecting exposures. The compressed air variety, in particular, tends to cling to exposures and do a better job of pre-wetting surfaces.

Evaluating a Direct Attack

Application of hose streams applied directly into the burning structure is an exposure protection tactic that has merit, but the fire intensity and depth may be such that it is impossible to effectively extinguish the fire or knock down the flame front. It is important to recognize a total loss when confronted with one. If a direct attack fails to control the fire or the effectiveness of a direct attack is questionable for any reason, first cover nearby exposures and then direct as much water as possible on the main body of fire from a safe distance.

Sometimes it is possible to place a master stream in a position where it can be operated either directly on the fire or on an exposure without moving the apparatus or master stream appliance **Figure 9-4**.

> The IC must consider the possibility of fires starting inside exposed buildings due to radiant heat igniting combustibles that are near windows facing the fire building. Radiant heat passes through intact panes of glass.

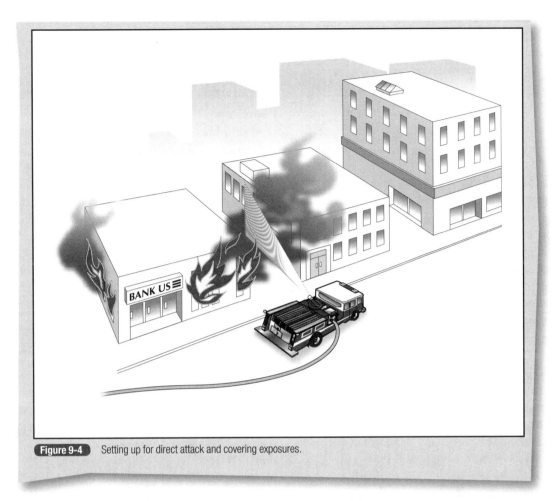

Figure 9-4 Setting up for direct attack and covering exposures.

Estimating the Number and Type of Master Streams Needed

There are times when master streams are used to augment an offensive attack. However, master streams are used primarily during defensive attacks. This chapter addresses defensive operations.

Master streams can be operated from ground level, the top of an apparatus, or an elevated position. However, it is important to remember that any tactic that hinders the upward and outward movement of heat and smoke is usually detrimental to the operation. Elevated master streams are particularly problematic. If improperly positioned, elevated master streams will push a venting fire back into the building, thus spreading the fire into uninvolved areas of the building and reversing the positive effects of ventilation.

An experienced IC knows that operating a stream into a vent opening is always an unwise and dangerous tactic during offensive operations. However, there may be circumstances during a defensive attack when these openings can be used to provide direct access to the fire.

Sometimes, enough of the building has been destroyed that the fire is not being pushed into unburned spaces, or there may not be any remaining unburned areas in the building of origin. For instance, if the IC considers the building a total loss, directing streams into vent openings may be necessary to protect nearby exposures. The IC must carefully consider the desired outcomes before operating lines into openings above the fire.

It is extremely difficult for a person to see where a stream is hitting from behind the nozzle. A fire fighter or officer should step back or to the side to get a better view of the stream, giving directions to the fire fighter who is operating the nozzle from that position.

Once master streams are in position, staffing needs diminish. Personnel inside the fire zone should be kept to a minimum. This is a good time to start rehabilitation. Handheld lines have little or no value during defensive attacks on large structures and often invite unsafe tactics. Anything less than a 2½" (64-mm) line would be ineffective. A common problem is that fire fighters who are not operating pumps or master streams are standing

> Master streams are the tools of choice for defensive operations. A master stream can apply more water from a greater distance and requires fewer personnel to operate as compared to a handheld hose line.

around at a big fire, and they may occupy themselves using handheld lines that serve no purpose. It is the IC's responsibility to manage this situation by placing fire fighters without a specific assignment in a safe location. If handheld lines were used for an abandoned offensive attack, it is recommended that these lines be disconnected from the apparatus during defensive operations to ensure that they are not used inappropriately. However, there are occasions when the fire is being fought defensively in one building and offensively in another. In this case, the use of smaller diameter hose may be acceptable on the interior of an exposure.

Estimating Water Supply Needs

Providing water at these large defensive operations may prove more challenging than most offensive attacks. Many master stream appliances are capable of requiring the total pump capacity of an apparatus. When several of these master stream appliances are operating, the water system can be exhausted even in areas with reliable water supplies. Supply lines should be large-diameter hose, and there may be a need to relay water.

Some water supply systems are made up of several separate systems that can be connected (cross-tied) when necessary. System connections can be accomplished by pumping from one system to the other. Typically, the water utility company will open street valves to cross-tie the systems. Cross-tie locations should be identified and pre-planned. The water utility company may also be able to bring additional pumps on-line, increasing the volume or pressure at the fire scene. Consider including the water utility in the multiple-alarm call-up list.

Selecting the Stream Position

The first consideration in placing exterior streams is safety. The second is the ability to apply water to exposures and to the interior of the building, preferably from a position that will not push the fire into uninvolved areas. Quite often, these two considerations are at odds. The best direct application positions are often in the collapse zone.

> When handheld hose lines are used as part of the defensive attack, the minimum size for an exterior attack line is 2½" (64 mm). When handheld lines are needed as part of a defensive attack, reduce fatigue by using hose loops whenever possible.

Unstaffed ground monitors are sometimes the answer to this dilemma. However, the effectiveness of these streams is often limited, which results in sending fire fighters back into the collapse zone to redirect the streams. Elevated streams have a natural advantage when the fire is in the upper stories of the building but require the apparatus to be parked fairly close to the building. Likewise, the apparatus-mounted master stream device is easier to place in service but places the pump operator and apparatus in the collapse zone when directing an attack on tall buildings. Remember, there is no justification to risk injury to fire fighters during a defensive operation, and there is seldom justification to risk damage to expensive apparatus and equipment.

> When evaluating the water supply, consider the flow capacity of individual hydrants as well as the total flow capacity of the water system.

Selecting the Nozzle Type

Is the solid-stream (smooth-bore), fog (or spray) stream, or straight-stream best for defensive operations? They all have their place, and it is important for the IC to know the advantages and disadvantages of all available tools. The solid-stream has the greatest reach and penetrating ability and is best suited to situations when the attack is on the main body of the fire in a large structure.

The fog (spray) pattern can be gently applied to an exposure, covering a wider area without breaking windows. A fog pattern has the maximum pushing effect, and sometimes the objective during an exterior attack is to push the fire away from occupants, fire fighters, or exposures.

Straight streams from a variable-stream nozzle are not generally considered equal to solid-stream (smooth-bore) nozzles, but most are capable of penetrating the main body of fire.

Calculating Staffing Needs

Unlike the offensive attack, in which several fire fighters are needed to advance a handheld line, the master-stream–oriented defensive attack typically requires only one fire fighter to direct the stream and one to operate the pumper, once a water supply has been established. The company officer should supervise the operation and stand to the side to provide direction to the fire fighter directing the stream. Unassigned personnel may be rotated through rehabilitation or assigned to staging, making companies available for other responses or to augment the attack.

Determining Apparatus Needs

During the offensive attack there may be available apparatus, whereas the defensive attack is apparatus intensive. Offensive attacks are people intensive. Where two or more fire fighters are required to advance a 2½″ (64-mm) line flowing 250 GPM (16 L/sec) in an offensive attack, one fire fighter can direct a 1000-GPM (63 L/sec) or greater master stream.

A master stream that is being supplied from a distant hydrant may require one or more pumpers to supply the water. Further, a large master stream appliance may require one or more pumpers to supply it. However, 1250-GPM (79 L/sec) or 1500-GPM (95 L/sec) pumpers connected to a reliable water supply with a large-diameter hose may be able to supply two master streams.

Elevated streams can also require additional pumpers, particularly if the aerial apparatus are not equipped with pre-piped waterways and pumps.

When water supplies are very limited, tanker shuttles or drafting from distant water sources will require a substantial commitment of personnel, apparatus, and time.

Conflagrations and Group Fires

<u>Conflagrations</u> occur less frequently now than in the past, but they remain a possibility. Fire departments should recognize the extraordinary challenge presented by a conflagration and determine the probability of a conflagration in their response area. Special tactics are needed to resolve a fire incident involving a large geographic area.

Early conflagrations, such as the burning of Rome, are not well-documented. More is known about later conflagrations that destroyed large tracts within urban centers. London had conflagrations in the years 798, 982, 1212, and 1666. Constantinople is the undisputed conflagration champion, with conflagrations recorded in the years 1729, 1745, 1750, 1756, 1782, 1791, 1798, 1816, 1870, 1908, 1911, 1915, and 1918. The first permanent American colony, founded at Jamestown, Virginia in 1607, was destroyed by fire in January of 1608. Plymouth, Massachusetts, was founded in 1620 and was substantially destroyed by fire three years later. Boston experienced nine conflagrations before the Revolutionary War. During the Revolutionary War, like all other wars,

Case Summary

On Monday, August 19, 2002 a conflagration that began in the Santana Row construction site of San Jose, California spread via burning embers to the Huff/Moorpark area, which is a half mile away.

The building of origin was a very large six-story structure covering six acres that was under construction within the Santana Row complex. This fire created burning embers as large as 2″ × 4″ (51 mm × 102 mm). The high wind carried embers to an apartment complex and townhouses in the Huff/Moorpark area where they ignited wood shake roofs. Some of the roofs had been replaced by composition roofing materials. The damage was largely confined to 10 large apartment and townhouse buildings with wood shake roofs. Buildings with composition roofs were, for the most part, undamaged.

Figure CS9-2 The fire at Santana Row construction site, from which burning embers traveled and created a conflagration half a mile away.

Source: John Lee Cook, Jr., USFA-TR-153, 1/2004, Santana Row Development Project Fire, San Jose, California.

fire was used as a weapon, destroying many villages in colonial America. Terrorists used fire as a weapon of mass destruction at both World Trade Center attacks. Conflagrations continued with great frequency in the New World until the early 1900s. Some conflagrations of notable significance are listed in Table 9-1.

The list in Table 9-1 is not all inclusive. Many group fires and some conflagrations are not investigated by national organizations and, therefore, go unnoticed outside the region where they occur.

The term conflagration is often misused. Many people refer to any large fire as a conflagration. It is not unusual to hear a news anchor refer to the conflagration in a downtown tenement building. The *NFPA Fire Protection Handbook* defines conflagration as a fire with major building-to-building flame spread over some distance.[3]

Group fires are similar to conflagrations; however, as explained in the *NFPA Fire Protection Handbook*, unlike a conflagration, a group fire is confined within a complex or among adjacent buildings. Earlier definitions describe a group fire as not extending beyond a complex or confined to the city block of origin. Most group fires have the potential to become conflagrations, but they are either confined early, or no exposures are readily available to allow fire spread. Group fires are generally smaller scale than a conflagration; however, there are exceptions. For

TABLE 9-1 Conflagrations of Significance

Year	Place	Involvement
1835	New York City, NY	700 buildings
1845	Pittsburgh, PA	1000 buildings
1851	San Francisco, CA	2500 buildings
1866	Portland, ME	1500 buildings
1871	Chicago, IL	17430 buildings (250 fatalities)
1871	Peshtigo, WI	17 towns (approx. 1200 fatalities)
1901	Jacksonville, FL	1700 buildings
1904	Baltimore, MD	2500 buildings
1906	San Francisco, CA	28,000 buildings (earthquake)
1908	Chelsea, MA	3500 buildings
1914	Salem, MA	1600 buildings
1916	Paris, TX	1440 buildings
1917	Atlanta, GA	1938 buildings
1918	Minnesota Forest	4000 buildings (559 fatalities)
1922	Ontario Forest	18 townships (44 fatalities)
1922	New Bern, NC	1000 buildings
More recent conflagrations and group fires:		
1961	Bel Air, CA	630 buildings
1973	Chelsea, MA	300 buildings
1979	Houston, TX	25 buildings
1981	Lynn, MA	28 buildings
1982	Anaheim, CA	51 buildings
1983	Dallas, TX	6 buildings (125 apartments)
1991	Oakland/Berkeley, CA	2886 buildings (25 fatalities)
2002	San Jose, CA	10 large buildings

instance, the main body of fire at the Santana Row fire in San Jose, California was contained within the complex, which would be defined as a group fire even though it did cover a large geographic area. However, this fire does not completely fit the definition of a group fire as flying brands traveled a considerable distance, igniting buildings well beyond the complex where the fire originated. From a strategic point of view, conflagrations and group fires present similar problems and have similar solutions.

Wildland/urban interface fires spreading from wildlands into an urban area, and destroying large numbers of buildings, are properly defined as conflagrations. The most notable wildland/urban interface fire occurred in Peshtigo, Wisconsin, on the same day as the Great Chicago Conflagration: October 7, 1871. The Peshtigo fire killed 1200 people (some estimate as many as 2000 fatalities) and destroyed 17 towns. A wildland fire that does not destroy a large number of buildings should not be labeled a conflagration.

The focus of this text is structural firefighting, thus tactics used to combat wildland/urban interface fires are beyond the scope of this book. However, in Table 9-1 we have listed several conflagrations that began as wildfires, because once a wildland fire enters a populated area, many of the tactics described in this text would apply. Therefore, discussion is limited to the tactics used when confronted with a large-area fire in a populated area.

When conflagrations are studied, common contributing factors are evident, including:

1. Closely built structures, especially combustible structures
2. Wood shingle roofs
3. Poor water supplies or fire-suppression weaknesses (automatic and manual)
4. Dilapidated structures, especially abandoned buildings in large numbers
5. Large-scale, combustible construction or demolition projects
6. Residential and/or commercial developments near wildlands
7. Built-up areas near high-hazard locations, where a transportation or industrial fire/explosion could quickly involve large numbers of buildings

Response areas with one or more of the seven conditions listed above have an increased probability of a conflagration, especially when experiencing dry and windy weather conditions. Most of the urban conflagrations that occur have two or more of the seven factors listed above combined with windy weather conditions. However, of the above, the wood shingle roof has historically posed the greatest problem and is most often cited as a conflagration factor. Codes allowing untreated wood shingle roofs ignore a real and present danger.

An understanding of how large fires spread is essential in order to effectively apply the tactics involved in controlling a conflagration or group fire. The wood roof shingles or other burning materials move upward via convection currents and then are carried by the wind to other areas. Also, the large flame front creates extremely high radiant heat that further spreads the fire.

The flying brand hazard is addressed by sending brand patrols into areas downwind, particularly areas containing materials that are likely to ignite. The 2002 Santana Row fire in San Jose, California, serves as an excellent example of the dangers associated with flying brands and wood shingle roofs. At the Santana Row fire, flying brands ignited wood shingle roofs a half mile away.

Understanding the methods of heat transfer is the first step in developing a conflagration strategy. The role of convection heat is described above; however, once the flame front widens, radiant heat is the primary means of fire extension from building to building and from groups of buildings to other groups of buildings. If the fire is extending along a wide, radiant flame front, the primary tactics are narrowing the flame front and setting up primary and secondary lines of defense while protecting exposures. In setting up lines of defense, take advantage of natural fire breaks, such as wide streets.

As always, life safety is the top priority followed by extinguishment. In large group fires and conflagrations, being proactive and evacuating people in the fire's path long before they are actually threatened will be the key to success. Remember that the evacuation area must be beyond the secondary line of defense, and the police department is typically the best agency to manage the actual evacuation. Isolating the evacuation area from well-intentioned would-be victims and traffic control are extremely important; thus, a wide fire perimeter should be maintained. Establishing a police branch, under police supervision, is generally a good organizational tactic at large-scale fires.

A fully staffed planning section is of great value when commanding a fire that has spread over a wide geographic area; however, many times the initial line of defense is overrun by the fire before Planning can be properly

> When setting up a primary and secondary line of defense for a conflagration, place apparatus so they can be rapidly re-deployed if the line of defense must be abandoned.

staffed. Figure 9-5 shows three lines of defense for a Houston, Texas, apartment complex fire. The first two lines of defense had to be abandoned before a successful stop was made at the third line of defense. A battle-line strategy established without allowing enough time to position apparatus and equipment results in the IC expending scarce resources in a losing battle.

Halting or preventing a conflagration can seriously challenge any water supply system. The use of water must be prioritized, with individual units realizing the importance of water conservation. In most cases, water being discharged into a flame front is of little value. The top priorities are maintaining the fire break at the lines of defense and protecting exposures. Once the decision has been made to write off an area, do not waste water trying to save what will surely be lost. Large cities are generally served by several water service areas, and the planning section chief should identify auxiliary water sources well in advance. For instance, when water mains were destroyed in the Marina District fire area during the 1989 San Francisco earthquake, the San Francisco Bay became an auxiliary water source as fireboats relayed water to the shore.

Collapsed structures, accumulations of combustible debris, utility problems (electrical shorts, gas leaks, etc.), and heating and lighting with open flames are only a few of the many reasons that areas struck by a natural disaster are prone to fires and conflagrations. Cincinnati, Ohio suffered a conflagration in 1937 during a major flood. Petroleum tanks floated off their bases, spreading flammable and combustible liquids on top of the water. The petroleum products were ignited and spread fire to many buildings over a wide area. Fire hydrants and access roads were underwater, hampering fire control efforts. Many buildings burned to the water line.

The five tactical elements of a successful conflagration strategy are:

1. Evacuate and rescue people in imminent danger.
2. Evacuate people in the endangered area beyond the secondary line of defense.
3. Set up a line of defense in an area with natural or artificial fire breaks.
4. Establish a secondary line of defense.
5. Narrow the flame front.
6. Provide flying brand patrols beyond the line of defense.

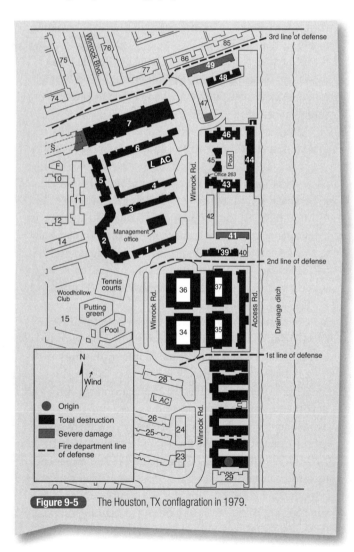

Figure 9-5   The Houston, TX conflagration in 1979.

Classifying as a Non-attack

It is difficult to stand by and watch a building being consumed by fire, but this is sometimes the only safe approach. Non-attack postures are underused because ICs often fail to recognize a total loss.

Case Summary

A conflagration that destroyed three city blocks in Perkasie, Pennsylvania, was believed to have been started by juveniles setting fires in grass next to buildings. The fire entered a lumber shed and appeared at first to be a small fire. First attempts at extinguishment seemed to be successful, but hidden fires reignited, and wind swept the flames through the lumber shed. The fire continued to burn out of control and eventually either destroyed or damaged seven buildings and two fire trucks.

Figure CS9-3 Buildings damaged or destroyed in the Perkasie fire.

Source: Jay Bradish, Conflagration destroys three city blocks. *Fire Command*, January 1989, pp. 22, 23, and 26.

Often, in non-attack operations, environmental concerns are paramount. In these situations, resources may be assigned to protect streams from contaminated run-off water or evacuate downwind neighborhoods. As an example, a Sherwin-Williams warehouse fire in Dayton, Ohio, threatened the community's water supply.[4] A non-attack posture avoided a major environmental catastrophe. If a safe offensive attack is not possible and there is little or nothing to be gained by initiating a defensive attack, choose the non-attack option.

Summary

There are three possible operations at a structure fire: offensive, defensive, and non-attack. A proper size-up will indicate which operation is appropriate. The offensive attack gives fire fighters the best chance of saving lives and property from fire; therefore, when a proper size-up justifies an offensive operation, it is the operation of choice. Rarely is the defensive or non-attack strategy an effective means of saving lives in the building of origin. However, under certain circumstances these may be the IC's only reasonable options.

If the IC's size-up indicates that a defensive operation is appropriate, operations are initiated from the exterior. The strategic objectives of the defensive attack are to protect internal and external exposures while extinguishing the fire.

If an offensive attack is not possible and the building and surroundings are going to be a total loss regardless of the actions taken, a non-attack posture may be best. The risk-management principle, "No risk to the safety of members shall be acceptable when there is no possibility to save lives or property" applies.

Wrap-Up

Key Terms

conflagration A fire spreading over a large area involving multiple structures.

defensive A fire attack conducted from the exterior of a building.

group fire A large fire involving several buildings that is confined within a complex or among adjacent buildings.

offensive A type of fire attack in which fire fighters advance into the fire building with hose lines or other extinguishing agents to overpower the fire.

Suggested Activities

1. The fire building located in the center of **Figure 9-6** is a fully involved one-story warehouse with fire visible at the roof. It is 2:00 AM, with cloudy skies and a temperature of 55°F with wind 10 MPH in the direction shown. Central City has three water supply systems that can be interconnected. This area is supplied by the business district system that has a total capacity of 10,000 GPM (631 L/sec). The other two systems can be cross tied to the business district system by opening street valves. The total water supply from all three systems is 30,000 GPM (1892 L/sec). Hydrants are spaced every 500′ (152 m) with a 1200 GPM (76 L/sec) capacity. Use the scenario in Figure 9-6 to answer the questions related to a defensive attack.

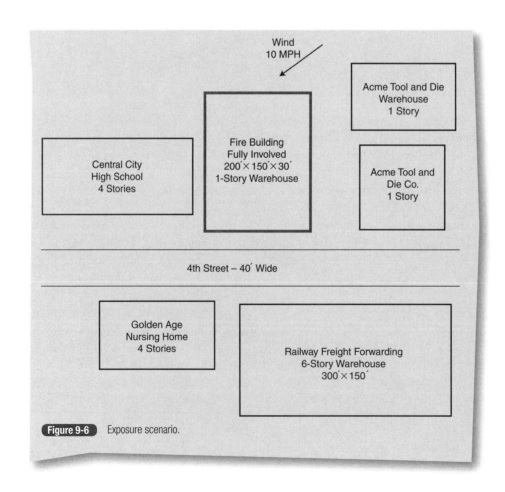

Figure 9-6 Exposure scenario.

A. Establish a collapse zone.

B. Prioritize exposures, indicating which exposure you would protect first, second, third, fourth, and fifth.

C. Develop an incident action plan for this scenario.

D. What is the probability that the exposures are occupied at the time of this fire? Explain.

E. Show the location of master streams. Explain the pros and cons of each placement.

F. What type and size of nozzle would you use on each master stream appliance?

G. Would you attack the main body of fire before or after protecting the exposures? Explain your decision.

H. Explain the effect elevated master streams would have on the main body of fire and on exposed structures.

I. Would the available water supply support your defensive operation?

J. How many apparatus would be needed to support your operation?

K. List the companies and staffing needed to conduct a safe and effective attack.

L. Develop a NIMS chart for your operation.

2. Use the scenario in #1, but change the time of day to 1:00 PM on a school day. How does this change your exposure priorities and tactics?

3. Use the scenario in #1, but change the Golden Age Nursing Home to a vacant building (time is 2:00 AM). How does this change your exposure priorities and tactics?

Wrap-Up, continued

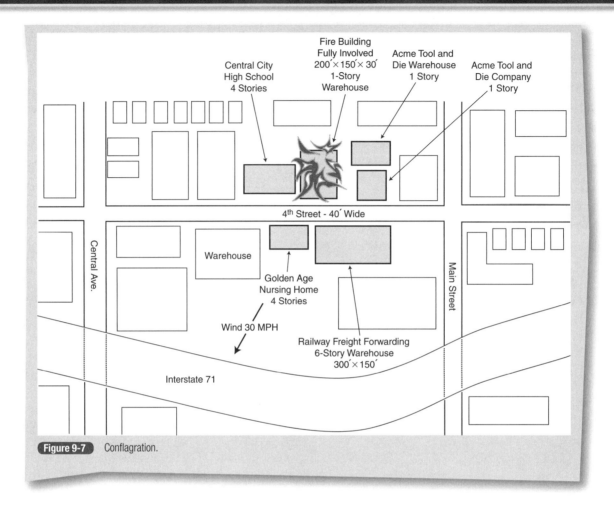

Figure 9-7 Conflagration.

4. The fire in scenario #1 has extended to surrounding buildings and across 4th Street **Figure 9-7**. The wind has increased to 30 MPH in the direction shown in Figure 9-7. The Central City business district extends 8 city blocks beyond Interstate 71 to a large river (approximately 1800′ [549 m] wide). A heavily wooded hillside rises sharply on the far side of the river. Rows of houses with wood shingle roofs are situated on this hillside. Develop an incident action plan for this fire based on the tactics discussed in this chapter.

5. Evaluate the probability of a conflagration or large group fire in your response area.

Chapter Highlights

- Defensive attacks are used when the building's structural integrity or other hazards prohibit entry, when resource needs outweigh capabilities, or when the risk to fire fighters is too great in terms of what can be saved.
- Solid fire-ground information, training, and experience are the keys to determining whether to use an offensive, defensive, or non-attack strategy.
- Defensive attacks seek to save intact property and/or to protect surrounding exposures.
- The collapse zone is the height of the building plus an allowance for debris scatter usually estimated as 1½ times the height of the building.
- Exterior streams can limit fire spread to exterior and interior exposures, but improperly placed streams can also push fire into unburned areas.
- A master stream can apply more water from a greater distance and requires fewer personnel to operate as compared to a handheld hose line.
- Water supply and apparatus positioning are key considerations during a defensive attack.
- A defensive attack is apparatus intensive but generally requires fewer fire fighters than an offensive attack.
- Factors that lead to conflagration include closely built and/or dilapidated structures, wood shingle roofs, poor water supply, large-scale construction/demolition, nearby wildlands, or industrial hazards.
- Apparatus at large-scale fires must be positioned for rapid re-deployment.
- A successful conflagration strategy includes evacuating civilians, creating fire breaks for a primary and secondary line of defense, narrowing the flame front, and patrolling for fire brands.

References

1. National Fire Protection Association, Francis Brannigan, and Glenn Corbett. *Brannigan's Building Construction for the Fire Service*, 4th edition. Sudbury, MA: Jones and Bartlett Publishers, 2008.
2. National Fire Protection Association, *NFPA 1670: Standards on Operations and Training for Technical Rescue Incidents*. Quincy, MA: NFPA, 2004.
3. National Fire Protection Association, *NFPA Fire Protection Handbook*, 19th edition. Quincy, MA: NFPA, 2003.
4. Michael S. Isner, $49 million loss in Sherwin-Williams warehouse fire. *NFPA Fire Journal,* March/April: 65–73, 93, 1988.

Property Conservation

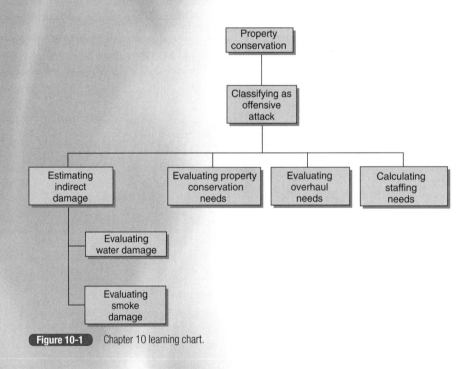

Figure 10-1 Chapter 10 learning chart.

Chapter 10

Learning Objectives

- List the three tactical priorities, in priority order, and explain how property conservation can be accomplished simultaneously.
- Define primary and secondary damage.
- To the untrained observer, ventilation and forcible entry may often appear to cause unnecessary property damage. Explain how proper ventilation and forcible entry actually reduce property damage while protecting occupants and fire fighters.
- Given a pressure and discharge orifice size, estimate the flow from a single and multiple sprinkler heads.
- List and explain six tactics used to reduce water damage.
- Explain why removing property from a building is not generally a good property conservation tactic.
- Calculate the weight of water from a nozzle discharging 350 GPM (22 L/sec) over a 10-minute period.
- Describe the importance of ventilation in property conservation.
- Discuss the importance of overhaul.
- Explain how thermal imaging cameras can be used to reduce overhaul damage and the precautions necessary when using thermal imaging cameras to find hidden fires.
- Enumerate safety issues related to overhaul and fire investigation.
- Calculate rate of flow for a given area, and relate the rate of flow to property conservation issues.
- Develop a fire scenario and apply property conservation tactics for a fire controlled by the sprinkler system or an accidental discharge from the sprinkler system.
- Develop a fire scenario and apply property conservation tactics for a fire controlled by hose lines using V/100 to determine the rate of flow.

Introduction

Proficient incident commanders (ICs) realize the importance of property conservation, and property conservation tactics can substantially reduce the property loss. Although property conservation is the lowest of the three operational priorities, being listed as the third priority does not mean that it must wait until all life safety and extinguishment tactics are completed. The IC would first do all that is necessary to save lives, and extinguishing the fire is often the best life safety tactic. Extinguishment is also the ultimate property conservation tactic. Once staffing permits, the IC should simultaneously attend to all three operational priorities during offensive operations.

> The three tactical priorities are:
> 1. Life Safety
> 2. Extinguishment
> 3. Property Conservation.

Good public relations results from professionalism shown in property conservation efforts. Few citizens (or politicians) recognize a good extinguishment effort, generally because expertly conducted offensive operations are hidden from view. Property conservation tasks are obvious, however, even to laypeople. Efforts directed at saving valuable property show the citizens that you care about them and their property. Remember, fire departments are expected to save both people and property. In a residential fire, the personal property of the occupants is being protected—property that is often irreplaceable. The occupants will appreciate close attention to salvaging what can be saved.

> The golden rule of property conservation is "Do unto the property of others as you would have them do unto yours."

Fire departments often receive unfair criticism when they conduct forcible entry, ventilation, and overhaul operations. The untrained observer does not understand why it is necessary to cut a hole in the roof or tear walls apart to check for extension. To prevent unfair criticism, the department must educate the media, the public, and politicians about the purpose and importance of these tactics. Most important, the IC should explain the overhaul process to the owner and occupants. Although most secondary damage is essential, training should emphasize keeping the damage to a minimum while stopping the forward progress of the fire and preventing a rekindle.

Property conservation is prioritized behind life safety and extinguishment, but property conservation activities need not be delayed until the fire is extinguished or until the primary and secondary searches are complete. Priority activities are assigned first, and property conservation is delayed when staffing is not sufficient to attend to all of the life safety and extinguishment tasks. However, proficient ICs include staffing for property conservation in their request for assistance or reassign forces to property conservation activities early in the battle.

The purpose of this chapter is to explain the primary components of property conservation operations. However, this is not a complete reference on how to complete the various tasks involved in property conservation. Other sources of information such as the NFPA's *Fundamentals of Fire Fighter Skills* are dedicated to the actual tasks involved in protecting property.[1]

In commercial and industrial occupancies, employees may be extremely helpful in locating valves, drains, and mechanical controls. Large industrial complexes sometimes have trained fire brigades, many of which are dedicated to property conservation duties.

Classifying as an Offensive Attack

Figure 10-1 is a learning chart for this chapter. Notice that property conservation generally is limited to offensive attacks. If fire crews have been withdrawn from the building because of a lack of resources or because of deteriorating fire and building conditions, property conservation is no longer an issue. It makes no sense to send fire fighters into a building to throw salvage covers or otherwise protect property when the entire building is at serious risk of being destroyed by fire. Remember, fire fighters should *never* be placed at risk attempting to save what is already lost, or ultimately will be lost.

Estimating Indirect Damage

A balance must be struck between doing necessary damage to halt the loss being caused by the fire and minimizing the damage caused by fire-ground activities. Categorizing damage helps in understanding this relationship.

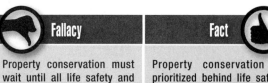

Fallacy	Fact
Property conservation must wait until all life safety and extinguishment activities are complete.	Property conservation is prioritized behind life safety and extinguishment. Be certain that resources are assigned to accomplish these priorities first. When staffing is adequate, property conservation can and should be conducted simultaneously.

Ventilation and forcible entry often cause some degree of property damage. However, delaying or not performing these tactics will cause a delay in extinguishing the fire, which places occupants and fire fighters at unnecessary risk while actually increasing the overall property damaged.

Primary damage will consume the entire property if fire forces are reluctant to force entry or ventilate. On the other hand, if the building is washed off its foundation, nothing has been saved. In a fire, both primary and **secondary damage** will occur. Using caution and good judgment, the IC must do everything necessary to limit all types of property damage. However, fire fighters should never be placed in grave risk to save property.

Fire suppression systems can also cause secondary damage. In earlier chapters, we emphasized letting the system control the fire and not shutting down the system prematurely. Smoke and steam reduce visibility, making it difficult to determine the proper time to close valves and shut down systems. Therefore, when the IC does decide to shut down a system, a fire fighter should remain at the control valve so that the system can be charged again if needed.

Primary damage is caused by the products of combustion. Secondary damage is the result of fire-ground activities or the operation of a fire suppression system.

Sprinkler systems have an excellent record of extinguishing or holding fires in check with minimal water damage. On occasion, system piping or sprinkler heads are damaged, causing an unwanted or accidental water flow. Although this situation is not a fire, the IC needs to be prepared to promptly mitigate it.

Flows from sprinkler heads can vary depending on the size of the orifice opening, the type of sprinkler head, and the water pressure in the system. As **Figure 10-2** shows, water flows of up to 80 GPM (5 L/sec) per head are possible.

Whether the head was knocked off accidentally or opened automatically because of a fire, the water flow will continue until action is taken to stop it. A 30-GPM (2-L/sec) head flowing during a 10-minute response time will discharge 300 gallons (1136 L) of water.

Once the fire is under control, it is necessary to stop the flow of water, either by shutting the system down at the riser, by closing division valves, or by using sprinkler stops. A two-part article in *Fire Engineering* magazine describes property conservation as having the following three components: locating, containing, and removing water.[2,3] The model presented in the article outlines an approach to property conservation operations, and a similar six-point model is offered in this chapter.

Other fire suppression agents can also harm equipment and stock. For example, the discharge of a dry chemical system will require considerable cleanup and can damage certain types of equipment, such as computers, telephone equipment, and other sensitive electronics. Generally, the cleanup should be the responsibility of the property owner, not the fire department.

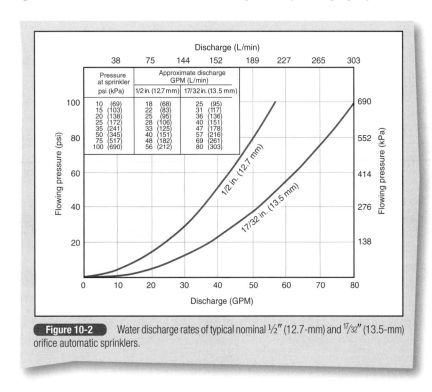

Figure 10-2 Water discharge rates of typical nominal ½" (12.7-mm) and 17/32" (13.5-mm) orifice automatic sprinklers.

Evaluating Water Damage

A fallacy exists regarding secondary damage and rate of flow. Many believe that larger-than-needed hose streams cause unnecessary water damage. In reality, the opposite is true. Larger-than-needed hose streams, if properly applied, will result in less water damage, as the fire is quickly extinguished and the time that water is flowing is reduced. Conversely, choosing a hose line that is too small to quickly extinguish the fire will result in both increased water and fire damage, as water will flow for an extended period of time without effectively extinguishing the fire.

Life safety and extinguishment activities concentrate on the fire floor and floors above, while efforts to control water damage are concentrated on the fire floor and floors below. Water damage is reduced by doing the following:

1. *Promptly extinguishing the fire, and avoiding wash downs*. In addition to reducing water damage, prompt extinguishment activities also reduce fire and smoke damage.

2. *Stopping the flow of water from sprinkler systems*. Promptly shutting down hose lines is one method of stopping the flow of water into the building. In the case of sprinkler systems, or other fixed fire suppression systems, it is important to locate control valves. Using sprinkler stops **Figure 10-3** or controlling the water at the main or divisional control valve is the primary way to reduce the flow from an automatic sprinkler system. Be careful never to shut a system down too soon, as the result will be additional water and fire damage.

3. *Containing run-off*. Capturing the water before it has a chance to run off or migrate to another location is another property conservation tactic. Catchalls that are made with salvage covers, or containers (drums, buckets, etc.) are placed under a ceiling leak below the fire floor. Also, pumps and water vacuums can be very effective in property conservation operations. Preventing further migration of water will reduce the damage.

4. *Channeling the water into drains or chutes or otherwise directing it out of the building*. Fire fighters have devised unique ways of channeling water out of buildings. Makeshift chutes are sometimes fashioned by using ladders, pike poles, and salvage covers. These chutes are then directed out a window, down a stairway, or to a drain.

Fire debris will tend to clog floor drains, so it is important to keep these drains clear. If floors do not have drains, a common way to provide one is by removing a toilet, creating a drain through the 4″ (102-mm) opening in the floor. If this is done carefully, the only damage to the toilet will be to the seal and bolts. As with any drain, however, it is critical to ensure that it does not become clogged with debris. Removing water from the floor is not only important to property conservation, it is also a safety issue. A large buildup of water on a floor can cause structural failure. The IC must determine whether it is safe to conduct interior operations once a large quantity of water has been applied. This is particularly true after long defensive attacks when it is time to move into the building for final extinguishment and overhaul. It is also important to ensure that any standing water in areas such as basements is not hiding hazards, such as openings, changes in elevation, or steps. If it is necessary to send fire personnel into a basement or other confined space that contains standing water, it is crucial to ensure that the electric power has been shut down to avoid any possibility of electrocution.

The nature of the commodities that are stored in a building can affect the weight loading on the floor. If absorbent materials are being stored in the fire area, the weight of the water that they will absorb will

Figure 10-3 A sprinkler stop, used to control water flow from a sprinkler head.

remain for a considerable time and these absorbent materials may expand, placing pressure against outside walls, which can also affect the structural integrity of the building. Water weighs 8.33 lb/gal (1 kg/L). When multiple master stream appliances are flowing water into a structure, the weight of the water can rapidly increase the collapse potential. Depending on conditions, the water may flow out of the building or be contained. Table 10-1 assumes that materials stored in the building or fire debris is containing the water being discharged from large master stream appliances.

It is difficult to estimate the amount of water remaining inside a building. Water streaming out of a building may represent only a fraction of the total amount of water being applied.

5. *Covering valuable property.* It is important for the IC, as well as all other members on the scene, to be aware of the need for property conservation. By placing salvage covers over exposed property, an alert fire fighter can often prevent water damage to valuable property. When using salvage covers, crews should group together furniture and other items to protect them from flowing water. This will allow several items to be protected by a single cover.

6. *Moving or removing valuable property.* Water flows downward through a building, following the path of least resistance. Stairs, elevator shafts, and drains provide paths of least resistance. At times the water is blocked by debris, or the volume simply overwhelms the natural flow out of the building. Even when water is flowing through the ceiling, the path of least resistance will be followed. This is generally around openings in the ceiling, such as light fixtures. If a ceiling is holding water run-off for a period of time, the weight of the water is likely to cause the ceiling to collapse. This ceiling collapse can injure fire fighters

assigned to the area and increase damage to the contents below. Using a pike pole to drain the ceiling can alleviate this problem.

As soon as possible, property should be moved away from natural flow paths, grouped together, and covered with salvage covers.

On occasion, property is completely removed from the building by fire crews to save it from water damage Figure 10-4 . This is a very labor-intensive

Figure 10-4 Museum employees look at art that was removed by fire fighters.

TABLE 10-1	Weight of Water from Master Streams		
Number of Master Streams at 1000 GPM (63 L/sec) Each	Weight Added Each Minute at 8.33 lb/gal (1 kg/L)	Total Weight at 30 Minutes, lb (kg)	Total Weight at 60 Minutes, lb (kg)
1	8330 lb (3785 kg)	249,900 lb (113,550 kg)	499,800 lb (227,100 kg)
3	24,990 lb (11,355 kg)	749,700 lb (340,650 kg)	1,499,400 lb (681,300 kg)
10	83,300 lb (37,850 kg)	2,499,000 lb (1,135,500 kg)	4,998,000 lb (2,271,000 kg)

way to conduct property conservation operations and has very limited application. If materials are moved outside, they must then be protected from inclement weather, vandalism, or theft.

Evaluating Smoke Damage

Smoke can cause considerable property damage. Quite often, the smoke damage greatly exceeds water damage. Smoke generally follows an upward path but can cause damage below the fire as well. In previous chapters, ventilation was discussed in terms of life safety and extinguishment. However, ventilation is also an important property conservation tactic. Blowers and fans are generally used to remove smoke during property conservation operations. Of course, ventilation tactics used to remove smoke for life safety and extinguishment purposes also help to prevent property damage.

Calculating Staffing Needs

Once critical life safety and extinguishment positions are staffed and a tactical reserve has been established, remaining personnel can be assigned to property conservation. The earlier property conservation is addressed, the more successful it will be. Property conservation that is delayed too long will be of little value, as the exposed materials will already be damaged by water and/or smoke.

Property conservation staffing needs vary greatly. For most offensive operations, at least one crew should be assigned to start property conservation activities on the floor below the fire. Several crews may need to be assigned to property conservation duties depending on the amount of damage, the size of the area to be protected, and the type of property that needs to be protected.

Evaluating Property Conservation Needs

Property conservation activities should not be delayed until other priorities have been completed. This idea needs to be emphasized. It is very easy to become overwhelmed with fire attack activities and neglect property conservation until the fire is completely extinguished. Implementing a planning section greatly assists the IC in defining necessary property conservation activity. Staffing and equipment are assigned to life safety and extinguishment first, but if property conservation tasks still remain to be assigned, then there is a need to call for additional resources.

Case Summary

An electrical fire in a four-story apartment building was very difficult to overhaul. The fire started in an electrical outlet on the third floor and extended into the wall. The fire department disrupted the power supply to the apartment, opened up the wall, and extinguished the fire. However, the fire had extended into the apartment above and the attic. Even after the wall and ceiling were opened up, some hidden fire was missed. Eventually, all of the fire was found, and the fire was completely extinguished.

Wires should be traced through the wall space, as heating could occur anywhere along the circuit. Thermal imaging cameras could also be a valuable tool for use during overhaul operations.

Figure CS10-1 A fire fighter uses a thermal imaging camera while another uses a pike pole, and a third awaits with a fire hose.

Source: Frank C. Montagna, Chasing down electrical fires. *Fire Engineering*, July 1999, pp. 83–86.

Crews generally view property conservation as an undesirable assignment, preferring to be involved in life safety and extinguishment activities. Companies generally do not freelance into property conservation tasks. The ability to simultaneously conduct life safety, extinguishment, and property conservation tasks is a sure sign that the IC is managing the incident effectively.

Evaluating Overhaul Needs

Overhaul is extremely important; some tactical priority models list it as a separate priority. Overhaul is the completion of the extinguishment priority. Overhaul can result in what might appear to be additional damage to the property, but this damage is warranted to prevent further primary damage. The use of a thermal imaging camera can substantially reduce damage caused by opening walls and ceilings to locate hot spots. However, it is important that members using the thermal imaging camera be well trained in using the device and interpreting readings. If a thermal imaging camera is used to examine walls and ceilings, and any doubt remains as to the presence of hidden fire, the wall and/or ceiling must be opened.

The purpose of overhaul is to ensure that the fire is completely extinguished and that the building is safe for personnel such as investigators or the property owner to re-enter. To permit people to re-enter a building, only to have fire rekindle, is a serious failure on the part of the IC. Overhaul is an activity that is appealing to few fire fighters, which sometimes leads to a decision to simply wet down everything in sight. This wash down is not effective in getting to hidden fires and may result in unnecessary property damage.

There is a tendency for fire scene discipline to deteriorate during the overhaul phase. Fire fighters are often injured as they rush to complete this distasteful chore so that they can return to the fire station or home. As the smoke clears, many fire fighters take the opportunity to remove their breathing apparatus and other items of protective clothing. Studies indicate that toxic gas levels are often greater during overhaul operations than during active firefighting. If the IC does not maintain discipline, overhaul can become an uncontrolled, dangerous activity.

To keep property damage to a minimum and to ensure that the fire is completely extinguished, overhaul must be accomplished in a well-planned, systematic manner. A good overhaul technique is to assign a fully protected crew inside as a fire watch while the building is being ventilated. This will allow other fire fighters on the scene to rest, thereby reducing the potential for injuries. If it is practical, keep crews outside at least 15 minutes. After the building is well ventilated, the rested troops can return to a much safer environment to complete the overhaul.

Full protective clothing, including breathing apparatus, must be worn during overhaul activities. Opening walls and ceilings will expose smoky, toxic, smoldering debris. In addition, ceilings and walls may contain asbestos or other materials that are capable of harming the unprotected fire fighter. Members operating inside the structure, including fire investigators, cannot be permitted to remove their facepieces. Many departments use carbon monoxide meters to determine when it is safe to remove the breathing apparatus facepieces. Carbon monoxide is usually the most prevalent fire gas, and low readings after a fire indicate that some of the smoke and toxic gas has been removed. However, there are many other inhalation hazards that are not read by carbon monoxide detectors. The best practice is to remain "on air" throughout the overhaul phase of the operation.

A Word About Fire Investigation

Improper overhaul can destroy any hope of properly conducting a fire scene investigation. A determination of what areas are to be left undisturbed should be made while fire suppression forces are resting between fire control and overhaul. The IC should attempt to make a preliminary determination as to what areas will be of interest to the investigators. If a fire investigator is on the scene during overhaul operations, he or she will be an invaluable resource to the IC. However, proper and complete overhaul should always be done to ensure that the fire is completely extinguished.

Summary

The overall goal of firefighting is to save lives and protect property. Extinguishing the fire is an important step in accomplishing this goal, but it is not the only step. Property conservation activities are also important elements of the IC's overall strategic plan. Further, the IC must remember that fire fighters have a tendency to let their guard down during property conservation and overhaul activities. Therefore, to guard against injuries, the IC must ensure that these tasks are accomplished in a well-planned, systematic manner.

Wrap-Up

Key Terms

primary damage Damage caused by the products of combustion.

secondary damage Damage that is the result of fire-ground activities or the operation of a fire suppression system.

Suggested Activities

1. Locate a multi-story sprinkler-protected building with high-value equipment or stock. Simulate a fire scenario on the floor above the high-value property. Compare and contrast a sprinkler-controlled fire and a manually controlled fire.

   A. List ways in which water and smoke damage could be minimized while conducting an effective offensive attack.

   B. Use the rate-of-flow calculations to determine the GPM required to manually extinguish the simulated fire. In the sprinkler-protected scenario, use the actual flow for the sprinkler heads used in the property. If the exact flow is not known, calculate one head controlling the fire at 30 GPM (2 L/sec) for the sprinkler operation. Assume that the fire is rapidly controlled by this one head but continues to flow for the 10-minute response time plus time needed to either chock the sprinkler head or control the flow through a division or main control valve.

   C. Discuss how each of the six strategies for water removal listed in this chapter, as well as venting, can reduce the property loss.

Chapter Highlights

- Property conservation is the third priority, but can be performed simultaneously with life safety and extinguishment when staffing permits.
- Property conservation is limited to offensive attacks.
- Ventilation and forcible entry generally cause less property damage than delaying extinguishment.
- Attempting to extinguish a fire with a hose line that does not meet or exceed the required rate of flow results in unnecessary water and fire damage.
- Floor drains must be kept clear; a toilet can be removed to provide a drain if needed.
- Covering or moving valuable property can help prevent water damage.
- Smoke damage generally occurs above the fire but can damage areas below the fire as well.
- Proper ventilation is a life safety, extinguishment, and property conservation tactic.
- Staffing and equipment are assigned to life safety and extinguishment first, but if property conservation tasks still remain to be assigned, call for additional resources.
- At times overhaul operations that cause property damage are necessary and warranted to prevent further primary damage.
- To ensure complete extinguishment and minimize property damage, overhaul must be accomplished in a well-planned, systematic manner.
- Improper overhaul can hinder a fire scene investigation.

References

1. National Fire Protection Association, *Fundamentals of Fire Fighter Skills*. Sudbury, MA: Jones and Bartlett Publishers, 2004.
2. Anthony J. Pascocello, Jr., Saving property: Salvage operations, part 1. *Fire Engineering*, September: 70–72, 1996.
3. Anthony J. Pascocello, Jr., Saving property: Salvage operations, part 2. *Fire Engineering*, October: 97–99, 1996.

The Role of Occupancy

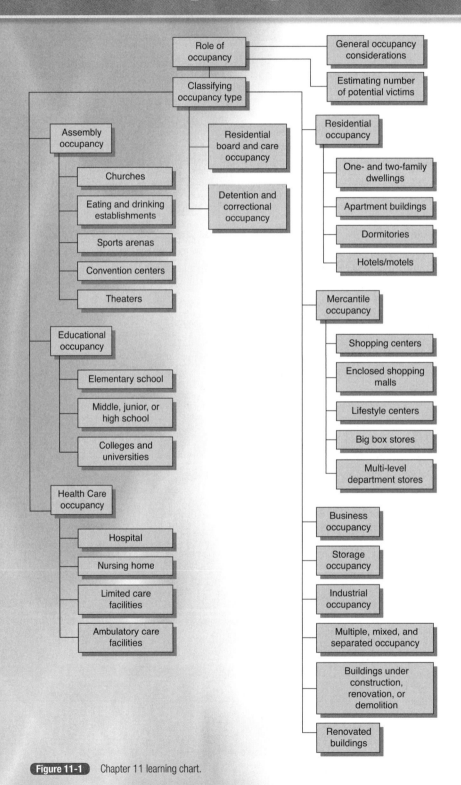

Figure 11-1 Chapter 11 learning chart.

Chapter 11

Learning Objectives

- Classify occupant ability to evacuate using an occupancy factor matrix.
- Explain what is meant by occupant density and how it affects occupant safety.
- Given the dimensions, number of floors, and occupant density, determine the maximum number of people who could be in a building.
- Define assembly occupancy and provide examples of different types of assembly occupancies.
- Compare and contrast the risk to fire fighters when fighting a fire in an assembly occupancy compared to a residential occupancy.
- Define educational occupancy and compare the life hazards in elementary schools, high schools, and colleges.
- Define health care occupancy and explain how evacuations in health care occupancies are different than in most other occupancies.
- Compare and contrast occupants in hospitals, nursing homes, and limited care facilities and describe how the occupant characteristics in each affect life safety during a structure fire.
- Define residential board and care occupancy and compare these facilities to nursing homes.
- Define detention and correctional occupancy and explain the special challenges associated with combating a fire in a large correctional facility.
- Define residential occupancy and compare various types of residential buildings in terms of life safety.
- Evaluate and discuss civilian fire deaths and fire fighter on-duty death rates in residential occupancies.
- Define mercantile occupancy and compare older style shopping centers to enclosed malls and lifestyle centers in terms of life safety and extinguishment.
- Examine life safety and extinguishment problems related to "big box" stores.
- Explain how search and rescue procedures in a large commercial structure differ from search and rescue in a typical residential building.
- Define business occupancy.
- Compare and contrast business and residential occupancies in terms of the risk to fire fighters during fire-ground operations.
- Define storage occupancy and evaluate the effect of fuel load on manual firefighting.
- Describe the problems associated with changing the commodities stored in a sprinkler-protected storage occupancy.
- Compare and contrast storage and residential occupancies in terms of the risk to fire fighters during fire-ground operations.
- Define industrial occupancy and the effect of hazardous materials related to life safety and extinguishment.
- Compare and contrast industrial and residential occupancies in terms of the risk to fire fighters during fire-ground operations.
- Define multiple, mixed, and separated occupancies and explain the difference between a mixed and separated occupancy.
- Explain the increased hazard to occupants in multiple-story buildings where the first floor is occupied by stores and shops with apartments above.
- Describe the fire hazards associated with buildings under construction, renovation, or demolition.

Introduction

As a part of the size-up process, the incident commander (IC) must consider the building's occupancy type. This information will assist the IC in determining the level of risk to occupants and fire fighters. Residential, high-rise, and assembly occupancies all require different strategies. Time of day and day of week are also critical factors. This chapter reviews the process of evaluating occupancies and formulating safe and effective fire-ground strategies. Figure 11-1 lists occupancy types discussed in this chapter. Once the occupancy type is known, the IC can apply a strategy applicable to the occupancy and can begin determining the life safety, extinguishment, and property conservation tactics required to bring the incident to a successful conclusion.

Classifying the Occupancy Type

The function or use of a building has much to do with life safety. Codes and standards are written with a focus on the special hazards presented by the occupancy. Some of the major factors related to occupancy type are as follows:

- Mobility of the occupants
- Age of the occupants
- Leadership (Will there be an organized evacuation?)
- Awareness (sleeping, mentally impaired, etc.)
- Occupant density and total number of occupants
- Familiarity (Do they know the building layout?)
- Time factors (When are people present in the structure?)

Occupancy factors affecting the ability to evacuate without assistance will be rated as part of the discussion of each major occupancy type. A 1-to-10 scale will be used to indicate whether each factor is positive (10), neutral (5), or negative (1) for the specific occupancy type. As an example, an awareness rating of 10 would mean that all occupants within this occupancy would be expected to be fully awake and mentally unimpaired. An awareness rating of 1 would be given to an occupancy where occupants are totally unaware, such as people who are under general anesthesia. The numerical ratings are general in nature and buildings of the same occupancy classification may have different ratings. As an example, some ambulatory care facilities anesthetize patients, which would result in a very negative awareness rating, while other ambulatory care facilities do not anesthetize or use medications that reduce patient awareness.

The following occupancy types will be addressed in this chapter:

- Assembly occupancies
  - Churches
  - Eating and drinking establishments
  - Sports arenas
  - Convention centers
  - Theaters
- Educational occupancies
  - Elementary schools
  - Middle, junior high, or high schools
  - Colleges/universities*
- Health care occupancies
  - Hospitals
  - Nursing homes
  - Limited care facilities
  - Ambulatory care facilities
- Residential board and care occupancies
- Detention and correctional occupancies
- Residential occupancies
  - One- and two-family dwellings
  - Apartment buildings
  - Dormitories
  - Hotels/motels
- Mercantile occupancies
  - Shopping centers
  - Enclosed shopping malls
  - Lifestyle centers
  - Big box stores
  - Multi-level department stores

*Colleges are technically categorized as business occupancies where classrooms accommodate fewer than 50 students and as assembly occupancies when the room accommodates 50 or more students.

> Occupant density is a measure of the number of people in a given area. Codes usually state occupant density as the number of ft² per person. In some assembly occupancies the maximum density is 7 ft² (0.65 m²) per person (occupant density factor). In other words, 100 people could occupy a 700-ft² (65 m²) area. Some business occupancies have a maximum density of 100 ft² (9 m²) per person or the same 700-ft² (65 m²) area would be limited to seven people. Most codes require the maximum number of people allowed in a given area to be posted in assembly occupancies and sometimes in other occupancies. If the area is not posted, a rough approximation of the number of people who could safely occupy an area can be determined by dividing the area in ft² by the occupant density factor.

- Business occupancy
- Storage occupancy
- Industrial occupancy
- Multiple occupancy
- Buildings under construction or demolition
- Renovated buildings

Each occupancy- and sub-occupancy type will be discussed in terms of the operational priority list: life safety, extinguishment, and property conservation.

Codes are promulgated to avoid the repetition of past mistakes. However, strengthening codes is not enough. Each of the major loss-of-life fires described in this chapter contains lessons for the would-be IC. The various occupancies are used to discuss common problems and solutions but also to offer history lessons. The NFPA's Fire Investigations Department publishes fire investigations and research reports. The U.S. Fire Administration (USFA), National Institute for Occupational Safety and Health (NIOSH), and National Institute for Science and Technology (NIST) produce technical reports that are available for recent large loss-of-life fires. Periodicals such as the *NFPA Journal* and *Fire Engineering* have offered technical reports to the fire service for over 100 years.

Assembly Occupancies

NFPA 101: Life Safety Code,[1] defines an assembly occupancy as:

> **3.3.168.2** Assembly Occupancy. An occupancy used for the gathering of 50 or more persons for deliberation, worship, entertainment, eating, drinking, amusement, awaiting transportation, or similar uses; or used as a special amusement building, regardless of occupant load.

The discussion here is limited to larger places of assembly and tactical difficulties related to the occupancy. Included under the places of **assembly occupancy** category are the following:

- Churches
- Eating and drinking establishments
- Sports arenas
- Convention centers
- Theaters

An occupancy factor matrix for assembly occupancies is shown in Table 11-1. Note that occupant familiarity and occupant density tend to be negative factors, but mobility and awareness tend to be positive factors for most assembly occupancies.

Places of assembly are not occupied at all times. However, when they are occupied, large numbers of people gather in a relatively small area. In some instances, a population density of 7 ft^2 (0.65 m^2) per person can be allowed; stated another way, nearly 1500 people can occupy a 100′ × 100′ (approximately 930-m^2) room, and even more people are allowed where there is fixed seating. This compares to 333 people in the same area of a grade-level mercantile occupancy, 100 in a business occupancy, or 50 in a hotel.[1] The primary challenge in managing an incident in an occupied assembly building is removing people to the outside or to a place of safe refuge.

Assembly occupancies are dangerous places, not only for the occupants, but also for fire fighters. The fire fighter fatality rate for residential fires is 3.8 fatalities per 100,000 fires compared to a rate of 11.6 fire fighter line-of-duty deaths per 100,000 fires in assembly

> The philosopher and poet George Santayana said "Those who cannot remember the past are condemned to repeat it." As practitioners of fire strategy and tactics, it is imperative that we learn from the past.

TABLE 11-1 Occupancy Factor Matrix: Assembly Occupancies

Assembly Occupancies	Mobility of Occupants	Age of Occupants	Leadership	Awareness	Occupant Density	Familiarity of Occupants
Churches	8	5	9	10	2	5
Eating and drinking establishments	8	7	5	4	2	3
Sports arenas	8	7	6	7	2	2
Convention centers	8	6	5	10	2	1
Theaters	8	6	5	10	2	2

Key: 1 = extremely negative factor, 10 = very positive factor.

occupancies. Fire fighters working at an assembly occupancy fire are three times more likely to be killed than when working at a residential fire Figure 11-2 .

The fuel load will vary by the assembly classification. The concern is not only for the total fuel load; flame spread rate is also critical. Decorative materials with high flame spread rates were significant contributing factors in the large loss of life at the Cocoanut Grove, Beverly Hills, and Station Night Club eating and drinking establishment fires. Fire control efforts must place a high priority on keeping the fire out of the exits. Aside from the main tactical consideration of removing victims during times of peak occupancy is the fact that these structures are unattended for long periods of time when they are not in use, allowing fires to gain considerable headway before discovery. The probability of total involvement on arrival in these situations dramatically affects fire fighter safety. Fires in places of assembly are fairly rare, accounting for approximately 3% of all structure fires,[2] but the potential for a large loss-of-life fire in these buildings requires pre-incident planning and special tactical considerations.

Churches

Churches often use open flames as part of religious services and tend to be overcrowded on religious holidays. Churches are also left unattended for long periods of time, and many church fires occur when the building is unoccupied. However, churches are occupied at times other than during religious services. Meetings, educational programs, fund-raisers, and social events take place within the church or attached rooms. Some churches provide a wide array of social services, such as child care, elderly programs, and shelter for the homeless. These services require supplies that add to the fuel load. Collecting clothing and food for poor or homeless people in the community is a common practice. These supplies—food and clothing—are often stored in places that are not separated from the main church and can pose a severe fire threat.

In the case of old gothic-style churches, such as the one shown in Figure 11-3 , fires quickly involve large concealed spaces. Further, many churches are attached to other buildings, such as rectories or schools, which can create an exposure problem Figure 11-4 . In addition to these traditional church structures, religious services are sometimes held in small stores and other buildings not originally designed as churches.

Statistically, fires and loss of civilian lives are rare in religious properties. On average only 1760 of 517,130 structure fires (0.3%) per year occurred in religious properties from 1999 to 2002.[2] The fire loss in places of worship is 1.1% of the total loss.[2] However, the potential for a large loss of life is a concern. Most fires in these

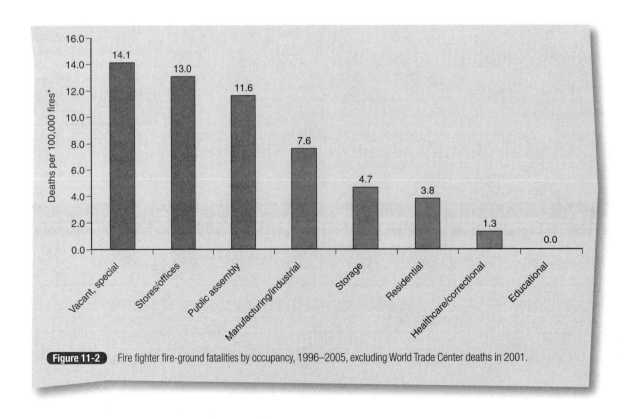

Figure 11-2   Fire fighter fire-ground fatalities by occupancy, 1996–2005, excluding World Trade Center deaths in 2001.

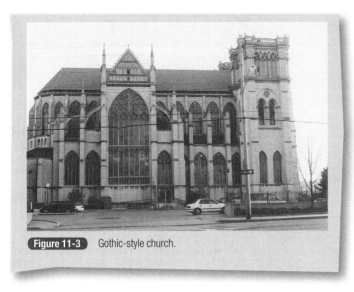

Figure 11-3 Gothic-style church.

Figure 11-4 Modern church.

occupancies take place when services are not being held, resulting in an extended alarm time that allows the fire to gain considerable headway prior to the fire department's response. Many times, fires that occur when a church is unoccupied are not reported until fire breaks through the roof. Recent fire fighter fatalities have occurred due to truss roof collapses with fire fighters working on the interior (see Chapter 5 for a detailed analysis). The roof structure should be noted during pre-incident planning. Bell towers typically contain large, heavy bells that are higher than the church and present a potential collapse hazard.

The large open spaces in places of worship can result in a large-volume fire that is extremely difficult to extinguish once the fire gains considerable headway. Large church fires usually dictate a defensive operation and often result in full or partial building collapse.

Places of worship tend to have priceless artifacts, stained glass windows, and other high-value property. Unfortunately, a large volume of fire and structural damage/collapse may render property conservation efforts unsafe and ineffective.

Eating and Drinking Establishments

Overcrowding is a common problem in eating and drinking establishments. If a fire occurs while the occupancy is overcrowded, exits may quickly become blocked. Many tragic fires have occurred in assembly occupancies during the past 100 years; most notable are the ones shown in **Table 11-2**. The largest loss of life in an eating or drinking establishment was in the Cocoanut Grove fire in Boston, where nearly 500 people were killed, many of them servicemen on their way to fight in World War II.

> Knowing the roof structure is crucial. Roof structure refers to the building components supporting the roof, not the roof covering.

TABLE 11-2 Notable Fires in Assembly Occupancies		
Fire	**Year**	**Deaths**
Iroquois Theater, Chicago, Illinois	1903	602
Rhodes Opera House, Boyerton, Pennsylvania	1908	170
Rhythm Club, Natchez, Mississippi	1940	207
Cocoanut Grove, Boston, Massachusetts	1942	492
Ringling Brothers Barnum and Bailey Circus, Hartford, Connecticut	1944	168
Indiana State Fairground, Indianapolis, Indiana	1963	75
Beverly Hills Supper Club, Southgate, Kentucky	1977	165
Happy Land Social Club, New York, New York	1990	87
Station Nightclub, West Warwick, Rhode Island	2003	100

Case Summary

An electrical fire was reported in the basement of a church on March 13, 2004 in Pittsburgh, Pennsylvania. Operations began as an interior attack with several hose lines. Deteriorating structural conditions and a backdraft caused the IC to withdraw and begin a defensive attack, which eventually brought the fire under control. During overhaul, the bell tower collapsed, killing two fire fighters.

Source: Fire Fighter Fatality Investigative Report F2004-17, CDC/NIOSH, Morgantown, WV, 2006.

Figure CS11-1 The remains of the Pittsburgh Church after the fire that occurred on March 13, 2004.

Case Summary

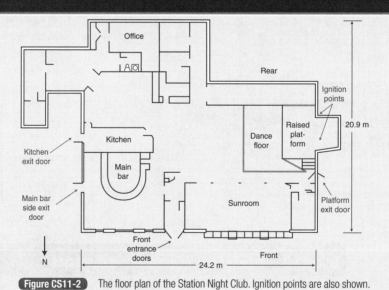

Figure CS11-2 The floor plan of the Station Night Club. Ignition points are also shown.

At the Station Night Club in West Warwick, Rhode Island on February 20, 2003, pyrotechnics used as part of a stage act ignited a highly flammable polyurethane foam insulation material used on the walls and ceiling around the stage, resulting in a rapidly advancing fire in a crowded nightclub. At first occupants failed to realize the need to evacuate, but 30 seconds after ignition, the band stopped playing and occupants began to evacuate. Smoke quickly filled the building and guests began stacking up at the main entrance approximately 100 seconds from ignition. Some occupants escaped via windows at the front of the building. Flame was visible at the front of the building 312 seconds after ignition. The foam covering and crowded conditions provided little time for evacuation, resulting in 100 civilian fire deaths.

Source: William Grosshandler, Nelson Bryner, Daniel Madrzykowski, and Kenneth Kuntz, NIST NCSTAR 2: Vol. I, Report of the Technical Investigation of The Station Nightclub Fire, National Institute of Standards and Technology, U.S. Department of Commerce, 2005.

An analysis of eating and drinking establishment fires reveals the killing potential of fires in assembly occupancies. Occupants will often be mobile, but they may be impaired because of alcohol consumption. Fire probability is fairly high, owing to smoking, cooking, and highly combustible decorations and furnishings. Fire fighters must quickly confine the fire while protecting the exits.

Most patrons will attempt to exit the same way they entered. Although alternative exits are available, they are often underutilized, and overcrowding occurs at the main entrance. Fire fighters should direct people to alternative exits while keeping the crowd at the main entrance moving. It is essential to identify alternative exits, including windows and other openings, which could be used to provide emergency egress during pre-incident planning for eating and drinking establishments. Pre-incident plans should also consider the possibility of using forcible entry tools to create alternative exits or to widen existing exits during an emergency evacuation.

When large numbers of people are involved in an evacuation, many times fleeing occupants stop as soon as they reach safety, thereby unintentionally blocking the exit for others. Someone should be stationed at exit discharges to keep people moving away from the building. This is best done by fire fighters, but police, security personnel, or property management can be assigned this task when fire department staffing does not permit assigning fire fighters. However, consideration must be given to fire and smoke conditions at the exit discharges. If heavy smoke or fire is present at the exit discharge, only fire fighters will be able to facilitate movement at the exit.

Panic has long been blamed for many deaths at nightclub fires. In reality, panic is a scapegoat. Often the real problem is a lack of proper fire protection features. Extensive human behavior research was done after the Beverly Hills Supper Club fire Figure 11-5 .[3] The findings indicated that people actually assisted one another as long as options were available. Only in desperation, when no chance of escape was available, did panic occur. As long as there is hope for survival, panic should not be a significant problem.

People are known to respect role models under emergency conditions: children will follow their parents, and workers will look to the manager. At the Beverly Hills Supper Club, patrons listened to their server while ignoring the busperson and others. Fire fighters most certainly represent an authoritative role model. People should be expected to follow any reasonable directions given by fire officials during an emergency. The challenge is not whether the occupants will follow orders; the challenge is formulating clear, concise, and correct instructions.

Figure 11-5   Beverly Hills Supper Club, Southgate, Kentucky, 1977.

As mentioned previously, extinguishment, the second tactical priority, often is the most effective means of accomplishing the top tactical priority of life safety. Extinguishing or confining the fire to areas away from exit pathways will do more to save lives than individual rescues when there are large numbers of victims and the fire is controllable. The first-arriving company at a major fire in an occupied eating or drinking establishment is faced with a critical life safety decision: Should efforts be directed toward removing endangered occupants, or should the operation begin with an offensive fire attack? If the fire can be controlled quickly and many occupants are in danger, extinguishment is generally the best choice.

Sports Arenas

In 1985, 56 people died and 300 were injured in a Bradford, England stadium fire.[4] The 77-year-old grandstand was an outside facility that resembled wooden grandstands of the past. Problems would be greatly magnified in an inside facility holding nearly 100,000 people, like many huge, domed stadiums of today.

Fortunately, newer facilities limit the amount of combustible material, although there is still plenty to burn, largely in the form of plastics. Stadium owners may point to their facility's modern construction and state that "it can't happen here." However, that type of thinking is a recipe for disaster. These occupancies pose an extreme life-hazard threat, and a close examination of evacuation tactics is a must. The logistics associated with evacuating 10,000 or 100,000 people are very complex. When conducting pre-incident planning at these facilities, the fire officer should keep in mind that fire is not the only type of incident that may require a rapid evacuation. Today, any place that attracts large crowds is a potential target for terrorists.

A fire and explosion occurred in Indianapolis, Indiana, in 1963, killing 74 people at the Indiana State Fairgrounds Coliseum. It can and did happen here in the United States. A situation of this nature should be handled like a nightclub fire: protecting egress while directing and facilitating evacuation. Of course, a stadium or coliseum fire presents a significant life safety hazard. On the positive side, open-air arenas allow smoke and toxic gases to dissipate. Even coliseums and covered stadiums have large open-air spaces that will allow more time for evacuation. Newer facilities, such as the sports arenas shown in **Figure 11-6** are equipped with sprinklers, which greatly improve life safety.

Convention Centers

The 1967 McCormick Place Convention Center fire in Chicago, Illinois, caused $52 million in damage, the largest loss of property in the United States at that time. Since the McCormick Place fire, automatic fire sprinklers have become commonplace in convention center construction. Sprinklers greatly diminish the potential loss of life and property. Large loss-of-life fires are unheard of in fully sprinkler-protected buildings when the systems are installed and maintained in compliance with nationally recognized codes and standards.

However, sprinkler systems are not the total solution. The sprinkler system may become overwhelmed if it is impaired or if the building has a fuel load exceeding the system's design capacity. Also, event organizers may be reluctant to cancel events when the sprinkler system is not operating. Fire companies that respond to convention centers should be aware of the event schedule and conduct on-site inspections of the property whenever large events are planned. It is also a good idea for fire companies to participate in the code enforcement process for buildings located in their response districts.

Large quantities of combustibles may be brought into convention centers, and hazardous demonstrations may be performed. Because of the combination of large fuel loads and large numbers of people, suppression efforts can be very difficult. As previously mentioned, it is important to ensure that the building's fuel load does not exceed the designed capacity of the sprinkler system. If a fire does overwhelm the sprinkler system and involves a large area, the rate-of-flow requirement may be beyond the fire department's resources and capabilities. In this case, extinguishment may not be a viable option. Fire departments must always be ready for the potential of large fires and high risk to life in these occupancies. Remember, tactical considerations for convention centers, such as protecting egress routes, are similar to those for nightclubs and similar public assembly properties.

> The largest loss of life in a single building fire in the United States, prior to the World Trade Center attack on September 11, 2001, took place in 1903 at the brand-new Iroquois Theater in Chicago, when 603 people (mostly children) lost their lives.

Theaters

Movie theaters built during the first half of the twentieth century were built to accommodate live shows or to present both live shows and movies. The construction of these facilities was similar to that of the Iroquois Theater in Chicago, the site of a large loss-of-life fire in 1903. Some of these older theaters have now been subdivided into several smaller units.

Modern movie theaters are housed in very large buildings. Generally, these are one story, but there is a movement toward mega-theaters that have multiple stories. These buildings are subdivided into several smaller theaters **Figure 11-7**.

Figure 11-6 Modern sports arena. A. Exterior. B. Interior.

CHAPTER 11 *The Role of Occupancy* 267

Figure 11-7 Modern movie theater. A. Exterior. B. Interior.

but smaller theaters located in a single, large building also increases the probability that extinguishment efforts will be effective. In addition to the multiple, smaller theaters allowing for a reduced rate of flow, the tendency for everyone to exit via the main entrance is also reduced in these buildings, as alternative exits should be well-marked and easily identified by patrons. Many newer theaters have plainly marked exits at several locations within the individual theater and throughout the building complex. Furthermore, modern movie theaters are often equipped with sprinkler systems.

While movie theaters are getting smaller, venues for live entertainment are not. The fire during a live stage show at the Iroquois Theater taught building designers some important lessons, but it was not until recently that theaters were required to be constructed with sprinkler systems. If an older theater has not been equipped with a fire sprinkler system and its exit design updated, it can be just as dangerous as the Iroquois Theater was in 1903. Today, many live performances, formerly staged in theaters, are presented in large arenas and stadiums.

Educational Occupancies

NFPA 101: Life Safety Code[1] defines an educational occupancy as:

> **3.3.168.6** Educational Occupancy. An occupancy used for educational purposes through the twelfth grade by six or more persons for 4 or more hours per day or more than 12 hours per week.

College classrooms are not technically considered **educational occupancies**. However, there are similarities between college and other classrooms in terms of life safety, extinguishment, and property conservation. Therefore, the discussion of college classrooms is included here. For purposes of this comparison three types of educational occupancies will be discussed:

- Elementary school
- Middle, junior high, or high school
- Colleges and universities

An occupancy factor matrix for educational occupancies is shown in **Table 11-3**. Mobility factors tend to be very

The current trend in movie theaters toward smaller theaters within very large buildings should make the fire fighter's job less difficult. Tactics would be much the same as for the eating and drinking establishments, except that there would generally be fewer people in an individual movie theater unit. The modern concept of having numerous

TABLE 11-3 Occupancy Factor Matrix: Educational Occupancies

Educational Occupancies	Mobility of Occupants	Age of Occupants	Leadership	Awareness	Occupant Density	Familiarity of Occupants
Elementary school	10	6	10	10	4	10
Middle, junior high, or high school	10	10	8	10	4	10
College/university	8	8	7	10	5	5

Key: 1 = extremely negative factor, 10 = very positive factor.

good in school settings. In the case of elementary and high schools, most of the occupants would be mobile. At the college level, there will be an older population, but this should not pose a problem. However, younger elementary school students may not be capable of taking independent action, hence the lower "Age" ratings for elementary school occupancies in the educational matrix. A high degree of leadership is provided by classroom teachers, especially at the elementary school level. At the high school and college levels, students move from area to area and from building to building; therefore, there are times when they are not under the direct control of a teacher.

Students at all levels should be aware of fire conditions and be able to hear the evacuation alarm. Fire drills held on a regular basis in elementary and high schools increase awareness and better establish teacher-to-student leadership. Occupant density minimums are 20 ft^2 (2 m^2) per person in elementary and high school classrooms.[1] There are fewer people in the same relative area in elementary and high school settings as compared to an assembly occupancy, but there are more people in these settings than in many other occupancies.

Intentionally set fires in remote areas are a problem in schools, with 38% of elementary and high school fires being intentionally set, and 19% of all educational occupancy fires originating in bathrooms and locker rooms.

Elementary Schools

Most tragic fires involving educational occupancies occur in primary or elementary schools, mainly because younger children are less likely to take appropriate action on their own. All other things being equal, the younger the students, the greater the danger.

Some notable school fires include the ones listed in Table 11-4. Figure 11-8 shows the scene at Our Lady of Angels School in Chicago in 1958.

Only recently have schools been protected with automatic sprinkler systems. Very few older school buildings

Figure 11-8 Our Lady of Angels School fire, Chicago, Illinois, 1958.

are sprinkler protected. Placing large numbers of children within a small, non-sprinkler-protected area creates the potential for a tragedy. In addition to not being sprinkler protected, older schools tend to be three or more stories high Figure 11-9.

The saving factors in elementary schools are the discipline gained through frequent fire drills, and the fact that teachers are natural role models for the children to follow in an emergency. In the event of a fire in a school, it is necessary to coordinate operations with the school principal to determine the evacuation status and the location of students who are missing. If rescue efforts are necessary, fire fighters can be directed to the most probable location for trapped students and teachers on the basis of this information.

Fire department personnel should observe school fire drills whenever possible. Observing a school fire drill provides invaluable training for fire fighters who should know the prescribed location for students and the school contact person. Being present during a fire drill also provides an opportunity to evaluate accountability and evacuation procedures. Most elementary and high schools maintain a visitor's log, but visitors may not be included in the accountability system. Knowing how long it takes to totally evacuate the school is important, but not nearly as important as knowing whether students are under control and whether the accountability system is reliable. Determining whether the school evacuation plan is

TABLE 11-4 Notable Fires in Educational Occupancies

Fire	Year	Deaths
Lakeview Grammar School, Collinwood, Ohio	1908	175
Cleveland School, Beulah, South Carolina	1923	77
Consolidated School, New London, Texas	1937	294
Our Lady of Angels, Chicago, Illinois	1958	95
Star Elementary School, Spencer, Oklahoma	1982	7

Figure 11-9 Buildings used as primary schools. A. Modern construction. B. One-story construction. C. Multi-story, older construction. D. Multi-story with temporary classroom.

consistent with fire department operations is also important. Sometimes students are placed in areas that could obstruct or slow fire department access. Other times, the school contact person is located in an area that is not in the line of travel for arriving fire units. Minor changes in the school evacuation plan can significantly improve fire department response. Many jurisdictions conduct unannounced fire drills that are timed and otherwise evaluated by the fire department.

Some newer, one-story schools have doors from each classroom leading directly to the outside, greatly reducing the life hazard. In older schools, heavy reliance is placed on the use of stairways. As with any building, the stairs in multi-story school buildings must always be controlled by the fire department during evacuations. Evacuating trapped victims through the interior stairs is typically the most effective and safest option.

When absolutely necessary, aerial ladders, elevated platforms, and ground ladders can be used to evacuate a multi-story school. Ladders allow pupils to form a steady evacuation stream with teachers assisting from inside. However, fire department ladders should be used only when fire fighters cannot gain control of the interior stairs. Establishing an incident accountability system that ensures complete evacuation and accounting for all students is essential. It is also important to make a sound evaluation as to whether the immediate priority is rescue or fire control.

As was mentioned above, many school buildings pose special firefighting challenges. Fire extension should be expected in school buildings of ordinary construction. Where mobile buildings are used for temporary classrooms, complete destruction may result in a short period of time. Many schools add to the fire control problem by storing clothing, furniture, and other collected materials within the school, and at times this storage is under stairwells and in corridors. This greatly increases the fuel load and the threat to life safety for any occupants.

Middle, Junior High, and High Schools

Fire drills are required in all schools for kindergarten through grade 12. As mentioned previously the accountability process may not be as efficient at the high school level.

Fire fighters are inclined to think schools are only occupied during school hours. When determining the probability of a school being occupied, it is important to remember that teachers, maintenance, and administrative personnel may occupy the school outside of normal classroom hours. There are also many sporting and social events that occur at schools outside the normal school day. At high schools and colleges, there will be more of these events than at the elementary school level. The fire department should obtain a list of scheduled events and include event information with the pre-incident plan. A basketball court or auditorium will likely be classified as an assembly occupancy within the educational occupancy. High schools may also have evening classes or permit citizens to use computers, gym equipment, and other school resources outside normal working hours.

Colleges and Universities

Colleges and universities usually have much larger campuses as compared to elementary or high schools, with many more evening and weekend activities. The adult population is better able to take independent action to evacuate a building; however, students move from building to building at different time intervals and are not under close supervision. Further, there typically is no occupant accountability system, and large universities may be situated on a very complex campus where a map is needed to find specific buildings. Most campuses will have one or more large classroom facilities that hold 200 or more students. These classrooms resemble an assembly occupancy and are classified as such. Any college classroom with 50 or more students is considered an assembly occupancy.

Fires in classrooms, except for the very large college classrooms or large laboratory and shop areas, will generally be within the flow capabilities of standard pre-connected hose lines. Rate of flow should be pre-calculated for any large undivided areas within a school.

Schools contain library collections, computers, and other high-value contents. Therefore, property conservation tactics should be carefully thought out and included in the pre-incident plan.

Health Care Occupancies

NFPA 101: Life Safety Code[1] defines a health care occupancy as:

> **3.3.371.6** Health Care Occupancy. An occupancy used for purposes of medical or other treatment or care of four or more persons where such occupants are mostly incapable of self-preservation due to age, physical or mental disability, or because of security measures not under the occupants' control.

Four types of health care facilities will be discussed:
- Hospitals
- Nursing homes
- Limited care facilities
- Ambulatory care facilities

An occupancy factor matrix for **health care occupancies** is shown in **Table 11-5**. Mobility is a major concern in hospitals and nursing homes. Some patients are considered ambulatory in these facilities, but ambulatory and mobile are not the same. An ambulatory person is able to move, but may move very slowly and require a walker or cane. Many patients in hospitals and nursing homes are non-ambulatory, meaning they are unable to self-evacuate and will need assistance if it is necessary to move to a place of safe refuge. The nursing staff generally provides this assistance, but the staff available may not be sufficient if large numbers of people need assistance. Occupants in limited care facilities are more mobile than in a nursing home, but many move slowly or need walkers, canes, wheelchairs, or other types of assistance.

Ambulatory care facilities treat minor illnesses and injuries. However, some of these facilities perform

TABLE 11-5 Occupancy Factor Matrix: Health Care Occupancies

Educational Occupancies	Mobility of Occupants	Age of Occupants	Leadership	Awareness	Occupant Density	Familiarity of Occupants
Hospital	2	4	10	3	7	2
Nursing home	2	1	8	3	7	7
Limited care facility	4	2	7	5	8	9
Ambulatory care facility	6	5	6	5	7	2

Key: 1 = extremely negative factor, 10 = very positive factor.

surgical or testing procedures that require the patient to be anesthetized. If this is the case, patients will be immobile and have limited or no cognitive ability. Part of the pre-incident planning process should include notations of ambulatory care facilities where patients are anesthetized.

Residents of nursing homes and limited care facilities are typically much older than the general population. The nursing staff provides leadership and assistance during an emergency. In a hospital or nursing home, patients will look to their nurse for direction in an emergency. Limited care facilities will generally have staff personnel who can provide leadership, but patients are more independent. Patients at ambulatory care facilities may not recognize the staff, and conscious patients will be more likely to take independent action in an emergency.

The nursing home and assisted living facility is home to the occupant, thus most patients will be more familiar with the layout of the building. However, residents may not remember details of the building under stress, and some patients may not be able to follow directions. Patients and visitors in hospitals and ambulatory care facilities may be unfamiliar with the building.

Large numbers of people located in a single building always represent life safety concerns. Physically or mentally challenged people present additional difficulties for the rescuers. Fire fighters should expect facility personnel to assist with evacuations; this should be reinforced through training. In the hospital or nursing home, nurses and attendants will be available to care for patients.

These occupancies often require fire forces to rely on other people who are not trained to work under fire conditions. Health care facilities use places of safe refuge within the facility during evacuations. Hospital and nursing home personnel move patients through fire doors to safe areas. Seldom are hospital or nursing home patients evacuated to the outside. Moving people on beds with life support equipment is difficult, to say the least; therefore, the **defend-in-place** strategy is the preferred option.

Hospitals

Building features assist in accomplishing the life safety mission at hospitals, with building areas being separated by substantial fire walls and fire doors. Making sure that fire separations are not compromised is an essential part of a hospital fire protection plan. Fire doors are often left open in hospital facilities, allowing smoke and fire to enter common hallways and other patient areas, as was the case in a hospital fire in Petersburg, Virginia Figure 11-10 , which killed five patients. The fire department may be needed

Figure 11-10 Hospital in Petersburg, Virginia.

to assist in moving patients from the fire area to another designated defend-in-place location. The IC must make an early determination as to progress being made in moving patients to an area of safe refuge and whether the place of safe refuge is defendable. Evacuation to a lower floor or to the outside is a mammoth undertaking and is rarely necessary. The nursing staff should be able to account for patients, but in most cases they will not be able to account for visitors or doctors. Therefore, a primary search should be conducted.

There are large open areas within hospitals and some of these areas have a heavy fuel load, such as laundries, storage areas, and refuse areas. Further, the presence of oxygen can increase flame intensity and fire spread. However, patient rooms in hospitals are usually within the flow capacity of a standard pre-connect or fire department standpipe hose line. It is essential to consider the possible smoke spread to the area of safe refuge when a fire door is opened to advance a hose line.

Hospitals will often have very expensive equipment. Some of this equipment is extremely sensitive to water and smoke damage. Given the high life safety problem, property conservation could be delayed; however, the IC should call for enough assistance to assign personnel to property conservation activities as soon as possible.

Nursing Homes

Nursing home residents are unable or only partially able to assist themselves. This special population presents many evacuation and rescue problems, which requires a significant commitment of fire department resources whenever a working fire is encountered in a nursing home.

During the past few decades, as the population aged, the need for nursing facilities increased. A larger percentage of the population moved into these occupancies, and large loss-of-life fires became common. Much of the United States now has codes requiring sprinklers in nursing homes, which has greatly reduced the number of large loss-of-life fires in health care facilities. Multi-fatality fires still occur in nursing homes, but not where sprinklers have been properly installed and maintained. When a properly installed and operating sprinkler system is in place, fires are generally limited to one room, and the primary tactic is to move patients out of smoky areas to places of safe refuge while supporting the sprinkler system. If the nursing home is not sprinkler protected or the sprinkler system is not operating properly, use fire doors to contain the fire whenever possible. Even though self-closing fire rated doors separate rooms from hallways, obstructions may prevent the door from closing, which allows fire and smoke to spread to the hallway and other rooms. Moving patients out of harm's way is the primary tactic, but extinguishment is essential to defend the place of safe refuge. If the fire is not extinguished, smoke and possibly fire will eventually enter the area of safe refuge, requiring further evacuation and rescue.

Some notable examples of fires in health care facilities (nursing homes and hospitals) are listed in Table 11-6. It should be noted that during the period 1999–2002, over 90% of the civilian fire fatalities in health care occupancies occurred in nursing homes.[2] The two most recent multi-fatality fires listed in Table 11-6 were in non-sprinkler-protected nursing homes.

TABLE 11-6 Notable Fires in Hospitals and Nursing Homes

Fire	Year	Deaths
Hospital Cleveland, Ohio	1929	125
Hospital, Effingham, Illinois	1949	74
Nursing Home, Hillsboro, Missouri	1952	20
Nursing Home, Warrenton, Missouri	1957	72
Nursing Home, Fitchville, Ohio	1963	63
Nursing Home, Marietta, Ohio	1970	31
Nursing Home, Buechel, Kentucky	1971	10
Nursing Home, Honesdale, Pennsylvania	1971	15
Nursing Home, Cincinnati, Ohio	1972	10
Nursing Home, Denmark, Wisconsin	1972	10
Nursing Home, Springfield, Illinois	1972	11
Nursing Home, Philadelphia, Pennsylvania	1973	11
Nursing Home, Wayne, Pennsylvania	1973	15
Nursing Home, Chicago, Illinois	1976	23
Nursing Home, Little Rock, Arkansas	1984	2
Hospital, Riverside, California	1986	5
Health Care Center, Memphis, Tennessee	1988	6
Nursing Home, Norfolk, Virginia	1989	12
Nursing Home, Dardanella, Arkansas	1990	3
Hospital, Petersburg, Virginia	1994	5
Health Center, Hartford, Connecticut	2003	16
Nursing Home, Nashville, Tennessee	2003	15

Limited Care Facilities

Limited care facilities, sometimes called assisted living facilities, are similar to nursing homes, but the occupants do not require continuous nursing care. Some limited care facilities provide services that are similar to comprehensive nursing home care, while others establish independent living criteria for occupants. Often a nursing complex will include both assisted living and full nursing care, such as the one shown in Figure 11-11. Elderly residents generally move from their homes into the assisted living facility with a degree of self-reliance. Over time these residents require additional nursing services and many eventually occupy a full nursing facility. Most codes now require automatic sprinkler protection for all new nursing homes, but they may not require existing homes to be retrofitted with sprinklers or sprinkler protection in limited care facilities. Part of the fire protection plan for a nursing home requires a minimum number of on-duty staff to assist patients in an emergency. Staffing will generally be less at a limited care facility, and there may not be places of safe refuge where a defend-in-place strategy can be implemented. Many limited care facilities resemble apartment buildings; therefore, a primary search and quick extinguishment are the principal tactics. Occupants of these facilities, who are mostly elderly, would have great difficulty descending a ladder; therefore, attempting to rescue residents using fire department ladders would be an option of last resort.

Ambulatory Care Facilities

As noted previously, these facilities provide a variety of services. Some are group practice physicians who provide emergency care for minor illnesses and injuries. Others perform outpatient surgery or testing that requires full anesthesia. The emergency

> NFPA records indicate that "no fire in a completely sprinkler-protected public assembly, educational, institutional, or residential building has killed more than two people."[5]

> Many apartment buildings rent exclusively to elderly occupants. These buildings will be classified as a residential occupancy and have less stringent fire protection code requirements. However, the firefighting challenges will be very similar to those in a limited care facility.

Figure 11-11 Limited care facility attached to a nursing home.

care facility where patients are conscious requires the same tactics as an office building. If patients are anesthetized, they will have to be physically removed from the building and provided medical care. These facilities should be pre-incident planned with notations regarding the types of service provided, with a special notation if patients are anesthetized.

Residential Board and Care Occupancies

NFPA 101: Life Safety Code[1] defines a residential board and care occupancy as:

> **3.3.168.12** Residential Board and Care Occupancy. A building or portion thereof that is used for lodging and boarding of four or more residents, not related by blood or marriage to the owners or operators, for the purpose of providing personal care services.

An occupancy factor matrix for <u>residential board and care occupancies</u> is shown in **Table 11-7**. Awareness is given a rating of three due to the diminished mental and/or physical capacity of many of the residents and resultant problems in evacuating the building. There may be one or more staff members on duty depending on the facility size and the mental/physical status of patients being housed. Thus, leadership is a positive factor, but not to the extent found in schools and hospitals.

Mentally challenged patients who were formerly housed in large, fire-resistive structures are now being housed in board and care facilities. This has created a problem similar to the problems experienced in nursing homes in the past. However, the board and care problem has been more difficult to regulate, as fewer patients are kept in a single structure. Staff requirements are less in board and care

TABLE 11-7 Occupancy Factor Matrix: Residential Board and Care Occupancies						
	Mobility of Occupants	Age of Occupants	Leadership	Awareness	Occupant Density	Familiarity of Occupants
Residential board and care occupancies	5	5	7	3	7	7
Key: 1 = extremely negative factor, 10 = very positive factor.						

TABLE 11-8 Notable Fires in Residential Board and Care Occupancies

Fire	Year	Deaths
Connellsville, Pennsylvania	1979	10
Farmington, Missouri	1979	25
Washington, District of Columbia	1979	10
Pioneer, Ohio	1979	14
Brady Beach, New Jersey	1980	24
Keansburg, New Jersey	1981	31
Eau Claire, Wisconsin	1983	6
Worcester, Massachusetts	1983	7
Gwinnett County, Georgia	1983	8
Cincinnati, Ohio	1983	7
Tennessee	1989	16
Georgia	1990	4
Texas	1990	4
Bessemer, Alabama	1990	4
Colorado Springs, Colorado	1991	10
Detroit, Michigan	1992	10
Alabama	1994	6
Broward County, Florida	1994	6
Oregon	1995	4
Laurinburg, North Carolina	1996	8
Shelby County, Tennessee	1996	4
Mississauga, Ontario, Canada	1996	8
Pennsylvania	1996	4
Ste. Genevieve, Quebec, Canada	1997	7
Harvey's Lake, Pennsylvania	1997	10
Arlington, Washington	1998	8
California	2003	4
Tennessee	2004	5

The NFPA has issued several fire investigation reports for recent fires in board and care facilities. Some of these facilities are occupied by mentally challenged residents; others are scaled down facilities for elderly patients who do not require full nursing care.

Some notable board and care fires include those listed in Table 11-8. Widespread use of residential board and care facilities is a relatively new phenomenon; therefore, the multi-fatality fires shown in Table 11-8 are fairly recent. The number of multi-fatality fires in these facilities seems to be trending downward. This is partially due to more stringent codes and better code enforcement, but also because fire departments are now aware that these occupancies exist in their communities.

Detention and Correctional Occupancies

NFPA 101: Life Safety Code[1] defines a detention and correctional occupancy as:

> **3.3.168.5** Detention and Correctional Occupancy. An occupancy used to house four or more persons under varied degrees of restraint or security where such occupants are mostly incapable of self-preservation because of security measures not under the occupants' control.

An occupancy factor matrix for **detention and correctional occupancies** is shown in Table 11-9. Residents of nursing homes and hospitals are characterized as having physical and/or mental challenges that hamper self-evacuation, whereas prisoners in detention facilities are generally physically restrained. During a prison fire, inmates are moved to other areas within the confines of the detention property.

The prison staff provides leadership for the inmates. Close cooperation between prison staff and the fire department is needed for successful operations in a jail or prison environment. Some detention and correctional occupancies are no more than cells within a police station designed to temporarily hold prisoners. Other correctional facilities are nearly self-sufficient mini-cities, with laundries, storage areas, large-scale food preparation areas, and many other facilities as part of the prison complex.

During a detention facility fire, maintaining security is extremely important; however, dealing with tight

facilities, and many resemble a residential occupancy. The board and care fire problem is substantial, and tactics must be planned in advance. It is essential that the fire department identify and inspect board and care facilities and develop pre-incident plans for these occupancies located within their jurisdiction. Tactics should take into account the limited ability of residents to escape on their own. Residents will be categorized as independent, but will usually be mentally or physically disadvantaged. Most of these buildings are residential buildings; therefore, the standard pre-connected hose line should be sufficient to extinguish most fires that occur.

TABLE 11-9 Occupancy Factor Matrix: Detention and Correctional Occupancies

	Mobility of Occupants	Age of Occupants	Leadership	Awareness	Occupant Density	Familiarity of Occupants
Detention and correctional occupancies	1	6	9	6	7	6

Key: 1 = extremely negative factor, 10 = very positive factor.

TABLE 11-10 Notable Fires in Detention and Correctional Occupancies

Fire	Year	Deaths
Columbus, Ohio	1930	320
Lancaster, South Carolina	1972	11
Sandford, Florida	1975	10
Williamsport, Pennsylvania	1975	3
Danbury, Connecticut	1977	5
St. John, New Brunswick, Canada	1977	18
Columbia, Tennessee	1977	42
Leavenworth, Kansas	1979	7
Biloxi, Mississippi	1982	29

security measures while implementing fire department tactics is sometimes exasperating for the responders. Close cooperation between the IC and prison administrators is an absolute necessity.

During the period 1999–2002, 52% of fires in prisons were intentionally set and 55% originated in bedrooms or cells.[2] The inmate arsonists will likely resent the fire department's interference. Prisoners present a physical threat to the fire fighter, so close coordination with prison officials is needed to enable the fire fighters to operate safely and suppress the fire.

Modern prisons are usually constructed with a fair degree of fire safety in mind, but older structures are still being used as prisons. In the older prisons and jails, the building will complicate operations by permitting the free circulation of smoke and toxic gases as well as allowing rapid fire extension. However, modern and old prisons alike tend to be overcrowded, holding more prisoners than they were designed to house. Any form of overcrowding increases the life hazard.

Several notable examples of fires in detention facilities are listed in Table 11-10.

Pre-incident planning is an absolute necessity for a detention or correctional facility. Securing a water supply and advancing hose lines may require special precautions. The details of fire department interaction with security are worked out during pre-incident planning. Questions that must be answered in advance include:

- How will fire fighters be provided access to secured areas? This generally involves a security person meeting the fire department at a pre-designated location and escorting them while inside the prison. Be sure to detail how security will provide access when smoke or fire conditions require fire fighters to advance ahead of security personnel into areas requiring self-contained breathing apparatus and full protective clothing.
- Are there areas within the prison complex that do not house prisoners and can they be accessed without assistance from security? Work areas such as laundries may be occupied by prisoners during certain times of the day but unoccupied at other times.
- What are the emergency procedures if a system malfunctions, automatic lockdown occurs, or when there is a power failure?
- What are the evacuation routes, and where are the places of safe refuge for prisoners located?

The goal is to conduct a safe and effective operation while avoiding direct contact with the prison population. There may be a need for special RICs and accountability and rehabilitation procedures for a detention and correctional occupancy.

Residential Occupancies

NFPA 101: Life Safety Code[1] defines a residential occupancy as:

> **3.3.168.13** Residential Occupancy. An occupancy that provides sleeping accommodations for purposes other than health care or detention and correctional.

Residential occupancies include:
- One- and two-family dwellings
- Apartment buildings
- Dormitories
- Hotels/motels

An occupancy factor matrix for residential occupancies is shown in Table 11-11. Note that mobility and age are

TABLE 11-11 Occupancy Factor Matrix: Residential Occupancies

Residential Occupancies	Mobility of Occupants	Age of Occupants	Leadership	Awareness	Occupant Density	Familiarity of Occupants
One- and two-family dwellings	5	5	7	4	8	10
Apartment buildings	5	5	5	4	6	9
Dormitories	7	7	5	3	4	8
Hotels/motels	6	5	2	3	4	1

Key: 1 = extremely negative factor, 10 = very positive factor.

near average (5) except in dormitories, where occupants tend to be young adults who are very mobile. There is little leadership in most hotels and motels except when provided by a parent traveling with children. In one- and two-family dwellings, the head of household should be in a strong leadership position, particularly if the family has practiced a home escape plan. Awareness is rated slightly negative in all residential occupancies because occupants are sleeping part of the time that they occupy the building. Except for dormitories, hotels, and motels, there are very few people in a fairly large area, thus exit capacity is not generally considered a problem. People living in a residence or dorm should be familiar with their surroundings. Hotels and motels are transient in nature; therefore, most occupants will not know the location of alternative means of egress.

Most fires, most fatal fires, and most property loss occur in residential fires. In 2005, 83.1% (3055) of civilian fire deaths and 77.5% of all fires occurred in residential occupancies.[6] Most of the deaths in residential fires occur one or two at a time. However, the annual U.S. Multiple-Death Fires report for 2005 includes 12 one- and two-family dwelling fires where 5 to 11 fatalities occurred.[7] This is not an unusual circumstance; every year, several multiple-death fires occur in single-family homes. If we were to provide a list of notable residential fires for the past 100 years, including those in homes and apartment buildings, the list would have over a thousand entries. According to the NFPA One Stop Data Shop, during the period 1996–2005, 41.5% of all fire fighter fatalities occurred in residential properties, mostly in one- and two-family dwellings.

In addition to the one- and two-family dwelling, apartments, hotels, motels, and dormitories are also classified as residential. Large loss-of-life fires in residential properties have occurred frequently in hotels. **Table 11-12** lists some significant residential fires, excluding one- and two-family homes and apartment buildings, over the past 100 years. **Figure 11-12** shows the Dupont Plaza Hotel fire in San Juan, Puerto Rico, which was the deadliest residential fire in the last 50 years.

In residential occupancies, as with most occupancy types, there is a direct correlation between life safety and the time of day. In residential occupancies, the hours when residents are sleeping are the most dangerous. Wide acceptance of smoke detectors has greatly diminished the number of deaths in residential properties. Many high-rise buildings are now constructed for use by senior citizens and family living. Expect to find many people throughout the structure in need of assistance, including the possibility of the **convergence cluster phenomenon**.[8]

Convergence cluster behavior was studied in depth after the MGM Grand Hotel and Casino fire in Las Vegas. Many guests attempted to use stairs to exit the building but encountered smoke as they descended. They then sought refuge in a room other than their own. This resulted in several people being in a single room, rather than the expected one to five guests per room. It would be logical to assume that guests had evacuated the floor successfully after finding several rooms empty, but this assumption could have serious, deadly consequences. The convergence cluster phenomenon should reinforce the importance of conducting a complete, systematic search.

TABLE 11-12 Notable Fires in Residential Occupancies Other Than One- and Two-Family Dwellings

Fire	Year	Deaths
Gulf Motel, Houston, Texas	1943	54
La Salle Hotel, Chicago, Illinois	1946	61
Winecoff Hotel, Atlanta, Georgia	1946	119
MGM Grand Hotel/Casino, Las Vegas, Nevada	1980	85
Hilton Hotel/Casino, Las Vegas, Nevada	1981	8
Dupont Plaza Hotel/Casino, San Juan, Puerto Rico	1986	96
High-Rise Apartment Building, Johnson City, Tennessee	1989	16
Motel, Hagerstown, Maryland	1990	4
Ramada Hotel, Indiana	1992	11
Paxton Hotel, Chicago, Illinois	1993	20
Fraternity House, Chapel Hill, North Carolina	1996	5
Off-Campus Student Housing, Columbus, Ohio	2003	5

Figure 11-12 Dupont Plaza Hotel fire, San Juan, Puerto Rico, December 31, 1986.

One- and Two-Family Dwellings

One- and two-family dwellings are an important and large part of the fire problem, accounting for 56.2% of all structure fires and 69.9% of civilian fire deaths in 2005.[6] Fire fighters generally feel confident attacking a fire in a one- or two-family dwelling "bread and butter" fire. However, more fire fighter line-of-duty deaths occur in these properties than any other occupancy, which is primarily due to the large number of fires that occur in these properties. The point that must be emphasized is that no fire should ever be considered routine. Proper precautions must be taken at every structure fire.

One- and two-family dwellings are primarily occupied at night and on weekends, but people can be home and asleep at any time. Whenever it is safe to conduct an offensive attack, a primary and secondary search should be conducted, unless there is a *credible* and *verifiable* report that everyone escaped. Reports that everyone is out of the building should be carefully scrutinized for reliability. When in doubt, conduct a primary and secondary search. When the fire occurs late at night, the primary search should normally begin in bedrooms.

Except for some very large, open-layout buildings or mansions, fires in one- and two-family dwellings should be within the capabilities of one or two pre-connected hose lines. As is most often the case, extinguishment is usually the best way to achieve the life safety objective.

The actual monetary value of the building's salvageable contents will vary, but it is important to remember that in many cases personal items cannot be replaced and are of great value to the occupants. Property conservation efforts must take into consideration that personal mementos are often invaluable and cannot be replaced.

Apartment Buildings

In terms of life safety, extinguishment and property conservation, apartment buildings are much like one- and two-family dwellings. However, a larger number of people are concentrated in a smaller area, and the probability of someone being able to provide a *credible* and *verifiable* report that everyone has successfully escaped is unlikely. The life safety problem is directly proportional to the number of units in a single building. Larger buildings with more apartment units require a larger staffing commitment to complete the primary and secondary searches.

With the large number of people potentially exposed to the fire and the lack of a reliable evacuation status report, extinguishment becomes even more important. Most apartment buildings will require flows that are within the capacity of one or two standard pre-connected hose lines. However, some apartment buildings have large open storage or common spaces that may require additional flow.

As is the case in other residential properties, the occupants' personal items will be of great value.

Exposure problems are more likely at apartment complexes; a pre-incident plan should, at minimum, include a map showing the building layout with an address for each unit within the complex, as well as areas requiring a rate of flow exceeding the flow capacity of two standard pre-connected hose lines. Pre-incident plans should also note access problems. Some apartment buildings have back-to-back units with fire separations between units or a center hallway separating units. Apparatus access may be limited to the front of the building with no road access to the rear units. A serious fire in a rear unit may not be visible from the front of the building, and access to rear apartments may be limited.

If a fire is being attacked in a defensive mode at an apartment complex where buildings are closely spaced, the primary concern is that other buildings will quickly become involved in fire, as was the case in the Houston, Texas, apartment complex conflagration discussed in Chapter 9.[9] It is important to get ahead of the fire, for both extinguishment and evacuation purposes, as it is fairly easy to alert and evacuate residents who are not in immediate danger. However, if the fire spreads and threatens the occupants, the rescue effort will be much more complex.

High-rise residential property would also fall into the category of apartment buildings. The special perils and

Case Summary

In a four-story apartment complex fire in Bremerton, Washington, four residents were killed in a fast-spreading, early morning fire. The fire originated in a third floor apartment that was unoccupied at the time and quickly spread out the apartment door, which was left open when the fire was detected. The fire moved into the attic space and then throughout the complex. There were four fire separation walls in the attic space, but openings in these walls allowed the fire to spread beyond them. This case shows how an attic fire can create a severe challenge for the fire department in attempting to contain the fire and protect the occupants.

Source: Edward R. Comeau, *Apartment Complex Fire, Bremerton, Washington*, Fire Investigation Report. Quincy, MA: NFPA.

Figure CS11-3 This apartment complex in Bremerton, Washington, was severely damaged by a rapidly spreading fire. The roof was burned off, and the upper floors were destroyed by this early morning fire.

---

tactics associated with high-rise buildings are discussed in Chapter 12.

Dormitories

College and university dormitory housing is supervised. University policies and jurisdictional codes require a specified level of safety in dormitories. However, there are other types of housing for college students. Students rent housing in apartments or homes that have no affiliation with the college or university. Fraternity and sorority housing may or may not be controlled by the college. College dormitories tend to be larger facilities than off-campus housing; some are high-rise buildings. The number of people in both on- and off-campus structures is fairly high compared to most residential properties. Dormitories tend to provide small living spaces for each student with common public areas. Many fraternity, sorority, and private houses rented by students are old residential properties that have been modified to hold more tenants than the original design intended.

The occupant load is increased substantially when guests are invited to social events. A 2003 fire in private housing for students near Ohio State University claimed the lives of five students. The three-story building had 12 residential units plus a large number of visiting students. In another deadly fire, three students perished in an off-campus housing unit near Miami University in Oxford, Ohio, in 2006.

Off-campus statistical information is difficult to obtain, because many fire reports categorized them as apartment occupancies, and many smaller fires go unreported. Approximately 75% of college students live in off-campus housing compared to 25% living in dormitories. A dormitory fire at Seton Hall University and a fraternity house fire at North Carolina University prompted officials to require sprinkler protection. The American Fire Sprinkler Association, Campus Firewatch, the National Fire Protection Association, the National Fire Sprinkler Association, the United States Fire Administration, and other organizations and fire safety advocates are working to require sprinkler protection in dormitory, sorority, and fraternity housing, but many on- and off-campus housing units remain unprotected.

Life safety is a critical issue in dormitory and off-campus housing. Alcohol consumption can result in a lack of awareness and reduces mobility. Many victims of fatal campus housing fires have high blood alcohol levels. When a fraternity or sorority house (or other private student housing unit) invites other students to social events, alcohol consumption is often excessive, and many of the invited students spend the night at the off-campus housing unit. The IC must consider the possibility that more students than expected may occupy an off-campus house, and occupants may be in areas other than normal sleeping areas. Many of the multi-fatality fires in off-campus housing occur after a party. Further, arson is a leading cause of fire in student housing; therefore, a heavily involved fire with multiple victims is not uncommon.

Most rooms in dormitory or off-campus housing will be within the flow capacity of standpipe or pre-connected

Case Summary

On May 12, 1996 a fire in an off-campus Phi Gamma Delta fraternity house at the University of North Carolina at Chapel Hill resulted in five fatalities and $475,000 in property damage. The fire occurred after a graduation party and alcohol was a factor in four of the five deaths.

On June 8, 1996, less than a month after the fire at Phi Gamma Delta, another blaze struck the UNC-Chapel Hill fraternity Sigma Chi. The Chapel Hill fire chief determined that there was no connection between the two fires. The second was an arson fire with no fatalities. Recent renovation that included noncombustible interior finish, smoke and fire doors on stairways, a full alarm system, fire escape stairs, and an escape plan contributed to a much different outcome.

On June 19, 1996 an ordinance was passed requiring all fraternity and sorority houses to be sprinkler protected within five years.

Figure CS11-4 The burned fraternity house at the University of North Carolina at Chapel Hill.

Source: Michael S. Isner, NFPA Fire Investigation Report, *Fraternity House Fire,* Chapel Hill NC, May 12, 1996.

hose streams. Common areas used for study and recreation will be larger, but most will still be within the capabilities of one or two standard hose lines. Irreplaceable personal heirlooms are not as common in dormitories and off-campus student housing as compared to some other residential occupancies; however, as with other residential properties, the importance of personal items should be considered during incident operations.

Hotels and Motels

Like dormitories and off-campus housing, hotels, and motels have a much greater occupant density than the typical home or apartment. Similar to the occupants of hotels and motels, guests in the dormitories and off-campus housing will not be familiar with building and exit facilities. Small motels may have doors that lead directly to the outside at grade level or doors leading to a walkway with direct access to ground level. Larger hotel and motel room doors typically lead to an interior hallway with well-marked exits. Motels and hotels are required to post evacuation information inside each unit. Unfortunately, many people fail to read the evacuation instructions or familiarize themselves with exits when they are assigned a room. Occupants in a rush to evacuate will seldom read the instructions in their room, and exit signage may be obscured by smoke at the time of the fire.

Many times fire apparatus access will be limited to only one or two sides of a hotel or motel; therefore, fire department aerial and tower ladders will not be able to reach occupants on the inaccessible sides of the building. Further, fires in units on the opposite side of back-to-back units or units with an interior hallway may not be visible upon arrival. Expect to find occupants still in their rooms, possibly with occupants from several rooms converged in a single unit, disoriented in hallways, in elevator lobbies, on exterior balconies, and in stairways. If the roof is accessible, occupants may try to seek refuge there as well.

When encountering heavy fire and smoke conditions, several fire companies will be needed to conduct search-and-rescue operations in a large hotel or motel property. The desk attendants may have a list of people occupying the hotel, but they will not know which occupants are in the facility or be aware of people who share rooms without notifying hotel management. Most large hotels will have assembly, food, and drinking establishment areas within the building. Because of the potential for multiple rescues and a large rate of flow, it might be necessary to assign multiple companies to these areas. Most of the large loss-of-life fires in Table 11-12 are in high-rise hotels. High-rise tactics will be addressed in Chapter 12.

Mercantile Occupancies

NFPA 101: Life Safety Code[1] defines a mercantile occupancy as:

> **3.3.168.9** Mercantile Occupancy. An occupancy used for the display and sale of merchandise.

An occupancy factor matrix for **mercantile occupancies** is shown in Table 11-13. Shoppers and employees should be fully alert and most will be able to evacuate with little assistance. However, there may be people who need assistance in an emergency, particularly in multi-story mercantile buildings. Leadership is not provided in most stores, and during busy shopping times the occupant density can be very high (as much as one person for every 30 ft² [3 m²]).[1]

As with the residential occupancy, time factors are critical in sizing up a mercantile occupancy. During times of peak occupancy the life hazard can be high. Property and content losses can be significant in these buildings. Property loss is typically the largest problem facing fire forces in mercantile occupancies, but don't be misled: the danger to life may be substantial. Consider the life safety problem in a crowded department store during the Christmas season.

Mercantile occupancies include shopping centers and malls, individual stores, and shops. Individual stores and shops can range from very small shops to mammoth "big box" stores. Similar tactics can be applied to the smaller mercantile building; however, the "big box" retail, multi-story department store, shopping center, lifestyle center, and enclosed mall each presents special challenges.

Any building with large, undivided spaces requires special search-and-rescue techniques. The right- and left-hand searches used in residential occupancies are not effective in a store that could be hundreds of feet in each direction with aisles, cross aisles, and various obstructions. These large areas require team searches, which are difficult to coordinate and are labor- and time-intensive. Team search techniques using ropes and thermal imaging cameras will be needed when vision is obscured. However, it takes longer for smoke to fill these large-volume compartments, allowing many evacuations to be completed in a relatively clear atmosphere. If smoke is filling the store, proper venting can restore visibility and may allow occupants to evacuate with little assistance.

Large, undivided spaces will require a substantial rate of flow; extinguishment may not be possible within the store of origin once it gains sufficient headway. Rate of flow should be computed during pre-incident planning for each store and storage area exceeding the flow capacity of two standard pre-connected hose lines. Consider using 2½″ (64 mm) hose lines or master streams as initial attack lines inside large stores.

Shopping Centers

Older shopping centers are usually configured as a line of stores with direct access to the outside from each store. Shopping centers characteristically will have large anchor stores on each end. Anchor stores are typically "big box" stores or multi-story department stores. Having direct access to the outside makes evacuation less confusing as compared to evacuating stores located inside enclosed malls. Most occupants of stores in a shopping center will be able to find the main entrance with little difficulty. However, secondary exits are sometimes limited to the front and rear of the building due to fire wall separations between stores on each side. The rear of the store is often used as a storage area; thus, rear exits will require occupants to go through unfamiliar surroundings to reach the outside.

Many shopping centers lack sprinkler protection. As stated previously, the rate of flow could be very large, and it may be beyond the fire department's ability to extinguish a fire in a large store. Having stores attached on both sides creates an internal exposure hazard. Exposure hose lines should be deployed to stores on each side of the store of origin as soon as critical life safety and initial extinguishment assignments are made or if a defensive attack becomes necessary. Fire can penetrate anywhere along the common fire wall, but the roof line and any place with utility penetrations are the most likely extension paths. Older strip malls may have a common attic, which could allow the fire to spread horizontally to all attached stores. If this is the case, hose lines must be

TABLE 11-13	Occupancy Factor Matrix: Mercantile Occupancies					
	Mobility of Occupants	Age of Occupants	Leadership	Awareness	Occupant Density	Familiarity of Occupants
Mercantile occupancies	7	5	3	9	3	2

Key: 1 = extremely negative factor, 10 = very positive factor.

quickly extended into these false spaces on each side of the fire. Some shopping centers have double fire walls between buildings with no penetrations. Extension is less likely with this higher degree of compartmentation.

Enclosed Shopping Malls

The enclosed shopping mall took shopping convenience one step further by enclosing the total shopping area with entrances to various shops from an interior walkway, thus providing access to shops in a climate-controlled environment. Shopping malls tend to be larger than shopping centers and may have several anchor stores located at the ends and off multiple wings. Shopping malls are usually two or more levels. The layout of these structures is much more confusing than in the shopping centers. It is virtually impossible to find a specific smaller store without a map and/or guidance from building management. Pre-incident plans should include exterior access areas for fire apparatus, and floor layout drawings showing the location of each store should be carried on the apparatus of the first-arriving units. Primary and alternative access locations for each general area inside the mall should also be noted. If pre-incident plans are computerized, it is possible to search electronically for specific stores and identify the best exterior access points.

Stores within a shopping mall will be separated by partition walls of noncombustible construction, but will be open to the front, which faces the walkway area. During business hours, large door openings are designed to provide easy customer access to the store from the mall walkway. At night metal grates are lowered and secured. These metal grates provide a barrier to criminals but will not stop the forward progress of a fire. Most modern shopping malls are sprinkler protected, greatly reducing the potential fire problem. Without a properly maintained and operational sprinkler system, these properties would present an unmanageable life hazard for occupants and fire fighters.

The size of enclosed shopping malls has increased dramatically. Some of the largest malls cover several blocks and are virtually small cities inside a single building. These larger malls contain many occupancy types, including places of assembly. One such mall in Edmonton, Alberta, Canada, has hundreds of stores, an amusement park, a hotel, and a wave pool inside the structure.

Along with sprinkler protection, smoke control systems are also an integral part of the fire safety system within a mall. If not controlled early, fire in one store will likely spread smoke and heat into the walkway area and other stores. Smoke control systems and large open walkways should allow enough time for a successful evacuation.

Lifestyle Centers

The latest trend in shopping is toward lifestyle centers. The lifestyle center is laid out as blocks of stores, much like a city street. Large anchor stores are usually surrounded by smaller shops at grade level, and each individual store can

Case Summary

A fire in Altoona, Pennsylvania demonstrates the problems associated with shopping malls that are not sprinkler protected. The fire began in an unprotected store and rapidly spread to the mall walkway area and other stores. This shopping mall fire also demonstrates another common fire problem in strip and enclosed malls: common concealed spaces. The fire began in a utility closet, then spread to the ceiling space, involving a large concealed area before breaking out into the mall area. The result was the total destruction of several stores and damage to other stores within the mall in the weeks before Christmas.

Source: Michael S. Isner, *Shopping Mall Fire, Township of Logan, Pennsylvania, December 16, 1994*, Fire Investigation Report. Quincy, MA: NFPA.

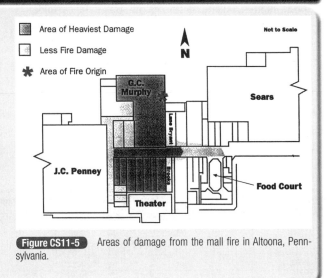

Figure CS11-5 Areas of damage from the mall fire in Altoona, Pennsylvania.

be accessed directly from the outside. The same basic tactics can be used as in the shopping center, but lifestyle centers will have fewer stores in each cluster and nearly all will be protected with automatic sprinklers. Finding individual shops and stores will normally be more difficult in the lifestyle mall than in a shopping center. Like the enclosed mall, a drawing showing the location of each store is needed as part of the pre-incident plan.

There is generally a fairly regular turnover of tenants in shopping centers, enclosed malls, and lifestyle centers. It is important to communicate with building management on a regular basis and update the pre-incident plan whenever a store's occupant changes.

"Big Box" Stores

Many "big box" stores are anchor stores at shopping centers, but there are also many stand-alone stores or "big box" stores with a few small stores nearby. The typical layout for a "big box" store is an open layout store to the front with an attached storage area to the rear. The "big box" store should be sprinkler protected. Therefore, as long as the sprinkler system is operating properly and store management has not increased the fuel load beyond the capabilities that the sprinkler system was designed to protect, following the operational guidelines provided in Chapter 7 will be sufficient. Many of these stores will have standpipe drops from the sprinkler system, but it may be better to lay hose directly from the apparatus to avoid robbing water from the sprinkler system.

The sprinkler system is the only positive factor in these buildings. Variances in construction methods, exit facilities, and size are permitted because of the sprinkler protection. This is a valid concept provided that the sprinkler system is properly maintained and is fully operational. If the sprinkler system is out of service or unable to control the fire, the "big box" store is a very dangerous place for fire fighters and occupants. Even with a properly operating sprinkler system, the height and configuration of storage can require substantial intervention to achieve final extinguishment. As an example, consider an indoor lumber yard where sheets of plywood are stacked with no in-rack sprinkler protection, such that the upper stacks of plywood shield the lower racks from overhead sprinkler water. The storage racks are typically metal; therefore, a serious fire in a lower rack could result in a collapse of the entire rack. Further, these stores may contain merchandise of an extra-hazardous nature (e.g., chemicals, compressed gas cylinders, aerosols, racked clothing, etc.), which are extremely difficult to extinguish and create a serious life hazard for occupants and fire fighters.

Multi-Level Department Stores

As noted previously, multi-level department stores can be part of a shopping center, enclosed mall, or lifestyle center. They can also be stand-alone buildings. Life safety and extinguishment are more difficult in these occupancies because of the multiple stories. Evacuation from upper floors requires the use of stairs or escalators. In many stores there is little separation from floor to floor, thus fire on a lower level is likely to extend to upper floors. Due to the large fuel load, extinguishment will be difficult. Property conservation is also more problematic in a multi-level structure. Property conservation is a major issue in all mercantile occupancies where high-dollar inventories are common and many items are extremely sensitive to smoke and water damage.

Business Occupancies

NFPA 101: Life Safety Code[1], defines a business occupancy as:

> **3.3.168.3** Business Occupancy. An occupancy used for account and record keeping or the transaction of business other than mercantile.

An occupancy factor matrix for business occupancies is shown in **Table 11-14**. The occupants of a **business occupancy** should be awake and alert while occupying the building. Occupant mobility is generally not a major problem, but there may be physically or mentally challenged people in the building who re-quire special assistance. Further, many business occupancies are located within high-rise buildings that will require the use of stairways for evacuation as well as places of safe refuge for occupants with special needs.

TABLE 11-14 Occupancy Factor Matrix: Business Occupancies

	Mobility of Occupants	Age of Occupants	Leadership	Awareness	Occupant Density	Familiarity of Occupants
Business occupancies	7	7	7	9	5	6

Key: 1 = extremely negative factor, 10 = very positive factor.

In business settings, the people occupying the premises are typically between the ages of 18 and 65. Leadership depends on the specific occupancy, but many business occupancies rehearse evacuation plans, particularly in high-rise buildings. Occupant density can be as high as one person per 100 ft² (9 m²),[1] which is considerably better than most places of assembly but more densely populated than many other occupancies. Employees in a one- or two-story office building with exits leading directly to the outside will probably be familiar with egress. People working in high-rise buildings usually take elevators to their work areas and may be unfamiliar with stairway locations.

Fire fighter line-of-duty death statistics for business and mercantile occupancies are combined in NFPA reports. The frequency rate for fire fighter fatalities in residential fires is 3.8/100,000 fires. In mercantile and business occupancies the frequency rate is 13.0/100,000 fires (see Figure 11-2). The large open areas and complex layouts place the fire fighter at greater risk.

Business occupancies have a relatively good history in terms of life safety for occupants. However, the trend toward more and larger high-rise structures increases the probability of a large loss-of-life fire in a business occupancy. For many years, fire professionals predicted that a future high-rise fire would result in the loss of a thousand or more lives. Unfortunately this prediction became a reality on September 11, 2001. However, the circumstances of the terrorist attacks on September 11, 2001 were much different than the fire the experts envisioned. The September 11, 2001 attack on the World Trade Center was the deadliest fire in U.S. history, but it was not the first terrorist attack on a high-rise building, nor even the first attack on the World Trade Center. The first World Trade Center attack in 1993[10] and the bombing of the Alfred P. Murrah Federal Building in Oklahoma City in 1995[11] were also the results of terrorist activity, and both included fire and explosions as part of the scenario.

Historically, fire has been used as a weapon, and most experts agree that high-occupancy buildings will continue to be targets of choice for terrorists, with explosives or fire used as the destructive force. The World Trade Center had approximately 50,000 occupants at the time of the first bombing in 1993. The Murrah Building is believed to have been occupied by 350 people at the time of that explosion. At nine stories, the Murrah Building was small in comparison to many high-rise office buildings. High-rise buildings create special problems and are discussed in detail in Chapter 12.

At night, fires in business occupancies can progress for a long period if the building is not protected by a detection and/or suppression system. Most new construction in business occupancies provide built-in automatic fire suppression and alarm systems.

In office occupancies, separation between individual offices is typically poor or nonexistent. Many times, movable partitions are used to separate work stations, allowing heat and smoke to spread quickly throughout a large, undivided area. Furthermore, this layout makes search-and-rescue operations very complex, since the configuration can resemble a maze. Wall separations are sometimes not full fire barriers as some walls separating offices do not extend beyond the suspended ceiling. The area above the suspended ceiling provides an easy path for fire extension.

Fire control efforts vary widely in business occupancies, depending on the fuel load and the size of the undivided area. Fire control will not always be possible or could be substantially delayed owing to large flow requirements. Business occupancies tend to have a light-to-moderate fuel load. Like other wide-open areas the rate-of-flow requirements will depend on the volume of the largest undivided compartment. If the sprinkler system does not control the fire, heavy-volume solid- or straight-stream appliances may be needed for suppression.

Property conservation in business occupancies must emphasize the importance of protecting sensitive electronic equipment and business records.

Storage Occupancies

NFPA 101: Life Safety Code[1] defines a storage occupancy as:

> **3.3.168.15** Storage Occupancy. An occupancy used primarily for the storage or sheltering of goods, merchandise, products, vehicles, or animals.

The factors affecting occupant evacuation in **Table 11-15** are mainly positive for a **storage occupancy**. Most employees

TABLE 11-15	Occupancy Factor Matrix: Storage Occupancies					
	Mobility of Occupants	Age of Occupants	Leadership	Awareness	Occupant Density	Familiarity of Occupants
Storage occupancies	9	8	6	9	9	8

Key: 1 = extremely negative factor, 10 = very positive factor.

in these occupancies are expected to perform physical labor, thus most will be mobile. Generally speaking, the very young and elderly are at greatest risk during a structure fire. Age should not be an evacuation factor at a storage occupancy with a working-age population. Some storage facilities have an emergency action plan that prescribes internal leadership, while others will lack leadership in the initial stages of an emergency. Employees should be awake and alert, as well as being familiar with the building layout and exit facilities. Some modern warehouses are automated with conveyors extending into remote and difficult-to-access locations. Occupant density is generally low in storage facilities, especially the automated warehouse.

The risk of a fast-moving fire or explosion may be high within the storage occupancy, which could have an immense fuel load. Yet the fire safety record in these occupancies has been good, primarily because of steps taken by responsible owners in protecting their workforce and property from fire.

Storage facilities vary in size from small self-storage modules to mammoth facilities like the K-Mart warehouse in Falls Creek, Pennsylvania. Most one-story modern warehouses are of non-combustible construction with metal truss roof structures. Like the "big box" stores described under mercantile occupancies, these storage buildings can have tremendously high fuel loads and the roof structures make them prone to early roof collapse. On the positive side, most of the large modern warehouses are sprinkler protected. Conversely many older warehouses will be of fire-resistive, heavy timber, or ordinary construction, but are often multi-story structures with limited egress and confusing layouts. Many of these older warehouse buildings do not have sprinkler protection.

Fighting a fire in a large commercial building is more challenging tactically than attacking a fire inside a small residential building. The size and complexity of the building, coupled with the possibility of an extremely high fuel load, makes these buildings more hazardous for fire fighters working inside. Escape time will be much longer from a building that may be a 1 million ft^2 (approximately 9 hectares) or greater as compared to a residential structure where dimensions seldom exceed 100′ (30 m). A fire in a vacant warehouse in Worcester, Massachusetts resulted in six fire fighter fatalities. The interior of this multi-story warehouse was very confusing. Fire fighters are more likely to be killed in these large commercial structures than in residential fires, but the danger increases exponentially in large vacant buildings, where the fire fighter fatality rate is nearly four times as great in vacant buildings as compared to residential structures. During the period 1996–2005, there were 3.8 fire fighter fatalities per 100,000 residential fires versus 14.1 fire fighter fatalities per 100,000 fires in vacant structures (see Figure 11-2).

If an emergency action plan is in place with provisions for employee accountability, the primary search and rescue will be much easier, because the search team can focus on a targeted area.

If the property is sprinkler protected, supporting and augmenting the sprinkler system will be the primary extinguishment tactic. However, facility management will sometimes increase the fuel load inside a warehouse beyond the design capacity of the sprinkler system. Additional water supplied by fire department pumpers could make the difference between a successful operation and a total loss of the building when the fuel load exceeds the design capacity of the system. However, if the fire overwhelms the sprinkler system and reaches an advanced stage, extinguishment will be difficult, if not impossible.

A common practice in warehouse facilities is to place large stacks of idle pallets in a staging area. This creates a tremendous challenge to the sprinkler system. Only an extra-hazard sprinkler system is capable of handling a fire in large quantities of wooden pallets. When a new occupant leases a storage facility, the new owner/occupant may store a different commodity, possibly exceeding the design capacity of the sprinkler system. As an example, a warehouse sprinkler system may be designed to protect metal parts stored in cardboard boxes. If a new owner or occupant stores rubber tires, flammable liquids, or plastics in this same area, the system, which was designed for the lower-hazard commodity, would be inadequate to extinguish the fire. The result would likely be a total loss.

The K-Mart fire in Falls Creek, Pennsylvania is a good example of the loss potential in a storage occupancy fire.[12] The K-Mart warehouse, covering nearly 1.3 million square feet (approximately 12 hectares), was fully sprinkler protected and compartmentalized, yet it was completely destroyed in a $100 million fire Figure 11-13. Aerosol cans of carburetor cleaner, which exceeded the design capacity of the sprinkler system, were first ignited at this incident. The burning aerosol cans rocketed from the compartment of origin to other compartments, eventually spreading the fire to all four building quadrants. Had the fire chief failed to see the danger to his personnel, this fire could well have resulted in several fire fighter fatalities. By developing a pre-incident plan for the structure, the chief recognized the collapse potential and the futility of manual firefighting operations.

Figure 11-13 K-Mart warehouse fire, Falls Creek, Pennsylvania, June 1982.

The rate of flow must be calculated during pre-incident planning. Standard V/100 calculations may not be sufficient for storage occupancies. Sprinkler system calculations should be used when dealing with extra-hazard storage. It is important to remember that sprinkler calculations assume the fire will be contained to a limited area due to quick operation of the system with water discharging directly onto the fire. These assumptions are invalid when manually attacking the fire with handheld hose lines, because sprinklers are much more efficient in applying water directly on the fire. Therefore, the calculated rate of flow for hose lines may result in a flow that is beyond the available water supply and fire department staffing and pumping capacities. With that said, when calculating rate of flow and developing pre-incident plans, remember that in many cases the fire will be limited to a small area of a very large building.

Many firms that suffer a large fire go out of business. Even with insurance, the loss of production and subsequent loss of customers are more than many businesses can overcome. Businesses apply risk-management principles to protect their profitability, and many of the risk-management measures involve installation of suppression systems or other built-in fire protection.

Most warehouses contain large quantities of valuable commodities. Some of these commodities are very susceptible to water and smoke damage. Storage in racks or on pallets reduces water damage from run-off. Storage of materials that absorb water, especially if the materials are stored on the floor, will greatly increase the water damage and add significant weight to the floor.

Industrial Occupancies

NFPA 101: Life Safety Code[1] defines an industrial occupancy as:

> **3.3.168.8** Industrial Occupancy. An occupancy in which products are manufactured or in which processing, assembling, mixing, packaging, finishing, decorating, or repair operations are conducted.

The occupant profile in an industrial setting is much like that of the storage occupancy with mobile, alert workers who are familiar with their surroundings **Table 11-16**. The occupant density is usually higher in an **industrial occupancy** facility as compared to a storage occupancy. Industrial plants are required to have emergency evacuation plans; therefore, leadership is usually better in the industrial occupancy than in many other occupancies. Emergency action plans include an accountability process that can assist the IC in facilitating an orderly evacuation and accounting for plant personnel.

Recognizing hazards and equipment limitations is paramount during an industrial fire. In the large industrial building or complex, pre-incident planning is critical to fire fighter safety and efficient operations. However, pre-incident planning is not enough. Cooperation between plant personnel and fire fighters is essential. In-plant employees are plant experts: They know valve locations, how to navigate the plant, and how to shut down processes. The fire fighter is the fire suppression expert. Successful industrial fire suppression activities involve plant staff assisting fire department personnel. Anything less than a cooperative effort invites total destruction and places fire fighters at unnecessary risk. The number of hazards and the potential for harm dictate

TABLE 11-16	Occupancy Factor Matrix: Industrial Occupancies					
	Mobility of Occupants	Age of Occupants	Leadership	Awareness	Occupant Density	Familiarity of Occupants
Storage occupancies	9	8	8	9	7	8

Key: 1 = extremely negative factor, 10 = very positive factor.

that special tactics be developed for each large manufacturing and storage property within a jurisdiction.

Fire fighter fatalities per 100,000 fires in industrial occupancies are twice the rate of residential fires (see Figure 11-2). Industries processing or storing hazardous materials are most hazardous.

Industrial occupancy fires are notable because they tend to cause a large loss of life and/or high dollar loss. Some of the more notable large loss-of-life fires in industrial occupancies are listed in **Table 11-17**. Several of the multi-fatality fires in industrial properties involved hazardous materials, such as ammunition, fireworks, explosives, chemicals, and flammables. In our research, we found several oil refinery fires that claimed three to five lives, but limited the list in Table 11-17 to larger loss-of-life fires.

In the industrial setting, the main tactical activity often involves controlling a manufacturing process or otherwise stabilizing the incident to protect life and property. Depending on the type of burning fuel, the use of water may be counterproductive. Using knowledgeable facility personnel as advisors to the IC during an incident is imperative, as they are familiar with the building, systems, and processes and can advise the IC on a course of action to take (or not take) to safely mitigate the problem.

Fire control efforts must be customized to the hazard involved. Pre-incident planning, which includes identifying and outlining the responsibilities of plant personnel, is key to successfully controlling fires in industrial occupancies.

In addition to the normal salvage activities, property conservation includes stabilizing manufacturing and operating processes. Many industrial facilities have interlocks that sense a drop in pressure in the water supply and/or fire protection system. These interlocks sometimes interrupt production or dump materials being processed. Flowing water from a fire hydrant located on the property may unintentionally interrupt a process, which could result in a significant property loss.

Multiple-and Mixed-Occupancy Buildings

NFPA 101: Life Safety Code[1], defines a **multiple occupancy** as:

> **3.3.168.11** Multiple Occupancy. A building or structure in which two or more classes of occupancy exist.

Multiple occupancies are further categorized as separated or mixed.

> **3.3.168.10** Mixed Occupancy. A multiple occupancy where the occupancies are intermingled.
> **3.3.168.14** Separated Occupancy. A multiple occupancy where the occupancies are separated by fire resistance–rated assemblies.

Separated multiple occupancy buildings can be handled as two buildings due to fire-resistive separations.

> Damage or destruction of high-value storage and manufacturing equipment at industrial properties results in many large-dollar losses. A million-dollar fire is commonplace in an industrial occupancy. The $100-million-plus fires at K-Mart in Pennsylvania,[12] Tinker Air Force Base in Oklahoma,[13] and Central Storage Warehouse in Wisconsin[14] are examples of large-loss fires eclipsed by the billion-dollar fire at a Texas oil refinery in 2005.

TABLE 11-17	Notable Fires in Industrial Occupancies	
Fire	Year	Deaths
R. B. Grover Shoe Factory, Brockton, Massachusetts	1905	50
Triangle Shirtwaist, New York City, New York	1911	145
Eddystone Ammunition, Eddystone, Pennsylvania	1917	133
Aetna Chemical Company, Oakdale, Pennsylvania	1918	193
Semet-Slovay, TNT Manufacturing, Split Rock, New York	1918	50
Hercules Powder Company, Kenvil, New Jersey	1940	52
Elwood Ordinance Plant, Joliet, Illinois	1942	54
East Ohio Gas, Cleveland, Ohio	1944	130
Munitions Depot, Port Chicago, California	1944	300
Chemical Plant, Texas	1989	23
Chicken Processing Plant, North Carolina	1991	25
Fireworks Manufacturing, Michigan	1998	7
Power Plant, Dearborn, Michigan ($1 billion property loss)	1999	6
Automobile Insulation Manufacturing, Kentucky	2003	7
Oil Refinery, Texas	2005	15

Mixed multiple-occupancy buildings are more problematic because fire can easily travel from one occupancy to another. Having an unoccupied area in the same building where other areas are occupied can result in a severe life safety threat. As an example, many buildings have a series of small stores or shops on the first floor with residential units above. Fuel loads in the shops will typically be greater than the fuel loads found in residential occupancies and often include hazardous processes. Further, the shops are usually unoccupied outside normal business hours allowing a fire to go undetected in an area below sleeping residents. By the time residents realize there is a fire, egress routes may be untenable and the fire may have already extended into the residential sections of the building.

Buildings Under Construction, Renovation, or Demolition

Buildings under construction, renovation, or demolition are not classified as specific occupancies, but these buildings are vulnerable to fire and warrant special attention. Buildings under construction or undergoing extensive renovation may lack fire protection equipment and structural features designed to impede fire until late in the construction process. Conversely, in buildings being demolished these same fire protection features and equipment are often the first to go, including disabling the sprinkler system and removing interior walls.

Buildings under construction will not have fire barriers or exterior coverings that retard fire growth, and large quantities of building materials may be stored within the interior framework of a partially constructed building. Like the building under construction, the building being renovated may have large stacks of building materials stored inside the building. When a building is being razed some interior materials are removed for salvage, then work begins to demolish the outer shell with large quantities of debris accumulating around and within the structure.

Buildings under construction, renovation, or demolition are typically unoccupied when work crews leave for the day; however, these structures may be illegally occupied by youths or vagrants. In either case, after-hour fires in these structures usually gain considerable headway before being reported, allowing the fire time to spread throughout the structure and involve external exposures. Free-burning fires involving wooden framing, combustible building materials, or debris can develop a large fireball with subsequent high levels of radiant heat. These buildings can also produce flying brands that can ignite fires far away, such as at the Santana Row fire in San Jose, California (see the case summary in Chapter 9). As discussed in Chapter 9, the radiant heat ignites nearby structures while flying brands can travel a considerable distance beyond the building of origin. The open status of buildings under construction, renovation, or demolition makes these structures vulnerable to flying brands from nearby fires.

A building that is only partially constructed or one that is partially razed is more likely to collapse than a finished structure. Many times, scaffolding is placed around buildings under construction, tempting fire fighters to use this scaffolding to gain rapid access to upper floors of the building. However, fire fighters should avoid using scaffolding, because water from extinguishment efforts can undermine the base of the scaffolding and the fire can destroy the scaffolding connections to the building. Either scenario could result in a total collapse of the scaffolding system, injuring or killing fire fighters who were working from the scaffolding at the time of collapse, while leaving fire fighters who used the scaffolding to access upper floors stranded, possibly with no means of escape.

Renovated Buildings

Once the renovation process is completed the occupancy classification may change. Many multi-story, downtown buildings that were once offices, stores, schools, or factories have been converted to other occupancy types. Many old school buildings are now apartments, condominiums, and restaurants; old department stores, lofts, and buildings that were built as factories are now offices. These are but a few examples of the kinds of conversions taking place in many urban areas.

In many cases, converted buildings are safer, because building and fire codes require upgrades when occupancy classifications are changed or when a significant portion of the building is renovated. However, these converted occupancies can create challenges. For instance, interior walls are often removed to create large open spaces to accommodate the new building use. In addition, false ceilings are added to reduce the ceiling height for aesthetic and energy conservation reasons.

The present trend of moving back into the central city core is expected to continue. As preservationists use their political clout to protect older buildings from being razed, much of the new residential and commercial space will be located in renovated structures. Many of the potential problems in renovating these older buildings can be avoided by enforcing building and rehab code requirements. Unfortunately local

Case Summary

On October 6, 2006, an early morning fire in a 113-year-old church resulted in the total destruction of the church. The collapsing church spewed burning embers, which ignited three other buildings several blocks away. One of the buildings ignited by the embers was a 22-story building being renovated into condominiums. Embers entered window openings to ignite materials on the interior, resulting in a fire that seriously damaged the high-rise building.

Source: Memphis Daily News.

and state government officials, in their zeal to revitalize the urban core while at the same time appeasing special interest groups, may bow to pressure to approve waivers to these codes. For fire fighters, this means combating fires in buildings that contain many false spaces that can hide fire and provide channels for fire extension, confusing floor plans, and large open areas that will complicate search-and-rescue efforts and require a large rate of flow.

General Occupancy Considerations

Special occupancy fires require fire departments not only to pre-plan for specific properties, but also to develop plans of action for general use within particular types of buildings and occupancies. Many of the fire fighter fatalities that have been recorded over the years were in residential properties, including single-family detached dwellings. However, the size and complexity of business and industrial buildings place the fire fighter at additional risk when combating fires in these structures. Extra precautions and more frequent accountability procedures may be necessary in combating assembly, mercantile, storage, business, and industrial fires.

It is fair to say that fire fighters are at much greater risk of dying when confronted with commercial fires than when working a residential structure fire. Obviously, the threat to a fire fighter's life is substantial at the residential fire, but the IC must recognize the even greater potential danger when working at a commercial fire.

Using operational priorities as a basis, fire suppression forces can wage a safe and effective operation at commercial and other special occupancies. Few ICs gain confidence through experience in fighting special occupancy fires;

therefore, it is imperative that they learn through training and from the experiences of others.

Most of the large loss-of-life fires in commercial and other special occupancies occurred in the fairly distant past. The number of civilian fire deaths and on duty deaths for fire fighters has shown much improvement in the recent past. Two important points should be made about these positive statistics:

1. The fire threat to human life is real and present. Maintaining a properly staffed, well-trained fire force is absolutely necessary.
2. Improvements in codes and standards and the widespread use of smoke detectors, fire sprinkler systems, and public education programs are directly related to life safety improvements.

Estimating the Number of Potential Victims

Determining the total number of occupants is closely related to the type of occupancy. The high-rise office building could have thousands of occupants, compared to one or two occupants in many single-family detached dwellings. Occupant load is also time sensitive. The office occupancy will most likely be fully occupied during normal working hours, while the residential building will most likely be occupied at night. A fire at the Peachtree Plaza office building in Atlanta, Georgia, occurred on a Friday morning, killing five occupants and injuring 20 others in the fully occupied office building.

The staffing requirements for a fire department response to a single-family detached dwelling is usually inadequate for larger building fires because of the additional search-and-rescue requirements combined with a need for larger fire flows. The more complex the structure, the greater the need for personnel. However, some of these buildings will be more substantially built and will be protected by fire suppression equipment. These fire protective features will lower the risk to fire fighters and occupants, provided that the fire safety features designed into the building are maintained and operating properly.

The first step in determining evacuation needs is an evaluation of the time of day, building size, and occupancy type. Classifying these variables will provide a rough estimate of how many people could be in the building; however, it is important to remember that this is at best only an estimate.

Summary

The occupancy type will directly affect the fire-ground strategy. Since the primary responsibility of any fire-ground operation is life safety (of both the occupants and the fire fighters), it is critical that the IC conduct a complete size-up based on a number of factors, such as building construction, occupancy type, time of day, and related factors that will guide the decision-making process.

Wrap-Up

Key Terms

assembly occupancy An occupancy used for the gathering of 50 or more persons for deliberation, worship, entertainment, eating, drinking, amusement, awaiting transportation, or similar uses; or used as a special amusement building, regardless of occupant load.

business occupancy An occupancy used for account and record keeping or the transaction of business other than mercantile.

convergence cluster phenomenon Groups that take shelter together to provide mutual support.

defend-in-place A tactic utilized during a structure fire when it is very difficult to remove occupants from the building. Occupants are either protected at their present location or moved to a safe location within the building.

detention and correctional occupancy An occupancy used to house four or more persons under varied degrees of restraint or security where such occupants are mostly incapable of self-preservation because of security measures not under the occupants' control.

educational occupancy An occupancy used for educational purposes through the 12th grade by six or more persons for four or more hours per day or more than 12 hours per week.

health care occupancy An occupancy used for purposes of medical or other treatment or care of four or more persons where such occupants are mostly incapable of self-preservation due to age, physical or mental disability, or because of security measures not under the occupants' control.

industrial occupancy An occupancy in which products are manufactured or in which processing, assembling, mixing, packaging, finishing, decorating, or repair operations are conducted.

mercantile occupancy An occupancy used for the display and sale of merchandise.

mixed multiple occupancy A multiple occupancy where the occupancies are intermingled.

multiple occupancy A building or structure in which two or more classes of occupancy exist.

occupant density A measure of the number of people in a given area; usually stated as the number of square feet per person.

residential board and care occupancy A building or portion thereof that is used for lodging and boarding of four or more residents, not related by blood or marriage to the owners or operators, for the purpose of providing personal care services.

residential occupancy An occupancy that provides sleeping accommodations for purposes other than health care or detention and correctional.

separated multiple occupancy A multiple occupancy where the occupancies are separated by fire resistance–rated assemblies.

storage occupancy An occupancy used primarily for the storage or sheltering of goods, merchandise, products, vehicles, or animals.

Suggested Activities

1. Classify specific occupancies in your jurisdiction using an occupancy factor matrix similar to the general occupancy factor matrices used in this chapter.

2. Compute the maximum number of occupants in a 100' × 100' (30 m × 30 m), two-story office building assuming 100 ft^2 (9 m^2) per person.

3. Use the plan view drawing in Figure CS11-2 showing the exit doors and ignition points from the Station Night Club Case Summary for this activity. Change the ignition sequence to a slower moving fire (one not involving foam plastic materials). When you arrive on the scene as the first-in engine company, there are heavy smoke and fire conditions with people jammed at the main entrance, blocking people attempting to escape.

   A. What is the best course of action for this first-arriving unit, assuming the company is staffed with an officer, apparatus operator, and two fire fighters? Explain the rationale for your decision.

   B. If you decide to lay a hose line, what would be the best place to enter the building to attack or control the fire?

   C. If you decide to perform manual rescues, how could you rescue the most people in the shortest period of time?

   D. Assume the role of a battalion chief arriving to take command. What are some immediate assignments that should be given to address the life safety hazard?

4. Review an evacuation plan for a local elementary or high school in your jurisdiction.

   A. Does the plan clearly spell out the location of the school contact person for a fire emergency?

   B. Are students assembled in areas that interfere with fire department operations?

5. Obtain a copy of a floor plan for a nursing home or hospital in your jurisdiction or mutual aid response area. Develop a fire scenario where the fire originates in a patient room and extends into the hallway.

   A. Describe what the nursing staff should do.

   B. Explain the cooperative effort between the nursing staff and fire department in moving patients to an area of safe refuge.

   C. Discuss the use of hose lines and what effect hose lines would have on the place of safe refuge. Be sure to consider the effect on smoke and fire movement when hose lines penetrate fire doors.

6. Obtain a list of residential board and care occupancies in your response area. What are the occupant characteristics in these facilities in terms of their ability to escape without assistance?

7. It is 2:00 AM and you are responding as the first-due engine company to a 30-unit apartment building. Upon arrival you can see fire in the main stairway; the fire appears to have originated in the basement storage area. Some residents are already out of the building, while others are at windows or on balconies calling for help. What is your best first action? Explain.

8. It is 4:00 AM, and there is a report of fire in a three-story building with a basement that is used as a multi-purpose dining/party room. This off-campus building is known to house students from the local university. You have had many fire and EMS calls in this area due to parties associated with homecoming week.

   A. What occupant load do you expect to find in this building?

   B. Where will the occupants most likely be located?

9. It is 2:00 PM when a fire alarm is received for a "big box" home improvement store. You are the battalion chief arriving with the first-due engine and truck companies. Upon arrival the store manager meets you at the front door and states that he thinks everyone is out of the store and the fire is located in the indoor lumber yard. The first-due engine company is located near the main entrance and has connected to a water supply. The apparatus operator is hooking up to the sprinkler intakes. The engine company officer checks the alarm panel and confirms sprinkler operation in the location of the indoor lumber yard. The truck company reports heavy and increasing smoke conditions inside the building.

- **A.** How credible is the report that everyone is out of the building?
- **B.** What are some of the reasons why the smoke would be increasing with the sprinkler system operating?
- **C.** What size hose line should the engine company advance into the building?
- **D.** Should the engine company personnel use the standpipe drop from the sprinkler system or manually lay a line from their apparatus?
- **E.** What tasks would you assign to the truck company?

10. An insurance company's claims adjustment unit is housed in a 250′ × 250′ (76 m × 76 m) building of non-combustible construction. Office space is provided in the form of 6′ (2-m) high cubicles with aisles and cross-aisle access to the cubicles. Describe a search method that could be used under conditions of low visibility in this building.

11. A sprinkler-protected building that formerly housed an auto parts warehouse storing auto parts in cardboard boxes (no tire or flammable liquid storage) was sold to a toy manufacturer, who is now storing plastic toys in cardboard cartons. How does this affect life safety and extinguishment inside the warehouse?

12. Select a pre-planned industrial property in your jurisdiction. Describe the role of plant employees and the role of the fire department for this facility.

Chapter Highlights

- Occupancy type, time of day, and day of week are critical factors in determining fire-ground strategy.
- Major factors related to occupancy type are number, mobility and age of the occupants; leadership; awareness; familiarity; and time of day/week/year.
- Types of occupancy include assembly, educational, health care, board/care, detention/correctional, residential, mercantile, business, storage, industrial, and mixed occupancy.
- Assembly occupancies include churches, eating/drinking establishments, stadiums, convention centers, and theaters where 50 or more persons gather.
- The number of potential victims in an assembly occupancy varies depending on the time of day/week; in general, assembly occupancies represent a significant threat to civilians and fire fighters.
- An educational occupancy includes elementary, middle, junior high, or high schools.
- Mobility factors are usually above average in school settings, but younger elementary school students may not be capable of taking independent action to safely evacuate.
- Leadership and preparedness are generally positive factors in educational occupancies.
- Educational occupancies should be pre-planned.
- Health care occupancies include hospitals, nursing homes, and limited and ambulatory care facilities.
- A defend-in-place strategy may be the preferred option for highly immobile occupants in a hospital or nursing home.
- Property conservation of expensive medical equipment is an important consideration in hospitals.
- Board and care residents may include individuals with diminished mental and/or physical capacity, which complicates rescue efforts.
- Operations in detention/correction occupancies are complicated by the fact that prisoners cannot self-evacuate and must be kept secure during firefighting operations.
- Pre-incident planning is an absolute necessity for a detention or correctional facility.
- Residential occupancies include single and multiple family dwellings, apartment buildings, dormitories, and hotels/motels.
- Most fires, fatalities, and property loss occur in residential fires.
- In many occupancies, including residential, there is a direct correlation between life safety and the time of day.
- More fire fighter on-duty deaths occur in residential properties than any other occupancy due to the large number of fires that occur in these properties.
- Most fires in one- and two-family dwellings should be within the capabilities of one or two pre-connected hose lines.
- Apartment buildings typically have a larger number of people concentrated in a smaller area as compared to single-family dwellings.
- Exposure problems are more likely at apartment complexes.
- Life safety is a critical issue in dormitory and off-campus housing.
- Social events at dormitories and off-campus housing may complicate rescue efforts, as alcohol consumption and extra occupants within the structure are significant factors.
- Hotels and motels generally have a much greater occupant density than the typical home or apartment.
- Occupants of hotels and motels will not be familiar with building and exit facilities and the IC should anticipate convergence cluster.
- Property loss is typically the largest problem facing fire forces in mercantile occupancies; however, depending on time factors, danger to life may be substantial.
- Mercantile occupancies include shopping centers and malls, individual stores, and shops, with large "big box" stores presenting special challenges.
- Many shopping centers lack sprinkler protection and require large rates of flow.
- Multiple exposures and large occupant loads should be expected in shopping centers and malls.
- Fire extension may occur quickly, especially in older strip malls.
- Pre-incident plans should include exterior access areas for fire apparatus and floor layout drawings showing the location of each store in shopping centers, malls and life-style centers.
- Most modern shopping malls are sprinkler protected and many have smoke control systems.

Wrap-Up, continued

- High turnover of tenants in shopping centers and malls makes it important to update pre-plans on a regular basis.
- A "big box" store should be sprinkler protected, but the height and configuration of storage areas can require substantial intervention to achieve final extinguishment.
- Stores may stock hazardous merchandise.
- Multiple-level department stores and malls allow rapid fire extension.
- Large fuel loads in mercantile occupancies make extinguishment difficult.
- Store occupants are alert and usually mobile, but some people may require special assistance.
- Business occupancies have a relatively good history in terms of life safety. In the past, terrorists using fire and explosives targeted high-rise business structures such as the Murrah Building and World Trade Center.
- Some businesses have large open areas with minimal separation between spaces (e.g., movable cubicles) and maze-like layouts with suspended ceilings creating a fire and smoke extension hazard.
- Occupants in a storage facility are generally mobile, alert, and few in number.
- The risk of a fast-moving fire or explosion is high in a storage occupancy, especially if the fuel load exceeds the design capacity of the sprinkler system.
- Most large, modern warehouses are sprinkler protected; older warehouses are often of fire-resistive, heavy timber, or ordinary construction, but often lack sprinkler protection.
- Occupant profile in an industrial occupancy is similar to a storage occupancy, but with higher occupant density.
- Emergency evacuation and pre-incident plans are required in most industrial facilities.
- Plant staff in an industrial occupancy can usually assist fire suppression forces by shutting down processes and equipment.
- Industrial fires often result in a large loss of life and/or high dollar loss.
- Property conservation in an industrial occupancy includes stabilizing manufacturing and operating processes.
- Mixed- and multiple-occupancies combine the fuel loads and life safety hazards of two or more occupancy types.
- Vacant buildings should not be occupied, but illegal occupancy is common and should be considered.
- Fires at construction sites often produce flying brands that can ignite fires elsewhere.
- Structural collapse and/or collapse of construction scaffolding is highly likely in buildings under construction.
- A change of occupancy from one type to another may result in updated safety features (e.g., sprinklers), but may also add features that create additional firefighting hazards (e.g., suspended ceilings).
- Occupant load is related to building size, type of occupancy, time of day, and day of week.

References

1. National Fire Protection Association, *NFPA 101: Life Safety Code*. Quincy, MA: NFPA, 2006.
2. Marty Ahrens, *U.S. Fires in Selected Occupancies 1999 to 2002*. Quincy, MA: NFPA Fire Analysis and Research Division, 2006.
3. R. L. Best, *Reconstruction of a Tragedy: The Beverly Hills Supper Club Fire, Southgate, Kentucky, May 28*. Quincy, MA: NFPA, 1977.
4. Thomas J. Klem, *56 Die in English Stadium Fire, Fire Journal*, May: 128–147, 1986.
5. Edward Comeau, *Board and Care Fire, Mississauga, Ontario, March 21, 1995, Fire Investigation Report*. Quincy, MA: NFPA, 1995.
6. Michael J. Karter, U.S. fire loss for 2005. *NFPA Journal* September/October: 46–51, 2006.
7. Stephen G. Badger, U.S. multiple-death fires for 2005. *NFPA Journal* September/October: 54–59, 2006.
8. John Bryan, A phenomenon of human behavior seen in selected high rise building fires: convergence clusters. *NFPA Fire Journal* November: 27–30, 86–90, 1985.
9. NFPA Fire Investigations Department, Wood shingle roofs fuel conflagration. *NFPA Fire Command* February: 22–24, 1980.
10. U.S. Fire Administration, *World Trade Center Bombing: Report and Analysis*. Emmitsburg, MD: U.S. Fire Administration, undated.
11. Ed Comeau and Stephen Foley, Oklahoma City, April 19, 1995. *NFPA Journal* July/August: 51–53, 1995.
12. Richard Best, $100 million fire in K-Mart distribution center. *NFPA Fire Journal* March: 36–42, 80, 1983.
13. James Goodbread, Fire in Building 3001. *NFPA Fire Command* July: 34–37, 1985.
14. Michael Isner, $100 million fire destroys warehouses. *NFPA Fire Journal* November/December: 37–41, 1991.

High-Rise Buildings

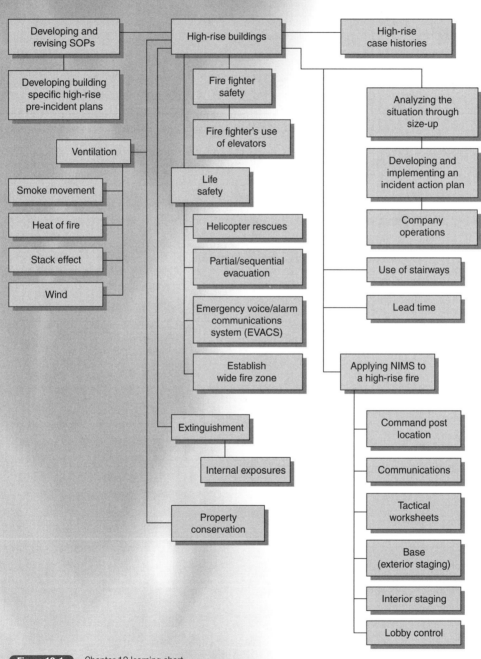

Figure 12-1 Chapter 12 learning chart.

Chapter 12

Learning Objectives

- Describe the magnitude of the high-rise fire problem in terms of number of fires, number of fire fatalities, and property loss.
- Explain the effect the loss of life and property at the World Trade Center on September 11, 2001 has on the statistical analysis of high-rise fires in relation to loss of life and property in high-rise buildings.
- Define a high-rise building from a fire department perspective.
- Explain elevator recall and its advantages during a fire emergency.
- Enumerate the responsibilities of lobby control.
- Explain why using an elevator during a fire emergency is a calculated risk.
- List and discuss the seven rules for safety when using an elevator during a high-rise structure fire.
- Describe the process of using an elevator under fire department control.
- Explain the duties of stairway support and when it should be used.
- Compute the number of fire fighters needed to staff stairway support for a fire on the 30th floor of a high-rise building where the elevators are unsafe to use.
- Identify negative and positive aspects of using helicopters at a high-rise building fire.
- Construct a chart showing the angle of deflection for fire streams operated from the exterior into an upper floor of a high-rise building.
- Compute the approximate pump discharge pressure needed to supply a hose stream operating on the 20th floor.
- Explain when the command post should be located in the lobby, outside the building, or at a remote location.
- List four occupancy types that are commonly found in high-rise buildings.
- List five special considerations when developing a pre-plan for a high-rise building.
- Compare old-style tower construction to modern, planar-style high-rise buildings.
- Define and explain the advantages of a smoke-proof tower.
- Discuss methods that can be used to maximize the limited stairway capacity in a high-rise building and how fire department operations can hinder evacuation.
- Describe an EVAC system and identify the advantages and disadvantages as compared with other notification systems.
- Explain the difficulties in determining the location of the fire from the exterior of a high-rise building.
- Develop a high-rise decision tree with decision points addressing fire location, size, and elevator availability.
- Explain how heat affects smoke movement in a high-rise building.
- Define stack effect and explain the conditions necessary for positive and negative stack effect.
- Explain how wind affects the neutral pressure plane.
- Discuss the importance of extinguishment to life safety in a high-rise fire.
- Explain what is meant by a "wrap-around" fire.
- Describe the hazards involved in ventilating upper floors by removing window glass and how to protect fire fighters and civilians on the street below.
- List pathways for floor-to-floor fire extension in a high-rise building.
- Define "lead time" and extrapolate the estimated lead time for a fire on the 40th floor of a high-rise building when elevators are unavailable.
- List practical forms of non-radio communication that can be used at a high-rise building fire.

Learning Objectives, continued

- Discuss when interior and exterior staging would be used at a high-rise fire.
- List the duties of the interior staging officer.
- List the duties of lobby control.
- Compare and contrast factors that affected fire operations at the One Meridian Plaza fire in Philadelphia and at the First Interstate Bank Building fire in Los Angeles.
- Evaluate the number of possible fire deaths in a high-rise building at the Peachtree Plaza fire in Atlanta based on the number of actual civilian fire deaths and probable deaths had the fire been on a higher floor.
- Explain the negative effects of operating an exterior stream into a high-rise building.
- Compare the designed "accidental aircraft" impact at the World Trade Center to the actual impact of the aircraft on September 11, 2001.
- Describe the conditions leading to structural collapse at the World Trade Center.
- Describe how a "convergence cluster" could affect search-and-rescue operations at a high-rise building.
- Evaluate operations at a simulated high-rise fire in your response area.
- Size-up, develop an incident action plan, assign units to carry out the plan, and develop a NIMS organization for a simulated high-rise fire scenario.

Introduction

The basic tactics and strategic objectives used in high-rise firefighting are the same as those that apply to any other structure fire, but with special considerations because of the height of the building. The differences are great enough to deserve this separate chapter.

Table 12-1 shows the average number of high-rise fires, as well as the average annual loss of life and property in high-rise buildings for the period 1985 to 2002,[1] which provides a good indication of the magnitude of the problem. As can be seen in Table 12-1, the September 11, 2001, attack on the World Trade Center skews the statistical analysis. The World Trade Center bombings in 1993 and attack in 2001 are not typical high-rise fire situations. But this is the nature of the high-rise problem. There are often intervals of several years between major loss-of-life or large property-loss fires in high-rise buildings in the United States, but the potential always exists. The overall number of fires, civilian fire deaths, and property loss seems to be trending downward. This is probably due to more high-rise buildings being sprinkler protected. In the mid-1970s, most jurisdictions required new high-rise buildings to be sprinkler protected, but very few required existing buildings to be retrofitted. The World Trade Center was fully sprinkler protected, but the system was severely damaged on impact, and the jet fuel provided a fire load beyond the design capacity of the system.[2] There is absolutely no doubt that a sprinkler-protected building is a safer building for fire fighters and occupants; however, fire departments should not place total reliance on the sprinkler system.

High-rise fires represent an extraordinary challenge to fire departments and are some of the most challenging incidents a fire department encounters. High-rise buildings can hold thousands of people well above the reach of fire department aerial devices, and the chance of rescuing victims from the exterior is near zero once the fire is above the operational reach of aerial ladders or elevating platforms.

Case histories provide an excellent way of learning. By studying past fires, it is possible to gain experience that may not be gained in any other way. Outside of New York, Los Angeles, Chicago, and a few other very large cities, not many chiefs experience enough working high-rise fires to gain confidence in managing such incidents. Communities with one or two high-rise buildings may never experience a serious fire, yet the fire department is expected to be prepared if a high-rise fire occurs. For this reason, several high-rise fires are discussed at the end of this chapter. Studying these and other high-rise fire case studies is strongly recommended.

Figure 12-1 is a learning chart showing the topics in this chapter. It also represents a logical method of first planning for high-rise fires in your jurisdiction, and then handling an incident in a safe and effective manner. This chapter ends with a discussion of a few notable high-rise fires and the lessons that have been learned from each.

Developing and Revising High-Rise Standard Operating Procedures

Pre-incident planning and code enforcement can reduce the scope of the high-rise problem. However, special tactics will be needed to control fire forces working in different areas within these large structures while providing the necessary logistical support. Using the national incident management system (NIMS) and developing high-rise SOPs can do much to ensure successful operations.

A difference of opinion exists as to the definition of a high-rise building. Most codes define the high-rise building in terms of height and/or stories. Fire departments tend to think of a high-rise building as being beyond the reach of the aerial fire equipment available to them. Since the focus of this text is on fire-ground tactics, the fire department definition is most appropriate. However, do not forget the obvious; an eight-story building will not present the same challenges as an 80-story high-rise building. Logistics and access problems increase with height. The more floors that are located above the fire, the more people are likely to

TABLE 12-1 High-Rise Fires 1985–2002		
Annual Average*	Including World Trade Center*	Without World Trade Center*
Number of high-rise fires	22,171	22,171
Number of civilian fire deaths	244	91
Property loss	$2,230,000,000	$372,000,000
*Note: These numbers are annual averages over 18 years, not exact numbers per year.		

need fire department assistance, and the more fuel there is to burn.

High-rise buildings were once found exclusively in larger cities, but today they are commonly found in small and mid-sized communities as well. In most cases high-rise buildings in these smaller communities are newer, lower in height, and protected by automatic sprinkler systems. Even if your department does not respond to a high-rise building at present, if urban sprawl continues as expected, it probably will in the future. If you have a mutual aid contract with a jurisdiction that contains high-rise buildings, they are likely to need your help in combating a working high-rise fire. Several larger cities that once contained just a few notable high-rise structures have experienced an incredible growth in high-rise construction. Given the special problems that these buildings present, each department that could reasonably be expected to respond to a high-rise fire should have a high-rise SOP and train accordingly.

Fire Fighter Safety

The risk to fire fighters and occupants increases in proportion to the height of the building and the height of the fire above grade level. Once fire fighters are operating above the reach of aerial devices, the only viable means of egress is the interior stairs; extra protection afforded by laddering the building is not possible.

Good tactics and fire fighter safety cannot be separated. The tactics that are explained throughout this chapter improve fire fighter safety. The safety considerations that were discussed in Chapter 5 apply to high-rise firefighting, but additional safety measures need to be taken during a high-rise operation.

It is common knowledge that occupants should not use elevators to escape a high-rise fire except under special circumstances, and then only under fire department supervision. For this reason, many modern building elevators have fire department controls. When the alarm system is activated, the elevators return to the ground floor and remain there for use by the fire department. Older buildings may have fire department controls but may not have the feature that automatically returns elevators back to the ground level.

> Elevator recall is a safety feature that is designed to send elevators to ground level when the fire alarm is activated. Elevators then remain locked out at ground level until fire fighters arrive and use a key to place the elevator on fire department service.

Case Summary

A fire on May 4, 1988, in the First Interstate Bank Building in Los Angeles destroyed four floors in a 62-story building. The fire required a total of 64 fire companies and 383 fire fighters; it took 3½ hours to control. Despite the magnitude of this fire, only one person was killed: a maintenance worker who took an elevator to the fire floor to investigate the alarm.

Source: Thomas J. Klem, *First Interstate Bank Building Fire, Los Angeles, California, Fire Investigation Report.* Quincy, MA: NFPA.

Figure CS12-1 The fire at the First Interstate Bank Building.

A responsibility of <u>lobby control</u> is to control, operate, and account for all elevators. Some fire service professionals say that elevators should never be used under fire conditions or suspected fire conditions until their safety can be verified from the fire floor. This may not be practical in situations where fire companies respond to alarms in high-rise buildings several times each day. Requiring fire fighters to ascend 20 flights of steps to check an odor of smoke is not a productive use of resources. A more reasonable approach is to develop procedures and conduct training to reinforce the safe use of elevators.

Fire Fighters' Use of Elevators

A critical variable in high-rise fire operations is the availability of reliable elevators. If fire fighters can safely use the elevators, fire-ground logistics are dramatically improved. When the fire is located many floors above ground level, there is a strong inclination to use the elevators. However, elevators often stall or act erratically under fire conditions. Fire fighters who are trapped in a stalled elevator become part of the problem, as other fire fighters are then needed to rescue the rescuers. Therefore, the department SOPs should address the safe use of elevators, including circumstances in which it is unsafe for fire fighters to use them. These procedures should include alternative measures for getting needed equipment to the fire floor when elevators cannot be safely used. Fire department service controls and other elevator safety features will vary by age and elevator manufacturer. Building pre-plans should include information and instructions for using the elevator system.

One of the dangers in using elevators is that the doors may open on the fire floor, exposing fire fighters on the elevator to smoke and heat before they are in position with hose lines to attack the fire. Once elevator doors open on the fire floor, sensing devices (electric eyes, etc.) may prevent the doors from closing, thus trapping the fire fighters. This is a potentially deadly mistake that should be avoided at all costs. If fire fighters are caught in this situation, they may be able to use an override switch to force the doors to close. If not, they may be forced to exit the elevator in an attempt to escape the fire floor via the stairway. Similarly, elevators can stall in the shaft at or above the fire floor, also trapping fire fighters in the elevator car. Fire fighters should never, under any circumstances, use an elevator if there is a chance that the elevator will travel to or above the fire floor. This is another, possibly fatal, mistake.

Elevators are equipped with redundant safety systems to prevent them from falling. In many cases, an elevator car has to be nearly destroyed before it will fall. The elevator shaft is a fire-protected enclosure, but

> Using elevators during a fire emergency is a calculated risk.

a fire of sufficient intensity can invade the shaft, and smoke and toxic gases will certainly enter elevators that are stalled above the fire floor. This could prove deadly for fire fighters, who should be fully equipped with personal protective clothing, including self-contained breathing apparatus

Case Summary

On April 11, 1996, a fire occurred on the ground floor in the airport terminal in Düsseldorf, Germany. It spread rapidly throughout the area and into the upper levels. Several people standing on the roof of the adjacent parking garage observed the smoke and decided to use the elevator to exit the garage. Unfortunately, the elevator opened into the fire area on the ground floor of the terminal. The smoke obscured the electric eye on the elevator door, so it could not be closed. Seven people died in this elevator. Ten others were killed throughout the terminal.

Source: Edward R. Comeau, *Düsseldorf Airport Terminal Fire*, NFPA Fire Investigation Report. Quincy, MA: NFPA.

Figure CS12-2 The elevator that brought seven people to the ground floor opened directly into the fire area, trapping and killing them.

(SCBA). Civilians without protective gear have even less chance of survival when trapped in an elevator on or above the fire floor. Maintenance and security people should never be taken into the elevator until it has been verified that the elevator is completely safe.

The first rule of elevator safety is to avoid the use of elevators unless they will substantially improve operations. Fires on lower floors do not warrant the use of an elevator unless someone on the fire floor can verify its safe use. The first-arriving companies should use the stairways for fires on lower levels. Department SOPs and pre-incident plans must address this issue specifically. Once the safety of an elevator has been established, then fire fighters can use it under the close supervision of lobby control. Two other factors to be considered are fire separations between the elevator bank and fire location and whether the elevator goes to or above the fire floor.

High-zone/low-zone (split bank) elevators are not uncommon, and some split bank elevators are divided into several zones. For example, a three-zone bank of elevators would have one bank serving lower floors only, another bank serving only the middle floors, and a third bank providing access to upper floors only. This is important information that should be included in the pre-incident plan. For example, for a confirmed fire on the 10th floor, it is fairly safe to use elevators that terminate at the eighth floor **Figure 12-2**.

When there is a fire separation between the elevators and the fire area, it may be possible to safely use the elevators in another building zone and then travel horizontally to the fire. This approach is much easier than ascending the stairs. Large high-rise buildings, especially hospitals, often have building zones. Fire walls and fire doors create a horizontal barrier to smoke and heat from the fire. Hospital patients are usually moved horizontally to another zone, rather than taken down to the ground floor or evacuated from the building. If another building zone is safe for the patients, it should certainly be safe for fire fighters.

When separate low-zone/high-zone elevators are unavailable or the building is not adequately zoned horizontally and the fire is on the upper stories of the building, extreme caution should be used if the IC decides to use elevators to transport fire fighters.

Listed below are several rules that should be observed in taking the calculated risk of using an elevator during a fire situation. These rules for elevator safety should be considered when writing department high-rise SOPs.

1. *Do not use an elevator of questionable safety or for a fire on a lower level.* Elevators should not be used for

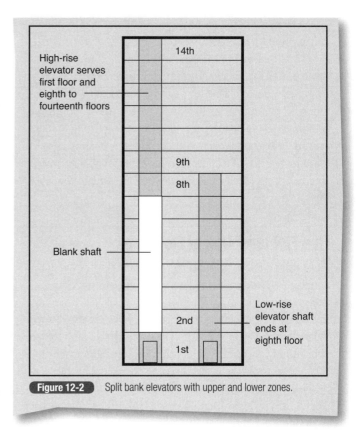

Figure 12-2 Split bank elevators with upper and lower zones.

fires on lower floors in the building or if there is any doubt about the safety of the elevator.

2. *Never take an elevator directly to the fire floor or above.* This is the cardinal rule of elevator safety and must be rigidly enforced. Department SOPs should state that the elevator should be taken to two floors or more below the fire, and then fire fighters should walk up the stairways to the fire floor. This rule also applies to split bank elevators that do not travel to the fire floor. If split bank elevators are available, use an elevator that does not travel to or above the fire floor. Again, exit the elevator at least two floors below the fire and use the stairway to reach the fire floor.

3. *Place the elevators under independent (fire department) control.* Keys should be made available to the fire department so that elevators can be placed under independent control. Newer elevators cannot be operated without a key once the fire alarm has sounded. This independent control greatly increases fire fighter and civilian safety, as the elevator will not be responding to calls from occupants.

4. *Control all elevator cars in multiple hoistways.* This not only provides the fire department access to upper

floors, but also prevents erratic response to other calls within the building. Controlling elevators is a responsibility of lobby control.

5. *Never overcrowd elevators.* This is doubly important for fire fighters, not only because it is unsafe to exceed the weight limit on an elevator, but also because fire fighters may need space to don SCBA or to use tools to force their way out of the elevator.

6. *Wear personal protective clothing, including SCBA, and bring forcible entry tools.* Air supply should be conserved by not donning the SCBA facepiece or connecting the breathing hose until approaching the fire area or when conditions indicate the need to be on air. Never forget that using an elevator is taking a calculated risk. Even when every reasonable precaution is taken, what lies at the top of the ascent is unknown. Forcible entry tools must be available to escape the elevator if necessary. Understanding elevator door operations can be extremely valuable if personnel become trapped in an elevator that stalls. In trying to escape a stalled elevator, always activate the emergency stop switch. Opening an emergency escape door or ceiling panel should also stop the elevator, as the interlocks are designed to prevent accidental movement when people are trying to escape.

7. *Send equipment rather than fire fighters on elevators.* Many times elevators that are considered unsafe for fire fighters can be used to transport tools and equipment to the interior staging area (usually two or more floors below the fire floor). Fire fighters can then safely ascend the stairs without the burden of heavy tools and equipment. It is not always possible to send an unstaffed elevator to a desired floor. When the fire alarm is activated it may be necessary to mount the elevator and hold a floor button or elevator key before the elevator will move. If this is the case, equipment cannot be sent above without a fire fighter. Freight elevators are usually less safe than passenger elevators, but they may be suitable for sending equipment above without placing fire fighters in the elevator car.

When ascending in an elevator, it is good practice to stop periodically, possibly every five floors, to check for smoke. This is accomplished by opening the top elevator escape panel and using a flashlight to see whether smoke is present in the shaft. This is also a test of the elevator's ability to stop on demand. It is further recommended that fire fighters stop the elevator three or more floors below the fire floor, step out into the hallway, and check the general floor arrangement to get a feel for the building layout. By examining an uninvolved floor, fire fighters can become familiar with landmarks and can quickly identify the location of secondary exits, floor configurations, and potential ventilation locations. The first and second floors are generally not typical floor layouts; therefore, it is better to check an upper floor that will likely be similar to the fire floor. Many high-rise office buildings have floor layouts that resemble a maze, making it very easy for a fire fighter to become disoriented and lost. Therefore, fire fighters should constantly ask themselves, "Where am I in relation to the stairways and elevators?"

The IC must always conduct a careful risk-versus-benefit analysis before placing fire fighters in an elevator. Furthermore, there is seldom justification for placing civilians in elevators before the fire is completely extinguished and the smoke is ventilated from the building. SOPs regarding the use of elevators should be kept current, and all fire fighters should be thoroughly familiar with them. These procedures should consider the height of buildings in your jurisdiction, automatic fire suppression equipment, type of occupancy, and the safety considerations outlined in this text.

Stairway Support

How do fire fighters, air cylinders, hoses, nozzles, first aid supplies, and forcible entry tools get to the fire floor? The easiest way is to use the elevator. What if it is not safe to use the elevator? As was previously discussed, many times elevators are out of service or unsafe to use. Moving supplies and staff up 10, 20, 30, or more stories is an arduous task. If it is not properly managed, no one will reach the fire floor with the physical stamina necessary to fight the fire. Imagine returning to the apparatus for a fresh air tank from the top floor of the 1454′ (443-m), 110-story Sears Tower in Chicago. When elevators cannot be used, gaining access to the upper floors of a building will take significant time. Getting fire fighters and equipment to an interior staging area will considerably reduce lead time for supporting a high-rise fire operation.

Stairway support is a procedure that is used to move supplies to the interior staging area when using elevators is not a safe option. Fire fighters

> Stairway support places one or more fire fighters on every other floor to shuttle equipment up through the building to an interior staging area or resource floor.

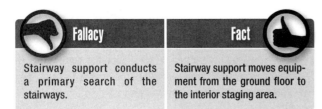

Fallacy	Fact
Stairway support conducts a primary search of the stairways.	Stairway support moves equipment from the ground floor to the interior staging area.

> Interior staging is set up in a safe area two or more floors below the fire floor when it is impractical for fire fighters to go outside at ground level to change SCBA cylinders and for rest and recuperation (REHAB).

assigned to stairway support ascend two stories with air cylinders and other equipment, where the next fire fighter picks up the equipment and relays it two additional floors. After moving the equipment up two stories, fire fighters descend two stories empty-handed, providing a rest period. During extended operations involving many companies in rescue and suppression activities, it may be necessary to place two fire fighters on every other floor or possibly recycle air cylinders to grade level for refill.

Moving equipment up through the building with the elevators out of service is a mammoth undertaking, but stairway support provides a reasonable alternative. Depending on conditions in the stairway, it may be possible to allow members who are assigned to stairway support to work without their SCBA and turnout gear. This will preserve their energy and allow them to transport more equipment faster. For example, experience indicates that wearing rubber boots while ascending the stairs is particularly fatiguing. However, always err on the side of caution. If there is any chance that members assigned to stairway support will encounter smoke or other hazardous conditions, they should be required to wear appropriate protective equipment. Stairway support should be one of the first assignments given when the fire is on the upper floors in a building without elevator service. Stairway support should be explained in the department's high-rise SOP.

Life Safety

As search-and-rescue teams proceed with a systematic search, they must provide status reports to their supervisor and mark areas that have been searched. A simple marking system is placing a chalk "X" on doors to rooms that have been searched and/or indicating that the whole floor has been searched by marking hallway doors or walls opposite the elevator. Door hangers or other marking devices are commercially available for this purpose as well. A method of indicating areas searched should be part of the department's high-rise SOP.

Many high-rise buildings lock the doors from the stairs to the hallway, further complicating search-and-rescue efforts. This was a major factor in the loss of life at the Cook County Administration Building fire in Chicago, Illinois,[3] where occupants encountered fire on the 12th floor while attempting to escape. Fleeing occupants then attempted to re-enter floors above the fire, where doors leading to the hallway were locked on the stairway side. Forcible entry tools should be carried by rescue and extinguishment teams to force entry when necessary. When selecting the tools to carry, remember that power saws might not operate because of heavy smoke conditions. Further, using gasoline-powered equipment inside of a building can create other hazards if the area is not adequately ventilated.

Once doors have been opened, it is important to prevent them from relocking behind fire fighters entering the floor. Some departments use a piece of rubber with two holes that cover the doorknob and lock to prevent the door from relocking. For most door types this method works very well. During pre-incident planning, examine the doors to determine which method would work best for the style(s) of doors in that building. However, whatever method you decide to use, be careful not to leave doors propped open, as this will most certainly allow smoke to move from the stairs into the hallway or from the fire floor into the stairs.

The primary search should also include a search of all elevators. Elevators should be brought down to the ground level and checked by lobby control. If elevators are stalled or otherwise located above the ground floor, they must be checked for victims. It is essential that all elevators be accounted for and checked for occupants.

Rescuing and Evacuating Occupants
Helicopter Rescues

There have been occasions when helicopters were successfully used to rescue occupants during high-rise fires. **Figure 12-3** shows a helicopter rescue operation in progress at the MGM Grand fire in Las Vegas. However, helicopter rescues are extremely dangerous and, in most cases, unnecessary. Few instances warrant the use of a helicopter in removing occupants from a roof, and many roofs make poor helicopter landing zones. A few fire departments have developed

Figure 12-3 MGM Grand helicopter rescues.

programs that use helicopters to place fire fighters on roofs, sometimes by having them rappel from the helicopter to the roof. Fire fighters placed on the roof can calm the occupants and protect the area from fire, and they can also descend to areas that are inaccessible from below.

These programs have merit, but require constant training on the part of the rappel team and the helicopter crews. At last count, New York City had 32 buildings over 600′ (183 m) in height. New York and other very large cities can no doubt justify the training and associated expenses related to such a program. However, the need or justification for such a program is doubtful for a department with only a few well-protected high-rise buildings within their response area. It is unlikely that members of such a department would have the expertise or equipment needed to safely operate from helicopters.

Helicopters can sometimes be used for reconnaissance. An aerial view can provide information that is not available from the interior or the exterior at ground level. At the World Trade Center on September 11, 2001,[2] police in helicopters could see signs of an impending structural collapse. Unfortunately, they were unable to communicate this message to the fire department.

Helicopter operations above a burning building can actually create additional risks owing to the thermal updraft. Helicopters flying near a burning building can create high winds that negatively affect operations. Remember, as when considering any tactic, before deciding to use helicopters to assist fire fighters or occupants or for reconnaissance, always conduct a risk-versus-benefit analysis. Is the risk to occupants, fire fighters, and the helicopter crew warranted? Or, could other less risky methods be used?

Partial or Sequential Evacuation

A decision for partial or sequential evacuation can be made in advance of a fire and incorporated into the alarm system. Alternately, the IC may make this decision at the time of the fire. When the fire is small or smoke has entered the stairway, the IC may decide that it would be best to leave occupants in their rooms rather than attempting a complete evacuation. A defend-in-place tactic places a great burden on the IC, as a successful outcome depends on the fire being promptly extinguished. However, many times occupants are placed in greater danger by attempting to evacuate them through smoke, using resources that could have been assigned to extinguishment duties.

If the fire is not quickly controlled, the people who are left in the building will be in great danger because of the non-evacuation decision. Much of this decision-making process has to do with fixed fire protection features that are designed into the building. For example, there is less risk associated with the non-evacuation decision if the building is sprinkler-protected.

Many departments permit the use of alarm systems that advise only the occupants in affected areas to evacuate the building. Rather than sounding a general alarm throughout the building, the evacuation alarm may sound only on the fire floor, one floor above, and one floor below. Occupants on the other floors may receive a pre-alarm notification advising them to remain on the floor until further notification.

> Any time a decision is made to leave occupants inside a burning building, the IC is taking a calculated risk. However, pre-incident plan information along with accurate, timely, and continuous status reports will provide the IC with the information necessary to make the correct decision.

This evacuation tactic reduces stairway traffic but also leaves occupants inside the building. Use of a partial or sequential evacuation system requires a sophisticated fire alarm system and/or manual control. The IC should have pre-incident plan information about the operation of the system readily available at the command post.

Relying on building management or security to notify occupants of the need to evacuate is generally a mistake. Maintenance or security people have a propensity to investigate alarms first, thus delaying notification of the fire department. This has been the case in many deadly high-rise fires, such as the First Interstate Bank Building in Los Angeles and One Meridian Plaza in Philadelphia. When a fire is discovered, many times in-house employees are not properly trained in how to initiate a safe, orderly evacuation. The answer lies in requiring direct and immediate fire department notification, regardless of the type of internal alarm system or method of evacuation.

Whenever occupants are left in the structure, the IC is depending on the construction features, fixed fire suppression systems, or manual suppression efforts to confine the fire and ventilate the smoke and toxic gases.

Emergency Voice/Alarm Communications System (EVACS)

Many high-rise buildings are equipped with an emergency voice/alarm communications system (EVACS). The EVACS uses recorded messages to notify occupants of a fire and provides specific directions for reaching places of safe refuge within the building. The EVACS sometimes directs people above the fire to higher floors within the structure, avoiding the need to pass the fire floor. People on the fire floor and the floor below are directed to an area two or three floors below the fire or to evacuate to the outside. When occupants are sent to other floors within the building, the occupants on the receiving floors are also notified of the evacuation. The EVACS notifies people on elevators that the elevator is responding to the ground floor, where they are to exit. By directing occupants away from the immediate fire area soon after the fire is detected, the EVACS helps to solve many high-rise evacuation problems.

Because a fire scenario presents many uncontrolled variables, caution is in order when using any automated system such as the EVACS. Stairways that would be perfectly safe if the fire were to occur in one floor area might not be safe in another circumstance. All evacuation variables should be thought out in advance, and a manual backup system must be in place. The EVACS will generally have manual overrides that allow fire fighters or building management to direct the evacuation. Manual directions could include which occupants are to evacuate and the stairway(s) to use. Most of these systems can limit notification to occupants on specific floors or notify all of the building's occupants. Sequential evacuation may be best to reduce stairway congestion. If fire companies are using a stairway to attack the fire, it is best to direct evacuating occupants to other stairways.

At the Cook County Administration Building fire in Chicago, Illinois,[3] security unintentionally notified all building occupants to evacuate; the evacuation message was intended for the fire floor only. In this case, the fire department was criticized for not using the internal communications system to notify occupants to avoid the stairway where fire operations were taking place in favor of the stairway on the opposite side of the building. However, occupants will often disregard verbal instructions, which was the case at the Cook County Administration Building fire, where many occupants ignored instructions to use the stairs and instead used the elevators. Another example is the World Trade Center, where occupants were trained to sequentially evacuate the building. When aircraft struck the buildings, many occupants self-evacuated and ignored the EVACS as well as what they had been taught during training.[2] In this instance, the decision to ignore the system proved to be the best decision, as a sequential evacuation would have resulted in more people being in the buildings when they collapsed.

Reliance on the EVACS should be limited to compartmentalized buildings with full sprinkler system coverage. If the sprinkler system fails to control the fire and fire forces are unable to quickly extinguish or confine the fire, all occupants above the fire floor should be evacuated. Further, the IC and lobby control officer should have a thorough knowledge of the EVACS.

Extinguishment

What makes high-rise firefighting different? In the lower portions of the building, the main difference is the exposure of many stories above the fire to vertical fire spread. Above the eighth floor most exterior defensive fire control tools

---

Case Summary

A New York City high-rise fire occurred in a 10-story apartment building. The fire started in an apartment on the 10th floor, and three fire fighters were killed during firefighting operations. Another fire occurred in an apartment in a 51-story high-rise building in Manhattan. Four civilians were killed in this fire. All of the fatalities were remote from the fire and occurred when the victims were overcome while trying to exit the upper stories.

Source: Robert Duval and Robert Solomon. Unpublished NFPA Fire Investigation Report. Quincy, MA: NFPA.

are no longer effective. Elevated streams can sometimes reach the fire floor, but the angle of deflection inside the window diminishes with each floor above the maximum height of the appliance. To improve fire stream penetration, it is necessary to move the nozzle further away from the building, thus reducing the angle of deflection. In most cases the width of the street is the limiting factor in determining how far away the appliance can be placed from the building. If the width of the street is not a limiting factor, the effective reach of the nozzle will determine the maximum distance the nozzle can be placed from the building. At the Cook County Administration Building fire in Chicago, Illinois, an elevated master stream was used to control a fire on the 12th floor; however, a delay occurred while the apparatus was repositioned.[3] In operating above the eighth floor, it is unlikely that ground-based appliances will deliver effective stream penetration to the fire area **Figure 12-4**. Furthermore, operating an exterior master stream into a building can push the fire into unburned areas, endangering fire fighters and civilians who are in the building. This is true in any building, but the danger is much greater in a high rise due to the size of the building and the possibility of pushing the fire into hallways or stairways and involving more than one floor. Above the eighth floor, the occupants are also beyond the reach of elevating platforms, aerial ladders, and ground ladders; therefore, attempting an exterior rescue would be extremely dangerous for both fire fighters and occupants. In fact, the building's stairways are the best means of egress, even when floors are within the reach of aerials. For these reasons, it is extremely important that fire forces prevent fire extension into the stairways, because stairways are the safest and most effective egress routes.

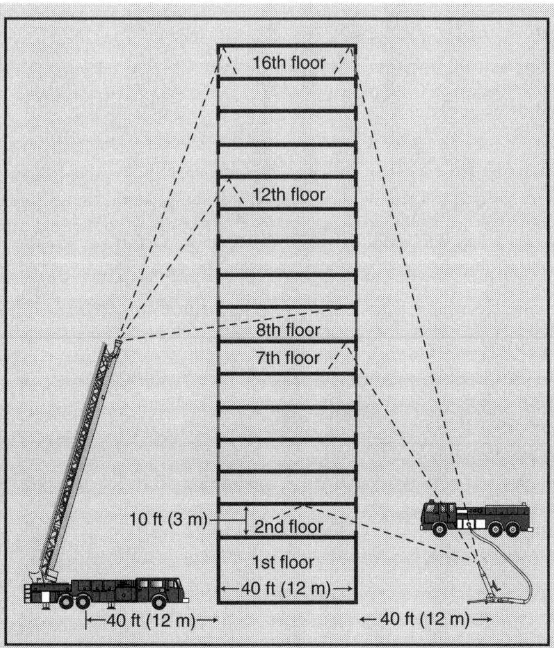

In Chapter 7, we discussed standpipe operations including pressure- or flow-reducing valves. Generally speaking, the pressure at the standpipe discharge will be lower than the pump discharge pressure used to supply hose streams connected to the pumper. Departments compensate for the lower pressure by using smooth-bore or low-pressure nozzles. Some departments have purchased hose that has less friction loss, or they use 2½″ (64-mm) hose for standpipe operations. Any method of reducing the friction loss in the hose will increase the nozzle pressure. However, 2½″ (64-mm) hose is more difficult to handle in small areas. As discussed in Chapter 8, flow requirements are lower in smaller rooms; therefore, 2½″ (64-mm) hose may not be the best choice in buildings subdivided into small compartments.

Taller buildings will tend to have pressure-reducing valves; however, not all high-rise buildings are equipped with these pressure-reducing or flow-reducing devices. If a building is not equipped with pressure-reducing valves, the pressure on the lower floors of the building will be higher than the pressure on the top floors. It is also possible to increase the pressure in the standpipe system by pumping into the fire department connection. Pumping into the fire department connection may be an effective means of boosting the pressure and increasing the water supply. Very high buildings have complex water supply systems, and pumping into the fire department connection may not increase pressure at the standpipe outlet. In these buildings, fire department pumpers alone cannot supply sufficient pressure for firefighting on upper floors, and the building's internal fire pumps must provide primary pressure.

Although the ceiling height of individual floors will vary in high-rise buildings, a fair estimate for the floor-to-ceiling height is allowing 20′ (6 m) for the first floor and 10′ (3 m) or more for each additional floor.

The loss of pressure due to gravity (elevation) is 0.434 psi per foot (3 kPa/305 mm). The pressure loss for a hose line operating at the top floor of a 40-story building with 10′ (3-m) floors would be calculated as follows:

$$\text{First floor} = 20' \ (6 \text{ m})$$
$$39 \text{ floors} \times 10' \ (3 \text{ m})/\text{floor} = 390' \ (117 \text{ m})$$
$$\text{Total elevation: } 410' \ (123 \text{ m})$$
$$410' \times 0.434 \text{ psi/ft} = 178 \text{ psi elevation loss}$$
$$(123{,}000 \text{ mm}/305 \text{ mm} \times 3 \text{ kPa} = 1210 \text{ kPa})$$

Figure 12-4 Stream deflection.

If the floor height is changed to 12′ (3.7 m), then the pressure loss would be

First floor = 20′ (6 m)
39 floors × 12′ (3.7 m)/floor = 468′ (144 m)
Total elevation: 488′ (150 m)
488′ × 0.434 psi/ft = 212 psi elevation loss
(150,000 mm/305 mm × 3 kPa = 1475 kPa)

In addition to the elevation loss noted above, friction loss in the standpipe system and the hose line, as well as nozzle pressure, must also be added. Friction loss will vary greatly depending on hose size and type, and nozzle pressure will vary from 50 to 100 psi (345 to 690 kPa) according to the type of nozzle used. However, to compensate for friction loss and nozzle pressure, it would be necessary to add an additional 75 to 150 psi (1034 to 1379 kPa). A quick look at these calculations makes it obvious that standard fire department pumpers alone could not support hose streams on the upper floors of these ultra-high-rise buildings.

Pumping into the fire department standpipe connections can assist internal fire pumps and is a good practice. If provisions are not made to provide an external backup supply and the internal pumps fail, the only protection will be the fire-resistive nature of the structure (fire enclosures). In high-rise building fires, when the fire is above the reach of aerial ladders and the building is not protected by an automatic sprinkler system, the fire can be expected to burn out all floors above the fire unless the fire can be brought under control by an offensive attack. Contingency plans for supplying the standpipe and/or reservoirs in case of breakdown, developed in cooperation with building management, could prevent a disaster. These contingency plans should be included in the pre-incident plan.

Command Post Location

Fire department SOPs often list the lobby as the preferred command post location. Many times this is true, but on occasion the lobby is a poor choice for a command post. When fire or the products of combustion threaten the lobby, the command post should be located elsewhere. In Chapter 1 it was mentioned that the larger the incident, the farther away the command post should be. The idea is to isolate the IC from disruption so that the command process can be carried out efficiently. When large numbers of people are attempting to exit through the lobby and fire fighters are gathering there in an effort to ascend toward the fire, the lobby is a poor location for the IC. The September 11, 2001, World Trade Center fire provides an example of the lobby being a poor choice for the command post. At this incident, the buildings ultimately collapsed, but even if the structures had not collapsed, the heavy traffic on the first floor of each building would have made the lobby a poor choice for the command post.

Newer high-rise buildings are sometimes equipped with a command center, which is usually near the lobby. This is often the ideal command post location, as these centers often provide good communications and the needed work space for command activities.

Once the command post has been established, its location should be communicated to all responding companies. Some department SOPs use a street name to notify companies of an exterior command post location, such as Main Street Command. A command post located inside a building may be identified using the building's name, such as Whitmore Command.

> All high-rise buildings should be pre-incident planned.

Developing Building-Specific High-Rise Pre-Incident Plans

Knowing the location of interior command rooms is just one more of the many issues that should be addressed in the high-rise pre-incident plan. A good high-rise pre-incident plan will address all of the issues covered in a standard building pre-incident plan, including occupancy type. Chapter 11 describes the importance of occupancy types, and Chapter 2 lists occupancy type as one of the major subjects to be addressed in a pre-plan. Knowing the building's use is essential to pre-planning. Most high-rise buildings are businesses, hotels, apartment buildings, or health care occupancies. Some high-rise buildings are mixed occupancies with assembly or mercantile occupancies on the first floor. In addition to the standard pre-incident planning factors, there are many special considerations in a high rise, including:

- Elevator operations and elevator key location
- Access/egress issues, such as stairway doors that automatically lock
- Standpipe operations, especially if field-adjustable, pressure-reducing valves are present
- Floor layouts for each floor that is different or special (e.g., when tenants occupy several contiguous floors, there may be internal stairs between floors)
- Ventilation, such as the use of the heating, ventilation, and air conditioning (HVAC) system; presence of

operable, tempered glass, or special windows (e.g., hurricane resistant)
- Procedures or operations that are unique to the building

High-rise buildings are generally of fire-resistive construction, with the older high-rise buildings often being superior to newer ones in many construction features Figure 12-5 . Older tower buildings have better compartmentation, more fire-resistive components, and generally better exit facilities. Newer buildings may have one significant advantage: They are often fully sprinkler protected. However, there was a period when high-rise buildings were constructed using the modern, lightweight construction methods but were not protected with automatic sprinklers. These new-style, non-sprinkler-protected high-rise buildings are likely to be most problematic.

Since the mid-1970s, codes used in the United States, with one notable exception, required high-rise buildings to be sprinkler protected. In New York City, until a change was made in March 1999, it was possible to construct a residential high-rise building without a sprinkler system. Two tragic fires led to a change in the building codes, and now all new high-rise buildings built in New York are required to have a sprinkler system installed. Also, within the past few years, many major cities such as New York and Chicago have implemented high-rise sprinkler retrofit programs for some existing high-rise buildings. For instance, the city of Chicago recently adopted an ordinance that requires high-rise buildings (above 80′ [24 m]) to be retrofitted with sprinklers. However, if residential properties and certain other high-rise buildings that are classified as historical pass a locally-developed life safety evaluation (LSE), they may opt out of the retrofit requirement.

In many new-style high-rise buildings where core construction methods are used, fire attack and evacuation tactics are further complicated. In these buildings,

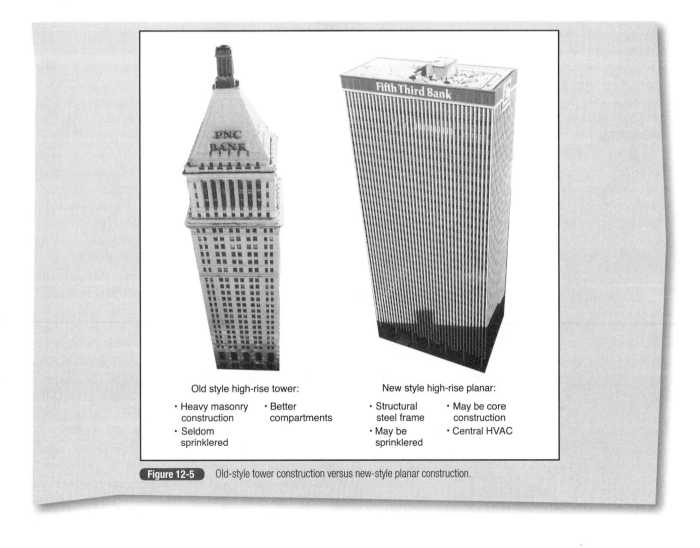

Figure 12-5 Old-style tower construction versus new-style planar construction.

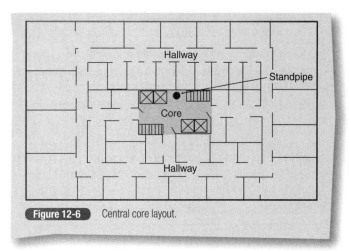

Figure 12-6 Central core layout.

stairways are usually located in the center of the building Figure 12-6. Therefore, if occupants are trapped or if the fire occurs near the core or between the occupants and the core, evacuation and fire attack positions are limited. Fire streams operated from the core may push the fire toward the victims at the periphery, creating a no-win situation. Pre-incident plan drawings will prove to be extremely valuable to the IC in developing evacuation and fire attack plans in these buildings. Some core construction methods move the core away from the center of the building, which is sometimes referred to as side core construction Figure 12-7.

Interior building configurations can be radically different in buildings that appear to be the same from the exterior. Notice that Figure 12-7 shows an open floor layout and standpipes in the stairs, whereas Figure 12-6 shows individual offices and the standpipe in the core area. These seemingly subtle differences can have a major impact on firefighting and rescue operations.

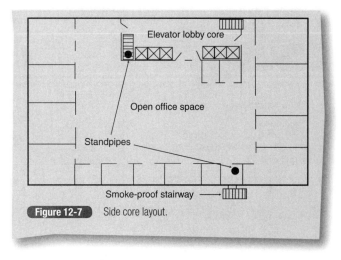

Figure 12-7 Side core layout.

The stairway at the periphery of the side core layout (Figure 12-7) is a **smoke-proof tower**. In smoke-proof construction the stairway is built as a separate structure, thus reducing the possibility of smoke entering the stairs. Also, the stairs could be pressurized to reduce smoke infiltration. Unless there is a compelling reason to use another stairway, the smoke-proof tower or the pressurized stairway should be the stairway of choice for occupant evacuation. This should be noted in the pre-incident plan.

Analyzing the Situation Through Size-Up

All of the factors discussed in Chapter 2 regarding size-up apply to high-rise fires. However, size-up at the scene of a high-rise fire will often be particularly difficult, as visual information may be nonexistent or misleading. Merely finding the fire can be a perplexing chore. A large structure can contain a substantial fire without displaying any external signs. If flame and smoke are visible, it is both a good sign and a bad sign. The good news is that the fire has self-vented, and the IC has some indication of the location and intensity of the fire. The bad news is that the fire has probably gained considerable headway before the arrival of fire fighters.

Observing a high-rise fire from the exterior could result in confusion in directing interior assignments. High-rise buildings often do not have a floor numbered 13, or the mezzanine may be counted as a floor. There may also be half-floor equipment rooms within the building. A heavy volume of fire that originated at a central core may appear at a single window, while a fire that originates near the periphery may cause several windows to break. Visible evidence gained from the exterior, many floors below the fire, may be unreliable. Occupants may report smoke or odors, but these reports typically come from occupants who are many floors above the fire. Information may be available from building occupants regarding evacuation status, but this information is seldom completely accurate in a large building. Even internal alarms are sometimes misleading, as smoke detectors sense smoke above the fire floor or in other areas where the HVAC is depositing smoke. For these reasons, even with fire and smoke showing outside the building, finding the seat of the fire may require assigning several fire companies to different areas of the building.

Matching the occupancy classification to time factors generally gives the IC a good idea of the life hazard potential. There is a tremendous life safety difference in a high-rise office building during working hours compared to the same building during non-business hours. There may be people in the building during non-business hours, but the occupant load during normal working hours will be much greater. How many people would you expect to find in an office building during normal working hours? As a rule of thumb, the average office building has one person for every 100 ft^2 (9 m^2).[4] The pre-incident plan would most certainly aid in determining the life safety problem.

The validity of dispatch information varies depending on the alarm type. A report of a smoke detector alarm on the 30th floor could be for a fire several floors below or for a localized problem on the 30th floor. A report of an odor of smoke from a building occupant may be no more reliable than the smoke alarm in determining the location and nature of the problem. However, a report of visible smoke or flames on the 30th floor in Suite 3012 gives the IC a wealth of information.

It is not unusual for the dispatch center to receive follow-up calls from the public during a high-rise fire. Many of these calls repeat and verify information already received. In many of the fires we studied, building occupants called dispatch after the initial alarm to report people trapped, request evacuation instructions or to report fire conditions in specific areas of the building. This information is critically important and should be immediately relayed to the IC.

The IC should notify the operations chief and/or divisions, groups, or companies working in these areas or deploy companies to evaluate the situation. Communications is often a problem; in some of the cases we reviewed, the IC had vital information but was unable to relay it to units working within the building.

In newer high-rise buildings of fire-resistive construction, structural stability will typically be very good. The First Interstate Bank Building and Meridian Plaza fires were both in new-style high-rise buildings, and both sustained large volumes of fire over several floors for an extended period of time. While there was structural damage, especially at the Meridian Plaza fire, neither building collapsed. Further, the old-style high-rise building can be expected to endure even more fire exposure than the more modern, lightweight construction. The Empire State Building serves as a good example of the structural stability of the older, tower-type high-rise buildings. On July 28, 1945, a bomber crashed into the Empire State Building. The aircraft's burning fuel cascaded down and through the building. The building, which sustained both the crash impact and the resulting fire, still stands as one of America's famous landmarks. Fire resistance in older high-rise buildings was achieved through massive structural members

Case Summary

In several fires the smoke has migrated to locations that are remote from the fire. At high-rise fires studied by the NFPA, victims have been killed in locations that were a considerable distance from where the fire originally started. In a hotel fire in Pattaya, Thailand, the fire started on the second floor, yet much of the smoke damage and a number of the fatalities were found on the top floors. Similar situations occurred at the MGM Grand fire, the Las Vegas Hilton fire, and a high-rise fire in 1998 at 124 West 60th Street in New York City that killed four people.

Figure CS12-3 Smoke from the fire that swept through the 36-story President Hotel in downtown Bangkok on February 23, 1997.

covered by concrete. The airplane that struck the World Trade Center was much larger, carried more fuel, and was traveling at a higher rate of speed upon impact compared to the bomber that crashed into the Empire State Building. Therefore, it is difficult to compare the two incidents. However, one of the major factors leading to collapse at the World Trade Center was aircraft debris removing sprayed-on fire-resistive coatings from steel structural members.[2] This sprayed-on insulation was not used in the construction of the older, tower-type high-rise buildings, although it is sometimes used when older buildings are renovated.

Evaluating resource needs is extremely difficult in a high-rise building. However, the staffing and resources necessary to combat a large fire are significant, as you will see in the analysis of high-rise case studies later in this chapter.

The high-rise decision tree shown in **Figure 12-8** considers a few major factors that could lead to a call for additional alarms at the scene of a high-rise fire. This decision tree by no means provides a complete size-up of the potential challenges that a high-rise fire presents. However, it does offer a thought process to assist in determining resource requirements. An examination of the decision tree demonstrates that a large fire, located beyond the reach of ground equipment and with elevators unavailable, will require extensive resources.

Some department SOPs require that the first-arriving company request an additional alarm for any working fire in a high-rise building. There is justification for this precaution, given the life hazard, extinguishment, and property conservation potential. In most instances additional personnel will be needed to augment the initial attack and to provide logistical support. The actual staffing requirements will vary depending on the circumstances, as demonstrated by the scenarios and case studies in this chapter.

Smoke Movement

Weather conditions, together with fire intensity, can have a significant effect on smoke movement within a high-rise structure. While the IC should consider these factors, a degree of unpredictability should be expected regarding the effects of weather and fire intensity on smoke movement.

Heat of the Fire

The intensity and size of the fire will determine the extent to which combustion gases are heated and how high they will rise inside the building. In lower structures there is

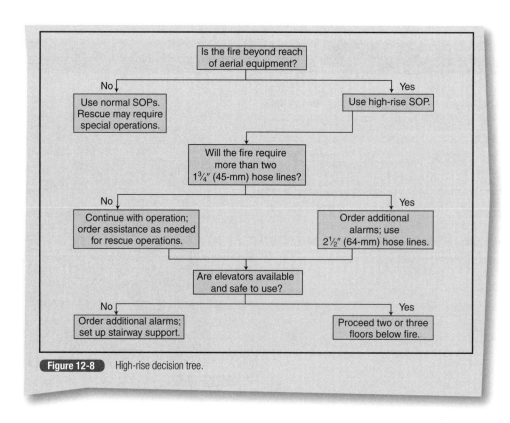

Figure 12-8 High-rise decision tree.

generally enough heat energy to cause the heated fire gases to rise to the highest level in the structure. In high-rise buildings the smoke and toxic gases will tend to rise until they reach temperature equilibrium, at which point they will stratify. It is not unusual to have heavy smoke on a mid-level floor and smoke-free floors above. This stratification can endanger occupants who enter a smoke-free stairway and then discover smoke several stories below. Many times doors leading back into a floor area are locked, forcing the fleeing occupants to seek refuge in the stairway or proceed through the smoke.

Stack Effect

The air tightness of the structure has much to do with **stack effect**—the vertical airflow caused by temperature differences within and outside the building. The unpredictable behavior of smoke within a high-rise is due, in large part, to stack effect. Many fire departments have been surprised when they tried to ventilate a high-rise building using a technique that was previously effective in that or a similar building, only to get an entirely different outcome. In some buildings the stack effect is so great that it interferes with the proper operation of the HVAC system. The colder it is outside, and the warmer it is inside, the greater the positive stack effect (upward movement). Conversely, the stack effect can be negative (downward) on a warm day within an air-conditioned building. The heat of the fire and stack effect are interdependent. The chances of smoke stratification are less on a cold day than on a warm day. There are formulas for calculating stack effect, but they have little application for field use.

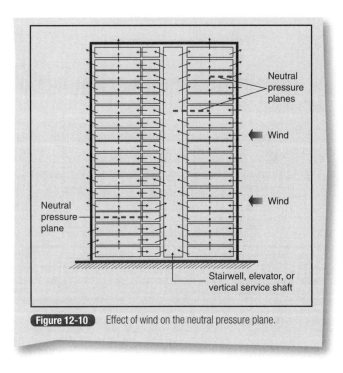

Figure 12-10 Effect of wind on the neutral pressure plane.

Wind

There is a point within a high-rise structure of sufficient height called the *neutral pressure plane (NPP)*. Below the NPP air is moving into the building; at the NPP forces are neutral (air is not moving in or out); above the NPP air is moving out of the building **Figure 12-9**. The NPP is affected by heat from the fire and the stack effect. Wind also plays a major role **Figure 12-10**.[5]

Ground-level winds are not always a good indication of wind direction and speed at higher elevations. Downtown areas of large cities containing large numbers of high-rise buildings are like giant canyons. Wind entering the high-rise canyon is redirected and becomes very turbulent. These wind gusts also prevail high above the ground but possibly in another direction and/or at a higher velocity.

Wind passing over a roof opening has a pulling or venturi effect. In addition, the wind will push smoke back

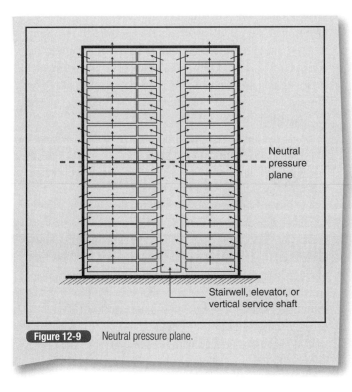

Figure 12-9 Neutral pressure plane.

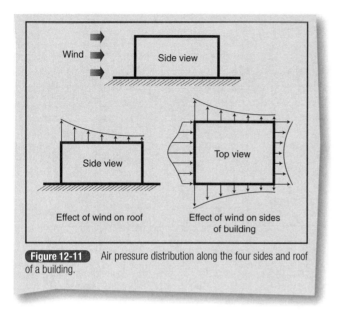

Figure 12-11 Air pressure distribution along the four sides and roof of a building.

into the building on the windward (upwind) side and help to vent smoke out of the building on the leeward (downwind) side **Figure 12-11**.

In reality, predicting the wind factor on the fire ground is more an art than a science. Wind direction and velocity can change dramatically and without warning, even when atmospheric conditions are not changing significantly.

Developing and Implementing an Incident Action Plan

Incident priorities (life safety, extinguishment, and property conservation) remain the same, regardless of the type of structure. Each type of building or occupancy presents a special set of hazards to the occupants and the property. When people and property occupy buildings that are 110 stories high, the risk to life safety and property is obvious.

> The primary rescue tactic in a high-rise building is a well-placed offensive attack.

In a high-rise, more than any other building, quickly confining and extinguishing the fire are critical, as ventilation and evacuation options are limited. Once the fire is extinguished, the toxic products of combustion are no longer being produced, and the operation becomes more manageable.

How much water flow will be needed to extinguish a well-involved fire in a large high-rise building? Compartmentation plays a major role in fire flow calculations. Some office occupancies are divided by substantial walls; others have glass panels or portable dividers separating work areas. Further, there are often large, undivided spaces within high-rise buildings. These undivided areas create a high hazard requiring large flows. In the case of the MGM Grand fire, which is described later in this chapter, the required rate of flow was beyond a reasonable flow capacity, and the size of the area made it impossible for fire streams to reach the opposite interior walls from the access points.

There have been cases in which master stream appliances were brought into upper stories of a high-rise building and operated on the interior, but few standpipes could flow a sufficient volume of water to support master streams. To compensate, 5″ (127-mm) hose was used in the building at One Meridian Plaza. However, even this tactic was unsuccessful in controlling or suppressing the fire.

Using master streams within a building would be an unusual tactic, but this could be the only realistic way to control a large-area fire on the upper levels of a high-rise building. Methods have been developed for advancing large diameter hose up interior stairways for use as a portable standpipe, but the time and staffing required to advance the hose to an upper story presents a significant challenge and becomes unrealistic at higher elevations. Master stream appliances could be supplied by fire department pumpers on the exterior, provided that the elevation pressure, friction loss, and required nozzle pressure do not exceed safe operating pressures. If master streams are operated on or into the interior, the structural integrity of the building must be closely monitored. Water from master streams could add several hundred thousand pounds of weight within the structure in a relatively short period of time. It is also easy to imagine the damage that internal master streams would do to property within the structure.

Until ventilation is accomplished, the floor area of a fire-resistive building can be very dangerous for fire fighters operating hose lines. Hose lines operated on the fire floor may push the fire. Like all other energy and matter, the fire will follow the path of least resistance. If the floor is not properly ventilated, fire, smoke, and heat will follow pathways through hallways or concealed spaces. This can be extremely dangerous in a closed high-rise building, as the fire can "wrap around," moving through these spaces and getting between fire fighters and their exit route. Fog streams exacerbate this problem and should be avoided. The floor should be ventilated as soon as possible; at the same time, fire fighters should be aware of the special hazards associated with breaking glass on the upper floors of a high-rise building. Backup lines should always be in place to protect exit routes.

Company Operations

In attacking a fire in a high-rise building, as in any structure fire, engine company and truck company operations must be coordinated. Truck company members will be needed to force entry to floors and rooms. Upon reaching the fire floor, truck company members should remove ceiling tiles to ensure that the fire is not getting behind the engine company crew and should ventilate the fire floor when possible. Suspended ceilings are common concealed spaces and a likely place for a "wrap-around" fire. This area sometimes contains cables and other equipment that can drop down, trapping fire fighters. Ventilation is crucial in a high-rise fire, though often very difficult to perform.

The attack line is used as the lifeline to safety. Members should stay within range of this protective line, not only for fire suppression purposes but also as a means of finding the stairway in heavy smoke conditions. It is also good practice to place a fire fighter at the stairway opening to the floor. This fire fighter will be needed to help extend the hose and can direct fire fighters to the exit if necessary. Some department SOPs require the use of lifelines (ropes) in these situations, which is a good safety precaution.

Use of Stairways

The occupants of an office building are generally mobile and able to escape on their own, provided that stairways are available for their use. When fire fighters are trying to advance up stairways while occupants are attempting to evacuate, the result is gridlock. What can the fire department or occupants do to alleviate the problem? Many high-rise building managers and fire officials recognize that it is not always the best policy to have all of the building's occupants in the stairways at the same time. The people on the fire floor and the floor immediately above are in the greatest danger and should be among the first groups evacuated. When the stairway is filled with people from other floors, those in greatest danger cannot safely evacuate the building.

Mass evacuations also complicate fire operations, as fire fighters are forced to "swim upstream" in a stairway full of evacuees moving downward. Furthermore, when there is a mass evacuation, occupants above the fire must pass by the fire floor to reach safety. Fire fighters tend to exacerbate this problem by blocking doors open with fire lines, thereby venting the products of combustion into the stairway. These were contributing factors to the loss of life at the Cook County Administration Building fire in Chicago, Illinois. There is no easy or, for that matter, best way to deal with this dilemma. Several ideas are presented here that may or may not work in buildings protected by your department.

First, search efforts must be systematic and include a complete primary search of the fire floor and floors above the fire. In searching above the fire it is important to check all stairways. One way to accomplish this is to enter the floor using one stairway and exit the floor using another stairway. If there are more than two stairways it may be necessary to assign a crew(s) to check the additional stairways. Some important questions that need to be answered include:

- Are people in the stairway?
- Are any stairways filled with smoke? Smoke may stratify; therefore, do not assume that the entire stairway is clear simply because it is clear on one floor.
- Are doors from the stairway to each floor level locked?

Forcible entry is often needed, so the search-and-rescue team must carry proper tools. Forcing doors not only will result in damage, but also will physically exhaust the team. Master keys supplied by building management can be invaluable in gaining access to rooms. It is necessary to check individual rooms to ensure that the occupants on any floor that is endangered by the fire or smoke have escaped.

As previously mentioned, another method of facilitating both evacuation and extinguishment is for the IC to designate separate stairways for occupant and fire department use. Fire companies must have control of the evacuation and/or good communications with occupants to successfully use this tactic. An internal public address system or EVACS can be used to direct occupants to evacuation stairways and away from the fire operations stairway. Even though occupants are directed to use stairways other than the fire operations stairs, the fire operation stairway must also be checked, as people do not always follow instructions. As was noted earlier, smoke-proof towers or pressurized stairs may be the best option for occupants. Controlling the stairs and dedicating one stairway for firefighting also reduces the possibility of opposing lines on the fire floor, as it is easier to control the entire attack if it is being made from one entry point. In most cases a stairway with standpipe outlets is the preferred fire operations stairway.

Ventilation

Ventilation and control of the HVAC system should be done by using reversible methods. Breaking windows high above grade level creates a serious hazard below. Plateglass windows tend to break into large, irregular pieces of glass, known as shards. These shards of glass create a very dangerous situation as they drop to the ground below. Tempered glass breaks into very small pellets and is less hazardous to people below. Even if windows are pulled into the building or are made of tempered glass, opening a window is much preferred to breaking a window. If a window is opened and the effect is negative, it can be closed. Unfortunately, most high-rise buildings have sealed windows, which cannot be opened. Sometimes fire codes require that a percentage of the windows be operable or made of tempered glass so that manual ventilation is possible. Codes that require these vent windows also generally require that windows designated for ventilation be marked as such. The location of tempered glass or operable windows should be noted on the pre-incident plan.

The operation of the HVAC system is usually reversible. If operating the HVAC system does not have the desired effect or spreads the smoke, immediately shut it down.

Interior Exposures

All floors above the fire are exposures. Companies need to check above the fire floor for extension and be equipped to fight the fire. In modern buildings with curtain walls fire can extend upward, inside the building near the exterior wall. Stairways and elevator shafts are vertical openings through which fire can spread. Additionally, pipe chases and utilities penetrate floors, creating openings through which fire can spread.

Fires can extend up the exterior of the building by "lapping" from floor to floor. While upward extension is the main concern, the fire can spread downward through melting expansion joints or burning materials dropping below via the HVAC system or by other means. Therefore, areas below the fire must also be checked. If sufficient numbers of fire fighters are available, the search-and-rescue team can check floors above the fire while fire fighters who are assigned to property conservation can check floors below the fire.

Property Conservation

As was previously stated, the first priorities are life safety and extinguishment. However, property conservation should also be considered early in the incident. High-rise office buildings typically contain valuable contents, such as computers and other office equipment and important files and documents. Also, high-rise residential buildings will contain personal items that cannot be replaced. Often, protecting these valuables is overlooked, owing to the labor-intensive nature of rescue and extinguishment challenges in a high-rise building fire.

When considering property conservation, the height of the building comes into play in an opposite way. In life safety and extinguishment, people and property on and above the fire floor are normally considered to be most at risk, because fire, smoke, and heat will first affect these floors. However, the greatest property conservation exposure is often downward as water flows through curtain walls, electrical fixtures, and other openings, damaging valuable property beneath the fire. In addition to moving and covering valuables, fire fighters should channel the flow of water down stairs or through drains. Like life safety and extinguishment, property conservation operations must be pre-planned and well executed to reduce the loss.

Lead Time

Experienced ICs understand that when an assignment is given, it takes time to see results. As an example, if a portable master stream appliance is to be used on the 10th floor it could take a considerable amount of time before fire fighters and equipment are in position to operate the master stream. The level where the fire occurs has the greatest impact on lead time, especially when elevators are not available. Table 12-2 displays the results of a training program conducted at a 48-story building. The results clearly demonstrate an increased lead time related to the height of the building. For purposes of this study, fire recruits and instructors were in full turnout gear, but not on air. Each five-person company (four recruits and an instructor) was assigned 150' (46-m) of 1¾" (44-mm) hose, nozzles, extra SCBA cylinders, and hand tools. The average time to walk up one floor for the first 10 floors was 20.8 seconds or a total of 3:07 to walk up to the 10th

TABLE 12-2 Climbing Stairs to the Top of a 48-Story Building

Floors	Average Time per Floor in Seconds
1–10	20.8
11–20	27.8
21–30	33.6
31–40	45.9
41–48	59.0

floor level. The last eight floors took 59.0 seconds per floor. On average, it took 28 minutes and 52 seconds to reach the top floor of this building. This does not include time needed to connect to the standpipe, place the SCBA in service, and advance on the fire. This building is less than half the height of some high-rise buildings, and the participants were in better-than-average physical condition. Our research indicates that fire fighters at some incidents were unable to reach the fire floor or needed to stop and rest during the ascent.

If the IC waits to give assignments until the need is obvious, he or she has waited too long. The IC must be able to anticipate needs and assign resources in advance.

Establishing a Wide Fire Zone

The area immediately adjacent to the exterior of a high-rise building fire is an unsafe place, particularly if glass is falling. A perimeter should be established to keep civilians out of danger and to provide a safe working area for fire fighters outside the building. Fire fighters who must work within the perimeter should be kept within vehicles or under protective structures. The importance of establishing zones changes with each incident. If there is danger from falling glass, a 200′ (61-m) perimeter should be enforced around the building. As was mentioned previously, plate glass falls to the street in large shards and at times will float a considerable distance from the base of the fire building. Opening windows by pulling the glass inward greatly increases safety at high-rise building fires. This technique has been tried with limited success, using tape, suction cups, and other methods. When there is a working fire in a high-rise building, fire fighters may be forced to break windows on the upper floors or the fire may self-vent without warning. Therefore, the IC should expect glass to fall at any time and direct that a safe perimeter be established early in the incident.

Applying NIMS to a High-Rise Fire

NIMS should be used at every structure fire, and a high-rise building fire is certainly no exception. Proper use of NIMS is a test of the department's training, pre-incident planning, and discipline.

A look at the logistical requirements and multiple staging areas makes it apparent that the span of control can be quickly exceeded at a major high-rise fire. In the case of a high-rise fire, the IC is well advised to hand off the operations and/or planning sections early on in the incident so that other priorities can be addressed. Record keeping is essential. If the IC has an assigned aide, this person usually acts as the initial planning officer and records situation reports and resource status and assignments. Units operating on the fire floor and staging area can be managed by a forward rescue/suppression branch director or operations section chief who documents which companies are in staging, in REHAB, and working on the fire floor.

Communications

Communication and accountability are simplified by using NIMS. Companies working on the fire floor may change; however, the geographic designation remains the same. For example, if Truck 1 and Engine 2 are assigned to the 15th floor for search and rescue, they would be identified as Division 15. If Truck 1 and Engine 2 are later reassigned to REHAB and replaced by Truck 3 and Engine 4, these companies now become Division 15. Remember that "division" is the NIMS term used to identify a geographic location or assignment.

Suppose a fire is on the 10th floor of a 20-story building. A rescue and evacuation group is doing the primary search on floors 12 to 20. Fire attack units are on the 10th and 11th floors. **Figure 12-12** is a NIMS chart and communications network for this operation. Communications for this fire would be as follows:

- The rescue and evacuation group would be communicating with the operations section as well as with companies in the rescue and evacuation group as to the status of various floors.
- The Division 10 supervisor (fire floor) would be communicating both with the operations section and with companies assigned to Division 10.
- The Division 11 supervisor would be doing much the same as Division 10.
- Companies assigned to stairway support would be communicating with each other and with logistics.
- Operations would be communicating with Divisions 10 and 11, the rescue and evacuation group, and the IC.
- The staging officer would be communicating with the operations section, which would then order the necessary staffing, equipment, and supplies for the staging area.
- Safety as well as the operations and logistics sections would be communicating with the IC.
- If staffed, the command staff positions of liaison and information, as well as the planning and administration/finance sections would most likely be at the command post location, communicating face to face with the IC.

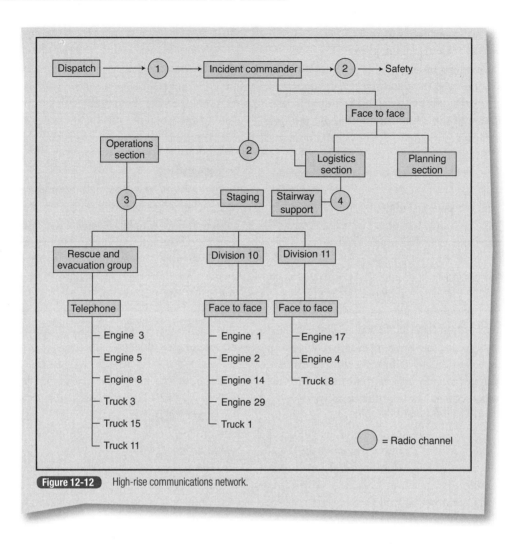

Figure 12-12 High-rise communications network.

Communications is the most frequently cited problem during major emergency operations. Communications can be very complicated within a high-rise building. If separate radio frequencies or hardwire communications are not provided at the incident scene, a breakdown in communications may occur.

Radio discipline *must* be maintained and alternative methods should be arranged (e.g., messenger, hardwire telephones). Note that the rescue and evacuation group in Figure 12-12 is communicating via telephone. Hardwire telephones are abundant in most high-rise buildings; consider using them as a means of emergency communications. In the Figure 12-12 example, the search-and-rescue group supervisor could be located in a room off the lobby, on a lower floor, or in another building where two or more telephones are readily available. All units working in this group would be given the contact numbers for the group supervisor, which would be used to call in status reports and receive new assignments. Alternate communications must also be available because the fire could damage the telephone system. Complicating the communications problem at a high-rise building is the fact that using radios is difficult within many structures. In fact, it is often impossible to receive radio transmissions from inside a structural steel building.

Tactical Worksheets

Fire departments have devised a wide assortment of tactical worksheets to be used in documenting and tracking activities on the fire scene, including some that are specifically designed for high-rise fires. Most worksheets contain checklists to remind the IC of important functions.

A good tactical worksheet will include sections to record the fire location and the location of companies at the incident scene. During high-rise fires companies will be working at different levels; therefore, the tactical worksheet should provide a generic sketch of a high-rise building that can be used to record activities on each floor.

Occupants on the fire floor and above are normally in the greatest danger, but it is necessary to search below the fire as well. A search-and-rescue group normally works above the fire to conduct a primary search and to assist in evacuation. This group is generally very active, moving from floor to floor. Therefore, it is very important for the group leader to keep the IC (or operations chief) current on the group's location in the building and the status of the primary search.

The high-rise worksheet in **Figure 12-13** lists various high-rise positions described in this chapter and provides space to enter companies working on each level.

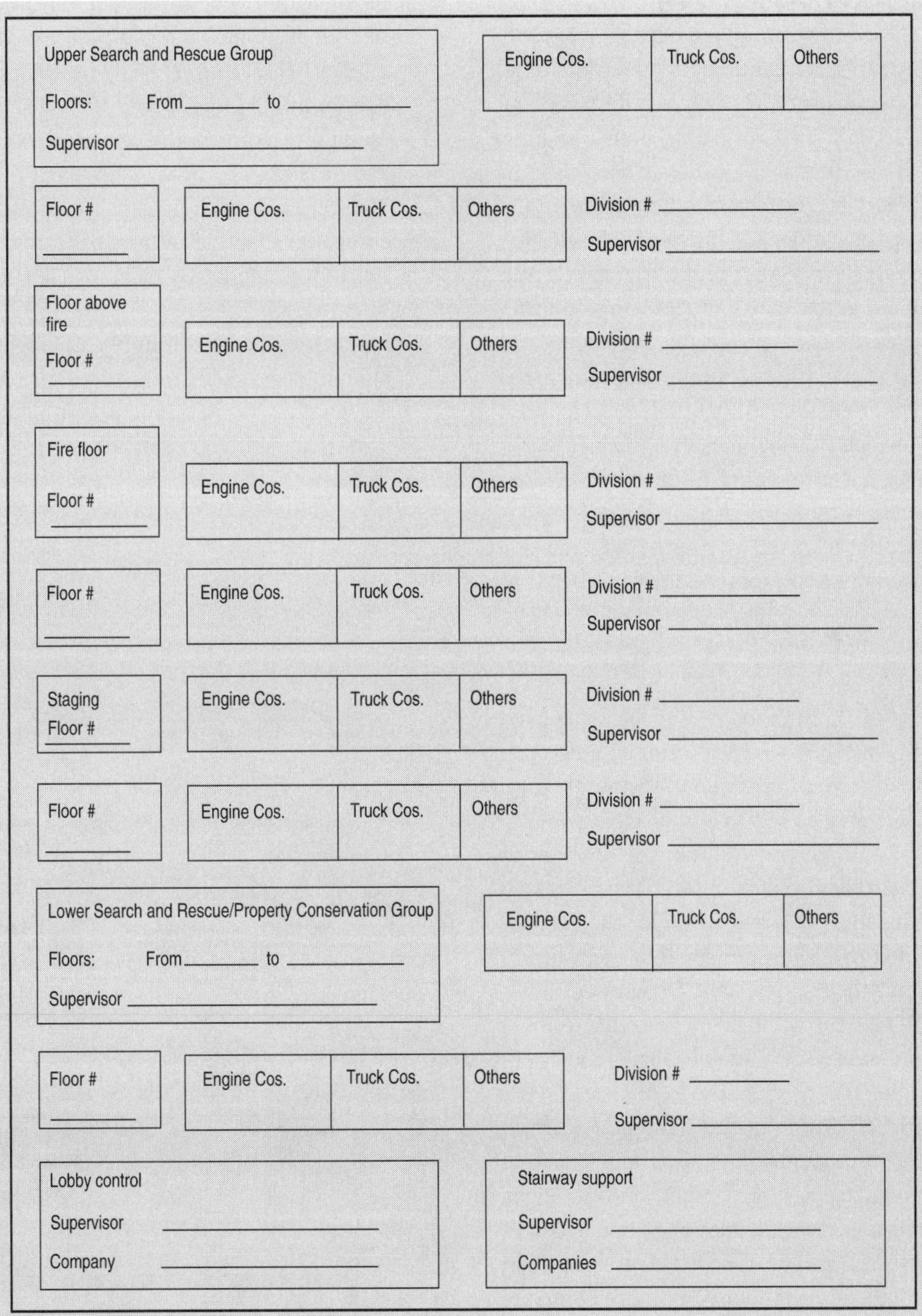

Figure 12-13 High-rise tactical worksheet.

The value of an operations section aide is clear. The operations section chief needs to know the location of all companies operating in the building but does not always have time to keep the worksheet up to date. An aide, acting in a planning capacity, can assist in tracking and recording this information.

Base (Exterior Staging for High-Rise Fires)
The term "base" is used by wildland fire fighters to identify a location housing reserve equipment and personnel (base camp). This term is not normally used for structural firefighting but has utility in managing high-rise fires. The base for high-rise fires is a location where support equipment and personnel are kept on the *exterior* of the building. Depending on interior conditions and the location of the fire, an exterior staging area may be established near the fire perimeter, but is generally limited to fires on lower floors. The reason for this distinction is that the staging area is normally moved inside the structure during a high-rise fire. Unless there is a possibility of moving to an exterior operation or the fire is involving other structures, it would be unusual to amass a large force on the exterior at a high-rise fire when the fire is on an upper floor. While "base" is the commonly used term for the exterior staging area at a high-rise fire, it would also be appropriate to refer to it as "exterior staging."

Staging (Interior)
In a high-rise situation, most of the reserve force is moved through the lobby and then to the interior staging area. This area is normally two or more floors below the fire. A rehabilitation area may be set up at the staging area or on another floor. The concept is the same as in exterior staging: to provide a readily available reserve force. To avoid the confusion of having a different name for the exterior staging area, the interior staging area could be identified as interior staging or as the resource area. Department SOPs should address this issue, using either the suggested terms of "base" and "staging" or other terminology that is consistent throughout the region.

The supervisor in charge of the staging area or an assistant can act as the accountability officer for crews that are working out of the staging area. Tracking crew rotations can be a challenging task. If an extended operation becomes necessary, three fire fighters will be needed for each position on a hose line. For example, a four-person fire company operating a 2½″ (64-mm) hose line on the fire floor will exhaust its air supply in approximately 15 minutes. To continually operate the hose line on the fire floor, this crew will need to be relieved by an available crew in staging. As the crew leaves the fire floor, it will move to REHAB, and another crew will move to the ready position **Figure 12-14**.

The person in charge of the interior staging area reports to the operations section but also communicates with the logistics section. The duties of the staging area officer are as follows:

- Report to the operations section if it is staffed. Otherwise, interior staging would communicate directly with the IC. The IC could elect to have the interior staging officer communicate directly with the logistics section.
- Maintain records of the companies that are in staging and REHAB.
- Maintain a minimum reserve of engine and truck company personnel as established by the IC.
- Request additional resources to maintain the established reserve force.
- Maintain an adequate supply of air cylinders and other equipment as needed.
- Supply first aid equipment and medical services for units that are involved in rescue and suppression.

Figure 12-14 Fire companies rotating from REHAB.

The concept of an interior staging area should not be reserved for high-rise fires. Any time an extensive interior attack is in progress where fire fighters expend more than one cylinder of air, staging and rehabilitation may be needed.

Lobby Control

The primary accountability officer may be part of the lobby control crew or co-located in the lobby. Lobby control is established regardless of whether the elevators are going to be used. The duties of lobby control include the following:

- Controlling, operating, and accounting for all elevators
- Assisting in incident command post operations
- Locating and controlling all interior stairs
- Directing incoming companies to the proper elevator or stairway
- Consulting with the building engineer
- Controlling or shutting down the HVAC system after consulting with the IC

Controlling the elevators and stairways is the only way to effectively gain access to the upper floors of a high-rise building. Many department SOPs assign these initial lobby control tasks to a member of the first-arriving truck company. Another important lobby-control duty is to gain control of the HVAC system. If operated properly, the HVAC system can prevent heat, smoke, and toxic gases from reaching occupants. Conversely, if it is operated incorrectly, it may spread the products of combustion well beyond the immediate fire area and into occupied areas.

In some buildings it is virtually impossible to know what is going to happen when the HVAC system is operated because of the many unknown variables, such as heat produced by the fire, stack effect, and wind. Furthermore, few building engineers are completely knowledgeable about the operation of the HVAC system. For these reasons many fire departments require an emergency shutdown switch near the lobby.

High-Rise Case Histories

The One Meridian Plaza Fire

The One Meridian Plaza fire in Philadelphia, Pennsylvania, occurred at 8:40 PM on Saturday evening, February 23, 1991 **Figure 12-15**.[6,7] Because the fire occurred outside normal working hours, the threat to occupants was minimal. However, this fire claimed the lives of three fire fighters.

Fire fighters provided a considerable occupant load in the fire building as the operation resulted in 316 fire fighters being called to the scene. The building was in total darkness and without elevator service.

Figure 12-15 One Meridian Plaza, Philadelphia, Pennsylvania.

Perhaps the One Meridian Plaza fire would not have gained the same headway had the building been occupied, because the fire probably would have been detected earlier. The reports do not provide the total number of occupants in the building during normal working hours, but the worst-case scenario was alarming. Each floor had 17,000 ft^2 (1579 m^2) of usable space. According to the 100-ft^2-per-person (9-m^2-per-person) rule from the *NFPA Life Safety Code*, if the building had been fully occupied, there could have been 170 people per floor.

The fire began on the 22nd floor and dropped down to the 21st floor. The top (38th) floor is a mechanical floor, meaning that the occupants on floors 21 to 37 (17 floors) would have been in immediate danger had the building been fully occupied. At 170 people per floor, 17 floors could be occupied by up to 2890 people. The eight floors that were destroyed could have held 1360 people.

The One Meridian Plaza fire most certainly points out the potential for a large-loss-of-life fire in a non-sprinkler-protected high-rise building. The building was being retrofitted with sprinklers at the time of the fire. Floors 22 to 29 were not sprinkler protected and were lost to the

Figure 12-16 First Interstate Bank Building, Los Angeles, California.

fire. It is interesting to note that the fire was suppressed on the 30th floor by ten sprinkler heads operating at the points of penetration. Can the value of sprinkler protection in high-rise buildings be underestimated? This fire serves as a testimonial to the value of automatic sprinkler systems in high-rise buildings.

The First Interstate Bank Building Fire

On May 4, 1988, at 10:25 PM, a fire similar to the One Meridian Plaza fire occurred in Los Angeles, California, at the 62-story First Interstate Bank Building Figure 12-16 .[8,9] At this fire four floors were completely destroyed, and a fifth floor was heavily damaged. The building was unoccupied except for security and maintenance personnel. The First Interstate fire burned floors 12 to 15 and was stopped at the 16th floor. Again, to consider the potential loss of life, we can look to the *NFPA Life Safety Code*. According to the *Life Safety Code*, the maximum number of occupants permitted per floor would have been 175. The fire was on the 12th floor of a 62-story building. Therefore, 51 floors were exposed to the fire and smoke conditions. This yields a potential occupant load of 8925 people above the fire. In this case the actual number of building occupants was listed as 4000. If the occupant load is averaged equally among the floors, as many as 3290 people could have been above the fire during normal working hours.

Comparing the One Meridian Plaza and First Interstate Bank Building Fires

Examining similarities between these two fires enhances the learning experience that would be gained by studying a single fire. Table 12-3 compares the two incidents.

When studying case histories you should note the effect of delayed alarms. A 16-minute delay occurred at the First Interstate Bank Building fire. Had the fire department been notified 16 minutes earlier, there is a high probability that the fire could have been extinguished without exposing fire fighters to prolonged, extreme danger, and with much less property loss and damage to the structure. Some would speculate that this delay is unlikely when the building is fully occupied, but Francis L. Brannigan, in one of his monthly *Fire Engineering* articles, took exception to this theory. Someone suggested that during normal working hours the employees at the First Interstate Bank Building would use the occupant-use hose line to extinguish the fire. Brannigan states:

> I think this answer is unrealistic. On a trading floor, operatives are dealing in multi-million-dollar split-

TABLE 12-3 Similarities Between the One Meridian Plaza and First Interstate Bank Fires

	One Meridian Plaza	First Interstate Bank
Deaths	3 fire fighters	1 maintenance person
Occupancy	Office building	Office building
Time of day	8:30 PM, unoccupied	10:25 PM, 40 occupants
Alarm	Delayed, 4 minutes+	Delayed, 16 minutes+
Sprinklers	Partial, being retrofitted	Partial, being retrofitted
Floor of origin	22	12
Area of involvement	8 floors, 17,000 ft^2/floor (1579 m^2/floor)	5 floors, 17,500 ft^2/floor (1626 m^2/floor)
Responding fire fighters	316	383

second transactions. I think it would be more like the Bradford Stadium fire in England, where the fans kept alternately watching the fire and the game until they died. In addition, to the typical computer nerd, the thought of water on his computer is anathema.[10]

There are many documented large-loss fires in occupied buildings in which delayed alarms played a significant role. In hotel and office occupancies, security or maintenance personnel may first attempt to investigate the fire before calling the fire department. Sometimes, they become involved in firefighting operations with occupant-use hose lines or fire extinguishers and delay calling the fire department until the fire is out of control.

A fire that occurred on January 6, 1995, in North York, Ontario, verifies Brannigan's assertion that delayed alarms should be anticipated in occupied buildings.[11] This fire also demonstrates that multiple-fatality high-rise fires are not limited to office and hotel occupancies. This fire, which occurred in a 29-story apartment building, killed six occupants. The occupant in the apartment of origin first attempted to extinguish a couch cushion fire with water. Other occupants smelled smoke and investigated, then assisted in trying to extinguish the fire. Finally, one of the occupants called 9-1-1. The door to the apartment of origin was left open, allowing the fire to spread into the hallway and extend into one of the two stairways located at the central core.

The high-rise apartment building fire discussed above was not a unique incident. According to Dr. John Hall, NFPA Assistant Vice President, between 1989–1998 high-rise apartment buildings and hotels averaged 10,500 reported structure fires a year, or 86% of the combination of high-rise apartment buildings, hotels, office buildings, and hospitals. Also, during this same reporting period, high-rise apartment buildings and hotels averaged 51 reported civilian fire deaths a year, or 96% of the combination of the four types of high-rise properties. The vast majority of these residential high-rise fires and deaths are specifically in apartment buildings. Also, the percentage of reported high-rise fires in buildings with sprinklers in 1998, the latest year in the range (and the one where the percentages were highest) was 36% for apartment buildings (compared to only 5% for apartment buildings that were not high-rise), 77% for hotels, 80% for hospitals, and 63% for office buildings. It is pretty clear that the high-rise buildings that present the largest problem are apartment buildings, where—no coincidence—the usage of sprinklers lags far behind the usage in other high-rise buildings.

Experienced fire officers have speculated that a working fire in a large, high-rise building during periods of full occupancy could require the services of 500 or more fire fighters. The fires in Philadelphia and Los Angeles make this estimate appear low, given that both occurred when the building was occupied by only a small staff of security and maintenance personnel.

Fire fighters in Los Angeles successfully mounted a manual attack on the 16th floor, whereas the One Meridian Plaza fire was not contained until it reached a sprinkler-protected floor. The difference was the standpipe system. At One Meridian Plaza pressure-reducing valves reduced the water pressure below the operating pressures needed to properly supply the automatic nozzles being used by the fire department. The codes governing standpipe installations required reducing valves to avoid overpressurization. The flow-reduction valves that were used to meet code requirements were incorrectly set at a pressure below the required pressure and could not be changed at the time of the fire. Manufacturers now provide nozzles that will effectively operate at lower pressures. Smooth-bore nozzles are another option when faced with low pressures.

The Peachtree Plaza Fire

On June 30, 1989, a much smaller fire in a 10-story office building in Atlanta, Georgia, occurred at 10:30 AM on the sixth floor while the building was occupied **Figure 12-17**.[12,13] Five of the 40 occupants on the sixth floor were killed, and at least six others where rescued by using aerial ladders. Eleven of forty occupants (28% of the occupants on the sixth floor) probably would have perished if the fire floor had been beyond the reach of aerials. Additional occupants were evacuated from other floors by using the interior stairways. Many of the occupants were able to survive by breaking windows and

Figure 12-17 Peachtree Plaza.

waiting for exterior rescue. If this fire had occurred above the reach of aerials, a fatality rate of at least 25% for the involved floor is a reasonable assumption. Translate that to the 170 and 175 people per floor in One Meridian Plaza and the First Interstate Bank Building, and the result would have been a major catastrophe.

Cook County Administration Building Fire

On October 17, 2003, a fire on the 12th floor of the 37-story Cook County Administration Building in Chicago, Illinois resulted in six civilian fatalities.[3,14] The fire was believed to have originated in a closet within a 2629 ft^2 (244 m^2) suite of offices on the east side of the 12th floor **Figure 12-18**.[15] Actual fire damage was contained to the office suite due to fire-resistive building features, but smoke and fire migrated to the entire 12th floor. The fire self-vented through eight exterior windows on the east side of the suite, thus creating the potential for floor-to-floor fire extension via the exterior **Figure 12-19**.

This fire was first noticed at approximately 5:00 PM; fire reports did not mention the total number of people in the building at the time of the fire. However, it is likely that some workers had left for the day, leaving the building below full occupancy. Building occupants were notified to evacuate the building via the stairways. However, some occupants used the elevators to escape, while others used the east and west stairways to evacuate. The fire department advanced a hose line into the 12th floor via the east stairway. When the door to the 12th floor was opened to advance the hose line into the fire area, smoke levels in the east stairway increased dramatically. As far as can be determined, there were thirteen occupants above the 12th floor in the east stairway when the fire department began attacking the fire. Doors from the hallways on each floor were locked on the stairway side trapping these occupants in the stairway above the fire. Seven of the occupants in the stairway found a door that did not completely latch on the 27th floor and were able to leave the east stairway and escape. Six others perished in the stairway.

Fire fighters were faced with an intense fire that they were unable to extinguish from their hallway position. Elevated master streams were used to knock down the fire from the exterior. Interior hose streams were then redeployed to achieve final extinguishment. Tight compartmentation and closed doors contained the fire to the suite where it originated.

Using exterior streams to attack a fire in a high-rise building is seldom recommended even on lower floors of

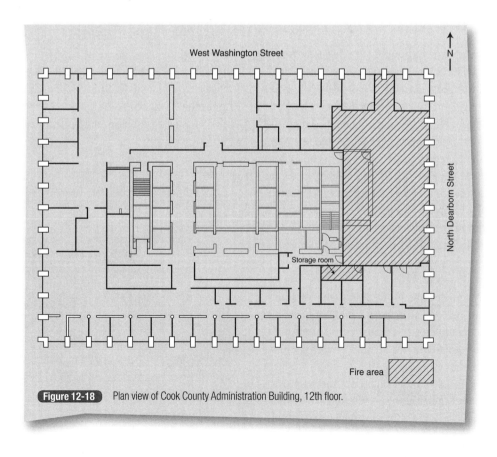

Figure 12-18 Plan view of Cook County Administration Building, 12th floor.

Figure 12-19 East side of 12th floor, Cook County Administration Building.

the building. If fire separations are left open or the compartmentation is compromised, exterior streams can rapidly spread the fire and threaten occupants and fire fighters.

At a 1994 fire on the ninth floor of the Regis Towers apartments in Memphis, Tennessee,[16] heavy smoke and fire were encountered on the ninth floor. Fire companies made several attempts to rescue occupants and extinguish the fire from the interior. When the fire self-vented a heavy volume of fire was visible from the exterior. An elevated master stream was operated into the apartment of origin to knock down the visible fire. In this case, the door between the apartment of origin and hallway was open. The exterior stream pushed the fire into the ninth floor corridor where a company was advancing toward the fire. Heat and smoke conditions immediately became worse on the ninth floor, forcing fire fighters to retreat. Two fire fighters who had been attacking the fire became disoriented and had to be rescued. Two fire fighters and two civilians perished at the Regis Tower fire.

Terrorist Attacks at the World Trade Center

The two largest high-rise incidents to date were the result of terrorist attacks on the World Trade Center in New York City in 1993 and 2001 **Figure 12-20**.

World Trade Center Bombing: February 26, 1993

The first attack involved a truck bomb.[17,18] Although this incident was the result of a terrorist's bomb, fire and smoke were the major concerns. A quote from Chief Anthony L. Fusco after the first attack identifies the fire danger and reinforces the contention made throughout this book: the most important life safety measure is often extinguishment. In his remarks, Chief Fusco stated, "The decision to attack the basement fires in the initial stages of the incident was the most important decision of the incident [commander] in that hundreds, maybe thousands, of lives were saved owing to timely extinguishment of the fires."

The first World Trade Center attack should have removed any doubts about the potential for death and injury in high-rise building fires. The World Trade Center was a seven-building complex that included two 110-story towers and a 22-floor hotel (the Vista Hotel), among other structures. The area of each floor was 40,000 ft^2 (3716 m^2). It was estimated at the time of the truck bomb incident that 60,000 people were working in the buildings, and there were also thousands of visitors. Six deaths and 1042 injuries were reported at this incident, and 50,000 people were evacuated from the building.

This incident was the equivalent of simultaneous multiple-alarm incidents. Five alarms were sounded for the Vista Hotel

Figure 12-20 New York City skyline view including the World Trade Center.

operation, five alarms were sounded for the Tower 1 operation, and four alarms were sounded for the Tower 2 operation. Another alarm was sounded for additional resources. Forty-five percent of the department's on-duty resources were required at this incident. In addition, a large contingent of police officers responded, and 174 ambulances treated and transported the injured. Few fire departments in the world have comparable resources. In fact, very few fire departments could provide the staffing needed to search the 99 elevators in each tower.

The World Trade Center was sprinkler protected, but the blast destroyed the sprinkler piping in the garage area where the fire occurred. With the sprinkler system out of service in the garage, the fire continued to burn and push smoke to the floors above. Once the fire was extinguished, the smoke production stopped. However, the smoke that had already entered the building lingered. As was previously mentioned, venting a high-rise is always problematic and takes considerable time.

World Trade Center Attack: September 11, 2001

The second attack on the World Trade Center resulted in the largest number of civilian and fire fighter fatalities ever recorded in a building fire. Lessons are still being learned from this disaster. Although aircraft flying into buildings is not a typical high-rise fire scenario, it occurs frequently enough to deserve consideration. Small aircraft have flown into high-rise buildings on several occasions. A B-25 bomber flew into the Empire State Building in 1945. World Trade Center building designers had actually planned for an accidental aircraft accident involving a Boeing 707 aircraft making a landing approach. World Trade Center structures were designed to withstand the impact and fire from this slightly lighter aircraft assumed to be low on fuel and traveling at approximately 180 miles per hour. This "planned for" accident would produce much less impact and fire than the Boeing 767s that were nearly full of fuel and traveling at an estimated 470 to 590 miles per hour when they were purposely flown into the buildings on September 11, 2001.

The Boeing 767 airplanes caused considerable structural damage on impact, but the two towers withstood these initial impact forces and remained standing. However, the aircraft debris moving through the building removed sprayed-on fire resistive coatings that protected steel structural components. The ensuing fire weakened the unprotected steel, resulting in the collapse of Towers 1 and 2, which killed 2749 people, including 340 fire fighters. The National Institute of Standards and Technology (NIST) theorizes that the two buildings would not have collapsed had the thermal insulation remained in place.[2] It is estimated that Buildings 1 and 2 at the World Trade Center were 33% to 50% occupied at the time of the attack. Had they been fully occupied, NIST estimates that it could have taken as long as three hours to evacuate the buildings, and as many as 14,000 people could have lost their lives.

Prior to the collapse of the two World Trade Center buildings, high-rise buildings had not been known to collapse even when subjected to intense fires on multiple floors. The Meridian Plaza and First Interstate Bank fires discussed earlier in this chapter both withstood severe fires on multiple floors without collapsing.

The MGM Grand Fire

During the MGM Grand (now Bally's) fire in Las Vegas, Nevada Figure 12-21, occupants displayed what we now call convergence cluster behavior.[19,20] They gathered in certain rooms as groups, gaining a feeling of safety in the presence

Figure 12-21 MGM Grand fire, Las Vegas, Nevada. A. Entire building. B. Close-up of people at windows.

of others. What implications does this have for the fire department? Searching fire fighters might not find anyone in several rooms or in an entire floor area, while one room may contain far more victims than anticipated.

The occupancy type has a direct bearing on the behavior and awareness level of occupants as well as the population density. The MGM Grand had 3400 registered guests at the time of the fire. The death and injury toll was due to smoke infiltrating the 21-story tower from a fire in the first-floor casino. The death toll of 85 people made this the second most deadly hotel fire in U.S. history (119 died in the Winecoff Hotel in Atlanta, Georgia, in 1946).

The fire in the MGM Grand started at the rear of the casino and eventually involved the entire 150′ × 450′ (46- × 137-m) casino floor. The ceiling height was not given in the reports, but the ceilings were probably a minimum of 20′ (6 m) high. Using the V/100 formula for 20′ (6-m) ceilings yields a rate of flow of 13,500 GPM (852 L/sec). The size of the fire area and the required rate of flow led to the conclusion that this fire was not going to be extinguished by manual means, and there were no sprinklers in the fire area. Therefore, the best tactic was to keep the fire out of the stairways and towers, and this was accomplished. However, smoke did spread throughout the structure, killing 85 people.

Summary

Recent fires at the World Trade Center and Cook County Administration Building demonstrate that the threat of a major high-rise fire continues. The chance of a serious fire is much greater in the non-sprinkler-protected, modern high-rise. Older high-rise buildings are constructed using massive structural members to support the structure and to provide fire barriers within the structure. Most high-rise buildings constructed in the 1970s and after are sprinkler protected. Buildings that are of newer lightweight construction that are not sprinkler protected present the greatest hazard. High-rise buildings have become a favorite target for terrorists. Previous to the attack on the World Trade Center on September 11, 2001, high-rise buildings had not been known to collapse.

A working fire in a high-rise building can threaten thousands of occupants and the most effective way to extinguish a high-rise fire is by mounting an offensive attack. In the vast majority of cases the best life safety tactic is extinguishing the fire. However, attacking the fire from the exterior places fire fighters and occupants who are still inside the building in great peril. In most cases, using fire-resistive structural features to contain the fire is preferable to an exterior attack.

Fires on the upper floors of a high-rise building will often be beyond the reach of aerial devices; therefore, occupant evacuation is usually limited to the interior stairs that can quickly become overcrowded. Sequential or partial evacuation can relieve the congestion in the stairway, but occupants who are not evacuated remain in a potentially hazardous location.

Fire fighters ascending the stairways can block occupants attempting to evacuate, and fire fighters opening a door from the stairway to the fire floor allow smoke to enter the stairway. These problems can be lessened by assigning a stairway exclusively for fire department use. However, a stairway dedicated to fire department operations decreases the building's total egress capacity.

As the level of the fire floor increases, reliance on the standpipe system also increases. Open-layout floor plans can challenge the standpipe's flow capacity and fire department resources. Getting resources to the places they are needed is complicated and, if the elevator is not safe to use, the lead time dramatically increases for a fire on an upper floor.

Wrap-Up

Key Terms

lobby control A high-rise assignment in which a crew is responsible for duties related to managing the stairways, elevators, and the HVAC systems.

smoke-proof tower A stairway designed to be separated from the building by a landing. This creates a separation that will limit the spread of smoke into the stairway and keep it clear for evacuation.

stack effect The vertical airflow within buildings caused by temperature differences between the building interior and the exterior; depending on conditions, stack effect could be positive or negative, causing smoke to move upward or downward.

Suggested Activities

1. Obtain an analysis of a recent high-rise fire. Critically review the operation in terms of manual fire suppression with emphasis on the following topics:
   A. Establishment of a command system (NIMS)
      i. Geographic and functional delegation
      ii. Command post location
   B. Identification and use of special high-rise components of the IMS
      i. Base
      ii. Lobby control
      iii. Staging
      iv. Stairway support
   C. Tactical priorities
      i. Determine what specific precautions were taken to enhance fire fighter safety.
      ii. Review the primary search-and-rescue efforts.
      iii. Review extinguishment tactics, especially those directed toward rescue operations.
      iv. Explain the ventilation tactics used.
      v. Calculate the required rate of flow.
      vi. Compare the flow applied to the calculated flow.
   D. Elevators
      i. Were elevators used?
      ii. Should they have been used?
      iii. Were the rules of elevator safety applied?
   E. Extension of fire and products of combustion
      i. Where did the fire go?
      ii. How did it get there?
      iii. Where did the smoke go?
      iv. How did it get there?
   F. How did building construction features affect the fire?
   G. What effect did the HVAC system have in smoke control efforts?
   H. Were fire streams effective? If not, why not?
   I. If the fire was in a tower-type high-rise, what effect did construction have on limiting the spread of the products of combustion? If it was in a newer (planar) style building, what effect do you think the heavier tower-type construction would have had on limiting the spread of the products of combustion?
   J. How would the fire have been different in a building protected by a properly operating automatic sprinkler system?

2. Use the following Las Vegas Hilton scenario to answer the questions listed in Question 1. The description of the Las Vegas Hilton fire presented here is a condensed version of an investigative report prepared by the National Fire Protection Association in cooperation with the Federal Emergency Management Agency, the United States Fire Administration, and the National Bureau of Standards.[21] The description provided here disregards non-tactical factors. In reading through the report, pay close attention to the topics discussed in Chapter 12.

Las Vegas Hilton Fire

At approximately 8:00 PM on February 10, 1981, a fire began at the Las Vegas Hilton Hotel in which eight people died and 350 more were injured. This was the largest hotel in the United States at the time. The fire was intentionally started in

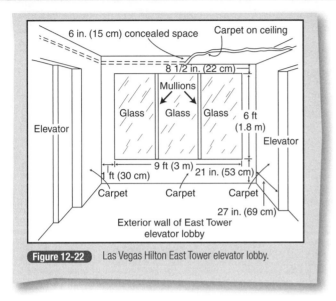

Figure 12-22 Las Vegas Hilton East Tower elevator lobby.

the eighth floor elevator lobby of the East Tower. This elevator lobby, shown in **Figure 12-22**, was completely covered with carpet, which was glued to the walls and ceilings. The elevator lobby also contained draperies, a small wooden bench with a foam cushion, and a table. Most elevator lobbies were similarly furnished; those on a few floors, including the 28th floor, contained larger pieces of furniture and a couch.

The fire spread via the building's exterior, eventually involving the 22 stories above the eighth floor fire.

The Las Vegas Hilton had 2783 guest rooms, which were constructed in three phases **Figure 12-23**:

- Central Tower: 1969
- East Tower: 1975
- North Tower: 1979

The ground floor and second floor contained the casino, restaurants, and showrooms and was larger in area than the hotel towers above. The top (30th) floor contained additional assembly areas. The entire building was of fire-resistive construction, but alarm systems, fire separations, and other protective features varied, owing to the differences in the codes enforced when the three towers were constructed. The guest rooms were decorated and furnished much like those in other luxury hotels. Parts of the first and second floors were sprinkler protected, but other areas were unprotected. There was a hotel fire brigade, and standpipes were located in the stairways. The HVAC system, located on the roof, supplied air to the corridors but did not return air from these corridors. Room air was supplied from the corridors. Smoke detectors, which were designed to close dampers under fire conditions, were installed in corridor ductwork. Toilet exhaust fans were also provided.

At 8:05 PM the hotel's security office received a telephone call from an employee reporting a fire on the eighth floor near the East Tower elevator lobby. Two security officers were sent to investigate via the Central Tower elevator. While the two security officers were responding, an annunciator transmitted an alarm for the eighth-floor elevator lobby. The hotel's fire brigade was notified. The two security guards reached the eighth floor, verified the fire, and radioed the security desk to call the fire department. The fire department received the alarm at 8:07 PM.

Clark County Fire Department Station 18 was adjacent to the hotel. While responding, Station 18 fire fighters saw flames coming from two floors on the building's exterior. During their short response time, the fire extended to involve two additional floors.

The first-arriving engine companies supplied the standpipe and proceeded inside with their high-rise packs. Additional alarms were sounded to muster additional forces for rescue, fire control, and emergency medical services support. A ladder pipe, used to knock down a portion of the fire, was also directed into the East Tower elevator lobbies from the fire floor up to the 16th and 17th floors. Hundreds of people were at their room windows, and fire fighters observed one person jump or fall from the 12th floor.

The fire chief established a command post in the parking lot near the East Tower. An interior command post and medical areas were set up on the seventh floor. Additional divisions were established on the 14th and 20th floors, as well as on the roof.

Fire fighters used the smoke-proof tower and stairway, located in the East Tower, for firefighting operations. Occupants were also using these stairways. Because of the number of fire fighters and equipment necessary for a multi-level attack, these stairways were inadequate. The interior stairway in the East Tower became filled with smoke, due to hose lines entering the hallway from the stairs, blocking the doors in the open position.

Alarm systems were activated, although many occupants reported that they did not hear the alarm. The alarm system in the East Tower may have failed because of electrical problems.

Wrap-Up, continued

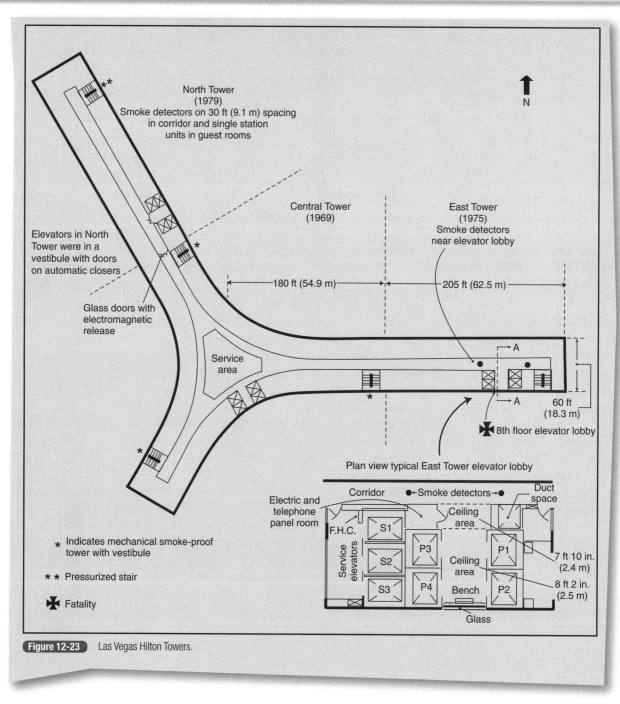

Figure 12-23 Las Vegas Hilton Towers.

The fire department used the voice annunciation component of the fire alarm system to notify guests that fire fighters were on the scene and advised the guests of what actions to take. People trying to evacuate via the stairways encountered smoke, especially those in the East Tower stairs. Some people in the stairways were able to reach the roof, where they were rescued by helicopters. Hotel operators advised guests who called to place wet towels around door openings and wait for the fire department. One person who was trapped in his room reported smoke coming in through the HVAC opening. He heard a sound like a door closing, and the smoke stopped. This undoubtedly was a smoke damper operating.

All East Tower and Main Tower elevators were brought to the first floor and placed on manual control. Three groups of fire fighters used elevators operated by security in the East Tower to reach the seventh floor. A guest using an elevator to evacuate, before it was controlled by the fire department, opened the elevator door on the eighth floor, which was filled with smoke. He crawled down the hallway and knocked on room doors. One door was opened, and he entered and waited out the fire there. Three other people who used the same elevator were found dead in the elevator lobby.

Fifteen hundred people were evacuated from a first-floor showroom without incident.

A total of 23 engine companies, 6 ladder companies, 2 snorkels, 9 rescue units, and 12 aircraft were used during the operation. Forty-eight fire fighters were injured, one suffering a severe heart attack.

Smoke spread throughout the East and Central Towers. The North Tower received little damage, as it was equipped with glass doors at the end of the Central Tower that automatically closed when the fire alarm system was activated.

Figure 12-24 shows the vertical spread of fire up the exterior of the building and the locations of the victims.

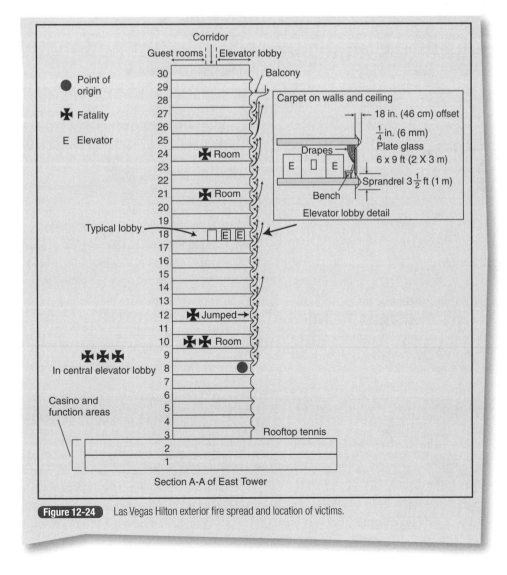

Figure 12-24 Las Vegas Hilton exterior fire spread and location of victims.

Wrap-Up, continued

TABLE 12-4 Las Vegas Hilton Flame Extension in East Tower Corridor

Floor	Extension of Flame Down Corridor, in ft (m)
8	299 (91)
9	247 (75)
10	181 (55)
11	162 (49)
12	161 (49)
13	174 (53)
14	88 (27)
15	147 (45)
16	143 (44)
17	104 (32)
18	126 (38)
19	90 (27)
20	131 (40)
21	83 (25)
22	75 (23)
23	68 (21)
24	63 (19)
25	65 (20)
26	169 (52)
27	38 (12)
28	302 (92)
29	limited
30	limited

Source: David P. Demers, *Investigative Report on the Las Vegas Hilton Fire*, Quincy, MA: NFPA, 1982, p. 16.

Note that Section A-A of this figure corresponds to Section A-A in the East Tower portion shown in Figure 12-23.

The fire spread varying distances down the hallways of the East Tower, as shown in Table 12-4. The differences were probably the result of the effectiveness of interior fire suppression, lead time, the effect of the exterior ladder pipe, and the nine-foot overhanging balcony on the 29th floor. It is estimated that the exterior vertical spread took 20 to 25 minutes to reach the top of the building.

3. Use the RGB high-rise fire scenario on the following pages to do the following:
   A. Size up the fire using the information from this chapter and the size-up factors in Chapter 2.
   B. Establish a strategic mode: offensive, defensive, or non-attack.
   C. Outline an incident action plan.
   D. Assign/reassign companies to tactics and tasks necessary to carry out the incident action plan.
   E. Organize the operation using NIMS. Use high-rise positions described in this chapter as necessary, such as stairway support, lobby control, staging, and base.

A sample answer is included. We suggest that you write out your answers and then compare them to the answers included in the text.

Figure 12-25 is the pre-incident plan narrative for the RGB high rise. Figures 12-26, 12-27, 12-28 and 12-29 are drawings showing the floor layout. Use the pre-plan information and scenario to complete suggested activity number 3.

RGB High-Rise Scenario

Assume the role of the IC and retain that role, even though command may change.

Using the alarm card, call whatever assistance you need. Start by using on-duty resources. Then request mutual aid assistance as needed and call back off-duty personnel. Background information:

- Fire report from dispatch:
  "Smoke detector alarms were received from floors 19, 20, 21, 22, and 23, and the 27th-floor equipment room. In addition, telephone calls were received from the 20th floor reporting a fire in the office area. A call from the 21st floor was also received reporting heavy smoke conditions."
- Day/date: Tuesday, August 1
- Time: 1600 hours
- Weather conditions:
  - 90°F (32.2°C)
  - Dew point 75 degrees
  - Partly cloudy
  - Wind from the southwest at 1 mph (1.6 kph)
- The report from Engine 1 is as follows:
  "Engine 1 and Truck 1 are on the scene of a 26-story office building with a working fire. Fire is visible from several windows on an upper floor on the Central Parkway side. We are proceeding to the 18th floor via the main elevator. Engine 1 is RGB command."

Engine 1 is carrying a high-rise pack including 1¾" (44-mm) hose to the fire area.

You arrive on the scene at the same time as Engine 2, Truck 2, and the heavy-rescue unit. You notice that Engine 1 has a water supply and that the pump operator is at the apparatus hooking up to the hydrant. The security people approach you and verify that people leaving the 20th floor saw visible flames on the Central Parkway side.

Truck 1's officer is now on the radio:

"Truck 1 reporting that Engine 1 and Truck 1 are hooking up to the standpipe on the 19th floor for a confirmed working fire on the 20th floor. The 20th floor appears to have an open layout and heavy fire. The main elevator is free of smoke and operational."

You ask security, "Who occupies the 20th floor?" Security replies, *"I will get back with you."* You request that security also find out as much about the floor layout as possible.

You set up command in the first-floor lobby alarm/command room. You notice on the annunciator panel that Floors 19 to the penthouse are in the alarm mode because of smoke detectors activating.

(continues on page 337)

Life Safety/Fire Fighter Safety

Occupancy Type
- Business occupancy

Occupant Status
- Estimated number of occupants
  - 98% occupancy rate for building on last inspection
  - As many as 9000 occupants during normal business hours (0700 to 1800 hours weekdays)
  - At least two security personnel and sometimes cleaning and maintenance people outside normal business hours
- Mobility of occupants
  - Most are highly mobile
  - Places of safe refuge in center core elevator lobbies floors 2 to 25
  - Floor wardens on each floor assigned to assist handicapped employees and visitors
- Primary and alternative egress routes
  - 3 stairways (see floor plans)
  - DOORS FROM STAIRWAYS TO HALLWAYS AUTOMATICALLY LOCK WHEN CLOSED–THERE IS NO RE-ENTRY FROM THE STAIRWAY SIDE.

Operational Status
- Rescue options
  - Stairways
  - Ladders will not reach beyond eighth floor
- Access to building exterior
  - Streets on three sides
  - Attached parking garage on "D" side
- Access to building interior (forcible entry)
  - Open during business hours
  - Front entry locked after 2000 hours, and on weekends and holidays
  - Security has keys to public areas and a few businesses
  - Security maintains log of people entering and exiting building from 2000 to 0600 hours and on weekends
  - Most doors leading to office areas are alarmed to an off-site security provider
  - Elevator keys located with standpipe pressure-regulating valve tool in special case in the first-floor lobby command room

Structure
- Collapse zone
  - Highly improbable collapse
  - Beware of debris and glass falling from upper floors
- Construction type
  - Fire-resistive core construction
- Roof construction
  - Builtup roof on concrete slabs supported by heavy metal trusses
  - Trusses protected with fire-resistive coating

Figure 12-25 RGB high rise pre-incident plan narrative.

- Condition
  - Last inspection revealed minor fire code violations including fire extinguisher inspections out of date
  - Building generally well maintained
- Live and dead loads
  - Offices tend to have a slightly higher than average live load
  - Storage area on Floor 26 has large live load
  - Mechanical room on Floor 27—large air handling equipment—a substantial dead load
- Enclosures and fire separations
  - Basement and sub-basement are open layout
  - First-floor lobby open layout with fire separated restaurant and bank
  - Monumental–open stairs between first- and second-floor lobbies
  - Floors 3 to 25 vary in layout. Most are open layout with six-foot cubicles separating work stations. Some floors have offices separated by walls that go from floor to suspended ceiling. Area between suspended ceiling and ceiling a common plenum on every floor.
- Extension probability
  - No fire separations except elevators and stairways to prevent fire from extending to involve entire floor on most floors
  - Heavy fire-resistive construction between floors with utility openings
- Concealed spaces
  - Area between suspended ceiling and actual ceiling
  - Telephone cables contained in cabinets on each floor—not sealed between floors
- Age
  - Built in 1970
- Height and area
  - 26 stories (no 13th floor)—Roof 316′ (96 m) above grade
  - Basement, sub-basement, and floors one to three 145′ × 185′ (44 m × 56 m)
  - Ceiling height 10′ (3 m) on all floors except basement and sub-basement (9′ [2.7 m]) and first floor (30′ [9 m])
  - Tower floors 4 to 27—60′ × 160′ (18 m × 49 m)
- Complexity and layout
  - See drawings
  - Core construction

Extinguishment
- Probability of extinguishment
  - Once fire involves a major part of open layout floors a large rate of flow will be required using multiple 2½″ (64-mm) hose lines (see rate of flow).
- External exposures
  - Exterior exposures are low-rise buildings with street separating buildings from RGB.
  - Primary extension is via burning debris.
- Internal exposures
  - The attached garage
  - Three-story buildings on Side "D"
  - Floor-to-floor extension possible via utility openings in floors
  - Floor-to-floor extension possible via fire venting out windows
- Manual extinguishment
  - Fuel load
    - Slightly above average fuel load in offices
    - Heavy fuel load in storage areas on 26th floor
    - Low to moderate fuel load in lobby, basement, sub-basement, and 27th floor
- Calculated rate-of-flow requirement
  - Flows are the maximum flow requirements with deductions for enclosures such as elevators and stairs.
  - Basement and sub-basement V/100 = 2000 GPM (126 L/sec)
  - First-floor lobby V/100 = 4500 GPM (284 L/sec)
  - First-floor bank V/100 = 480 GPM (30 L/sec)
  - First-floor restaurant V/100 = 320 GPM (20 L/sec)
  - Second-floor lobby V/100 = 920 GPM (58 L/sec)
  - Floors 4 to 27 with open layout V/100 = 960 GPM (61 L/sec)
  - Third floor has compartmented offices and some tower floors are sub-divided into smaller compartments. Areas not listed are within the capacity of two 125-GPM (8 L/sec) standpipe hose lines (250 GPM [16 L/sec]).
  - Garage (all levels) V/100 = 760 GPM (48 L/sec)

Figure 12-25 RGB high rise pre-incident plan narrative, *continued*.

- Water supply
  - 20,000 GPM (1262 L/sec) available from primary water supply
  - Can be cross-tied to the Eastern Hills supply for an additional 20,000 GPM (1262 L/sec)
  - Hydrants are located on each side of the building
  - Hydrants in the business district are located on each corner with an additional hydrant in the middle of the block
- Apparatus pump capacity
  - Pumpers have a minimum 1000 GPM (63 L/sec) pump, some have 1250 GPM (79 L/sec) pumps
  - Aerial apparatus are not equipped with fire pumps
- Manual fire-suppression system
  - Standpipe system with discharge on every floor in central core (see drawing)
  - 1000 GPM (63 L/sec) standpipe diesel pump located in basement
  - Fire department connection on Vine Street side (see drawing)
  - Field-adjustable pressure-regulating valves set to 100 psi (690 kPa)
  - Pressure-regulating valve tool located with elevator keys in special case in the first-floor lobby command room
- Automatic fire suppression equipment
  - No automatic sprinkler system
  - Dry chemical kitchen hood system in restaurant

Property Conservation
- Salvageable property
  - Computer equipment, business records, and furnishings located throughout building
- Water damage
  - Probability of water damage
  - Susceptibility of contents to water damage
    - High probability
    - Computers and business records very susceptible to water damage
- Water pathways to salvageable property
  - Utility holes in floors, stairs, elevators
- Water removal methods available
- Water protective methods available
  - Per standard SOPs
- Smoke damage
  - Probability of smoke damage
    - Smoke will migrate upward, but automatic dampers and strong compartmentation will retard spread
  - Susceptibility of contents to smoke damage
    - Office furniture and computer equipment very susceptible
- Damage from forcible entry and ventilation
  - Secured areas will require considerable force to enter
  - HVAC not reliable under fire conditions (shut down)
  - Venting will involve breaking windows which will endanger fire fighters and civilians at grade level

General Factors
- Total staffing and apparatus available
  - See alarm card—each company staffed by an officer, driver, and two fire fighters

Alarm	Engine Companies	Truck Companies	Others
First	1, 2, 3, and 4	1 and 2	Heavy Rescue, District Chief 1
Second	5, 6, 7, and 8	3 and 4	District Chief 2, ALS 1
Third	9, 10, 11, and 12	5	Chief of Department, 4 Assistant Chiefs
Fourth	13, 14, 15, and 16	6	Five additional District Chiefs from Administrative staff, ALS 2, BLS 1
Fifth	17, 18, 19, and 20	7	Air Supply Truck, Command Vehicle

Fire department has a total of 25 engine companies, 13 ladder companies, 2 heavy rescue companies, 7 BLS and 4 ALS units. Mutual aid is available from 50 fire departments in the county with an additional 123 engines, 22 aerials/platforms, and 41 heavy-rescue companies. Approximately 500 off-duty fire fighters can be recalled via a telephone-relay system.

- Utilities (water, gas, electric)
  - Main electric in sub-basement DO NOT ATTEMPT TO DE-ENERGIZE
  - Gas and water street valves on Court Street
  - Gas and water meter equipment in basement

Figure 12-25 RGB high rise pre-incident plan narrative, *continued*.

Wrap-Up, continued

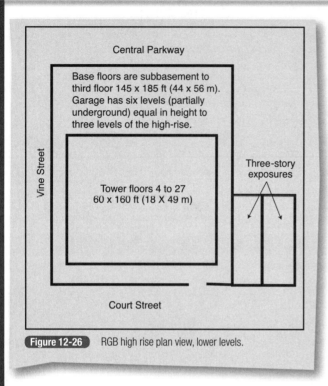

Figure 12-26 RGB high rise plan view, lower levels.

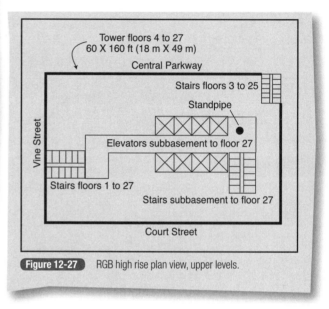

Figure 12-27 RGB high rise plan view, upper levels.

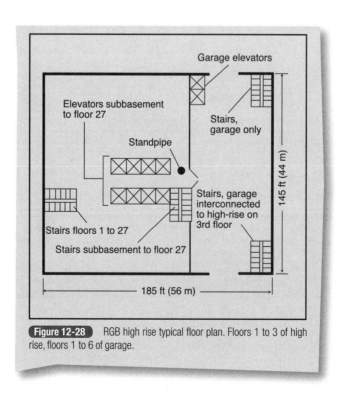

Figure 12-28 RGB high rise typical floor plan. Floors 1 to 3 of high rise, floors 1 to 6 of garage.

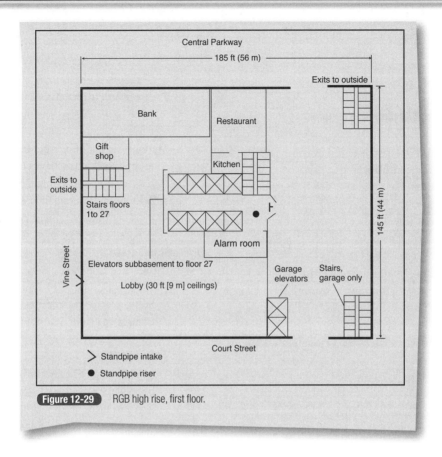

Figure 12-29 RGB high rise, first floor.

It is now 1608 hours. You strike the second alarm, and Engines 3 and 4 are approaching the command post. The heavy-rescue unit is in the elevator lobby, and all of the elevators have been recalled to the first floor. The heavy-rescue unit is assuming its normal role as the rapid intervention team.

Engine 1 is now reporting back to the command post via radio:

"Engine 1 and Truck 1 are now on the 20th floor. We estimate the fire to cover the entire Central Parkway side of an open-layout floor plan. We have heavy smoke throughout the floor. Our 1¾" (44-mm) line is not making progress on the fire. We need 2½" (64-mm) fire lines on the 20th floor."

Truck 1 is now reporting back to the command post via radio:

"On the way up to the 20th floor we ran into at least 10 people evacuating via the stairway. The occupants looked okay, but a few were coughing. The stairway has moderate smoke from the 19th floor to at least the 21st floor."

Now that you have reviewed the RGB high-rise building pre-incident plan and fire problem information, answer suggested activities 3A through 3E (page 332). After you complete suggested activity 3, compare your written responses to the sample answers on the following pages.

4. Use the RGB high-rise scenario, but change the situation and occupancy to see what effect these changes would have:

- Place the fire in a residential occupancy for the elderly.
- Place the fire in a hotel.
- Change the situation:
  – Elevators are out of service.
  – The fire area is subdivided into small office areas.
  – The fire occurs on Sunday afternoon in an office building.

- The building is protected by a working sprinkler system.
- There are heavy smoke conditions on the fire floor and above.

Sample Answer to the RGB High-Rise Problem
Size-Up
- Life safety/fire fighter safety
  - **Smoke and fire conditions**
    - Fire location
      - 20th floor
    - Direction of travel
      - Will involve remainder of 20th floor if not extinguished
      - Possibility of extension upward via utility openings or via building's exterior
  - **Ventilation status**
    - Self-vented on 20th floor
    - Smoke may migrate upward through building
  - **Occupancy type**
    - High-rise office building
  - **Occupant status**
    - Estimated number of occupants
      - The office building is probably fully occupied, given the time of day and the day of the week.
  - **Evacuation status**
    - Self-evacuation has begun, but the status and number of occupants remaining in the building are unknown.
    - Office occupancy density can be one person per 100 ft^2 (9 m^2), or 96 people per floor on the 20th (fire floor) through 25th floors, or nearly 600 people.
    - Floors 26 and 27 are equipment and storage areas.
    - The elevators have been recalled to the ground floor and are under fire department control.
  - **Occupant proximity to fire**
    - Occupied floor
  - **Awareness of occupants**
  - **Mobility of occupants**
    - Occupants should be alert, and most are mobile
  - **Occupant familiarity with building**
    - Occupants should be somewhat familiar with the building but probably do not routinely use the stairs
  - **Primary and alternative egress routes**
    - Three stairways are the only viable means of egress
  - **Medical status of occupants**
    - It appears that the few occupants encountered have minor smoke inhalation.
    - Additional EMS will be needed to treat and possibly transport many potential occupants exposed to smoke and toxic gases.
- Operational status
  - **Adherence to SOPs**
    - On-scene companies appear to be following SOPs related to standpipe operations
    - Using an elevator that reaches/penetrates the fire floor, without first confirming that the 18th floor is clear of fire and smoke, could prove to be deadly. Dismounting the elevator two floors below the fire is proper, but given the circumstances, the fire could be on a lower floor or the heavy fire conditions could be affecting the elevators.
    - Engine Company 1's officer is acting as the IC.
  - **Fire zone/perimeter**
    - PPE needed on fire floor and above
    - Beware of falling glass and debris.
    - Place exterior fire fighters in sheltered areas.
    - Maintain a fire perimeter of at least one additional block in all directions.
  - **Accountability**
    - Accountability has not yet been established but is needed.
  - **Rapid intervention**
    - A rapid intervention crew (RIC) has been formed.
  - **Organization and coordination**
    - Engine 1's officer established command.
    - Command transferred to District Chief 1.
    - Need to establish tactical level management units per NIMS.

- Rescue options
  - The only realistic rescue option available is evacuation via the three interior stairways.
- Staffing needed to conduct primary search
- Staffing needed to conduct secondary search
- Staffing needed to assist in interior rescue/evacuation
  - Primary and secondary searches are needed for the fire floor and all floors above.
  - A complete evacuation may be ordered.
  - Companies should be assigned as follows:
    - One company to the floor below the fire
    - One to the fire floor
    - One to each floor above
  - The actual staffing needed to accomplish the life safety priority is highly dependent on how many people have escaped and smoke conditions on the upper floors. Moderate smoke conditions have been reported in the stairway. Initially allow five to nine companies to search the fire floor and floors above.
- Staffing needed for exterior rescue/evacuation
  - None at this time, low priority
- Apparatus and equipment needed for evacuation
  - Interior evacuation, no apparatus or rescue equipment needed
- Access to building exterior
  - Access on three sides
  - Keep non-operating apparatus outside fire perimeter
- Access to building interior (forcible entry)
  - Building open for business, little need for forcible entry except possibly from stairways to floor areas

Structure
- Signs of collapse
- Collapse zone
  - Structural stability is not an immediate problem
- Construction type
  - New high-rise (fire-resistive) central core construction
  - Will withstand a considerable fire
- Roof construction
  - Sound roof structure
  - Not a factor; fire is many floors below the roof
- Condition
  - Well maintained
- Live and dead loads
  - Above average live load in 26th-floor storage area
  - Large dead load in 27th-floor mechanical room
  - 26th and 27th floors are not an immediate concern
- Water load
  - Possibly a factor
- Enclosures and fire separations
- Extension probability
  - Fire-resistive construction should contain main body of fire to the fire floor.
  - Need at least one precautionary hose line on the floor above the fire.
- Concealed spaces
  - Between suspended ceiling and actual floor/ceiling assembly
- Age
  - New-style planar construction
- Height and area
  - 26 stories
  - Large, undivided area on fire floor
- Complexity and layout
  - Elevators, stairs, and standpipe locations uniform throughout building
  - Work areas separated by cubicles—large floor area could be confusing and requires team search

Extinguishment
- Probability of extinguishment
  - The heavy volume of fire means that it will take time to amass the force necessary for extinguishment.
  - Further extension and the continued production of smoke and toxic gases are likely.
- Offensive/Defensive/Non-attack
  - A sizable offensive attack is warranted.
- Ventilation status
  - Self-vented in fire area
  - May need to ventilate other floors

- External exposures
  - The fire is self-venting; external exposures are well below the fire floor.
  - Falling debris could ignite buildings in area.
- Internal exposures
  - There is a possibility of internal extension.
- Manual extinguishment
- Fuel load
  - Above average fuel load in fire area
- Calculated rate-of-flow requirement
  - The pre-incident plan indicates that 960 GPM (61 L/sec) is needed to extinguish the fire as reported by Engine 1. Pressure-regulating valve tool located with elevator keys in special case in the first-floor lobby command room.
- Number and size hose lines needed for extinguishment
  - The only line being used at present is a 1¾" (44-mm) line at approximately 125 GPM (8 L/sec), leaving a deficit of 835 GPM (53 L/sec).
  - This is a definite "big water" fire; therefore, 2½" (64-mm) lines are needed. Three 2½" (64-mm) lines plus the 1¾" (44-mm) line already deployed should be enough to provide the 960-GPM (61 L/sec) fire flow.
  - The use of straight- or solid-stream nozzles are needed for reach and to avoid wrap-around fires. However, the fire appears to be self-vented, reducing the possibility of a wrap-around situation occurring.
- Additional hose lines needed
  - Additional hose lines are needed to back up the attack on both the 20th and 21st floors.
- Staffing needed for hose lines
  - Staffing should be at least three fire fighters per attack line (assign a full company to each attack line).
  - For each attack position there should be two crews in reserve to maintain a continuous attack.
- Water supply
  - More than adequate
- Apparatus pump capacity
  - Only need one pumper to supply fire department connection
- Manual fire suppression system
  - Connecting to the standpipe system below the fire floor will provide a marginal water supply with little reserve.
  - The building's 1000-GPM (63 L/sec) fire pump should be augmented by pumping into the fire department connections.
- Automatic fire suppression equipment
  - No sprinkler protection.

Property conservation
- Salvageable property
- Location of salvageable property
  - Computers and records storage are high-value items that are located throughout the structure.
- Water damage
  - Probability of water damage
    - Heavy water damage is likely on the 19th floor. Tarps should be used to cover equipment and channel water from the building.
  - Susceptibility of contents to water damage
    - Computers and records are highly susceptible to water damage.
  - Water pathways to salvageable property
    - Utility openings in floors
    - Stairs
    - Elevators (do not use for water removal)
  - Water-removal methods available
    - Chutes to stairway
    - Using toilet as temporary drain
    - Water vacs
  - Water-protective methods available
    - At least two companies will be needed to cover equipment and to channel water from the building.
- Smoke damage
  - Ventilation status
    - Self-vented on fire floor
    - Will need to vent other areas
  - Probability of smoke damage
    - Smoke damage is likely on the floors above the fire.

- Susceptibility of contents to smoke damage
  - Office furniture and electronic equipment sensitive to smoke
- **Damage from forcible entry and ventilation**
  - Little forcible entry needed
  - If necessary to vent via windows, expect window damage.
  - Considering the life safety problem, damage from venting or forcible entry is justified.

General factors

- **Total staffing available versus staffing needed**
  - Four engine companies, two truck companies, one heavy rescue, and 1 district chief responded on the first alarm.
  - Engine 1 and Truck 1 are already committed to the fire floor for fire extinguishment.
  - One company per floor is needed to conduct a primary search of the 19th and 20th floors.
  - One additional company is needed to back up the fire attack team on the 20th floor.
  - At least one company is needed on the 21st floor to protect internal exposures. This company may also be able to conduct a primary search of the 21st floor.
  - Additional search-and-rescue teams should be assigned to the lower floors (17 and below).
  - A property conservation team should be assembled.
  - A minimum of 13 companies are needed to provide initial primary search and extinguishment activities.
  - Personnel are needed to fill staff positions: safety, planning, division 20, search-and-rescue group, and staging (interior staging).
  - Search and rescue on floors below the fire floor, staging, and property conservation may require additional staffing, depending on initial success.
  - Additional EMS response of at least five units is needed. (Only one unit is responding.)
- **Total apparatus available versus apparatus needed**
  - Only need one pumper to supply standpipe

- **Staging/tactical reserve**
  - Staging needs to be established. (This could require as many as ten companies to provide the fire fighters needed to provide a continuous fire flow on the 20th floor.)
- **Utilities (water, gas, electric)**
  - Located in basement and sub-basement
  - Do not attempt to interrupt power supply.
- **Special resource needs**
  - Police to maintain fire perimeter.
  - Additional EMS units
  - Large supply of air cylinders
- **Time**
  - The time of day is 1600 hours.
  - The day of the week is Tuesday. Anticipate the building being occupied at full capacity.
  - Traffic will be moderate to heavy at this time of day.
  - Daylight will provide light for operation should the power fail.
- **Weather**
  - It is hot and humid with a light breeze.
  - Negative stack effect is possible.
  - If the HVAC system is shut down, conditions will become very uncomfortable inside.

Incident Action Plan
Primary Objectives

1. Initiate an aggressive interior fire attack on the fire floor.
2. Conduct a primary search on the fire floor and the floors above.
3. Check for fire extension on the floors above the fire.

Secondary Objectives

4. Conduct a primary search on floors below the fire.
5. Conduct a secondary search of fire floor and floors above the fire.
6. Conduct a secondary search of floors below the fire.
7. Initiate property-conservation operations.

Wrap-Up, continued

Assignments

- Engine Company 1 initially assumed a fast-attack incident command until relieved by District 1. On arrival of the fire chief, District 1 transferred command to the fire chief. The fire chief assigned District 1 as the planning section chief.
- Engine Company 1's officer was initially the Division 20 leader. This assignment was transferred to District 2.
- Engines 1, 2, 3, and 4 and Truck 1 are attacking the fire on the 20th floor. Engine 9 is staffing the backup line on the 20th floor. These companies, under the direction of District 2, make up Division 20.
- The heavy-rescue company is assigned as the RIC and located on the 18th floor.
- A search-and-rescue group is being supervised by Assistant Chief 1.

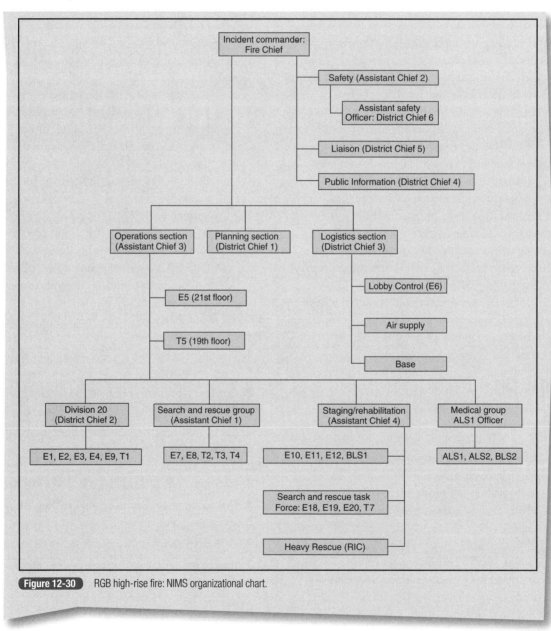

Figure 12-30 RGB high-rise fire: NIMS organizational chart.

- Truck 2 begins search-and-rescue activities at the 21st floor and works upward.
- Engines 7 and 8 and Trucks 3 and 4 are also assigned to the search-and-rescue group.
- Engine 5 is assigned to the 21st floor for extinguishment. Engine 5 is Division 21.
- Engine 6 is assigned the dual duties of lobby control and accountability. Engine 6 is located in the lobby.
- ALS 1 and additional EMS units are providing medical treatment in the lobby. ALS 1 is the Medical Group leader.
- Truck 5 is assigned to search-and-rescue duties on the 19th floor. Truck 5 is Division 19.
- Assistant Chief 2 is assigned as the safety officer.
- Assistant Chief 3 is assigned as the operations section chief.
- Interior staging is set up on the 18th floor with Assistant Chief 4 in charge of staging. Engines 10, 11, and 12 and BLS 1 are located in the 18th-floor staging area.
- District Chief 3 is assigned as the logistics chief with the following subordinate units reporting:
  - Lobby control
  - Air supply
  - Base (exterior staging)

If the fire is under control when three alarm companies are in place, reassignment of companies to REHAB and property conservation is possible. The IC may also elect to re-assign personnel to check conditions on lower floors.

If necessary, augment the primary search effort by rotating the original search-and-rescue companies through REHAB with at least four additional companies handling the secondary search assignment: Engines 18, 19, and 20 and Truck 7.

Air truck personnel are assigned to work with lobby-control personnel to move air cylinders to the 18th floor via elevators, provided the elevators are found to be safe for fire department use.

The four remaining district chiefs are assigned to command staff, groups, and divisions as needed. District Chief 4 is assigned as the public information officer, District Chief 5 as the liaison officer, District Chief 6 as the assistant safety officer, and District Chief 7 as the property conservation group supervisor. Property conservation is not yet assigned; therefore, the property conservation group is not shown in the NIMS chart in **Figure 12-30**.

There are many correct ways to organize this operation using NIMS. It is important to use the correct terminology, maintain a reasonable span of control, and account for all companies.

Wrap-Up, continued

Chapter Highlights

- Tactics and strategic objectives used in high-rise firefighting are the same as those that apply to any other structure fire, but with special considerations because of the height of the building.
- High-rise fires are some of the most challenging incidents a fire department encounters because of the high occupant load and building height/configuration.
- Pre-incident planning, code enforcement, use of NIMS, and developing high-rise SOPs improve the chance for successful operations in high-rise buildings.
- Logistics and access problems increase with building height.
- Elevators should not be used for evacuation except under special circumstances and only with fire fighter supervision.
- Avoid using elevators unless they will substantially improve operations, and even then, use with extreme caution.
- During a structure fire, never use an elevator to travel to or above the fire floor; exit the elevator at least two floors below the fire.
- All elevators should be under the control of fire fighters.
- When in doubt about the safety of elevators a possible option is to use the elevator to send equipment (rather than personnel) to the floors above.
- Getting fire fighters and equipment to an interior staging area via stairs is essential when elevators are unsafe or unavailable and returning to ground level is impractical.
- Stairway transport of equipment is done via relay (stairway support) to conserve personnel.
- When the fire is on an upper floor of a high-rise building interior staging should be set up two or more floors below the fire.
- Simple marking techniques indicating areas where the primary search is complete are essential to an efficient search-and-rescue operation.
- Tools for forcible entry should be carried to assist in eliminating barriers to evacuation (e.g., locked doors) and to facilitate access to locked areas.
- Helicopter rescues are extremely dangerous and usually unnecessary; alternate, superior tactics are usually available.
- Partial or sequential evacuation can reduce stairway congestion, but could also delay evacuation.
- Emergency Voice/Alarm Communications Systems (EVACS) are available in some buildings, but occupants do not always follow the instructions given.
- Most exterior defensive fire control tools are ineffective above the eighth floor.
- Fire forces must prevent fire extension into the stairways, which are the safest, most effective, and sometimes the only available, egress routes.
- Pressure-reducing valves and fire pumps both affect standpipe hose pressures.
- Pumping into the fire department standpipe connections can assist internal fire pumps and is a good practice.
- Due to large numbers of people and other distractions in the lobby, a high-rise lobby may be a poor choice for a command post.
- A command post's location should be transmitted to all responders regardless of whether it is inside or away from the building.
- Special considerations during a high-rise fire include elevator operations and elevator key location; access/egress issues; logistics, staging, and standpipe operation; floor layouts; ventilation; and procedures and operations unique to the building.
- Subtle differences in interior building configurations can have a major impact on firefighting and rescue operations.
- Weather conditions, fire intensity, stack effect, wind, and atmospheric pressure can have a significant effect on smoke movement within a high-rise structure.
- Ventilation is crucial in a high-rise fire, though often very difficult to perform.
- Dedicating a stairway with standpipe outlets for firefighting reduces the possibility of opposing lines on the fire floor.
- To avoid exposing people and equipment at ground level to falling hazards, a wide fire perimeter is essential.
- Communications is the most frequently cited problem during major emergency operations.
- High-rise tactical worksheets can be used to assist with planning functions and accountability.
- Controlling the lobby, elevators, stairways, and HVAC, facilitates both fire fighter access to, and successful evacuation of, the upper floors of a high-rise building.

References

1. John R. Hall, Jr., *High-Rise Building Fires*. Quincy, MA: NFPA, 2005.

2. National Institute of Standards and Technology, *Final Report NIST NCSTAR 1: Federal Building and Fire Safety Investigation of the World Trade Center Disaster*. NIST, 2005. Available online at http://wtc.nist.gov/reports_october05.htm (accessed July 26, 2007).

3. *Report of the Cook County Commission Investigating the 69 West Washington Building Fire of October 17, 2003*. Available online at www.co.cook.il.us/fire_report.htm (accessed July 26, 2007).

4. National Fire Protection Administration, *NFPA 101: Life Safety Code*. Quincy, MA: NFPA, 2006.

5. National Fire Protection Administration, *NFPA Fire Protection Handbook*, 19th edition. Quincy, MA: NFPA, 2003, pp. 12–119.

6. Thomas J. Klem, *One Meridian Plaza, Three Fire Fighter Fatalities, Philadelphia, Pennsylvania, Fire Investigation Report*. Quincy, MA: NFPA, 1991.

7. Gordon J. Routley, Charles Jennings, and Mark Chubb, *High-Rise Office Building Fire, One Meridian Plaza, Philadelphia, Pennsylvania*. Emmitsburg, MD: U.S. Fire Administration, undated. Available online at www.interfire.org/res_file/pdf/Tr-049.pdf (accessed July 26, 2007).

8. Thomas J. Klem, *First Interstate Bank Building Fire, Los Angeles, California, Fire Investigation Report*. Quincy, MA: NFPA.

9. Gordon J. Routley, *Interstate Bank Building Fire, Los Angeles, California*. Emmitsburg, MD: U.S. Fire Administration, undated. Available online at www.interfire.org/res_file/pdf/Tr-022.pdf (accessed July 26, 2007).

10. Francis L. Brannigan, Paranoia may save your life. *Fire Engineering*, July: 123–126, 1998.

11. NFPA Fire Investigations Department, *Residential High-Rise, North York, Ontario, Canada, January 6, 1995, Fire Investigation Report*. Quincy, MA: NFPA, 1995.

12. Michael S. Isner, *Fatal Office Building Fire, Atlanta, GA, NFPA Fire Investigation Report*. Quincy, MA: NFPA.

13. Charles Jennings, *Five-Fatality Office Building Fire, Atlanta Georgia*. Emmitsburg, MD: U.S. Fire Administration, undated. Available online at www.interfire.com/res_file/pdf/Tr-033.pdf (accessed July 26, 2007).

14. James Lee Witt Associates, *Cook County Administration Building Fire Review, Executive Summary*, undated. Available online at http://www.writer-tech.com/pages/ccab/CCAB%20Exec%20Summary%2093004%201550%205.0.pdf (accessed July 26, 2007).

15. D. Madrzykowski and W. D. Walton, *NIST SP 1021, Cook County Administration Building Fire, 69 West Washington, Chicago, Illinois, October 17, 2003*. Heat Release Rate Experiments and FDS Simulations. NIST, 2004.

16. Mark Chubb and Joe E. Caldwell, Tragedy in a residential high-rise, Memphis, Tennessee. *Fire Engineering*, March: 49–66, 1995.

17. Michael S. Isner, and Thomas J. Klem, *World Trade Center Explosion and Fire, New York, New York, Fire Investigation Report*. Quincy, MA: NFPA, 1993.

18. William A. Manning, *The World Trade Center Bombing: Report and Analysis*. Emmitsburg, MD: U.S. Fire Administration, undated. Available online at www.firetactics.com/wtc-93.pdf (accessed July 26, 2007).

19. Richard Best and David P. Demers, *MGM Grand Hotel Fire, Las Vegas, NV*, NFPA Fire Investigation Report. Quincy, MA: NFPA.

20. NFPA Fire Investigations Department, "The MGM Hotel Fire, Part 1", *Fire Service Today*, January 1982, pp. 18–23.

21. David P. Demers, *Investigative Report on the Las Vegas Hilton Hotel Fire*. Quincy, MA: NFPA 1982.

Appendix A

Sample ASETBX Data

This appendix contains sample data from the ASETBX applications within FPETOOL.

ASETBX is an application within a computer program called FPETOOL, produced by the National Institute of Standards and Technology (NIST) Building and Fire Research Laboratories. ASETBX estimates the rise in temperature and descent of the smoke layer based on a single compartment's size. Such programs are of limited value at an incident scene or as a pre-planning tool, but can be useful in reconstructing a fire. The data sheets in this appendix are intended to compare fires in large and small enclosures.

Many assumptions are made about fire conditions with the FPETOOL programs. These assumptions are necessary to calculate the scenario with a minimum of input from the user. However, the assumptions may not be true for the situation presented, thus leading to inaccuracies. FPETOOL output must be used with a great deal of skepticism. The purpose of presenting these FPETOOL readouts here is *not* to make engineering judgments, but to gain an appreciation of fire and smoke conditions related to the size of the compartment where the fire occurs.

FPETOOL is a product of the NIST Building and Fire Research Laboratories, which is a U.S. government agency. Other, more accurate, fire models exist that provide more in-depth information. These programs generally require inputting large amounts of data and considerable computer expertise. One such program is the NIST Fire Dynamics Simulator and Smokeview, which predicts smoke and/or air flow movement caused by fire, wind, and ventilation systems. The NIST Fire Dynamics Simulator and Smokeview simulator is included in several NIST fire reports, which are available at no charge either as a download or on a CD.

This appendix contains four sample scenarios to demonstrate the deterioration of conditions within a fire enclosure over a period of time.

1. 12′ × 15′ (3.7 m × 4.6 m) bedroom with a 10′ (3 m) ceiling (BedRm)
2. 15′ × 30′ (4.6 m × 9 m) family room with a 20′ (6 m) average height cathedral ceiling (FamRm)
3. 60′ × 120′ (18.3 m × 36.6 m) carpet store with 15′ (4.6 m) ceilings (Carpet)*
4. 120′ × 240′ (36.6 m × 73 m) warehouse with 30′ (9 m) ceilings (Warehouse)

The TEMP and LAYER columns contain the most useful information. The last statement at the bottom of the first three readouts estimates the time to flashover. The fourth chart does not contain a flashover statement because the model predicts that flashover will not occur in this very large compartment. The LAYER column is the smoke layer or the distance from the floor to the bottom of the smoke layer. The information presented is for comparison only. Several assumptions are made for each of these scenarios to allow a comparison, but these assumptions may not be true in a real-life incident. The assumptions made include:

- Each fire is a "Slow Fire." Other options within the FPETOOL ASETBX model are "Moderate Fire," "Fast Fire," or a user-created fire.
- Ignition was assumed to have been at 1′ above floor level.

*The carpet store data is based on an actual incident using data from the NFPA Fire Investigation report *Carpet Store Fire, One Firefighter Fatality, Bradford, CT; November 28, 1996*. The wood truss roof at this fire collapsed approximately 17 minutes after the arrival of the fire department.

- The fire was not vented, and natural air intake and heat loss were constant.
- The fire continued to burn after there was a possibility of self-extinguishment and was unaffected by the combustion air being mixed with fire products.

Change any condition on which these assumptions are based and the information changes. As an example, FPETOOL assumes the room to be unvented. If venting occurs, the smoke layer will be lifted and the heat will be reduced, with little chance of progression to flashover.

Figures A-1 through A-4 show the data sheets for each room. Figure A-5 is a graph comparing the progression to flashover for the four scenarios.

```
                         05-16-1999
                         FPETOOL V3.2
------------------------- ASETBX -----------------------------
```

Run title: BedRm

Heat loss fraction =   0.9
Fire height =   1.0'   0.3 m
Room height =   10.0'   4.6 m
Room area =   180.0 ft²   668.9 m²

TIME sec	TEMP °F	TEMP °C	LAYER ft	LAYER m	FIRE kW	FIRE BTUs
0	70	21.2	10.0	3.0	0.1	0.1
60	72	22.3	6.6	2.0	10.5	10.0
120	79	25.9	3.9	1.2	42.2	40.0
180	94	34.3	2.4	0.7	94.9	90.0
240	123	50.5	1.5	0.5	168.8	160.1
300	172	78.0	0.8	0.2	263.7	250.1
360	247	119.4	0.0	0.0	379.7	360.2
420	362	183.5	0.0	0.0	516.9	490.2
480	546	285.3	0.0	0.0	675.1	640.3
540	842	499.9	0.0	0.0	854.4	810.4

The drop of the upper level to the top of the burning item and the rise in the upper level temperature indicate vitiation of the combustion air with fire products. It is likely that the burning rate will be depressed, possibly smothered.

| 600 | 1337 | 725.1 | 0.0 | 0.0 | 1054.8 | 1000.5 |

Upper level temperature indicates that flashover has probably occurred.

Figure A-1 FPETOOL data sheet for BedRm scenario.

```
                    05-16-1999
                    FPETOOL V3.2
------------------- ASETBX -------------------
```

Run title: FamRm

Heat loss fraction =   0.9
Fire height =   1.0′   0.4 m
Room height =   20.0′   4.6 m
Room area =   450.0 ft²   668.9 m²

TIME sec	TEMP °F	TEMP °C	LAYER ft	LAYER m	FIRE kW	FIRE BTUs
0	70	21.1	20.0	6.1	0.1	0.1
60	71	22.4	14.7	4.5	10.5	10.0
120	72	22.2	9.3	2.8	42.2	40.0
180	75	23.9	6.0	1.8	94.9	90.0
240	80	26.9	4.0	1.2	168.8	160.1
300	89	31.7	2.8	0.9	263.7	250.1
360	102	39.0	2.0	0.6	379.7	360.2
420	121	49.2	1.2	0.4	516.9	490.2
480	146	63.1	0.4	0.1	675.1	640.3
540	178	81.0	0.0	0.0	854.4	810.4
600	220	104.6	0.0	0.0	1054.8	1000.5
660	276	135.6	0.0	0.0	1276.3	1210.6
720	349	176.1	0.0	0.0	1518.9	1440.7
780	444	229.1	0.0	0.0	1782.6	1690.8
840	570	298.9	0.0	0.0	2067.4	1960.9
900	737	391.6	0.0	0.0	2373.3	2251.1

The drop of the upper level to the top of the burning item and the rise in the upper level temperature indicate vitiation of the combustion air with fire products. It is likely that the burning rate will be depressed, possibly smothered.

TIME sec	TEMP °F	TEMP °C	LAYER ft	LAYER m	FIRE kW	FIRE BTUs
960	961	516.1	0.0	0.0	1054.8	1000.5
1020	1266	685.4	0.0	0.0	3048.4	2891.4

Upper level temperature indicates that flashover has probably occurred.

Figure A-2 FPETOOL data sheet for FamRm scenario.

```
                            05-16-1999
                           FPETOOL V3.2
--------------------------- ASETBX --------------------------------
Run title: Carpet

Heat loss fraction =        0.9
Fire height =               1.0'           0.3 m
Room height =              15.0'           4.6 m
Room area =              7200.0 ft²      668.9 m²
```

TIME sec	TEMP °F	TEMP °C	LAYER ft	LAYER m	FIRE kW	FIRE BTUs
0	70	21.1	15.0	4.6	0.1	0.1
120	72	22.1	14.2	4.3	42.2	40.0
240	75	23.8	12.8	3.9	168.8	160.1
360	79	26.2	11.2	3.4	379.7	360.2
480	86	29.7	9.6	2.9	675.1	640.3
600	94	34.5	8.2	2.5	1054.8	1000.5
720	106	40.9	6.8	2.1	1518.9	1440.7
840	121	49.4	5.7	1.7	2067.4	1960.9
960	141	60.3	4.6	1.4	2700.3	2561.2
1080	166	74.4	3.7	1.1	3417.6	3241.5
1200	198	92.2	2.8	0.8	4219.2	4001.9
1320	238	114.6	1.9	0.6	5105.2	4842.3
1440	288	142.2	0.9	0.3	6075.6	5762.8
1560	347	175.2	0.0	0.0	7130.4	6763.2
1680	420	215.8	0.0	0.0	8269.6	7843.7
1800	513	267.3	0.0	0.0	9493.2	9004.3
1920	631	332.8	0.0	0.0	10801.2	10244.9
2040	782	416.6	0.0	0.0	12193.5	11565.5

The drop of the upper level to the top of the burning item and the rise in the upper level temperature indicate vitiation of the combustion air with fire products. It is likely that the burning rate will be depressed, possibly smothered.

TIME sec	TEMP °F	TEMP °C	LAYER ft	LAYER m	FIRE kW	FIRE BTUs
2160	977	524.9	0.0	0.0	13670.2	12966.2
2280	1231	666.1	0.0	0.0	15231.3	14446.9

Upper level temperature indicates that flashover has probably occurred.

Figure A-3 FPETOOL data sheet for Carpet scenario.

```
                          05-16-1999
                         FPETOOL V3.2
- - - - - - - - - - - - - - ASETBX - - - - - - - - - - - - - - - - - - -
```

Run title: Warehouse

Heat loss fraction = 0.9
Fire height = 1.0' 0.3 m
Room height = 30.0' 9.1 m
Room area = 28800.0 ft² 2675.6 m²

TIME sec	TEMP °F	TEMP °C	LAYER ft	LAYER m	FIRE kW	FIRE BTUs
0	70	21.1	30.0	9.1	0.1	0.1
180	71	21.6	28.8	8.8	94.9	90.0
360	72	22.4	26.4	8.0	379.7	360.2
540	74	23.6	23.5	7.2	854.4	810.4
720	77	25.2	20.6	6.3	1518.9	1440.7
900	81	27.3	17.9	5.4	2373.3	2251.1
1080	86	30.1	15.4	4.7	3417.6	3241.5
1260	93	33.6	13.1	4.0	4651.7	4412.1
1440	101	38.1	11.2	3.4	6075.6	5762.8
1620	111	43.6	9.4	2.9	7689.5	7293.5
1800	123	50.5	7.9	2.4	9493.2	9004.3
1980	138	58.8	6.4	2.0	11486.8	10895.2
2160	156	68.8	5.1	1.6	13670.2	12966.2
2340	177	80.8	3.8	1.2	16043.5	15217.3
2520	201	94.2	2.6	0.8	15244.8	14459.7
2700	223	105.9	1.7	0.5	12933.7	12267.6
2880	241	116.0	0.9	0.3	10812.5	10255.7
3060	256	124.3	0.3	0.1	8881.2	8423.8
3240	268	131.1	0.0	0.0	7139.7	6772.0
3420	278	136.6	0.0	0.0	5588.1	5300.3
3600	286	140.9	0.0	0.0	4226.3	4008.7
3780	291	144.1	0.0	0.0	3054.4	2897.1
3960	295	146.3	0.0	0.0	2072.4	1965.7
4140	298	147.8	0.0	0.0	1280.3	1214.3
4320	300	148.6	0.0	0.0	678.0	643.1
4500	300	149.0	0.0	0.0	265.5	251.9
4680	301	149.2	0.0	0.0	43.0	40.8

Figure A-4 FPETOOL data sheet for Warehouse scenario.

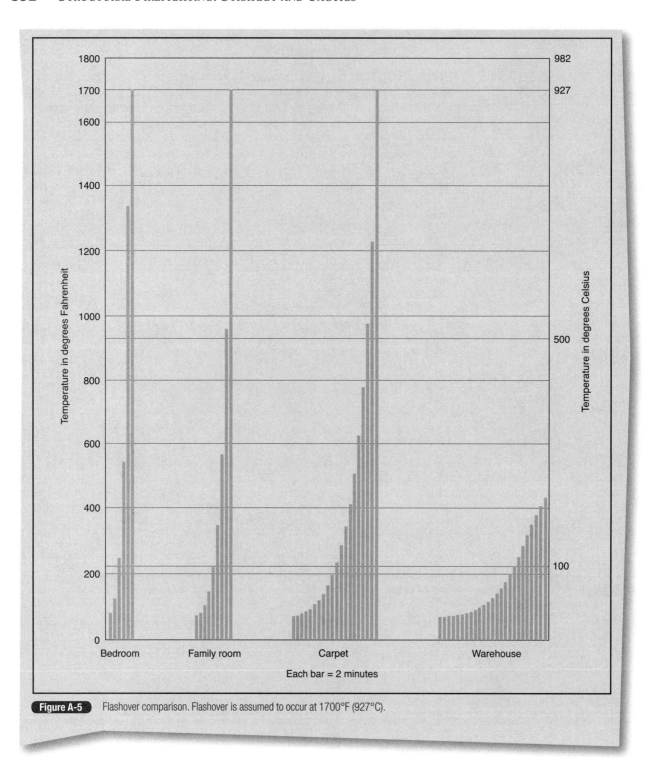

Figure A-5 Flashover comparison. Flashover is assumed to occur at 1700°F (927°C).

Appendix B

Imperial and Metric Conversions

TABLE B-1 Length

1 inch = 0.08333 foot, 1000 mils, 25.40 millimeters
1 foot = 0.3333 yard, 12 inches, 0.3048 meter, 304.8 millimeters
1 yard = 3 feet, 36 inches, 0.9144 meter
1 rod = 16.5 feet, 5.5 yards, 5.029 meters
1 mile = (U.S. and British) 5280 feet, 1.609 kilometers, 0.8684 nautical mile
1 millimeter = 0.03937 inch, 39.37 mils, 0.001 meter, 0.1 centimeter, 100 microns
1 meter = 1.094 yards, 3.281 feet, 39.37 inches, 1000 millimeters
1 kilometer = 0.6214 mile, 1.094 yards, 3281 feet, 1000 meters
1 nautical mile = 1.152 miles (statute), 1.853 kilometers
1 micron = 0.03937 mil, 0.00003937 inch
1 mil = 0.001 inch, 0.0254 millimeters, 25.40 microns
1 degree = 1/360 circumference of a circle, 60 minutes, 3600 seconds
1 minute = 1/60 degree, 60 seconds
1 second = 1/60 minute, 1/3600 degree

TABLE B-2 Area

1 square inch = 0.006944 square foot, 1,273,000 circular mils, 645.2 square millimeters
1 square foot = 0.1111 square yard, 144 square inches, 0.09290 square meter, 92,900 square millimeters
1 square yard = 9 square feet, 1296 square inches, 0.8361 square meter
1 acre = 43,560 square feet, 4840 square yards, 0.001563 square mile, 4047 square meters, 160 square rods
1 square mile = 640 acres, 102,400 square rods, 3,097,600 square yards, 2.590 square kilometers
1 square millimeter = 0.001550 square inch, 1.974 circular mils
1 square meter = 1.196 square yards, 10.76 square feet, 1550 square inches, 1,000,000 square millimeters
1 square kilometer = 0.3861 square mile, 247.1 acres, 1,196,000 square yards, 1,000,000 square meters
1 circular mil = 0.7854 square mil, 0.0005067 square millimeter, 0.0000007854 square inch

TABLE B-3 Volume (Capacity)

1 fluid ounce = 1.805 cubic inches, 29.57 milliliters, 0.03125 quarts (U.S.) liquid measure
1 cubic inch = 0.5541 fluid ounce, 16.39 milliliters
1 cubic foot = 7.481 gallons (U.S.), 6.229 gallons (British), 1728 cubic inches, 0.02832 cubic meter, 28.32 liters
1 cubic yard = 27 cubic feet, 46,656 cubic inches, 0.7646 cubic meter, 746.6 liters, 202.2 gallons (U.S.), 168.4 gallons (British)
1 gill = 0.03125 gallon, 0.125 quart, 4 ounces, 7.219 cubic inches, 118.3 milliliters
1 pint = 0.01671 cubic foot, 28.88 cubic inches, 0.125 gallon, 4 gills, 16 fluid ounces, 473.2 milliliters
1 quart = 2 pints, 32 fluid ounces, 0.9464 liter, 946.4 milliliters, 8 gills, 57.75 cubic inches
1 U.S. gallon = 4 quarts, 128 fluid ounces, 231.0 cubic inches, 0.1337 cubic foot, 3.785 liters (cubic decimeters), 3785 milliliters, 0.8327 Imperial gallon
1 Imperial (British and Canadian) gallon = 1.201 U.S. gallons, 0.1605 cubic foot, 277.3 cubic inches, 4.546 liters (cubic decimeters), 4546 milliliters
1 U.S. bushel = 2150 cubic inches, 0.9694 British bushel, 35.24 liters
1 barrel (U.S. liquid) = 31.5 gallons (various industries have special definitions of a barrel)
1 barrel (petroleum) = 42.0 gallons
1 millimeter = 0.03381 fluid ounce, 0.06102 cubic inch, 0.001 liter
1 liter (cubic decimeter) = 0.2642 gallon, 0.03532 cubic foot, 1.057 quarts, 33.81 fluid ounces, 61.03 cubic inches, 1000 milliliters
1 cubic meter (kiloliter) = 1.308 cubic yards, 35.32 cubic feet, 264.2 gallons, 1000 liters
1 cord = 128 cubic feet × 8 feet × 4 feet × 4 feet, 3.625 cubic meters

TABLE B-4 Weight

1 grain = 0.0001428 pound
1 ounce (avoirdupois) = 0.06250 pound (avoirdupois), 28.35 grams, 437.5 grains
1 pound (avoirdupois) = the mass of 27.69 cubic inches of water weighed in air at 4°C (39.2°F) and 760 millimeters of mercury (atmospheric pressure), 16 ounces (avoirdupois), 0.4536 kilogram, 453.6 grams, 7000 grains
1 long ton (U.S. and British) = 1.120 short tons, 2240 pounds, 1.016 metric tons, 1016 kilograms
1 short ton (U.S. and British) = 0.8929 long ton, 2000 pounds, 0.9072 metric ton, 907.2 kilograms
1 milligram = 0.001 gram, 0.000002205 pound (avoirdupois)
1 gram = 0.002205 pound (avoirdupois), 0.03527 ounce, 0.001 kilogram, 15.43 grains
1 kilogram = the mass of 1 liter of water in air at 4°C and 760 millimeters of mercury (atmospheric pressure), 2.205 pounds (avoirdupois), 35.27 ounces (avoirdupois), 1000 grams
1 metric ton = 0.9842 long ton, 1.1023 short tons, 2205 pounds, 1000 kilograms

TABLE B-5 Density

1 gram per millimeter = 0.03613 pound per cubic inch, 8345 pounds per gallon, 62.43 pounds per cubic foot, 998.9 ounces per cubic foot
Mercury at 0°C = 0.1360 grams per millimeter basic value used in expressing pressures in terms of columns of mercury
1 pound per cubic foot = 16.02 kilograms per cubic meter
1 pound per gallon = 0.1198 gram per millimeter

TABLE B-6 Flow

1 cubic foot per minute	= 0.1247 gallon per second, 0.4720 liter per second, 472.0 milliliters per second = 0.028 m³/min, lcfm/ft² = 0.305 m³/min/m²
1 gallon per minute	= 0.06308 liter per second, 1440 gallons per day, 0.002228 cubic foot per second
1 gallon per minute per square foot	= 40.746 mm/min, 40.746 L/min · m²
1 liter per second	= 2.119 cubic feet per minute, 15.85 gallons (U.S.) per minute
1 liter per minute	= 0.0005885 cubic foot per second, 0.004403 gallon per second

TABLE B-7 Pressure

1 atmosphere	= pressure exerted by 760 millimeters of mercury of standard density at 0°C, 14.70 pounds per square inch, 29.92 inches of mercury at 32°F, 33.90 feet of water at 39.2°F, 101.3 kilopascal
1 millimeter of mercury (at 0°C)	= 0.001316 atmosphere, 0.01934 pound per square inch, 0.04460 foot of water (4°C or 39.2°F), 0.0193 pound per square inch, 0.1333 kilopascal
1 inch of water (at 39.2°F)	= 0.00246 atmosphere, 0.0361 pound per square inch, 0.0736 inch of mercury (at 32°F), 0.2491 kilopascal
1 foot of water (at 39.2°F)	= 0.02950 atmosphere, 0.4335 pound per square inch, 0.8827 inch of mercury (at 32°F), 22.42 millimeters of mercury, 2.989 kilopascal
1 inch of mercury (at 32°F)	= 0.03342 atmosphere, 0.4912 pound per square inch, 1.133 feet of water, 13.60 inches of water (at 39.2°F), 3.386 kilopascal
1 millibar (1/1000 bar)	= 0.02953 inch of mercury. A bar is the pressure exerted by a force of one million dynes on a square centimeter of surface
1 pound per square inch	= 0.06805 atmosphere, 2.036 inches of mercury, 2.307 feet of water, 51.72 millimeters of mercury, 27.67 inches of water (at 39.2°F), 144 pounds per square foot, 2304 ounces per square foot, 6.895 kilopascal
1 pound per square foot	= 0.00047 atmosphere, 0.00694 pound per square inch, 0.0160 foot of water, 0.391 millimeter of mercury, 0.04788 kilopascal
Absolute pressure	= the sum of the gauge pressure and the barometric pressure
1 ton (short) per square foot	= 0.9451 atmosphere, 13.89 pounds per square inch, 9765 kilograms per square meter

TABLE B-8 Temperature

Temperature Celsius	= 5/9 (temperature Fahrenheit − 32°)
Temperature Fahrenheit	= 9/5 × temperature Celsius + 32°
Rankine (Fahrenheit absolute)	= temperature Fahrenheit + 459.67°
Kelvin (Celsius absolute)	= temperature Celsius + 273.15°
Freezing point of water:	Celsius = 0°; Fahrenheit = 32°
Boiling point of water:	Celsius = 100°; Fahrenheit = 212°
Absolute zero:	Celsius = −273.15°; Fahrenheit = −459.67°

TABLE B-9 Sprinkler Discharge

1 gallon per minute per square foot (gpm/ft²) = 40.75 liters per minute per square meter (Lpm/m²) = 40.75 millimeters per minute (mm/min)

Appendix C

FESHE Correlation Guide

Strategy and Tactics FESHE Course Outcomes	*Structural Firefighting: Strategy and Tactics, Second Edition* Chapter Correlation
1. Demonstrate (verbally and written) knowledge of fire behavior and the chemistry of fire.	2, 5, 8
2. Articulate the main components of pre-fire planning and identify steps during a pre-fire plan review.	2
3. Recall the basics of building construction and how they interrelate to pre-fire planning.	2, 5
4. Recall major steps taken during size-up and identify the order in which they will take place at an incident.	2
5. Recognize and articulate the importance of fire-ground communications.	1, 5
6. Identify and define the main functions within the ICS system and how they interrelate during an incident.	1
7. Given different scenarios, the student will set up an ICS, call for appropriate resources, and bring the scenario to a mitigated or controlled conclusion.	1
8. Identify and analyze the major causes involved in line-of-duty fire fighter deaths related to health, wellness, fitness, and vehicle operations.	5

Glossary

20-minute rule When a heavy volume of fire is burning out of control on two or more floors for 20 minutes or longer in a building of ordinary construction, structural collapse should be anticipated.

accountability system Established on the fire ground to ensure that everyone entering the area has a specific assignment, to track all personnel at the scene, and to identify the location of any missing personnel if a catastrophic event should occur.

area of refuge A floor area with at least two rooms separated by smoke-resisting partitions in a building protected by a sprinkler system, or a space located in an egress path that is separated from other building spaces.

assembly occupancy An occupancy used for the gathering of 50 or more persons for deliberation, worship, entertainment, eating, drinking, amusement, awaiting transportation, or similar uses; or used as a special amusement building, regardless of occupant load.

attack pumper The first-arriving engine company goes directly to the fire building without securing a water supply.

backdraft Occurs when oxygen (air) is introduced into a superheated, oxygen-deficient compartment charged with smoke and pyrolytic emissions, resulting in an explosive ignition.

back-pressure Also known as elevation pressure, this is the pressure required to overcome the weight of water in a piping or hose system. Each vertical foot of water in a pipe, hose, or tank exerts a pressure of 0.434 psi (3 kPa) at the base.

balloon-frame construction An older type of wood frame construction in which the wall studs extend vertically from the basement of a structure to the roof.

branches Immediately subordinate to section in NIMS hierarchy. These units are subordinate to the logistics, finance/administration, and planning sections. Divisions and groups are subordinate to these in the operations section. These are used to reduce the span of control at very large operations or to manage a particular function/agency.

business occupancy An occupancy used for account and record keeping or the transaction of business other than mercantile.

Class A foam Foam for use on fires involving Class A fuels such as vegetation, wood, cloth, paper, rubber, and some plastics.

cold zone An area where personal protective clothing is not required, where the command post, rehabilitation, and medical treatment are based.

collapse zone The area endangered by a potential building collapse, generally considered to be an area 1½ times the height of the involved building.

combination attack A type of attack employing both direct and indirect attack methods.

complex pre-incident plan A plan of a property with more than three buildings or when it is necessary to show the layout of the premises and relationship between buildings on the site.

conflagration A fire spreading over a large area involving multiple structures.

convergence cluster phenomenon Groups that take shelter together to provide mutual support.

critical time The time available until the structure becomes untenable.

dead load The weight of a building; consists of the weight of all materials of construction incorporated into a building, including but not limited to walls, floors, roofs, ceilings, stairways, built-in partitions, finishes, cladding, and other similarly incorporated architectural and structural items, as well as fixed service equipment.

decay phase The phase of fire development in which the fire has consumed either the available fuel or oxygen and is starting to die down.

defend-in-place A tactic utilized during a structure fire when it is very difficult to remove occupants from the building. Occupants are either protected at their present location or moved to a safe location within the building.

defensive A fire attack conducted from the exterior of a building.

deluge system A sprinkler system in which all sprinkler heads or applicators are open. When an initiating device, such as a smoke detector or heat detector, is activated, the deluge valve opens and water discharges from all of the open sprinkler heads simultaneously.

detention and correctional occupancy An occupancy used to house four or more persons under varied degrees of restraint or security where such occupants are mostly incapable of self-preservation because of security measures not under the occupants' control.

direct attack Firefighting operations involving the application of extinguishing agents directly onto the burning fuel.

division Tactical level management unit in charge of a geographic area.

division-of-labor principle Breaking down an incident or task into smaller, more manageable tasks and assigning personnel to complete those tasks (developing job skills in a concentrated area to allow for more productivity).

draft curtain A wall designed to limit horizontal spread of the fire and which extends partially down (usually no more than 20% of the height of the compartment) from the underside of the roof.

dry pipe sprinkler system A system in which the pipes are normally filled with compressed air. When a sprinkler head is activated, it releases the air from the system, which opens a valve so the pipes can fill with water.

educational occupancy An occupancy used for educational purposes through the 12th grade by six or more persons for four or more hours per day or more than 12 hours per week.

exclusion zone An area that is unsafe regardless of the level of personal protective equipment and which must be cleared of all personnel, including emergency response personnel.

extension Fire that moves into areas not originally involved, including walls, ceilings, and attic spaces; also, the movement of fire into uninvolved areas of a structure.

external exposures Buildings, vehicles, or other property threatened by fire that are external to the building, vehicle, or property where the fire originated.

finance/administration section The section that tracks and provides financial and administrative services required to compensate people or organizations providing goods and services at the incident scene.

fire perimeter A wide area beyond the working zones, usually staffed by police to keep unauthorized people away from the scene.

fire zone The area where emergency responders are working and which can be subdivided into exclusion, hot, warm, and cold zones.

flashover An oxygen-sufficient condition where room temperatures reach the ignition temperature of the suspended pyrolytic emissions, causing all combustible contents to suddenly ignite.

foam house A fixed facility consisting of an enclosure housing a foam concentrate supply tank, a foam solution proportioning system, a pump, and sometimes an extra supply of foam concentrate that can be added to the proportioning system.

formal pre-incident plan A plan for a property with a substantial risk to life and/or property; includes a drawing of the property, specific floor layouts, and a narrative describing important features.

freelancing Performing tasks outside the incident organization structure.

frequency As related to fire fighter injuries, a measure of how often an injury occurs; e.g., sprains and strains are the most frequent fire fighter injuries.

fuel load Fuels provided by a building's contents and combustible building materials; also called fire load.

fuel-controlled fire A fire in which the heat release rate and growth rate are controlled by the characteristics of the fuel, such as quantity and geometry, and in which adequate air for combustion is available.

fully developed phase The phase of fire development where the fire is free-burning and consuming much of the fuel.

Glossary, continued

group Tactical Level Management Unit in charge of a function.

group fire A large fire involving several buildings that is confined within a complex or among adjacent buildings.

growth phase The phase of fire development where the fire is spreading beyond the point of origin and beginning to involve other fuels in the immediate fire area.

health care occupancy An occupancy used for purposes of medical or other treatment or care of four or more persons where such occupants are mostly incapable of self-preservation due to age, physical or mental disability, or because of security measures not under the occupants' control.

hot zone An operating area considered safe only when wearing appropriate levels of personal protective clothing; established by the IC and safety officer.

ignition phase The phase of fire development where the fire is limited to the immediate point of origin.

immediate danger to life and health (IDLH) Describes any atmosphere or condition that poses an immediate hazard to life or is capable of producing immediate, irreversible, debilitating effects on health.

incident action plan The objectives for the overall incident strategy, tactics, risk management, and member safety that are developed by the IC; these are updated throughout the incident.

incident commander (IC) The person in command of the entire incident and all related activities.

incident management team An incident management team is the incident commander (IC) and appropriate command and general staff personnel assisting the IC in managing an incident.

incident safety officer The command staff position assigned to monitor the scene for safety hazards or unsafe operations, enforce safety practices, and establish a safety plan.

indirect attack Firefighting operations involving the application of extinguishing agents to reduce the buildup of heat released from a fire without applying the agent directly onto the burning fuel; ventilation is kept to a minimum while a fog stream is directed at the ceiling; most useful in unoccupied tightly enclosed spaces.

industrial occupancy An occupancy in which products are manufactured or in which processing, assembling, mixing, packaging, finishing, decorating, or repair operations are conducted.

internal exposures Areas within the structure where the fire originates or within buildings directly connected to the building of origin that were not involved in the initial fire ignition.

liaison officer The command staff position responsible to communicate and coordinate with agencies not directly involved in meeting objectives identified in the incident action plan.

live load The weight of the building contents, people, or anything that is not part of or permanently attached to the structure.

lobby control A high-rise assignment in which a crew is responsible for duties related to managing the stairways, elevators, and the HVAC systems.

logistics section The section that obtains needed supplies, equipment, and facilities.

mercantile occupancy An occupancy used for the display and sale of merchandise.

mixed multiple occupancy A multiple occupancy where the occupancies are intermingled.

multiple occupancy A building or structure in which two or more classes of occupancy exist.

National Incident Management System (NIMS) A Department of Homeland Security system designed to enable federal, state, and local governments and private-sector and nongovernmental organizations to effectively and efficiently prepare for, prevent, respond to, and recover from domestic incidents, regardless of the cause, size, or complexity, including acts of catastrophic terrorism. Adopted by the U.S. Department of Homeland Security, this system uses Incident Command System (ICS) organizational terminology and structure, thus the terms ICS and NIMS can be used interchangeably when referring to organizational structure and terminology.

notation A piece of information about the premises, such as the building has damage from a previous fire. This information may accompany a pre-incident plan or may be available when the building does not have a pre-incident plan.

occupant density A measure of the number of people in a given area; usually stated as the number of square feet per person.

offensive A type of fire attack in which fire fighters advance into the fire building with hose lines or other extinguishing agents to overpower the fire.

operations section The section that manages all tactical units deployed at an incident scene.

overhaul Examination of all areas of the building and contents involved in a fire to ensure that the fire is completely extinguished.

oxygen- (ventilation-) controlled fire A fire in which the heat release rate and/or growth of the fire are controlled by the amount of air available to the fire.

personal alert safety system (PASS) A device that continually senses a lack of movement of the wearer and automatically activates the alarm signal indicating that the wearer is in need of assistance. The device can also be manually activated to trigger the alarm signal.

planning section The section that gathers and evaluates information, assists the incident commander (IC) in developing the incident action plan and tracks progress; also tracks resource status.

platform-frame construction Construction technique using separate components to build the frame of a structure (one floor at a time). Each floor has a top and bottom plate that act as fire stops.

pre-action sprinkler A dry sprinkler system that uses a deluge valve instead of a dry-pipe valve and requires activation of a secondary device before the pipes fill with water.

pre-fire conditions Factors that can be contributing factors to a collapse in a building that is heavily involved in fire, and which include construction types, weight, fuel loads, damage, renovations, deterioration, support systems, and related factors such as lightweight truss ceilings and floors.

pre-incident plan A written document resulting from the gathering of general and detailed information to be used by public emergency response agencies and private industry for determining the response to reasonably anticipated emergency incidents at a specific facility.

primary damage Damage caused by the products of combustion.

primary factors The most important factors, assessed during size-up, which change from incident to incident and depend on specific incident conditions.

public information officer (PIO) The command staff position responsible for relaying information to the public.

pyrolysis The chemical decomposition of a compound into one or more other substances by heat alone; the process of heating solid materials until combustible vapors are emitted.

Quad Multi-function apparatus that are equipped to provide for the following four functions: water, pumps, hose, and ground ladders. In other words, this is a Quint minus the aerial ladder.

Quint Multi-function apparatus that are equipped to perform both engine and truck company operations. These are equipped to provide for the following five functions: water, pumps, hose, ground ladders, and an aerial ladder.

rapid intervention crew (RIC) A minimum of two fully equipped personnel on-site, in a ready state, for immediate rescue of injured or trapped fire fighters.

rational decision making (RDM) A form of decision making in which input is obtained from diverse sources and careful analyses of all options is considered.

recognition primed decision making (RPD) A form of decision making in which the incident commander (IC) must decide on the proper course of action with limited information available in a relatively short period of time.

rehabilitation The process of providing rest, rehydration, nourishment, and medical evaluation to members involved in extended incident scene operations and/or extreme weather conditions.

residential board and care occupancy A building or portion thereof that is used for lodging and boarding of four or more residents, not related by blood or marriage to the owners or operators, for the purpose of providing personal care services.

residential occupancy An occupancy that provides sleeping accommodations for purposes other than health care or detention and correctional.

risk-versus-benefit analysis The process of weighing predicted risks to fire fighters against potential benefits for owners/occupants and making decisions based on the outcome of that analysis.

Royer/Nelson formula A rate-of-flow calculation that calculates the rate of flow as the volume in the fire area in cubic feet divided by 100 (V/100). This formula is based on the assumption that structural fires are primarily oxygen-controlled.

secondary damage Damage that is the result of fire-ground activities or the operation of a fire suppression system.

secondary factors Less important factors in an incident, which change from incident to incident and depend on specific incident conditions.

separated multiple occupancy A multiple occupancy where the occupancies are separated by fire resistance–rated assemblies.

severity The extent of an injury's consequence, usually categorized as death, permanent disability, temporary disability, and minor.

single command One person is designated as the incident commander (IC). This person is responsible for the development and implementation of the incident action plan. The IC can delegate staff and command positions as needed to assist in command and control functions.

smoke-proof tower A stairway designed to be separated from the building by a landing. This creates a separation that will limit the spread of smoke into the stairway and keep it clear for evacuation.

span of control The number of people reporting to a supervisor. The span of control should not exceed seven people reporting to a single supervisor under emergency conditions. As an example, a fire captain supervising an apparatus operator and three fire fighters would have a four-to-one span of control.

sprinkler system calculations Specific rate-of-flow calculations for sprinkler systems based on the fuel load; can be found in various publications including NFPA documents and Factory Mutual Data Sheets.

stack effect The vertical airflow within buildings caused by temperature differences between the building interior and the exterior; depending on conditions, stack effect could be positive or negative, causing smoke to move upward or downward.

standard operating procedures (SOPs) Written rules, policies, regulations, and procedures intended to organize operations in a predictable manner.

standpipe system A piping system with discharge outlets at various locations; in high-rise buildings an outlet will normally be located in the stairway on each floor level; most are connected to a water source and the pressure is boosted by a fire pump.

storage occupancy An occupancy used primarily for the storage or sheltering of goods, merchandise, products, vehicles, or animals.

strike teams A set number of resources of the same kind and type that have an established minimum number of personnel, e.g., five, four-person engine companies. Strike teams always have a leader (usually in a separate vehicle) and have common communications among resource elements.

Superfund Amendments and Reauthorization Acts (SARA) Enacted in 1986, Title III of this federal law is also known as the Emergency Planning and Community Right to Know Act. Title III requires businesses that handle or store hazardous chemicals in quantities above specific limits to report the location, quantity, and hazards of those chemicals to the State Emergency Response Commission (SERC), Local Emergency Planning Committee, and the local fire department. Some of this information has been classified since the attack on the World Trade Center on 9/11/2001.

task forces Any combination of resources that can be temporarily assembled for a specific mission. All resource elements within a task force must have common communications and a leader. Task forces should be established to meet specific tactical needs and should be demobilized as single resources. A typical task force would be two engine companies and a truck company under the supervision of a chief officer.

ten-codes A numeric code used to communicate predefined situations or conditions. For example "10-4" generally means "finished communicating." The use of ten-codes is discouraged.

trench cut A cut made from bearing wall to bearing wall to prevent horizontal fire spread in a building.

two-in/two-out rule Fire fighters working inside the hazard area must work in crews of at least two people (two-in). The two people working in the hazard area must be backed up by at least two people outside the hazard area (two-out) who are properly equipped and immediately available to come to the aid of the inside crew.

Type I construction Construction in which the structural members, including walls, columns, beams, girders, trusses, arches, floors, and roofs, are of approved noncombustible or limited-combustible materials and which have the highest level of fire resistance ratings.

Type II construction Construction in which the structural members, including walls, columns, beams, girders, trusses, arches, floors, and roofs, are of approved noncombustible or limited-combustible materials but the fire resistance rating does not meet the requirements for Type I construction.

Type III construction Construction in which exterior walls and structural members that are portions of exterior walls are of approved noncombustible or limited-combustible materials, and interior structural members, including walls, columns, beams, girders, trusses, arches, floors, and roofs, are entirely or partially of wood of smaller dimensions than required for Type IV construction or of approved noncombustible, limited-combustible, or other approved combustible materials.

Type IV construction Construction in which exterior and interior walls and structural members that are portions of such walls are of approved noncombustible or limited-combustible materials. Other interior structural members, including columns, beams, girders, trusses, arches, floors, and roofs, are of solid or laminated wood without concealed spaces. Wood columns supporting floor loads are not less than 8″ (203 mm) in any dimension; wood columns supporting roof loads only are not less than 6″ (152 mm) in the smallest dimension and not less than 8″ (203 mm) in depth. Wood beams and girders supporting floor loads are not less than 6″ (152 mm) in width and not less than 10″ (254 mm) in depth; wood beams and girders and other roof framing supporting roof loads only are not less than 4″ (102 mm) in width and not less than 6″ (152 mm) in depth. Specifics for other structural members are required to be large dimension lumber as well.

Type V construction Construction in which exterior walls, bearing walls, columns, beams, girders, trusses, arches, floors, and roofs are entirely or partially of wood or other approved combustible material smaller than material required for Type IV construction.

U.S. National Fire Academy formula A rate-of-flow calculation that calculates the rate of flow as the area in feet squared divided by 3 (A/3).

unified command Application of the National Incident Management System when there is more than one agency with incident jurisdiction or when an incident crosses political jurisdictional boundaries. Agencies work together through the designated members of the unified command, often the senior person from an agency participating in the unified command, to establish a common set of objectives making up the incident action plan.

unity of command A pyramidal command system ensuring that no one reports to more than one supervisor.

warm zone Intermediate area between the hot and cold zone where personal protective equipment is required, but at a lower level than in the hot zone.

wet pipe sprinkler system A sprinkler system in which the pipes are normally filled with water.

wood truss An assembly made up of small dimension lumber joined in a triangular configuration that can be used to support either roofs or floors.

Index

A

A/3 water flow rate formula, 226
absorbent materials and excess water, 252–253
access points, 121. *See also* egress routes; entry to buildings, gaining
   apartment buildings, 277
   collapse zone, establishing, 59, 121–122, 234–236
   exclusion zones, establishing, 121–122
   identifying, 59. *See also* egress routes
   laddering the building, 132–134
     elementary schools, 269
     for rescue egress, 153–154
     scaffolding, 287
accountability, 6, 58, 130–132. *See also* command staff; incident commanders (ICs)
   defined, 142
   fire zones and perimeters, 122–123
   high-rise building fires, 317–318
   rehabilitation and, 140–141
administrative decision-making, 2
aerial ladders, for rescue, 153, 269
agencies, multiple, and unified command approach, 3–4
air protection. *See* protective clothing
airborne health hazards, 138
airport fire (Denver 1990), 184
airport fire (Düsseldorf 1996), 301
alarms. *See* fire alarms
ALC unit for rehabilitation, 140
Alfred P. Murrah Federal Building bombing (Oklahoma City, 1995), 3–4, 23, 85. *See* rehabilitation area
alphanumeric naming systems, 21–22
alternative egress, 132–134, 135
Altoona, Pennsylvania shopping mall fire, 281
ambulatory care facilities, 270–271, 272–273
apartment buildings, 277–278. *See also* high-rise buildings; residential occupancies
   Bremerton, Washington fire, 156, 158, 278
apparatus. *See* equipment, apparatus, and tools
area of buildings, evaluating, 61–62
area unit conversation formulas, 353
areas of refuge
   defined, 72
   identifying during size-ups, 57
   shelter needs, 164, 254
   unconscious victims, 156
   for valuable property, 254
arena occupancies, 265–266. *See also* assembly occupancies
arson. *See* fire scene investigation
ASETBX data sample, 347–352
assembly occupancies, 261–267
   defined, 290
   Memphis, Tennessee church fire (2006), 288
   Pittsburgh, Pennsylvania church fire (2004), 264
   Station Night Club fire (2003), 264
assisted living facilities, 272
attack pumpers, 100, 102
attack strategies, determining, 62. *See also* defensive operations; incident action plans; offensive operations
   life safety concerns, 82–83
   offensive vs. defensive operations, 83–84, 86–87, 234
     rate of flow and, 194
     risk-versus-benefit analysis, 59
     structural stability and, 113
automatic dry standpipe systems, 179
automatic fire suppression equipment, 64–65. *See also* fire protection systems
automatic wet standpipe systems, 179
automobiles/trucks as command posts, 7

B

back-pressure, fire pumps, 179, 188
backdraft conditions, 128. *See also* flashover, correlation with elapsed time
   assessing risk of, 53
   defined, 72, 142
backing up fire protection systems
   deluge systems, 179
   non-water-based extinguishing systems and, 186–187
   sprinkler systems, 177
backup lines, offensive operations, 216
balloon-frame construction
   defined, 72
   features, 46
   lack of fire separations in, 61
   pre-incident planning for, 45
Bangkok, China hotel fire (1997), 311
base (high-rise fires), 320
BASF plant fire (1990), 20
   IMS charts, 19
   incident management, 6
   span of control for, 18
benefit, in risk-versus-benefit analysis, 86
"big box" stores, 280, 282
Boston, Massachusetts restaurant file, 186
branches and branch structure, 17–19
   defined, 27
Brannigan's Building Construction for the Fire Service, 60
breaking windows, 152
breathing protection. *See* protective clothing
Bremerton, Washington apartment complex fire, 156, 158, 278
buddy system, 126
building access. *See* access points
building collapse. *See entries at* collapse; structural factors

364

building conservation. *See* property conservation
building construction. *See also* construction sites
  balloon-frame
    defined, 72
    features, 46
    lack of fire separations in, 61
    pre-incident planning for, 45
  concealed spaces, 61, 121
  construction types, 118. *See also* design loads
    identifying during pre-planning and size-ups, 60
    pre-incident planning and, 45
  fire extension, 118–121
    defined, 142
    in educational occupancies, 269
    probability of, evaluating, 61
  materials for. *See* construction materials for buildings
  platform-frame construction
    defined, 72
    features, 46
    pre-incident planning for, 45
  roofs. *See* roofs and roof operations
  truss construction, 114–115, 117
    floor assemblies, 118
    roofs and roof operations, 115, 116, 119, 128–129
building design loads, 112
  collapse potential, 116
  defined, 72, 142
  identifying during pre-planning and size-ups, 60
building fires vs. structural fires, 124
building materials, 113–116
  high-rise buildings, 309–310
business occupancies, 282–283
  Cook County Administration Building fire (2003), 304, 306, 307, 315, 324–325
  defined, 290
  First Interstate Bank Building fire (1988), 300, 311, 322–323
  Peachtree Plaza fire (1989), 323–324

C

capacity unit conversion formulas, 354
carbon dioxide extinguishing systems, 184–187
  property damage from, 251
cardiac deaths, fire fighters, 109
catastrophic collapse. *See entries at* collapse; structural factors
catchalls, to contain water run-off, 252, 316
ceilings, suspended, 118
cellular telephones for communications, 23
channeling water run-off, 252, 316
Chapel Hill, North Carolina fraternity building fire (1996), 157, 279
Charleston, North Carolina store fire (2007), 120
chemical extinguishing systems, 184–187
  property damage from, 251
chemical hazards, airborne, 138
chief aide, role and responsibilities, 13–14
children's behavior during fires, 268

chronology for size-ups, 68
church occupancies, 45, 262–263
  example incident action plan, 88–91
  Memphis, Tennessee fire (2006), 288
  Pittsburgh, Pennsylvania church fire (2004), 264
Class A foam, 184, 225. *See also* foam extinguishing systems
  defined, 188
  property damage from, 251
  protecting exposures, 236
clean agent extinguishing systems, 184–187
  property damage from, 251
cleanup responsibilities, 251
clothing, protective, 137–138
  elevator use, 303
  overhaul operations, 255
cold weather rehabilitation, 139–140
cold zone, 122–123
  defined, 142
  medical units in, 162
collapse. *See* structural factors
collapse, signs of, 59, 117
collapse zone
  defined, 142
  establishing, 59, 121–122, 234–236
college occupancies, 270
  Chapel Hill, North Carolina fraternity building fire (1996), 157
  dormitory housing, 278–279
combination attacks, 212–214, 226
Command Adjunct position, 14
command center, high-rise buildings, 308
command function, 40
command modes, company-level, 5
command post, 6–7
  high-rise buildings, 308
command staff, 9–11
  status reports, 5, 84, 130
    on initial incident conditions, 4–5
  truck company tasks, 97–98
command transfer procedures, 5–6
commercial buildings. *See* business occupancies; industrial occupancies; mercantile occupancies; storage occupancies
commercial property fires
  Altoona, Pennsylvania shopping mall fire, 281
  First Interstate Bank Building fire (1988), 300, 311, 322–323
  Hackensack, New Jersey building fire (1988), 115
  Phoenix, Arizona supermarket fire, 136
  Sofa Super Store fire (2007), 120
  Washington, D.C. store fire, 132
communications, 22–26, 84, 129–130
  common terminology in, 25–26
  establishing command, 4
  high-rise building fires, 317–318
  IMS-based networks, 24–25
  radios, 5, 23
communications unit (logistics section), 12–13

company officers, 4–5
   role during structural fires, 96
   safety obligations, 110
company operations. *See* standard operating procedures (SOPs)
compensation/claims unit (finance section), 12
complex pre-incident plans, 43–44, 72
computers
   for communications during incidents, 23
   for pre-incident plans, 50
   as property. *See* valuable property, covering or removing
concealed spaces, 121
   evaluating during size-ups, 61
condition of building, identifying, 60, 287
conflagrations, 239–242
   defined, 240, 244
conflagrations and group fires
   Perkasie, Pennsylvania conflagration, 243
   San Jose, California construction site fire (2002), 239
conservation of property. *See* property conservation
construction materials for buildings, 113–116
   high-rise buildings, 309–310
construction sites, 287
   San Jose, California construction site fire (2002), 239
   visiting, during pre-incident planning, 45–46
construction types, 118. *See also* design loads
   identifying during pre-planning and size-ups, 60
   pre-incident planning and, 45
containing run-off, 252, 316
control valves. *See* main control valves
convention center occupancies, 266. *See also* assembly occupancies
convergence cluster phenomenon, 276, 326
   defined, 290
converted buildings, 287
Cook Country Administration Building fire (2003), 304, 306, 307, 315, 324–325
correctional occupancies, 274–275
   defined, 290
cost unit (finance section), 12
covering valuable property, 253
crew size. *See* staffing needs
critical incident stress management, 141
critical time (time before flashover), 157. *See also* flashover
   defined, 166
   high-rise buildings, 316–317

D

damage to buildings. *See also* property conservation
   collapse and. *See also* structural factors
   from fire-ground operations, 250–252
   renovations, 113, 117
      occupancies and, 287–288
damage to buildings, collapse and, 116
dead loads, 112
   collapse potential, 116
   defined, 72, 142
   identifying during pre-planning and size-ups, 60
dead victims, separating walking wounded from, 163

deaths of fire fighters. *See* critical incident stress management; fire fighter safety
debris
   falling glass, high-rise buildings, 315–316, 317
   during overhaul operations, 255
   scatter in building collapse, 234. *See also* collapse zone
   water drainage and, 252
decay phase, 124, 128, 142
decision-making
   evacuation. *See* evacuation plans and drills
   at incident scene vs. administrative, 2
   offensive vs. defensive operations, 83–84, 86–87, 234
      rate of flow and, 194
      risk-versus-benefit analysis, 59
      structural stability and, 113
   organizational charts in, 2
defend-in-place tactics, 150, 152
   defined, 166, 290
defensive operations, 232–246
   apartment buildings, 277
   attack strategies, 83, 194
   changing from offensive operations to, 113
   collapse and exclusion zones, 121–122, 123
   conflagrations and group fires, 239–242
   deciding on, offensive operations vs., 83–84, 86–87, 234
      rate of flow and, 194
      risk-versus-benefit analysis, 59
      structural stability and, 113
   defined, 234, 244
   incident action plans, 86–88
dehydration (fire fighters), 139
delegating authority, 6
   to division/group level responders, 16
   during operations, 15
deluge systems, 173–174, 178–179
   defined, 188
deluge valve operation, 178
demobilization unit (planning section), 13
demolition, buildings under, 287
density unit conversion formulas, 354
Denver airport fire (1990), 184
department stores. *See* mercantile occupancies
deployment
   division-of-labor principle, 96
   incident action plans, 84, 86, 88–91
design loads, 112
   collapse potential, 116
   defined, 72, 142
   identifying during pre-planning and size-ups, 60
detection of fire, time needed for, 125
detention occupancies, 274–275
   defined, 290
deterioration of buildings, collapse and, 117
direct attacks, 211
   defensive operations, 236
   offensive operations, 226
direct hose streams, 136

direction of travel, identifying during size-ups, 54
discharge, standpipe systems, 182–183
dispatch centers, tracking elapsed time, 113
dispatch information, high-rise fires, 311
dispatch time, 125
division-of-labor principle, 96, 102, 165
divisions and division structure, 16–19
    defined, 27
    incidents needing sectoring, 16–17
    use of during BASF fire, 18–19
documentation unit (planning section), 13
Dodson, David W., 54
door locking, high-rise buildings, 304
dormitory housing, 278–279
    Chapel Hill, North Carolina fraternity building fire (1996), 157
draft curtains, 178
    defined, 188
drains for water run-off, 252, 316
drinking establishment occupancies, 263–265. See also assembly occupancies
dry chemical extinguishing systems, 185, 186–187
    property damage from, 251
dry pipe sprinkler systems, 173
    defined, 188
Düsseldorf, Germany airport terminal fire (1996), 301

E

eating establishment occupancies, 263–265. See also assembly occupancies
educational occupancies, 267–270
    defined, 266, 290
    Pangnirtung school fire (1997), 69
Edwards, Bruce, 203
egress routes, 152–154
    alternative (multiple), 132–134, 135
    directing occupants to, 155
    fire extension and, 118
    high-rise buildings, 300
    identifying during size-ups, 57
    laddering the building, 132–134
        elementary schools, 269
        for rescue egress, 153–154
        scaffolding, 287
    protecting, 150
electric power, water drainage and, 252
elementary schools, 268–269
elevated master streams, 236, 237, 239
    angle of deflection, 307
elevated platforms, for rescue, 153
    elementary schools, 269
elevator recall, 300
elevator use, 154
    high-rise buildings, 301–303
    lobby control, 321
emergency contact information, 48, 172
emergency evacuation signal, 130

emergency medical services. See EMS
Emergency Operations Centers (EOCs), 7
emergency voice/alarm communications systems (EVACS), 306
employees. See business occupancies
EMS (emergency medical services)
    evaluating need for, 162
    rehabilitation of fire fighters, 140
    triage, 163
enclosed shopping malls, 281
enclosures and fire separations
    backdraft and, 128
    concealed spaces, 61, 121
    evaluating during size-ups, 61
    hose streams and, 137
engine companies, 96–101
    first-arriving
        apparatus positioning, 100
        division of labor by, 98
        officers, 222
        staffing needs and attack approach, 220–221
    high-rise buildings, 315–317
    second-arriving, 221
Engine-RICs, 136
entry to buildings, 121. See also access points; egress routes
    apartment buildings, 277
    collapse zone, establishing, 59, 121–122, 234–236
    exclusion zones, establishing, 121–122
    forcible. See forcible entry
    identifying, 59. See also egress routes
    laddering the building, 132–134
        elementary schools, 269
        for rescue egress, 153–154
        scaffolding, 287
    sprinkler-protected buildings, 175, 177
equipment, apparatus, and tools
    apparatus positioning, 100–101, 236
    assessing needs for during size-up, 59
    Class A foam, 225
    defensive operations, 238
    fire fighter fatalities and, 106
    guidelines for, 66
    offensive operations, 225–226
    positioning by first-in truck company, 100
    pre-assigning, 98
    for property conservation, 254–255
    protective clothing, 137–138
        elevator use, 303
        overhaul operations, 255
    pump capacity for manual extinguishment, 64
    Quad apparatus, 102
    Quint apparatus, 99, 102
    for rapid intervention crews (RICs), 135
    relationship to SOPs and training, 41–42
    for search-and-rescue tasks, 158
    sending by elevator, 303
    sending up stairs, 304
    standpipe systems, 181–182

escape routes, 152–154
   alternative (multiple), 132–134, 135
   directing occupants to, 155
   fire extension and, 118
   high-rise buildings, 300
   identifying during size-ups, 57
   laddering the building, 132–134
      elementary schools, 269
      for rescue egress, 153–154
      scaffolding, 287
   protecting, 150
EVACS (emergency voice/alerts communications system), 306
evacuation plans and drills, 57, 150. *See also* occupancy, specific occupancy types
   alternative egress, 132–134, 135
   classifying evacuation status, 154–158
   conflagrations and group fires, 241
   emergency evacuation signal, 130
   high-rise buildings, 305–306, 315
   occupant self-escape, 154–155
   rescue options, analyzing, 152–154
   surveying floor layout and size, 159–160
   unconscious victims, 155–156, 158, 162–164
exclusion zones, defined, 142
exclusion zones, establishing, 121–122
Exit Drill in the Home (EDITH) educational program, 57
exposure evaluations
   apartment buildings, 277
   defensive operations, 236
   offensive operations, 215–216
extended operations, 136
extension. *See* fire extension
exterior access. *See* building access
exterior attacks. *See* defensive operations
exterior staging, high-rise fires, 320
external exposures, 62, 226, 236
extinguishment
   evaluating probability of, 150–151
   high-rise buildings, 306–308
extinguishment factors
   attack strategy. *See* attack strategies, determining
   automatic fire suppression equipment, 64–65
   external exposures, 62, 226, 236
   high-rise evacuations and, 315
   internal exposures, 62, 236
      high-rise buildings, 316
      shopping centers, 280
   manual extinguishment, 62–64
   probability of extinguishment, 62
   ventilation status, 62

F

face protection. *See* protective clothing
facilities unit (logistics section), 12–13
Falls Township, Pennsylvania, K-Mart fire, 209
Falls Township, Pennsylvania warehouse fire, 209
false alarms, from automatic systems, 187
fatalities to fire fighters. *See* fire fighter safety

fatalities to public. *See* life safety
Federal Emergency Management Agency (FEMA), 15
FESHE correlation guide, 357
Field Incident Technicians, 14
finance/administration section, 12, 27
fire alarms
   false, from automatic systems, 187
   time to transmission, 125
fire conditions, safety and, 117, 157
fire department accountability system model, 58
fire department connections
   buildings with multiple, 176
   supplying, 177, 180–181
fire doors, 150, 158
fire drills in schools, 268–269
fire escapes, 132, 152–153
fire extension, 118–121
   defined, 142
   in educational occupancies, 269
   probability of, evaluating, 61
fire fighter safety, 104–145, 127–128. *See also* fire zones and perimeters; life safety
   critical incident stress management, 141
   electricity and water drainage, 252
   fire conditions, 117
   fire extension, 118–121
   fire-ground operations, 129–138
      risk management, 111–116
   high-rise buildings, 300–301
   injury and fatality statistics, 106–111
      assembly occupancies, 261
      business occupancies, 283
      storage occupancies, 294
   on-site medical treatment, 163
   overhauls, 138
   pre-incident evaluation, 53–64, 116–117
      extinguishment factors, 62–64
      occupancy-related factors, 56–57
      overview, 53
      smoke and fire condition assessments, 53–54
      structural factors, 60–62
      ventilation status assessments, 54–56
   rehabilitation, 139–141
   response time to ignition, 124–127
      flashover, correlation with elapsed time, 124, 128–129
   roof operations, 116
   stream position and, 214
   tactical reserve, 127–128
fire-ground command system, 2, 62–64
fire-ground operations
   command system, 2, 62–64
   damage from, 250–252. *See also* property conservation
   safety and, 129–138
      risk management, 111–116
   safety issues in *NFPA 1500*, 107
fire intensity, 111–112
   high-rise buildings, 312–313
   structural stability evaluations, 113

fire lines (hose streams), 271
fire location, identifying, 54
fire protection systems, 172–187
    automatic fire suppression equipment, 64–65
    false alarms, 187
    non-water-based extinguishing systems, 183–187
        property damage from, 251
    pre-incident planning with, 172, 183
    sprinkler systems. *See* sprinkler systems
    standpipe systems, 179–183
        defined, 188
        high-rise buildings, 307–308
fire pumps, checking, 176–180
    deluge systems, 178
    sprinkler systems, 176, 177
    standpipe systems, 179, 180
fire scene investigation, 255
fire suppression water load, identifying, 60–61
fire walls, extending evacuation times, 150
fire zones and perimeters, 121–123
    collapse zone, establishing, 59, 121–122, 234–236
    defined, 142
    establishing during size-ups, 58
    exclusion zones, establishing, 121–122
    high-rise buildings, 317
    protective clothing, 137–138
        elevator use, 303
        overhaul operations, 255
firefighting tasks, 96–98
FIRESCOPE Incident Command System (ICS), 2
first-alarm fires, using SOPs for, 84
first-arriving (first-in) companies
    apparatus positioning, 100
    division of labor by, 98
    officers, 222
    staffing needs and attack approach, 220–221
first-in units, 4
First Interstate Bank Building fire (1988), 300, 311, 322–323
flaming, open, 54
flashover, 124
    assessing risk of, 53
    correlation with elapsed time, 124, 128–129
    defined, 72, 142
    rescue potential and, 157
floor layout and size, surveying, 159–160
floor numbering inconsistencies, 159, 310
floor volume, rate of flow calculations and, 197–198
Florence, Kentucky condominium fire (1999), 235
flow rate, sprinkler systems, 201–203, 226
    hydraulically designed systems, 201
flow-reducing valves in standpipe systems, 181–182, 307
flow unit conversion formulas, 355
flying brand hazards, 241
foam extinguishing systems, 183–184
    property damage from, 251
foam houses, 183, 188
food unit (logistics section), 12–13

forcible entry
    automatic alarms and, 187
    high-rise buildings, 304, 315
    need for flexibility in, 98
    property conservation and, 250
Ford dealership fire (New Jersey 1988), 115
formal pre-incident plans, 43–50
    defined, 44
formal rehabilitation of fire fighters, 139
FPETOOL programs, 347
freelancing, 5, 27
friction loss, 205, 308
fuel-controlled fires, 226
    water flow rate needs, 196, 197
fuel loads, 112–113
    assembly occupancies, 262
    collapse potential, 116
    defined, 72, 142
    identifying and updating in pre-incident plan, 49–50
    with manual extinguishment, 62
    pyrolysis and, 211
fully developed phase of fire, 124
    defined, 142
Fundamentals of Fire Fighter Skills, 196

G

gaining entry. *See also* access points
    forcibly. *See* forcible entry
    sprinkler-protected buildings, 175, 177
gases, behavior of, 54
general staff, 11–16
Geographic Information Systems (GIS), 50
gravity-related pressure loss, 307–308
ground ladders, 132–134
    elementary schools, 269
    for rescue egress, 153–154
    scaffolding, 287
ground-level winds at high-rise building fires, 313
ground support unit (logistics section), 12–13
group fires, 239–242
    defined, 240, 244
group/group structure, 16–17, 18–19, 27
growth phases of fires, 124, 142
guidelines, SOPs vs., 42
gusset plates, 115

H

Hackensack, New Jersey building fire (1988), 115
Halon extinguishing systems, 184–187
    property damage from, 251
hazardous materials, on property
    in convention centers, 266
    delegation of authority and, 6
    industrial occupancies, 285–286
    overhaul operations, 255
    pre-incident planning, 42–43
health care occupancies, 270–273
    defined, 290

heat intensity, 111–112
　high-rise buildings, 312–313
　structural stability evaluations, 113
heat protection. *See* protective clothing
heat transfer, conflagration and, 241
Heavy Rescue units, 223
height of buildings, evaluating, 61–62
helicopter rescues, 154, 304–305
helicopter transport of medical patients, 163
high-expansion foam systems, 183–184
　property damage from, 251
high-hazard occupancies, staffing needs for, 66
high-rise buildings, 277–278, 296–344
　case histories, 321–327
　command post location, 308
　Cook Country Administration Building fire (2003), 304, 306, 307, 315, 324–325
　defined, 299
　elevator use, 301–303
　evacuation, 150–151
　example incident action plan, 86–88
　extinguishment, 306–308
　fire fighter safety, 300–301
　First Interstate Bank Building fire (1988), 300, 311, 322–323
　floor numbering inconsistencies, 159
　incident action plans, 314–317
　life safety (general public), 304
　MGM Grand fire (1980), 153, 305, 314, 326–327
　New York City fires, 306
　NIMS application, 317–321
　offices and business occupancies, 283
　One Meridian Plaza fire (1991), 305, 311, 321–323
　Peachtree Plaza fire (1989), 323–324
　pre-incident planning, 308–310
　rescue and evacuation, 304–306
　size-ups, 310–314
　stairway support, 303–304
　standard operating procedures (SOPs), 299–300
　victim triage, 159
　World Trade Center bombings. *See* World Trade Center bombing (1993); World Trade Center bombing (2001)
high schools, 270
hose lines. *See also* master streams
　angle of deflection, 307
　direct, 136. *See also* direct attacks
　extending, RIC operations and, 136
　high-rise buildings, 307, 314
　improper venting and, 152
　with manual extinguishment, 63
　nozzles. *See entries at* nozzle
　number of, with offensive operations, 214–216, 220–224
　opposing, 136–137
　preconnected to standpipe systems, 180
　protecting egress routes, 150
　size of, 204–205
　water flow rates and, 194–195. *See also* water supply
　weight of water from, 252–253
hospital occupancies, 271

hospitals, patient transport to, 163
hot weather rehabilitation, 139–140
hot zone, 122–123, 142
Hotel Vendome fire (1972), 120
hotels and motels, 276
　Hotel Vendome fire (1972), 120
　MGM Grand fire (1980), 153, 305, 314, 326–327
　President Hotel fire (1997), 311
hydration (fire fighters), 139

I

I-beams, 116
IDLH (immediate danger to life and health), 123
ignition, time since
　effective actions and, 124–127
　flashover, correlation with elapsed time, 124, 128–129
　high-rise buildings, 316–317
　structural stability evaluations, 113
　tracking elapsed time, 113
ignition phase, defined, 142
immediate danger to life and health (IDLH), 123
immobile victims, evacuating, 155–156, 158, 162–164
imperial and metric conversions, 353–355
in-place suppression devices. *See* fire protection systems
incident action plans, 80–91
　deployment planning, 84, 86, 88–91
　division-of-labor approach, 96
　examples of
　　church fire, 88–91
　　high-rise apartment building, 86–88
　　single-family detached dwelling, 84–86
　high-rise buildings, 314–317
　life safety needs, evaluating, 82–83
　offensive vs. defensive approaches, 83–84, 86–87, 234
　rapid intervention crews (RICs), 58, 134–136
　　defined, 72
incident commanders (ICs), 4–7, 9
　changing of, during incident, 5
　command post, 6–7, 308
　defined, 27
　establishment of goals and objectives during incidents, 83–84
　offensive operations, 220
　progress reports to, 5, 84, 130
　　on initial incident conditions, 4–5
　references to, over radio, 23
　role and responsibilities, 5–6, 9
　　safety obligations, 107, 110
　　tracking elapsed time, 113
　use of SOPs, 41
incident management system, 2–3. *See also* NIMS
incident management team, 15, 27
incident safety officers, 9–10
incident stress management, 141
indirect application theory (water flow calculations), 195
indirect attacks, 211, 226
indirect damage, estimating, 250–251
industrial occupancies, 285–286
　defined, 290

informal rehabilitation of fire fighters, 139
initial attack staffing, 126–127
injuries to fire fighters. *See* fire fighter safety
injuries to public. *See* life safety
inmates. *See* detention occupancies
intelligence sections, 13
intensity of fire, 111–112
    high-rise buildings, 312–313
    structural stability evaluations, 113
interior access, 59. *See also* building access
interior hose lines, 136–137
interior spaces, evaluating size of, 61–62
interior staging, high-rise fires, 304, 320–321
interior stairs, 132, 152
interlocks, checking
    deluge systems, 178–179
    non-water-based extinguishing systems, 185–186
internal exposures, 62, 236
    high-rise buildings, 316
    shopping centers, 280
investigation. *See* fire scene investigation
ISO formula for response time, 126

J

junior high schools, 270
jurisdictions, multiple, 3–4

K

K-Mart warehouse fire, 209
kitchen hood systems, 185, 187

L

laddering the building, 132–134
    elementary schools, 269
    for rescue egress, 153–154
    scaffolding, 287
ladders, aerial, 153, 269
Las Vegas, Nevada Hilton fire (1981), 328–332
Las Vegas, Nevada hotel fire (1980). *See* MGM Grand fire (1980)
law enforcement. *See* police
Layman, Louis, 195
layouts of floors and buildings, including, 62
lead time, high-rise building fires, 316–317
length unit conversion formulas, 353
liaison officer, 9
    defined, 27
    role and responsibilities, 10
life safety (general public), 148–165. *See also* fire fighter safety; search and rescue operations
    conflagrations and group fires, 241
    dormitory housing, 278
    evacuation status, 154–158. *See also* evacuation plans and drills
    extinguishment, probability of, 150–151
    floor layout and size, 159–160
    high-rise buildings, 87, 304
    incident action planning, 82
    medical status of occupants, 57, 162–164. *See also* health care occupancies; unconscious victims
        nursing homes, 271–272
        residential board and care occupancies, 273–274
    number of people needing assistance, 158–159
    potential victims, estimating number of, 289
    pre-incident planning and size-ups, 51–57
        extinguishment factors, 62–64
        occupancy type assessments, 53, 56–57
        structural factors, 60–62
    rescue options, analyzing and prioritizing, 152–154, 160–162
    responsibility for, 96
    shelter needs, 164
    single-family dwelling fires, 85
    staffing requirements, 164–165
    stream position and, 212–214
    ventilation issues, 98–99
    ventilation profile, 151–152
lifestyle centers, 281–282
limited care facilities, 272
live loads, 112
    collapse potential, 116
    defined, 72, 142
    identifying during pre-planning and size-ups, 60
loads (building design), 112
    collapse potential, 116
    defined, 72, 142
    identifying during pre-planning and size-ups, 60
lobby control, 301, 308, 321
    defined, 328
local flooding systems, 184, 186
lock boxes, 172, 187. *See also* gaining entry
locked doors, high-rise buildings, 304
logistics section
    defined, 27
    functions and subsections, 12
    hierarchy, 13
low-hazard occupancies, staffing needs for, 66
low-rise elevators, 302

M

main control valves
    deluge systems, 178
    sprinkler systems, 175–176, 177
    standpipe systems, 180
main pump, sprinkler systems, 176–180
    deluge systems, 178
    sprinkler systems, 176, 177
    standpipe systems, 179, 180
maintenance, equipment, 206
malls (shopping), 280–281
manual dry standpipe systems, 179
manual extinguishment, 62–64
manual wet standpipe systems, 179
marking areas searched, 154–158, 161
Marrs, Gary, 3–4
mass-casualty incidents, evaluating victims in, 163–164

master streams
   angle of deflection, 307
   defensive operations, 237–238
   direct hose streams, 136
   high-rise buildings, 307, 314
   manual extinguishment, 63
   offensive operations, 199, 204, 214
   opposing fire lines, 136–137
   weight of water from, 253
materials of building construction, 113–116
   high-rise buildings, 309–310
medical status of occupants, 57, 162–164. *See also* health care occupancies; unconscious victims
   nursing homes, 271–272
   residential board and care occupancies, 273–274
medical unit (logistics section), 12–13
medium-hazard occupancies, staffing needs for, 66
Memphis, Tennessee church fire (2006), 288
mentally challenged occupants, 273
mercantile occupancies, 280–282
   Altoona, Pennsylvania shopping mall fire, 281
   defined, 290
metal trusses. *See* truss construction
metric and imperial conversions, 353–355
MGM Grand fire (1980), 153, 305, 314, 326–327
middle schools, 270
military command operations, 2
mixed occupancies, 286–287, 290
Mobile Data Computers (MDCs), 23
Mobile Data Terminals, 23
mobile RICs, 135
motels and hotels, 276
mouth protection. *See* protective clothing
movie theater occupancies, 266–267. *See also* assembly occupancies
moving valuable property, 253
multi-level department stores, 280, 282
multi-story buildings, enclosures and fire separations, 61. *See also* high-rise buildings
multiple conversations, handling, 25, 129
multiple fire department connections, buildings with, 176
multiple-occupancy buildings, 286–287, 290

N

naming systems, intuitive, 21–22
National Incident Management System (NIMS), 2–4, 9
   as accountability system, 58
   branches, divisions, and groups, 16–21
   chart for sectored incidents, 17
   chart for small incidents, 16
   command staff, 9–11
   defined, 27
   general staff, 11–20
   high-rise buildings, 317–321
   incident organization and coordination using, 58
   as safety system, 130
   total staffing vs. incident staffing needs, 66
   unity of command, 24–25
National Incident Management System (U.S. Homeland Security), 16

negative-pressure ventilation, 178. *See also* ventilation
   life safety and, 151–152
New York City
   World Trade Center bombing (1993), 6
   World Trade Center bombing (2001), 3–4, 7
New York, New York high-rise fires, 306, 311
NFPA Fire Protection Handbook
   NFPA 13, 172, 201
   NFPA 14, 179, 208
   NFPA 25, 173
   NFPA 220, 60
   NFPA 1142, 202
   NFPA 1223, 125
   NFPA 1410, 126
   NFPA 1500, 58, 106–107
      initial attack staffing, 126–127
      on relationship of training to SOPs, 42
      risk management principles, 111
      SCBA (self-contained breathing apparatus), 138
   NFPA 1521, 10
   NFPA 1561, 2
      deletion of word "sector" from, 20
      on SOPs, 41
   NFPA 1584, 139–140
   NFPA 1620, 43, 49
   NFPA 1710, 66, 125–127, 220–224
   NFPA 1720, 126
   NFPA 1962, 179–180
   NFPA 5000, 60
   sprinkler system calculations, 202
   staffing guidelines, 66
NIMS. *See* National Incident Management System (NIMS)
9/11. *See* World Trade Center bombing (2001)
non-water-based extinguishing systems, 183–187
   property damage from, 251
nonattack operations, 242–243
noncombustible buildings, 118–119
notation
   defined, 72
   in pre-incident plans, 45
nozzle flows, 206, 206–207
nozzle pressure, 205, 206–207
nozzle reaction force, 208
nozzles, automatic, 182
nozzles, types of, 210–211
NPP (neutral pressure plane), 313
numbering of building floors, 159, 310
nursing home occupancies, 271–272

O

occupancy, 258–293
   classifying occupancy type, 56–57, 260–261
   defined, 57
   general considerations, 288–289
   high-rise buildings, 308
   incident staffing needs and, 66
   indirect attacks and, 211
   number of potential victims, estimating, 289

pre-incident planning, 43, 45
SOPs for, 40
specific occupancy types, 260
　assembly occupancies, 261–267
　buildings under construction, renovation, or demolition, 287
　business occupancies, 282–283
　detention and correctional occupancies, 274–275
　educational occupancies, 267–270
　health care occupancies, 270–273
　industrial occupancies, 285–286
　mercantile occupancies, 280–282
　multiple- and mixed-occupancy buildings, 286–287
　renovated buildings, 287–288
　residential board and care occupancies, 273–274
　residential occupancies, 275–280
　storage occupancies, 283–285
occupant, defined, 57. *See also* life safety (general public)
occupant density, 260
　defined, 290
　dormitory housing, 278
　hotels and motels, 279
occupant escape and survival. *See also* evacuation plans and drills; life safety
　children, 268
　number of people needing assistance, 158–159
　panic and role models, 265
　potential victims, estimating number of, 289
　prioritizing rescues, 160–162
　time since ignition and, 124
　victim triage, 157–158, 163–164
occupant shelter and fire fighter rehabilitation (REHAB) stations, 67
offensive operations, 83, 192–231
　apparatus needs, 224
　changing to defensive operations, 113
　deciding on, defensive operations vs., 83–84, 86–87, 234
　　rate of flow and, 194
　　risk-versus-benefit analysis, 59
　　structural stability and, 113
　defined, 194, 234, 244
　hose lines, number needed, 212–216
　hose size selection, 204–208
　incident action plans, 86–88
　nozzle types, 210–211
　one- and two-family dwellings, 277
　property conservation and, 250–255
　rate of flow calculations
　　calculation approaches, 194–199
　　comparisons/choosing, 201–204
　　and percentage of area on fire, 200–201
　staffing needs, 220–224
　stream position, 210–214
　　combination attacks, 212–214
　ventilation needs, 218–220
　water supply needs, 216–218
office occupancies, 282–283
Oklahoma City, Oklahoma. *See* Alfred P. Murrah Federal Building bombing (Oklahoma City, 1995)

on-duty deaths, 106–111
　assembly occupancies, 261
　business occupancies, 283
　storage occupancies, 294
one-family dwellings, 277. *See also* residential occupancies
　floor numbering inconsistencies, 159
　life safety (general public), 85
One Meridian Plaza fire (1991), 305, 311, 321–323
124 West 60th Street high-rise fire, 311
operations section, 14–16, 27
opposing fire lines, 136–137
organizational charts, using to make decisions, 2
overhaul operations, 138, 255
　defined, 142
　property conservation and, 250
oxygen- (ventilation-) controlled fires, 186, 226
　carbon dioxide extinguishing systems, 184
　water flow rate needs, 196

P

PAHs (polycyclic aromatic hydrocarbons), 138
Pangnirtung, Northwest Territories school fire (1997), 69
panic during escape attempts, 265
PAR tags, 131–132
paramedics. *See* EMS
partial evacuation, high-rise buildings, 305–306
PASS (personal alert safety system), 134
　defined, 142
Pattaya, Thailand hotel fire, 311
Peachtree Plaza fire (1989), 323–324
people (potential victims). *See entries at* occupant; life safety (general public)
percentage-of-involvement modifier, 200–201
perimeter areas. *See* fire zones and perimeters
Perkasie, Pennsylvania conflagration, 243
personal alert safety system (PASS), 134
　defined, 142
personal items of occupants. *See* valuable property, protecting
personal protective clothing, 137–138
　elevator use, 303
　overhaul operations, 255
personnel. *See* staffing needs
Phi Gamma Delta fraternity house fire (1996), 157, 279
Philadelphia, Pennsylvania high-rise fire (1991), 305, 311, 321–323
Phoenix, Arizona supermarket fire, 136
PIO. *See* public information officer (PIO)
Pittsburgh, Pennsylvania church fire (2004), 264
Pittsburgh, Pennsylvania residential fire, 129, 160
planning section, 12–14
　defined, 27
plateglass windows, 315–316
platform-frame construction
　defined, 72
　features, 46
　pre-incident planning for, 45
police (law enforcement)
　delegation of authority to, 6
　role during structural fires, 3–4

positions, preassigning, 98–99
positive-pressure ventilation, 178. *See also* ventilation
　life safety and, 151–152
pre-action sprinkler systems, 173
　defined, 188
pre-burn time, 125
pre-fire collapse potential, 116–117
pre-fire conditions, defined, 142
pre-incident planning, 40, 42–53. *See also* size-ups
　collapse zones, 235
　defined, 72
　detention and correctional occupancies, 275
　with fire protection systems, 172, 183
　formal plans, 43–50
　high-rise buildings, 308–310
　　elevator use, 301–302
　preparation and time involved, 49–50
　storage occupancies, 294
preconnected hoses with standpipe system, 180
preplanning. *See* pre-incident planning
preservation (historical) of buildings, 287
President Hotel fire (1997), 311
pressure-reducing valves (PRVs), 181–182, 307
pressure unit conversion formulas, 355
pressure, ventilation, 178
pressure, water. *See* water flow rates
primary damage (property), 251
　defined, 256
primary factors, defined, 53, 72
primary search and rescue, 161–162
　organizing during size-ups, 58
prioritizing medical assistance. *See* victim triage
prioritizing rescues, 160–162
prisoners as occupants. *See* detention occupancies
private fire protection, addressing in SOPs, 40
procurement unit (finance section), 12
progress reports (status reports), 5, 84, 130
　on initial incident conditions, 4–5
property conservation, 65–66, 248–256
　evaluating needs for, 254–255
　fire scene investigation, 255
　offensive operations and, 250–255
　pre-incident planning for, 42–43, 51
　as reason for defensive attacks, 234. *See also* defensive operations
　salvageable property, 65, 286
　at specific occupancy types
　　colleges and universities, 270
　　high-rise buildings, 316
　　industrial occupancies, 286
　　mercantile occupancies, 280
　　at religious properties, 263
　　residential properties, 277
　sprinkler systems, 178
　water damage, 65, 251
　　evaluating, 252–254
protective clothing, 137–138
　elevator use, 303
　overhaul operations, 255

PRVs (pressure-reducing values), 181–182, 307
PSAP (Public Safety Answering Point), 125
public information officer (PIO), 9–11
　defined, 27
public property. *See* property conservation
public relations from property conservation, 250
Public Safety Answering Point (PSAP), 125
pump, sprinkler systems
　deluge systems, 178
　sprinkler systems, 176, 177
　standpipe systems, 179, 180
pump discharge pressure, 208
pyrolysis, 128, 211, 226
　defined, 142

Q

Quad companies/apparatus, 99, 102
Quint companies/apparatus, 99, 102

R

radiant heat, exposure levels for, 236
radio communications, 5, 23, 129–130, 318
　handling multiple communications, 25, 129
RAND average speed, 126
rapid intervention crews (RICs), 58, 134–136
　defined, 72
rapid removal operations, 136
rate of flow. *See* water flow rates
rational decision making (RDM), 2
　defined, 28
reactivating sprinkler systems, 178
recognition primed decision making (RPD), 2, 28
"Red-Card" system (forest service), 15
refuge areas
　defined, 72
　identifying during size-ups, 57
　shelter needs, 164, 254
　unconscious victims, 156
　for valuable property, 254
regional planning, in development of SOPs, 41
REHAB (rehabilitation) shelters, 67
rehabilitation area, 140
rehabilitation of fire fighters, 139–141
　defined, 143
　overhaul and, 138
religious properties, fires at. *See* church occupancies
removing valuable property, 253
renovations to buildings, 113, 117
　occupancies and, 287–288
replacing sprinkler system heads, 178
reports during incidents, 5, 84, 130
　on initial incident conditions, 4–5
rescue operations. *See* search and rescue operations
residential board and care occupancies, 273–274
　defined, 290
residential fires
　Bremerton, Washington apartment complex fire, 156, 158, 278
　Chapel Hill, North Carolina fraternity building fire (1996), 157, 279

Florence, Kentucky condominium fire (1999), 235
Pittsburgh, Pennsylvania fire, 129, 160
residential occupancies, 275–280
    defined, 290
    fire fighter fatalities, 107
    high-rise. *See* high-rise buildings
    incident action plans, 84–88
    property. *See* property conservation; salvageable property
residential sprinkler systems, 172
resource capabilities and needs
    equipment. *See* equipment, apparatus, and tools
    high-rise apartment building fires, 87–88
    incident action plans, 83, 86, 89
    staffing. *See* staffing needs
response time (to event), 125, 126
responsibility, 6
RESTAT (Resource Status) Departments, 13, 14
restaurant occupancies, 263–265. *See also* assembly occupancies
    Boston, Massachusetts restaurant file, 186
restoration of non-water-based extinguishing systems, 186, 187
reversible venting, 152
RICs (rapid intervention crews), 58, 134–136
    defined, 72
risk-benefit analysis
    building condition and, 60
    defined, 86
    and firefighting strategy decisions, 59
    incident action plans, 83, 85, 87, 88
risk management, 111–116, 285
role models during escape, 265
roofs and roof operations
    building construction, 60, 114–116
    helicopter rescues, 154
    safety, 116
    ventilation
        high-rise apartment building fires, 87
        for life safety, 151–152
    wind passing over, 313–314
rope rescue, 154
routes of egress or escape, 152–154
    alternative (multiple), 132–134, 135
    directing occupants to, 155
    fire extension and, 118
    high-rise buildings, 300
    identifying during size-ups, 57
    laddering the building, 132–134
        elementary schools, 269
        for rescue egress, 153–154
        scaffolding, 287
    protecting, 150
Royer/Nelson formula (V/100) (water flow rate), 226
    comparison with other approaches, 201–204
    and percentage of area on fire, 200–201
    when to use, 195–196
RPD. *See* recognition primed decision making (RPD)
run-off, containing, 252, 316
rural operations, staffing needs for, 53–59

S

safety
    of fire fighters. *See* fire fighter safety
    importance of incident commander, 4, 5
    incident safety officers, 9–10
    of public. *See* life safety
safety officers, 132
salvageable property, 65, 286
San Jose, California construction site fire (2002), 239
Santana Row construction site fire (2002), 239
saving lives. *See* life safety
saving property. *See* property conservation
scaffolding, 287
SCBA (self-contained breathing apparatus), 138. *See also* protective clothing
    elevator use, 303
    mandates for use of, 42
schools. *See* educational occupancies
scuttles, opening, 55. *See also* ventilation
search and rescue operations. *See also* life safety (general public)
    analyzing rescue options, 152–154
    assessing staff and equipment needs, 58–59
    fire suppression staffing and, 130
    flexibility in, 98
    high-rise buildings, 304–306, 315. *See also* high-rise buildings
    hotels and motels, 279
    mercantile occupancies, 280
    non-water-based extinguishing systems and, 185
    prioritizing rescues, 160–162
    surveying floor layout and size, 158–160
    as truck company task, 97–98
Seattle, Washington warehouse fire (1995), 119, 130, 160
second-arriving companies, 221
secondary damage (property), 251
    defined, 256
secondary factors, 53, 72
    organizing during size-ups, 59
secondary search and rescue, 161–162
section and sections chiefs, 12–16
sectoring incidents, 16–17
sectors, 20
security issues, detention and correctional occupancies, 275
self-contained breathing apparatus (SCBA), 138. *See also* protective clothing
    elevator use, 303
    mandates for use of, 42
semiautomatic dry standpipe systems, 179
separated multiple occupancies, 286
    defined, 290
sequential evacuation, high-rise buildings, 305–306
set-up time, 125, 126–127
severity, defined, 143
shards of glass, high-rise building fires, 315–316
shelter needs, 164, 254. *See also* areas of refuge
shipboard fires, 211
shopping centers and malls, 280–281
    Altoona, Pennsylvania shopping mall fire, 281

shutting down
   deluge systems, 179
   non-water-based extinguishing systems, 185
   sprinkler systems, 174, 177
      minimizing water damage, 251, 252
single command
   defined, 28
   unified command vs., 3
single-family detached dwellings, 277. *See also* residential occupancies
   floor numbering inconsistencies, 159
   life safety (general public), 85
SITSTAT (Situation Status) Departments, 13, 14
situation unit (planning section), 13
size-ups, 51–62
   checklist for, 52
   chronology for, 68
   high-rise buildings, 310–314
   structural factors, 59–62
   time of day as factor in, 67
   weather conditions and, 67–68
skin protection. *See* protective clothing
small incidents, NIMS chart for, 16
smoke and fire condition assessments, 128
smoke damage, 65–66
smoke movement, high-rise buildings, 310, 312–314
   elevators and, 301, 303
   reports of, 310
   smoke-proof tower, 304, 328
smoke removal, 254
smoke velocity, fire location and, 54
Sofa Super Store fire (2007), 120
SOPs. *See* standard operating procedures (SOPs)
span of control
   approaches to reducing, 17–21
   defined, 28
special occupancy, 40
special operations, 40
splitting crews, 131
sports arena occupancies, 265–266. *See also* assembly occupancies
sprinkler heads, replacing, 178
sprinkler systems, 172–174
   buildings with, exterior evidence of fire, 177–178
   buildings with, no evidence of fire, 175–176
   calculations of rate of flow, 201–203, 226
      hydraulically designed systems, 201
   in convention centers, 266
   deluge systems, 173–174, 178–179
      defined, 188
   discharge unit conversion formulas, 355
   flow rate calculations, 197
   high-rise buildings, 309
   in large noncombustible buildings, 119
   mercantile occupancies, 280–282
   nursing homes, 272
   property damage from, 251
   shutting down, 174, 177, 251, 252
      deluge systems, 179

   storage occupancies, 294
   theater occupancies, 267
stability, structural. *See* structural factors
stack effect, 313, 328
staffing needs
   accountability. *See* accountability
   addressing in SOPs, 66
   defensive operations, 237–238, 238
   life safety requirements, 164–165
   with manual extinguishment, 63
   occupancy and, 289
   offensive operations, 220–224
   property conservation, 254
   rapid intervention crews (RICs), 58, 134–136
      defined, 72
   set-up time and, 126–127
   tactical reserve, 67, 127–128
   total staffing vs. incident staffing needs, 66
staging function
   high-rise buildings, 304, 320–321
   standard operating procedures (SOPs), 40
staging reserves, 67, 127–128, 163
stairways
   as egress routes, 152, 155
   factors that affect smoke levels in, 55
   high-rise buildings, 303–304, 315
   in schools, 269
   venting, 151
standard operating procedures (SOPs), 4–5, 40–42, 49–51, 62–71
   defined, 72
   extinguishment and attack strategies, size-ups for, 62–64
   first-alarm situations, 84
   high-rise buildings, 299–300
      elevator use, 301–302
   importance of following, 96
   life safety/fire fighter safety size-ups, 53–59
   life safety staffing, 165
   modifying during operations, 68–71
   overhaul operations, 255
   structural factors, size-ups for, 59–62
standpipe systems, 179–183
   defined, 188
   high-rise buildings, 307–308
Station Night Club fire (2003), 264
stationary command posts, 4
statistics on fire fighter safety, 106–111
   assembly occupancies, 261
   business occupancies, 283
   storage occupancies, 294
Status Message Terminals, 23
status reports (progress reports), 5, 84, 130
   on initial incident conditions, 4–5
storage occupancies, 283–285
   defined, 290
stores (shopping). *See* mercantile occupancies
stream force. *See* water supply

stream position. *See also* hose lines
    angle of deflection, 307
    direct attacks, 211–212
stress management, 141
strike teams
    defined, 28
    reducing span of control using, 20–21
structural factors, 59–62
    absorbent materials and excess water, 252–253
    buildings under construction, 287
    collapse, signs of, 59, 117
    collapse potential, evaluating, 116–118
    collapse zone, establishing, 59, 121–122, 234–236
    evaluating structural stability, 113–116
    floor volume and rate of flow calculations, 197–198
    high-rise buildings, 311
    in incident action plans, 83, 85–89
    risk-benefit analysis, 59
    scaffolding, 287
    time since ignition and, 124
structural fires vs. building fires, 124
structure access. *See* access points
students. *See* college occupancies; dormitory housing
Superfund Amendments and Reauthorization Act (SARA), 42–43, 72
supply of extinguishing agent, 186
supply of water. *See* water supply
supply unit (logistics section), 12–13
supplying fire department connections, 150, 158
suppression devices, in-place. *See* fire protection systems
surveying floor for evacuation, 159–160
suspended ceilings, 118, 315
    backdraft and, 128

T

tactical (staging) reserves, 127–128
    addressing in SOPs, 67
tactical worksheets for high-rise building files, 318–320
task forces
    defined, 28
    reducing span of control using, 20–21
tasks. *See* firefighting tasks
technical specialists (planning section), 13
temperature unit conversion formulas, 355
tempered-glass windows, 315–316
ten-codes, defined, 28
testing equipment, 206
theater occupancies, 266–267. *See also* assembly occupancies
thermal imaging camera, 135
third-arriving companies, 222
thirst (fire fighters), 139
time since ignition
    effective actions and, 124–127
    flashover, correlation with elapsed time, 124, 128–129
    high-rise buildings, 316–317
    structural stability evaluations, 113
    tracking elapsed time, 113

time unit (finance section), 12
toilet removal for water drainage, 252
tools. *See* equipment, apparatus, and tools
total collapse. *See entries at* collapse; structural factors
total flooding systems, 184, 185–186
total loss, recognizing, 236, 242–243
tracking fire fighters, 131
training
    incident management team training, 15
    relationship to SOPs and equipment, 41–42
    use of hoses and nozzles, 208
transmission of fire alarm, time for, 125
transport of victims needing medical treatment, 163–164
trapped fire fighters. *See* rapid intervention crews (RICs)
trauma care centers, transport to, 163
travel time (to event), 125, 126
trench cut, 226
triage, victim, 157–158, 163–164
truck companies, 96–101
    first-arriving
        apparatus positioning, 100
        division of labor by, 98
        officers, 222
        staffing needs and attack approach, 220–221
    high-rise buildings, 315–317
    second-arriving, 221
truss construction, 114–115, 117
    floor assemblies, 118
    roofs and roof operations, 115, 116, 119
        backdraft and, 128–129
turnout time, 125–126
20-minute rule, 113
    defined, 142
two-family dwellings, 277. *See also* residential occupancies
two-in/two-out rule, 16, 126
    defined, 28
types of construction, 60, 61
    defined, 72–73

U

unconscious victims, 155–156, 158, 162–164
unified command, 3–4
    defined, 28
university occupancies, 270
    dormitory housing, 278–279
U.S. Fire Administration, 15
U.S. Homeland Security. *See* NIMS
U.S. National Fire Academy formula (A/3) (water flow rate), 226
    comparison with other approaches, 202–203
    percentage-of-involvement modifier, 200–201
    underlying assumptions, 196–197
utilities (water, gas, electricity, etc.), 67

V

V/100 water flow rate formula, 226
valuable property, covering or removing. *See also* property conservation

valuable property, protecting, 253, 263, 270
   high-rise buildings, 316
   residential properties, 277
valve, main. *See* main control valves
ventilating sprinkler systems, 178
ventilation, 151–152. *See also* oxygen-controlled fire
   backdraft. *See* backdraft conditions
   combination attacks, 212–214
   high-rise buildings, 315–316
      stack effect, 313
   offensive operations, 218–220
   property conservation and, 250
   status assessments, 54–56, 62
ventilation, building
   master streams into openings, 237
   minimizing smoke damage, 65
   need for flexibility in, 98–99
   proper vs. improper, 99
victim triage, 157–158, 163–164
victims. *See* life safety (general public)
volume unit conversion formulas, 354

W

walkaround, 135
warehouse fires. *See also* industrial occupancies; storage occupancies
   K-Mart fire (Falls Township), 209
   Seattle, Washington fire (1995), 119, 130, 160
   Worcester, Massachusetts fire, 112
warm zone, 122–123
   defined, 143
wash downs, avoiding, 252
Washington, D.C. store fire, 132
water apparatus. *See* equipment, apparatus, and tools
water damage, 65, 251. *See also* property conservation
   evaluating, 252–254
water flow rates (pressure), 62–63, 203, 294
   attack strategy and, 194
   calculating, 195–199, 202–204
      estimating size of largest area, 197–198
      sprinkler systems, 197, 201–202
   field-expedient area and volume estimations, 199–200
   gravity-related loss, 307–308
   large undivided spaces, 280
   with manual extinguishment, 62–63
   nozzle pressure, 205. *See also entries at* nozzle
   and percentage of area on fire, 199–200
   pumper discharge pressures, 181–182, 208–210
   standpipe discharge pressure, 181–182, 208–210
      high-rise buildings, 307–308

   stream force and nozzle size, 210–211
   unit conversion formulas, 355
water run-off capture or channeling, 252, 316
water supply
   addressing in SOPs, 40
   conflagrations and group fires, 242
   defensive operations, 238
   extinguishment and, 63–64
   for high-rise building fires, 314
   industrial occupancies, 286
   offensive operations, 216–218
   opposing fire lines, 136–137
   potential water load, evaluating, 60–61
   set-up staffing, 126–127
   sprinkler systems, 174, 175
      deluge systems, 178
      shutting down, 174, 177, 179, 251, 252
   standpipe systems, 181
water weight, concerns over, 252–253
weather conditions
   assessing during size-ups, 67–68
   hot and cold weather rehabilitation, 139–140
   shelter from, 163
weight of water, concerns over, 252–253
weight unit conversion formulas, 354
West Warwick, Rhode Island nightclub fire (2003), 264
wet chemical extinguishing systems, 185
wet pipe sprinkler systems, 173
   defined, 188
wildland fires, spreading of, 241
windows, high-rise buildings, 315–316
windows, opening vs. breaking, 152
winds at high-rise building fires, 313–314
wood shingle roofs, 241
wood truss, 45, 73
wooden construction. *See* construction materials for buildings
Worcester, Massachusetts warehouse fire, 112
workplace property. *See* business occupancies
World Trade Center bombing (1993), 325–326
   delegation of authority during, 6
World Trade Center bombing (2001), 299, 305, 306, 308, 312
   case summary, 326
   command post for, 7
   interagency communications problems, 3–4
wrap-around fires, 314, 315

Z

zones. *See* fire zones and perimeters; rehabilitation area

Photo Credits

Chapter 1
Case Study 2: Courtesy of FEMA.

Chapter 2
2.11: Courtesy of District Chief Chris E. Mickal/New Orleans Fire Department, Photo Unit; 2.17: Courtesy of the U.S. Fire Administration, Major Fires Investigation Series; 2.18: Courtesy of the U.S. Fire Administration, Major Fires Investigation Series; 2.19: Courtesy of the U.S. Fire Administration, Major Fires Investigation Series.

Chapter 3
3.5: Courtesy of Glen E. Ellman and Texas Engineering Extension Service, Texas A&M System.

Chapter 5
5.13: © David Huntley/ShutterStock, Inc.; 5.14: Photograph by Robert Duval, © National Fire Protection Association, Inc. used with permission; 5.18: Courtesy of Captain David Jackson, Saginaw Township Fire Department; 5.23: Courtesy of the Phoenix Fire Department; Case Study 1: © Paul Connors/AP Photos; Case Study 2: © Al Paglione/The Record; Case Study 3: © Gary Stewart, File/AP Photos; Case Study 4: © Boston Globe/George Rizer/Landov; Case Study 5: © Alexander Fox/AP Photos; Case Study 6: Reprinted with permission copyright © 2007, National Fire Protection; Case Study 7: Photograph by Robert Duval, © National Fire Protection Association, Inc. used with permission.

Chapter 6
Case Study 1a: © 1995, National Research Council Canada; Case Study 1b: © AP Photos; Case Study 2: Courtesy of the U.S. Fire Administration, Major Fires Investigation Series; Case Study 3: © The News & Observer; Case Study 4: Courtesy of the U.S. Fire Administration, Major Fires Investigation Series.

Chapter 7
7.2: Courtesy of Veltre Engineering; 7.3: Courtesy of Viking Sprinkler Company; 7.6: Courtesy of Ed Comeau, writer-tech.com; Case Study 1: Reprinted with permission copyright © 2007, National Fire Protection Association, Quincy, MA. All rights reserved.

Chapter 8
8.20: Courtesy of Chief Kelly Aylor, Florence Fire Department, Florence, Kentucky.

Chapter 9
9.2: © Bloomsburg Press Enterprise, Bill Hughes/AP Photos; 9.3: © Brian Kersey/AP Photos; Case Study 2: © Justin Sullivan/AP Photos.

Chapter 10
10.3: Courtesy of Scott Dornan, ConocoPhillips Alaska; 10.4: © The Messenger-Inquirer, John Dunham/AP Photos.

Chapter 11
11.5: © AP Photos; 11.6b: © Gheorghe Roman/ShutterStock, Inc.; 11.7b: Courtesy of Ed Comeau, writer-tech.com; 11.8: © AP Photos; 11.10: Courtesy of the U.S. Fire Administration, Major Fires Investigation Series; 11.12: © AP Photos; 11.13: © AP Photos; Case Study 1: © Reuters/Jason Cohn/Landov; Case Study 3: Courtesy of the U.S. Fire Administration, Major Fires Investigation Series; Case Study 4: © The News & Observer; Case Study 5: Reprinted with permission copyright © 2007, National Fire Protection Association, Quincy, MA. All rights reserved.

Chapter 12
12.3: © Master Sergeant Richard Diaz/AP Photos; 12.15: © AP Photos; 12.17: Courtesy of the U.S. Fire Administration, Major Fires Investigation Series; 12.19: Courtesy of NIST, Special Publication; 12.20: © Jemini Joseph/ShutterStock, Inc.; 12.21a: © AP Photos; 12.21b: © Master Sergeant Richard Diaz/AP Photos; Case Study 1: © Alan Greth/AP Photos; Case Study 2: Reprinted with permission copyright © 2007, National Fire Protection Association, Quincy, MA. All rights reserved; Case Study 3: © Sakchai Lalit/AP Photos.

Unless otherwise indicated, photographs are under copyright of Jones and Bartlett Publishers, courtesy of the Maryland Institute of Emergency Medical Services Systems, or have been provided by the authors.